Transactions on Computer Systems and Networks

Transactions on Computer Systems and Networks is a unique series that aims to capture advances in evolution of computer hardware and software systems and progress in computer networks. Computing Systems in present world span from miniature IoT nodes and embedded computing systems to large-scale cloud infrastructures, which necessitates developing systems architecture, storage infrastructure and process management to work at various scales. Present day networking technologies provide pervasive global coverage on a scale and enable multitude of transformative technologies. The new landscape of computing comprises of self-aware autonomous systems, which are built upon a software-hardware collaborative framework. These systems are designed to execute critical and non-critical tasks involving a variety of processing resources like multi-core CPUs, reconfigurable hardware, GPUs and TPUs which are managed through virtualisation, real-time process management and fault-tolerance. While AI, Machine Learning and Deep Learning tasks are predominantly increasing in the application space the computing system research aim towards efficient means of data processing, memory management, real-time task scheduling, scalable, secured and energy aware computing. The paradigm of computer networks also extends it support to this evolving application scenario through various advanced protocols, architectures and services. This series aims to present leading works on advances in theory, design, behaviour and applications in computing systems and networks. The Series accepts research monographs, introductory and advanced textbooks, professional books, reference works, and select conference proceedings.

Sanjeev Sharma · Kuntal Deka · Dharmendra Dixit
Editors

Next-Generation Wireless Systems

Fundamentals and Applications

Springer

Editors
Sanjeev Sharma
Department of Electronics Engineering
Indian Institute of Technology Varanasi
Varanasi, Uttar Pradesh, India

Kuntal Deka
Department of Electronics and Electrical Engineering
Indian Institute of Technology
Guwahati, Assam, India

Dharmendra Dixit
Department of Electronics and Communication Engineering
Motilal Nehru National Institute of Technology Allahabad
Prayagraj, Uttar Pradesh, India

ISSN 2730-7484 ISSN 2730-7492 (electronic)
Transactions on Computer Systems and Networks
ISBN 978-981-96-3184-1 ISBN 978-981-96-3185-8 (eBook)
https://doi.org/10.1007/978-981-96-3185-8

This Springer imprint is published by the registered company Springer Nature Singapore Pte Ltd.
The registered company address is: 152 Beach Road, #21-01/04 Gateway East, Singapore 189721, Singapore

Contents

Contributors

Anuj Abraham Technology Innovation Institute, Abu Dhabi, United Arab Emirates

Cédric Adjih Research Scientist, TRiBE Team, Inria Saclay, Paris, France

Nayan Moni Baishya Department of Electronics and Electrical Engineering, Indian Institute of Technology Guwahati, Assam, India

Vimal Bhatia SaSg, Department of Electrical Engineering, Indian Institute of Technology Indore, Indore, India;
Faculty of Informatics and Management, University of Hradec Kralove, Hradec Kralove, Czech Republic

Siddhartha S. Borkotoky Indian Institute of Technology Bhubaneswar, Khorda, Odisha, India

Jia Cao ENSTA Paris, Paris, France

Soumendu Das National Institute of Technology Jamshedpur, Adityapur, Jamshedpur, Jharkhand, India

Kuntal Deka IIT Guwahati, Guwahati, India

Dharmendra Dixit Motilal Nehru National Institute of Technology Allahabad, Teliarganj, Prayagraj, Uttar Pradesh, India

Maulesh N. Gadani Indian Space Research Organisation, Bengaluru, India

Iman Hmedoush Senior Radio Research Specialist, Nokia, Paris, France

Jayesh Jayarajan Indian Space Research Organisation, Bengaluru, India

Salil Kashyap Department of Electronics and Electrical Engineering, Indian Institute of Technology Guwahati, Guwahati, India

Prakhar Keshari SaSg, Department of Electrical Engineering, Indian Institute of Technology Indore, Indore, India

Ondrej Krejcar Faculty of Informatics and Management, University of Hradec Kralove, Hradec Kralove, Czech Republic

Nagendra Kumar National Institute of Technology Jamshedpur, Adityapur, Jamshedpur, Jharkhand, India

Vinit Kumar Department of Electronics Engineering, Indian Institute of Technology (BHU), Varanasi, India

B. R. Manoj Department of Electronics and Electrical Engineering, Indian Institute of Technology Guwahati, Assam, India

Priya Ranjan Muduli Department of Electronics Engineering, Indian Institute of Technology (BHU), Varanasi, India

Sangeeta Nalluru IIT Guwahati, Guwahati, India

Enukonda Venkata Pothan Department of Electronics and Electrical Engineering, Indian Institute of Technology Guwahati, Guwahati, India

Harsh Raj Department of Electronics and Electrical Engineering, Indian Institute of Technology Guwahati, Assam, India

Jyotsna Rani IIT Guwahati, Guwahati, India

Anupma Sharma SaSg, Department of Electrical Engineering, Indian Institute of Technology Indore, Indore, India

Mohit Kumar Sharma Digital Science Research Center, Technology Innovation Institute, Abu Dhabi, United Arab Emirates

Payal Sharma Indian Space Research Organisation, Bengaluru, India

Sanjeev Sharma Department of Electronics Engineering, IIT (BHU) Varanasi, Varanasi, India

Vaishali Sharma IIT, Indore, India

Anna Thomas Yenepoya Institute of Technology, Moodabidri, Mangalore, India

Chapter 1
Introduction to Next-Generation Communications

Sanjeev Sharma, Kuntal Deka, and Dharmendra Dixit

Abstract This book provides an overview of the evolution from previous-generation to next-generation communication networks, highlighting the key drivers and requirements for this transformative shift in wireless networks. Covering a wide range of topics including 5G and beyond, Internet-of-Things (IoT), Artificial Intelligence (AI), and emerging communication technologies such as visible light communication (VLC) and terahertz communication, orthogonal time-frequency space (OTFS) modulation, intelligent reflecting surfaces (IRS), this book offers insights into the latest advancements and innovations driving the development of next-generation communication systems. The book begins by examining the fundamental concepts and principles underlying current communication networks, providing readers with a solid understanding of the technological landscape. It then delves into the key challenges and opportunities presented by next-generation communication technologies, exploring topics such as multiple access, network architecture, security, and privacy.

Keywords 6G communication · Massive MIMO · Beamforming · Waveform design

S. Sharma (✉)
IIT (BHU) Varanasi, Varanasi, India
e-mail: sanjeev.ece@iitbhu.ac.in

K. Deka
IIT Guwahati, Guwahati, India

D. Dixit
MNNIT Prayagraj, Prayagraj, India

S. Sharma et al. (eds.), *Next-Generation Wireless Systems*,
Transactions on Computer Systems and Networks,
https://doi.org/10.1007/978-981-96-3185-8_1

1.1 Book Objectives and Outlines

The objective of the book "Next-Generation Wireless Systems: Fundamentals and Applications" is likely to provide comprehensive coverage of the fundamental concepts, latest developments, and practical applications in the field of next-generation wireless communication systems. Key objectives of the book are include:

Providing readers with a solid understanding of the underlying principles and theories behind next-generation wireless systems such as multiple access schemes, different waveforms design, massive multiple-input multiple-output (MIMO)-based system, etc.
Exploring the cutting-edge technologies and advancements shaping the landscape of wireless communication systems, including 5G, 6G, and beyond.
Addressing performance evaluation metrics, optimization techniques, and challenges associated with deploying and managing next-generation wireless networks.
Discussing practical applications, use cases, and real-world implementations of next-generation wireless systems across various industries and domains.
To foster innovative thinking and encourage readers to explore novel approaches and solutions in the design, implementation, and optimization of next-generation communication systems. Further, this book serves as a valuable educational resource for students, researchers, engineers, industry professionals, policymakers, and anyone interested in understanding the intricacies of 6G communication technologies.

The subsequent sections delve deeper into the intended scope of this book. Initially, we offer a historical retrospective on the progression of wireless communications spanning from 1G to 5G. Following this, we offer insights into the anticipated landscape of 6G technology. Subsequently, we elaborate on the principal 6G research endeavors globally. Finally, we conclude this chapter by providing an overview of the content covered in the subsequent chapters of this book.

1.2 Evolution of Wireless Communications

1.2.1 1G (First Generation)

Introduced in the 1980s, 1G systems were analog cellular networks tailored for voice calls. Provided basic voice services with limited coverage and low-quality audio. The most prominent standard was Advanced Mobile Phone System (AMPS). These are bulky mobile phones that have very little in common with modern communication devices. These networks encountered various constraints, with analog technology offering restricted capacity and subpar call quality. Among the foremost limitations of these early systems was the absence of security measures. From a hardware standpoint, 1G was not standardized globally, necessitating the need to modify equipment when making mobile phone calls from one county to another.

1.2.2 2G (Second Generation)

Emerged in the early 1990s, 2G systems marked the transition to digital communication. Introduced digital voice encoding (e.g., GSM[1]) offering better voice quality and increased security. Enabled basic data services such as SMS (Short Message Service). Introduced the concept of roaming and international compatibility.

GSM pioneered digital data services. Initially, speeds of 9.6 kb/s were offered; subsequently, the introduction of General Packet Radio Service (GPRS) and later Enhanced Data Rates for GSM Evolution (EDGE) increased the bit rates to several tens of kilobits per second and 1–200 kb/s, respectively. However, due to various factors such as high costs, latencies of several hundred milliseconds, and relatively low bit rates, these data services achieved only moderate success.

1.2.3 3G (Third Generation)

Rolled out in the early 2000s, 3G networks brought significant improvements in data rates and introduced mobile internet access. Supported higher data speeds, enabling multimedia services like video calling, mobile TV, and mobile internet browsing. Standards such as UMTS (Universal Mobile Telecommunications System) and CDMA2000 were prominent.

While the GSM air interface relies on narrowband frequency division multiple access/time division multiple access (FDMA/TDMA), UMTS utilizes code division multiple access (CDMA), which operates on a broader bandwidth. However, one of the numerous challenges associated with the UMTS air interface is its coherence bandwidth relative to the system's requirements.

1.2.4 4G (Fourth Generation)

Introduced in the late 2010s, 4G networks revolutionized mobile communications with high-speed data transmission. Provided significantly faster data rates, enabling seamless streaming of HD video, online gaming, and other bandwidth-intensive applications. Employed technologies like LTE (Long-Term Evolution) and WiMAX (Worldwide Interoperability for Microwave Access).

---

[1] It known as Global System for Mobile Communications.

LTE continues to undergo rapid evolution. It delivered high-speed downlink (up to 300 Mb/s) and uplink (up to 50 Mb/s) in Europe. Prototype trials aiming for speeds of 1 Gb/s and beyond are currently underway. Additionally, there is a shift toward transitioning voice services to Voice over IP, also known as Voice over LTE. This transition enables the utilization of wide-band speech codecs, unlocking a new dimension of voice quality. LTE is based on orthogonal frequency division multiplexing (OFDM). Moreover, it undergoes continuous enhancement, with advancements such as the evolution from MIMO to massive MIMO, further reductions in latencies, continuous increases in bit rates, and the readiness of the device-to-device (D2D) standard for implementation in products.

1.2.5 5G (Fifth Generation)

Began commercial deployment in the 2020s, 5G represents a paradigm shift with ultra-fast data rates, low latency, and massive connectivity. Offers multi-Gbps peak data rates, enabling transformative applications like Internet-of-Things (IoT), smart cities, autonomous vehicles, and virtual reality (VR). Utilizes technologies such as mmWave, Massive MIMO, and network slicing. 3GPP identified three high-level use cases for 5G as Enhanced mobile broadband (eMBB), Massive MTC (mMTC), and ultra-reliable and low-latency communications (URLLC).

Numerous projects focusing on 5G research are currently ongoing, making it premature to determine the definitive key innovations that will define this generation. Among the potential innovations is the utilization of significantly higher frequencies, reaching up to 60 GHz, enabling ultra-high-speed transmission. Discussions at various 5G conferences and events suggest that other innovations may encompass network slicing, aimed at virtualizing network resources to meet application-specific needs, as well as software-defined networking. Furthermore, it's argued that the 5G ecosystem holds significant potential for delivering important benefits to vertical industries.

Since the commencement of 5G initiatives within 3GPP in early 2016, significant advancements have been achieved in both standardization and commercial deployments. Currently, 3GPP is embarking on the second phase of 5G standardization, termed 5G-Advanced, which builds upon the foundational 5G technologies introduced in 3GPP Releases 15, 16, and 17. The onset of 3GPP Release 18 marks the initiation of 5G-Advanced, encompassing a diverse array of features that encompass both device and network enhancements. This comprehensive approach facilitates a balanced evolution of mobile broadband and further expansion into vertical domains, addressing immediate and long-term commercial requirements. 5G-Advanced is poised to significantly enhance 5G capabilities, cater to numerous new use cases, revolutionize connectivity experiences, and serve as a pivotal milestone in the progression toward 6G mobile communications.

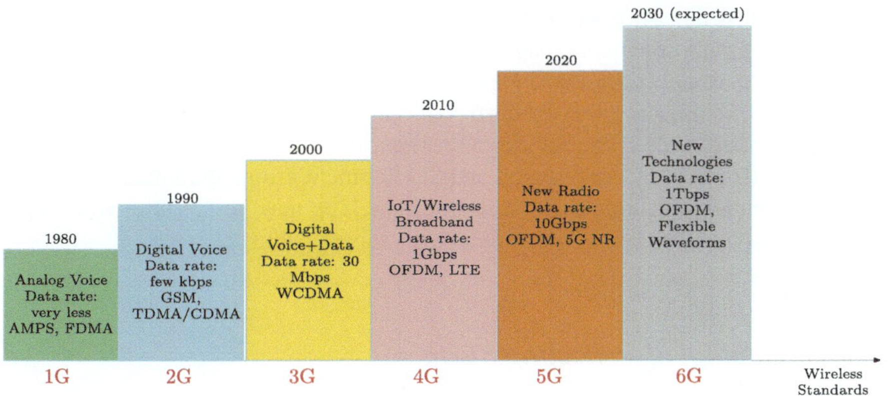

Fig. 1.1 Evolution of wireless standards

1.2.6 6G (Sixth Generation)

Still in the research and development phase, 6G aims to push the boundaries of wireless communication further. Envisions even higher data rates, URLLC, and ubiquitous connectivity. Expected to leverage technologies like terahertz frequencies, artificial intelligence (AI)-driven networks, holographic communication, and quantum communication.

The perpetual challenge of wireless capacity persists, urging continuous enhancement from the standards set by 3G. As the industry moves forward, performance targets for 6G are currently under examination. These targets aim to enable unparalleled services, including fully immersive Extended Reality (XR), high-fidelity mobile holograms, and digital replicas. Such services may demand data rates reaching up to 1 Terabit per second (Tbps). Consequently, the network necessitates access to spectrum ranging from hundreds of MHz to tens of GHz to meet these ambitious performance requirements (Fig. 1.1).

1.3 Key Technologies for 6G Communications

6G communication technologies represent the next frontier in wireless communication systems, aiming to revolutionize connectivity and enable new applications and services. Building upon the foundations laid by previous generations, 6G is expected to offer unprecedented data rates, ultra-low latency, and massive connectivity. Key technologies envisioned for 6G include terahertz communication, leveraging frequencies above 100 GHz to enable extremely high-speed transmission; intelligent reflecting surfaces (IRS) to enhance signal propagation and coverage; integrated satellite-terrestrial networks for ubiquitous connectivity; and AI and machine learning (ML) to optimize network management and resource allocation. Additionally, 6G is expected to support advanced applications such as holographic communication, quantum communication, and bio-inspired communication, paving the way for transformative innovations across various industries, including healthcare, transportation, and manufacturing [1]. As research and development efforts continue, 6G communication technologies hold the promise of unlocking new possibilities and shaping the future of wireless communication in the years to come [2].

1.3.1 Intelligent Reflecting Surfaces

Intelligent Reflecting Surfaces (IRS) represent a groundbreaking paradigm in wireless system design, offering transformative capabilities to enhance communication performance. By strategically deploying passive elements, such as reflective surfaces or metasurfaces, IRS-enabled systems can manipulate electromagnetic waves to optimize signal propagation [3]. This novel approach holds immense promise across diverse applications, including indoor and outdoor wireless networks, satellite communications, and Internet-of-Things (IoT) deployments. In IRS-assisted wireless systems, these passive elements are intelligently controlled to adjust signal strength, mitigate interference, and improve overall system efficiency. Furthermore, IRS technology enables dynamic adaptation to changing environmental conditions and user requirements, ensuring flexibility and scalability in wireless network design. With its potential to significantly increase coverage, capacity, and energy efficiency, IRS-assisted wireless system design represents a cornerstone of future communication networks, driving innovation and enabling new generations of high-performance wireless technologies.

1.3.2 Massive MIMO and Beamforming

Massive MIMO and beamforming are advanced antenna technologies that play a crucial role in enhancing the performance of wireless communication systems. Massive MIMO involves deploying a large number of antennas at the base station, allowing for simultaneous communication with multiple users using spatial multiplexing techniques [4, 5]. This enables significant improvements in spectral efficiency, throughput, and coverage, making it well-suited for next-generation wireless networks. Beamforming, on the other hand, focuses on directing radio signals toward specific users or regions of interest, rather than broadcasting signals uniformly in all directions. By steering signals toward intended recipients and minimizing interference, beamforming enhances signal quality, increases range, and improves overall network capacity. When combined, Massive MIMO and beamforming technologies offer synergistic benefits, enabling even higher data rates, improved reliability, and better utilization of available spectrum. These technologies are instrumental in meeting the ever-increasing demand for high-speed, low-latency wireless connectivity in today's data-driven world, and are expected to play a central role in shaping the future of wireless communication systems.

1.3.3 Hybrid Multiple Access Techniques

Hybrid multiple access techniques (HMAT) combine different access methods to efficiently utilize wireless resources and enhance network performance. Two prominent examples of HMAT are orthogonal multiple access (OMA) and non-orthogonal multiple access (NOMA) [6].

OMA assigns orthogonal resources, such as time slots, frequency bands, or codes, to different users to avoid interference. In TDMA and FDMA, users are allocated non-overlapping time slots or frequency bands, respectively. Orthogonality ensures that users can transmit and receive data simultaneously without causing interference. While OMA simplifies receiver design and provides predictable performance, it may underutilize resources, especially when the number of users is small or dynamic.

On the otherhand, NOMA allows multiple users to share the same time-frequency resources non-orthogonally [7]. In NOMA, users are separated by assigning different power levels or using advanced receiver techniques, such as successive interference cancellation (SIC), to decode overlapping signals. This enables better spectrum efficiency and higher throughput compared to OMA. NOMA is particularly beneficial in scenarios with a large number of users or varying channel conditions. However, NOMA introduces challenges such as increased receiver complexity and the need for sophisticated power control algorithms.

1.3.4 Terahertz Communication

Terahertz communication operates within the frequency range of 0.1–10 terahertz (THz), corresponding to wavelengths between 30 ms and 3 mm. This spectral region, also known as the terahertz or submillimeter wave band, offers immense potential for high-speed wireless communication due to its wide bandwidth and short wavelengths [8]. However, one of the significant challenges in terahertz communication is molecular absorption, particularly by atmospheric gases such as oxygen and water vapor. These molecules exhibit strong absorption characteristics at specific terahertz frequencies, leading to signal attenuation and limiting communication range. Oxygen absorbs radiation in the frequency range of about 60 and 100 GHz, while water vapor absorbs radiation around 22 GHz and between 180 and 325 GHz. Overcoming these absorption effects is crucial for realizing the full potential of terahertz communication systems. Researchers are exploring various techniques such as frequency-selective surfaces, atmospheric compensation algorithms, and adaptive modulation schemes to mitigate the impact of molecular absorption and enable reliable terahertz communication for future wireless networks.

1.3.5 Orthogonal Time Frequency Space

Orthogonal time-frequency space (OTFS) modulation is emerging as a transformative technology in wireless system design, promising significant advancements in communication performance [9]. OTFS leverages the unique properties of time-frequency scattering to overcome many of the challenges faced by conventional modulation techniques. In OTFS-assisted wireless systems, data symbols are spread across both time and frequency domains, enabling robust communication in multipath environments and overcoming the limitations of Doppler spread and delay spread. This innovative approach offers exceptional resilience to fading and interference, making it particularly well-suited for challenging propagation conditions encountered in urban environments, indoor deployments, and high-speed vehicular communications. Furthermore, OTFS technology provides inherent compatibility with existing modulation schemes, facilitating seamless integration into current and future wireless networks. As research and development efforts continue to advance, OTFS-assisted wireless system design holds great promise for revolutionizing communication capabilities and enabling the next generation of high-performance wireless technologies.

1.3.6 Integrated Sensing and Communication

Integrated Sensing and Communication (ISAC) technology represents a cutting-edge approach that seamlessly combines sensing and communication functionalities within a unified framework [10]. This integration enables devices to simultaneously gather real-time data from the environment while transmitting information to other devices or centralized systems. ISAC technology holds immense potential for various applications, including environmental monitoring, smart infrastructure, healthcare, and industrial automation. By integrating sensing and communication capabilities, ISAC systems can efficiently collect, process, and transmit data, leading to enhanced situational awareness, improved decision-making, and more effective resource management. Moreover, ISAC technology facilitates seamless coordination and collaboration among distributed sensors and devices, enabling scalable and resilient systems. As research and development in ISAC technology continue to advance, it is poised to revolutionize a wide range of industries, driving innovation and enabling the development of smarter, more interconnected environments.

1.3.7 Integrated Satellite-Terrestrial Networks

Integrated Satellite-Terrestrial Networks (ISTN) represent a pioneering approach to wireless communication systems, seamlessly integrating both satellite and terrestrial infrastructures. This integration offers unprecedented connectivity and coverage, particularly in remote and underserved areas where traditional terrestrial networks may be limited. By leveraging the strengths of both satellite and terrestrial networks, ISTN can provide ubiquitous and reliable connectivity across diverse geographic regions. This convergence enables efficient resource utilization, improved quality of service, and enhanced resilience to network failures. Moreover, ISTN opens up opportunities for a wide range of applications, including emergency response, disaster management, IoT deployments, and global connectivity solutions. As technology continues to advance and ISTN deployments become more widespread, they have the potential to bridge the digital divide, empower communities, and drive socioeconomic development on a global scale.

1.3.8 Artificial Intelligence (AI) and Machine Learning (ML)

AI and ML are revolutionizing wireless communication systems by enabling intelligent decision-making and optimization processes [11]. With AI and ML techniques integrated into wireless networks, systems can autonomously adapt to dynamic environments, optimize resource allocation, and enhance overall performance. These technologies enable predictive maintenance, spectrum management, interference mitigation, and self-optimization, leading to improved network efficiency and reliability. Furthermore, AI and ML facilitate the development of advanced features such as intelligent beamforming, adaptive modulation, and predictive analytics, which optimize signal transmission and enhance user experience. As AI and ML algorithms continue to evolve, they will play an increasingly critical role in shaping the future of wireless communication, unlocking new capabilities and applications that were previously unattainable.

1.3.9 Visible Light Communication

Visible Light Communication (VLC) is emerging as a promising technology for next-generation communication systems, offering several advantages over traditional wireless communication methods. VLC utilizes visible light spectrum, typically from light-emitting diodes (LEDs), to transmit data wirelessly. By modulating the intensity of light at high speeds, VLC enables data transmission rates comparable to or even higher than those achieved by radio frequency-based technologies. Additionally, VLC offers inherent security benefits as light signals cannot penetrate opaque

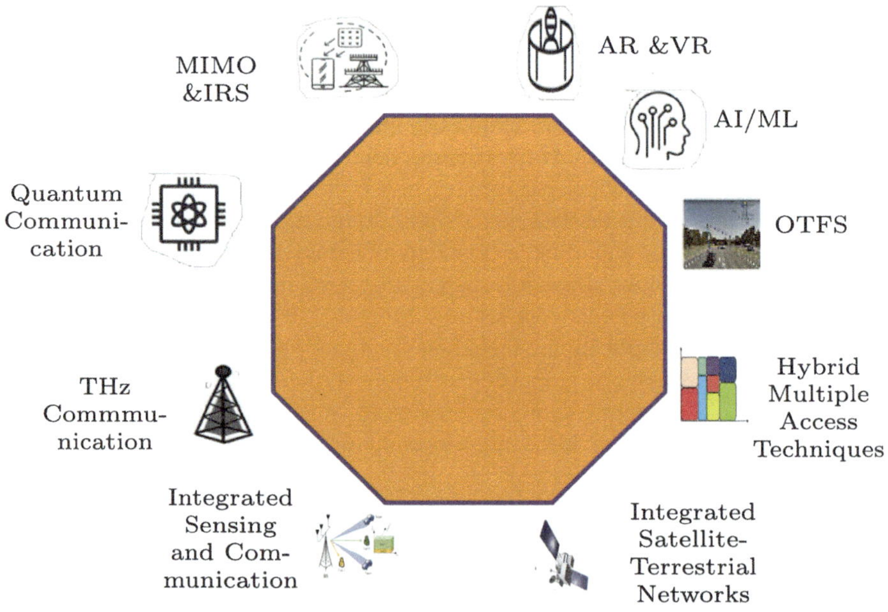

Fig. 1.2 6G enabling technologies [1, 8]

obstacles, providing a secure communication channel for sensitive data transmission. Furthermore, VLC can be seamlessly integrated into existing lighting infrastructure, making it a cost-effective solution for indoor communication, especially in environments where radio frequency interference is a concern. With its potential to deliver high-speed, secure, and energy-efficient communication, VLC is poised to revolutionize the way we transmit data in the next generation of communication systems (Fig. 1.2).

1.4 Applications

The advent of 5G technology has unlocked a multitude of new application scenarios and demands, shaping the landscape of connectivity and communication systems. One prominent application scenario enabled by 5G is eMBB, which promises significantly faster data rates and lower latency compared to previous generations. This opens up opportunities for high-definition video streaming, virtual reality (VR), augmented reality (AR), and immersive gaming experiences, revolutionizing entertainment and media consumption [1–3]. Additionally, 5G's URLLC capabilities cater to mission-critical applications such as autonomous vehicles, remote surgery, and industrial automation, where reliability and real-time responsiveness are paramount.

Another emerging application scenario for 5G is mMTC, facilitating the connectivity of billions of devices in the IoT ecosystem. This enables smart cities, smart grids, and smart homes, where interconnected sensors and devices gather and exchange data to optimize resource utilization, improve efficiency, and enhance quality of life. Furthermore, 5G technology and beyond enables seamless connectivity in remote and underserved areas through fixed wireless access, bridging the digital divide and providing broadband connectivity to previously inaccessible regions.

In addition to these application scenarios, the transition toward 5G and beyond brings forth new demands and challenges for communication systems coming from diverse field. Spectrum management becomes increasingly critical as the demand for spectrum resources intensifies, requiring efficient allocation and utilization strategies to accommodate diverse applications and services. Network densification is another key demand, necessitating the deployment of small cells, distributed antennas, and network slicing to enhance coverage, capacity, and reliability, particularly in dense urban areas and indoor environments.

1.5 Book Outline

Book chapters are outlined as follows

1. Chapter **Non-orthogonal Multiple Access (NOMA) System: Design and Analysis** delves a comprehensive overview of NOMA, focusing on its principles, advantages, and applications in the context of 5G networks and future communication. systems.
2. Chapter **Co-located and Cell-Free Massive MIMO Based Underlay Spectrum Sharing Networks** presents modeling and analysis to compare performance of co-located and cell-free massive MIMO (mMIMO) based underlay spectrum sharing networks in terms of interference outage probability and spectral efficiency.
3. Chapter **Intelligent Reflecting Surface Assisted Wireless System: Analysis and Design** introduces intelligent reflecting surface (IRS) assisted wireless system design and analysis.
4. Chapter **SFFT-based Two-stage OTFS: Fundamentals and Applications** presents the fundamentals of delay-Doppler-based signal processing for wireless system design.
5. Chapter **Zak-OTFS: Concepts of Delay-Doppler Communication** discusses the concepts and principles of the discrete Zak transform (DZT)-based orthogonal time-frequency space (OTFS) for doubly dispersive channels.
6. Chapter **Visible Light Communication: An Emerging Technology** provides an overview of the fundamentals of visible light communication (VLC) by thoroughly exploring its technologies, challenges, and future prospects.

7. Chapter **Advancements in Wireless Communication with Zero Energy Devices** explores the recent approaches toward achieving zero energy consumption in wireless communication systems, focusing on the integration of zero energy devices (ZED) in the wireless communication system.
8. Chapter **Integrated Sensing and Communications: Fundamentals, Recent Developments, and Prospects in 6G** presents research enthusiasm toward investigating the coexistence, collaboration, and integrated design of communication and sensing systems. ISAC encapsulates a design paradigm and its associated enabling technologies that intricately merge sensing and communication systems.
9. Chapter **Unmanned Aerial Vehicle Assisted Wireless Communications: Enabling Connectivity and Applications** presents the role of UAV communication in enabling connectivity and applications in these evolving systems.
10. Chapter **Machine Learning for Next-Generation Wireless Communications** presents the application of machine learning in communication system design and analysis.
11. Chapter **Rateless Coding for Reliable Long Range IoT** describes the integration of erasure-correcting rateless codes with Long Range (LoRa) for reliable and energy-efficient transfer of small files. An analytical framework is presented for the performance evaluation of rateless-coded LoRa.
12. Chapter **An Introduction to Modern Random Access Protocols for IoT Communications** presents a perspective on a category of grant-free communication methods, specifically those based on random access. They are well-suited for scenarios involving a large number of devices transmitting sporadically with short messages, commonly encountered in IoT communication scenarios.
13. Chapter **Terahertz Communication-Bridging the Gap in the Next-Generation Wireless Systems** provides fundamental aspects of THz communication, including spectrum characteristics, wave propagation, and advanced modulation and coding techniques tailored for THz frequencies.
14. Chapter **Ambulatory ECG Measurement Denoising and Implementation for Smart Healthcare Applications** presents a robust approach for motion artifact and baseline wander suppression using empirical wavelet transform (EWT) and nonlocal means (NLM) estimation.

References

1. Chafii M, Bariah L, Muhaidat S, Debbah M (2023) Twelve scientific challenges for 6G: rethinking the foundations of communications theory. IEEE Commun Surv & Tutor 25(2):868–904
2. Serghiou D, Khalily M, Brown TW, Tafazolli R (2022) Terahertz channel propagation phenomena, measurement techniques and modeling for 6G wireless communication applications: A survey, open challenges and future research directions. IEEE Commun Surv & Tutor 24(2):1957–1996

3. Chen X, Tan J, Kang L, Tang F, Zhao M, Kato N (2024) Frequency selective surface towards 6G communication systems: a contemporary survey. IEEE Commun Surv & Tutor PP(PP):1–41
4. Gong T, Gavriilidis P, Ji R, Huang C, Alexandropoulos GC, Wei L, Zhang Z, Debbah M, Poor HV, Yuen C (2024) Holographic MIMO communications: theoretical foundations, enabling technologies, and future directions. IEEE Commun Mag 61(1):40–46
5. Sharma S, Deka K, Beferull-Lozano B (2021) Low-complexity detection for uplink massive MIMO SCMA systems. IET Commun 15(1):51–59
6. Al-Dweik A, Alsusa E, Dobre OA, Hamila R (2023) Multi-symbol rate NOMA for improving connectivity in 6G communications networks. IEEE Commun Mag PP(PP):1–7
7. Deka K, Sharma S (2022) Hybrid NOMA for future radio access: design, potentials and limitations. Wireless Pers Commun 1–16
8. Petrov V, Kurner T, Hosako I (2020) IEEE 802.15. 3d: first standardization efforts for sub-terahertz band communications toward 6G. IEEE Commun Mag 58(11):28–33
9. Deka K, Thomas A, Sharma S (2021) OTFS-SCMA: a code-domain NOMA approach for orthogonal time frequency space modulation. IEEE Trans Commun 69(8):5043–5058
10. Demirhan U, Alkhateeb A (2023) Integrated sensing and communication for 6G: Ten key machine learning roles. IEEE Commun Mag 61(5):1–7
11. Singh A, Sharma S, Deka K, Bhatia V (2023) Deep learning-assisted OFDM detection with hardware impairments. J Commun Inf Netw 8(4):378–388

Chapter 2
Non-orthogonal Multiple Access (NOMA) System: Design and Analysis

Sanjeev Sharma, Kuntal Deka, and Dharmendra Dixit

Abstract Non-orthogonal multiple access (NOMA) has received significant attention due to its capability to enhance spectral efficiency and support diverse services with distinct quality-of-service requirements. This book chapter aims to provide a comprehensive overview of NOMA, focusing on its principles, advantages, and applications in the context of 5G networks and future communication systems. It delves into the fundamental concepts underlying NOMA, such as power domain multiplexing and successive interference cancellation (SIC), elucidating its benefits in terms of increased connectivity, enhanced throughput, and better user experience. Further, code domain techniques such as sparse code multiple access (SCMA) and pattern division multiple access (PDMA) are also introduced. Moreover, this chapter discusses the challenges and open research issues associated with deploying NOMA in practical scenarios. The exploration of NOMA's potential enhancements and its role in the evolution toward 6G networks will be elucidated, offering insights into the future trajectory of wireless communication systems.

Keywords 6G network · Multiple access techniques · NOMA · SCMA · PDMA · RSMA · Design and analysis

2.1 Introduction

In recent years, the demand for wireless communication services has grown exponentially due to the proliferation of mobile devices and the internet-of-things (IoT). To meet these escalating demands, novel communication techniques have been

S. Sharma (✉)
IIT (BHU) Varanasi, Varanasi, India
e-mail: sanjeev.ece@iitbhu.ac.in

K. Deka
IIT Guwahati, Guwahati, India

D. Dixit
MNNIT Prayagraj, Prayagraj, India

S. Sharma et al. (eds.), *Next-Generation Wireless Systems*,
Transactions on Computer Systems and Networks,
https://doi.org/10.1007/978-981-96-3185-8_2

developed to enhance spectrum efficiency, support a massive number of connected devices, and improve overall system performance. One such cutting-edge technology that has gained considerable attention is non-orthogonal multiple access (NOMA). NOMA is also analyzed for massive multiple-input multiple-output (MIMO)-based systems [1].

Traditionally, in wireless communication systems, multiple users access the network using orthogonal multiple access (OMA) schemes, such as time division multiple access (TDMA) or frequency division multiple access (FDMA). However, these methods suffer from limited spectral efficiency and struggle to accommodate the immense diversity in traffic demands and user requirements. NOMA, on the other hand, presents a paradigm shift by breaking away from the orthogonality principle in multiple access.

In NOMA, multiple users can simultaneously share the same time-frequency resources, enabling a more efficient utilization of the available spectrum. The key principle behind NOMA lies in the allocation of different power levels to each user, resulting in varying signal strengths at the receiver [2]. As a result, NOMA enables the concurrent transmission of multiple signals with different power levels, creating overlapping signal spaces, and hence, achieving a higher degree of multiplexing.

The implementation of NOMA offers several compelling advantages. First and foremost, NOMA substantially enhances spectral efficiency, as it allows multiple users to access the network simultaneously, occupying the same time-frequency resources. This enables a significant increase in the system's capacity and supports a higher number of connections. Additionally, NOMA provides superior user fairness, as it allocates more power to users with weaker channel conditions, ensuring that all users receive an adequate level of service. Moreover, NOMA holds immense promise in accommodating massive machine-type communications (mMTC) for the rapidly expanding IoT ecosystem. With the exponential growth in connected devices, efficient and reliable communication is essential with low latency, and NOMA's capabilities align perfectly with the stringent requirements of mMTC applications.

In this chapter, we consider the fundamental principles, design considerations, and implementation aspects of NOMA system. We explore the various NOMA techniques, including power domain NOMA (PD-NOMA) and code domain NOMA (CD-NOMA), and their applications in 5G, and beyond wireless networks. Additionally, we address the challenges and potential solutions in deploying NOMA, while highlighting its contributions to spectrum efficiency, network capacity, and user experience [3].

2.2 Fundamentals of Multi-user Communications

Multi-user communications form the backbone of modern wireless networks, enabling multiple users to share and access resources simultaneously. The efficient exchange of information among multiple users while minimizing interference and ensuring fair allocation of resources is a key challenge in the design and operation of such systems. The two primary multiple access schemes widely used in wireless systems are OMA and NOMA.

Furthermore, interference management plays a critical role in multi-user communications. Techniques like interference alignment, beamforming, and advanced signal processing are employed to mitigate interference and improve the overall system performance. Overall, mastering the fundamentals of multi-user communications is essential for designing and optimizing wireless networks to cater to the increasing demands of diverse applications and the growing number of connected devices. As technology continues to evolve, the efficient exchange of information among multiple users will remain a central theme in shaping the future of wireless communications.

The received signal y in uplink multi-user system is expressed as [4]

$$y = \sum_{i=1}^{K} h_i \sqrt{P_i} x_i + w, \tag{2.1}$$

where K is the total number of users in the system. P_i and $x_i \in \mathcal{A}$ denote the ith user transmit power and data symbol, respectively. $\mathcal{A}$ is the constellation diagram used for information bits to symbol mapping. $M=|\mathcal{A}|$ is the modulation order and each symbol has $\log_2 M$ bits. w is the additive white Gaussian noise (AWGN), and h_i is the channel impulse response (CIR) of ith user.

We explain system capacity in (2.1) using OMA schemes, while the capacity of a single user using FDMA (OMA scheme) can be written as [5]

$$C_i = \frac{W}{K} \log_2 \left[1 + \frac{P_i K |h_i|^2}{W N_0}\right], \quad \text{bits/sec}, \quad \mathrm{i} = 1, 2, \ldots, \mathrm{K} \tag{2.2}$$

where $\frac{W}{K}$ is the bandwidth (BW) allocated to a single user, when total BW is W Hz. N_0 is the noise power spectral density (PSD). $|h_i|^2$ is the channel gain at the receiver of the ith user. Thus, total capacity of the K users system can be expressed as

$$\sum_{i=1}^{K} C_i = \sum_{i=1}^{K} \frac{W}{K} \log_2 \left[1 + \frac{P_i K |h_i|^2}{W N_0}\right] \quad \text{bits/sec}. \tag{2.3}$$

It's worth noting that, with a fixed bandwidth W, the total capacity increases indefinitely as the number of users grows linearly with K. However, as K increases, the allocation of bandwidth to each user decreases (W/K), leading to a reduction in the capacity per user.

In TDMA scheme, each user transmits for $1/K$ time duration over the BW W, and a single user capacity is given as [5]

$$C_i = \left(\frac{1}{K}\right) W \log_2\left[1 + \frac{P_i K |h_i|^2}{W N_0}\right], \quad \text{bits/sec}, \quad \mathrm{i} = 1, 2, \ldots, \mathrm{K} \tag{2.4}$$

Here, we consider the average power of each user to be $P_i K$ unit. This capacity is identical to that of an FDMA system. Nevertheless, it's essential to highlight that, in TDMA, it may become impractical for the transmitters to maintain a transmitter power of K times P when K is very large. Consequently, there exists a practical constraint where the transmitter power cannot be increased further as K increases.

Example 2.1 In Fig. 2.1, the sum-rate of FDMA users is plotted while varying the number of users (K) and SNR in an AWGN channel. As the number of users increases, the sum-rate of each user decreases due to the $1/K$ factor mentioned in the FDMA sum-rate expression. Additionally, the total sum-rate of the FDMA system for $K = 32$ is depicted in Fig. 2.1, which increases as the SNR rises. If K becomes very large, the individual user sum-rate approaches close to zero as shown in Fig. 2.1.

Code-Division Multiple Access (CDMA): In CDMA system, each user transmits average power P_i over BW W Hz. At one end of the spectrum is non-cooperative CDMA, where each user's receiver is unaware of or intentionally disregards the codes and spreading waveforms used by other users during the demodulation process. As a result, the signals from other users are perceived as interference at the receiver of each user. Single user capacity in CDMA is given as

$$C_i = W \log_2\left[1 + \frac{P_i |h_i|^2}{\sum_{j=1, j\neq i}^{K} P_j |h_j|^2 + W N_0}\right], \quad \text{bits/sec}, \quad \mathrm{i} = 1, 2, \ldots, \mathrm{K} \tag{2.5}$$

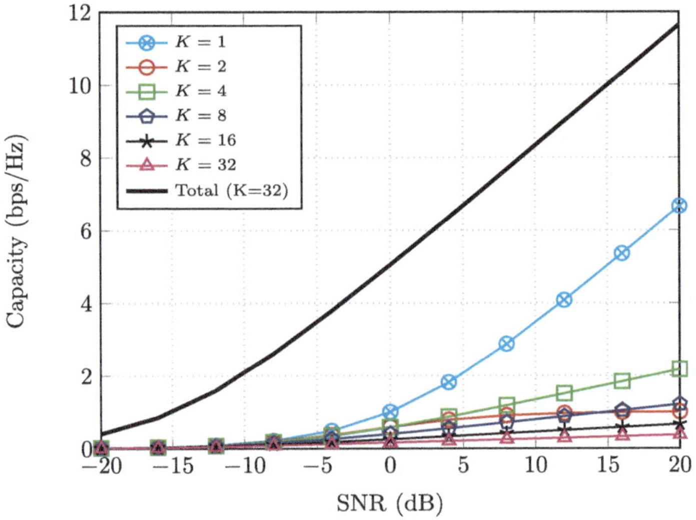

Fig. 2.1 Normalized capacity as a function of SNR for FDMA system

Further, if we consider AWGN channel only with average power P and BW W Hz, then capacity is expressed as

$$C_i = W \log_2 \left[1 + \frac{P}{(K-1)P + WN_0} \right], \quad \text{bits/sec}, \quad \text{i} = 1, 2, \ldots, \text{K} \tag{2.6}$$

As $K \to \infty$, total system capacity per unit BW is expressed as

$$\approx K \frac{P}{(K-1)P + N_0} \log_2 e \approx \log_2 e = 1.442 \text{ bits/sec} \tag{2.7}$$

Hence, in this scenario, the overall spectral efficiency is constrained, primarily due to the increasing interference, which ultimately becomes the limiting factor. This rate is often described as being interference-limited.

Example 2.2 The sum-rate of individual users is plotted in Fig. 2.2 for the K-user system. With an increase in K, the individual user's sum-rate decreases due to higher interference levels. Consequently, the capacity of the CDMA system becomes interference-limited and approaches zero as the number of users increases. Furthermore, it's observed that the sum-rate increases as the SNR rises, as depicted in Fig. 2.2.

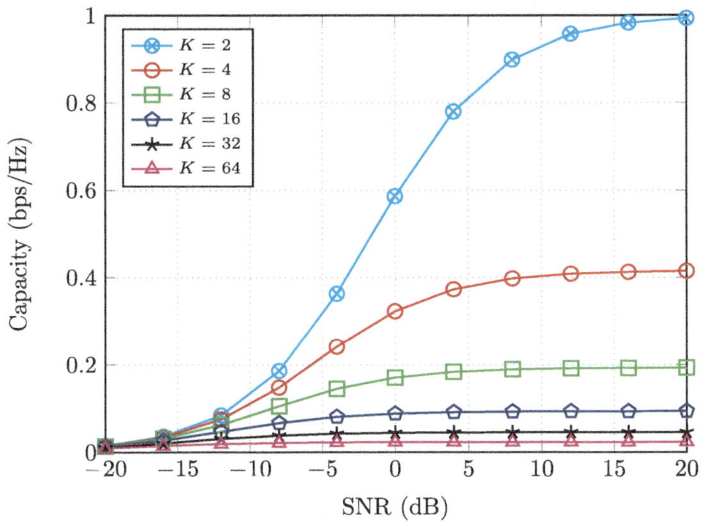

Fig. 2.2 Normalized capacity as a function of SNR for non-cooperative CDMA system

2.3 Multiple Access Techniques: Orthogonal Versus Non-orthogonal

In OMA, each user is assigned exclusive time, frequency, or code resources for transmission. TDMA divides time into discrete time slots, each allocated to a different user for transmission, as shown in Fig. 2.3. FDMA allocates unique frequency bands to different users, ensuring non-overlapping frequency regions. CDMA assigns unique spreading codes to each user, allowing simultaneous transmission over the same time (t) and frequency (f) band, as mentioned in Fig. 2.3. Further, space division multiple access (SDMA) focuses on spatial separation rather than time or frequency. It's commonly employed in technologies like MIMO systems, where multiple antennas are used to enhance communication by directing signals spatially. SDMA provides improved spectral efficiency and reduced interference, making it essential for modern wireless communication systems and technologies like 4G and 5G. OMA schemes ensure orthogonality among user signals, avoiding interference between users. However, they may lead to underutilization of resources when user traffic is sporadic or when the number of users is large.

In contrast, NOMA relaxes the orthogonality constraint by allowing multiple users to share the same time-frequency resources simultaneously. NOMA employs power domain or code domain techniques to multiplex users in the same resource block. The power domain NOMA (PD-NOMA) allocates different power levels to users, while code domain NOMA (CD-NOMA) assigns unique signature sequences to users, as shown in Fig. 2.4. NOMA significantly enhances spectrum efficiency, enabling more users to be supported with improved throughput and reduced latency due to superimposed users. By exploiting the varying channel conditions of users, NOMA achieves overlapping signal spaces, thereby increasing the overall system capacity.

Further, by considering three users, $\mathcal{U}_1$, $\mathcal{U}_2$ and $\mathcal{U}_3$ in the system by considering OMA and NOMA is also shown in Fig. 2.5, where each user uses orthogonal resources in OMA and overlapped in NOMA system. Users can have different power levels for data transmission.

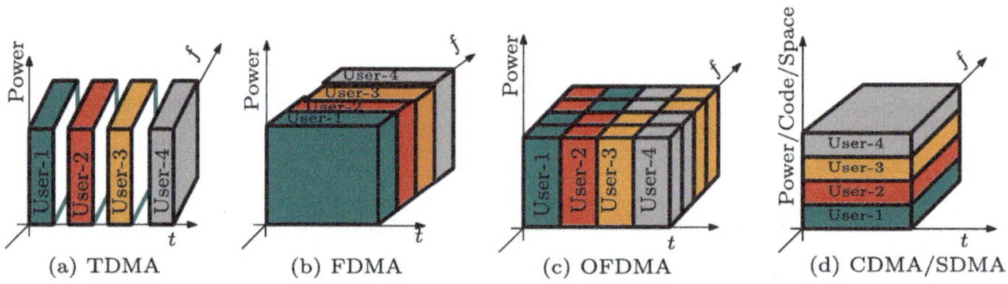

Fig. 2.3 Illustration of different OMA techniques

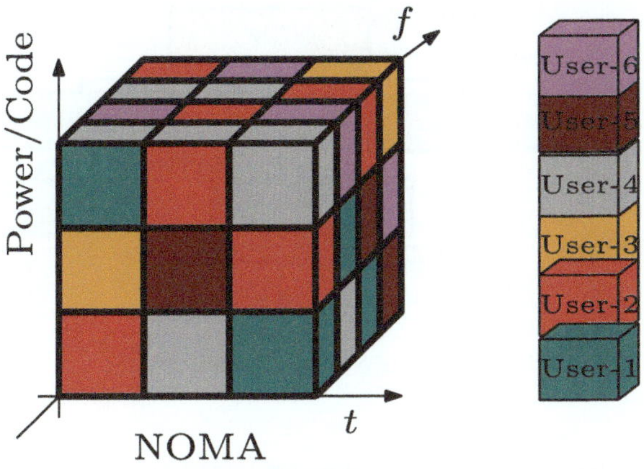

Fig. 2.4 Illustration of NOMA technique

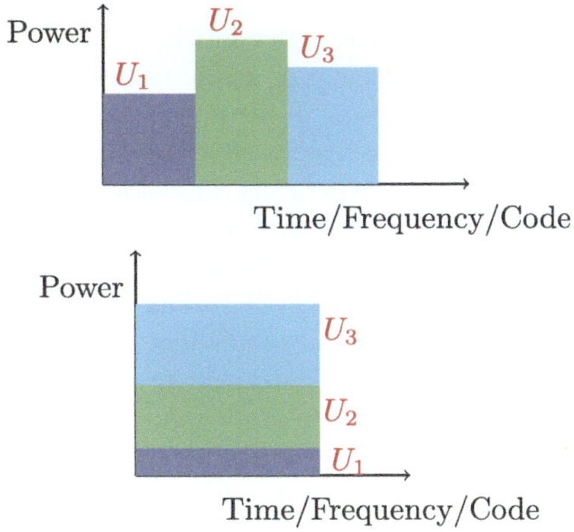

Fig. 2.5 Three users access in a system using OMA and NOMA

2.4 Power Domain NOMA (PD-NOMA)

There are several types of NOMA systems, each offering unique advantages. PD-NOMA allocates different power levels to users, allowing them to share the same time and frequency resources. Code Domain NOMA, on the other hand, assigns distinct codebooks to users, enabling them to share the same time and frequency slots. Hybrid NOMA combines both PS and CD-NOMA for more flexible resource allocation. Each of these NOMA variants contributes to the optimization of multi-user communication systems, catering to diverse requirements and providing greater capacity and efficiency in future wireless networks. Classification of NOMA in also shown in Fig. 2.6, where sparse code multiple access (SCMA) and low density spreading (LDS) use the sparse codes for codebook design and spreading, respectively. However, dense codebook multiple access (DCMA) and overloaded CDMA use the dense

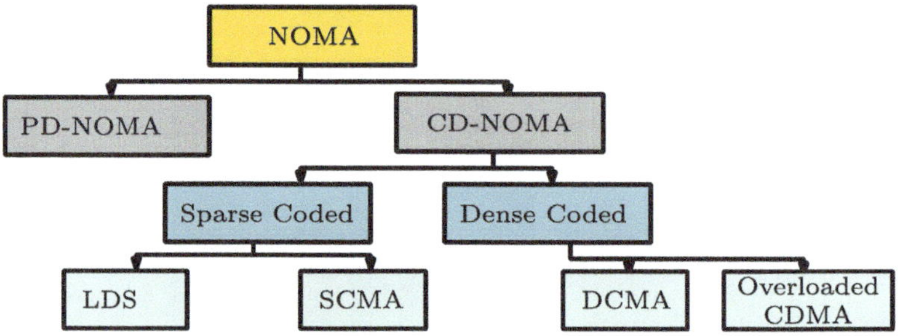

Fig. 2.6 Classification of NOMA schemes

codes for codebook design and spreading, respectively. In PD-NOMA, the users are allocated different power levels for their transmitted signals, allowing them to coexist within the same resource block.

The fundamental principle behind PD-NOMA is to exploit the diverse channel conditions of users in the power domain. Specifically, users with stronger channel conditions are assigned lower power levels, while users with weaker channels are allocated higher power levels. This power imbalance results in varying signal strengths at the receiver, enabling successful decoding of the weaker signals at the expense of slightly degrading the stronger signals. This concept is known as "successive interference cancellation" (SIC). The SIC process at the receiver involves decoding the strongest signal first, subtracting its contribution from the received signal, and then attempting to decode the next strongest signal, and so on. This successive decoding allows PD-NOMA to recover the signals of all users within the same resource block, effectively achieving multi-user multiplexing.

In a K users system, SIC is employed $(K - 1)$ times to successfully decode all users at the receiver, as shown in Fig. 2.7. The received signal y of K-users in uplink is expressed as

$$y = \sum_{i=1}^{K} \sqrt{P_i} h_i x_i + w. \tag{2.8}$$

Here, we assume that users gains are arranged as $|h_K|^2 > |h_{K-1}|^2 > \cdots > |h_1|^2$ at the receiver. Thus, Kth strongest user's symbol is detected first by considering remaining users as interference using maximum likelihood detector (MLD) as

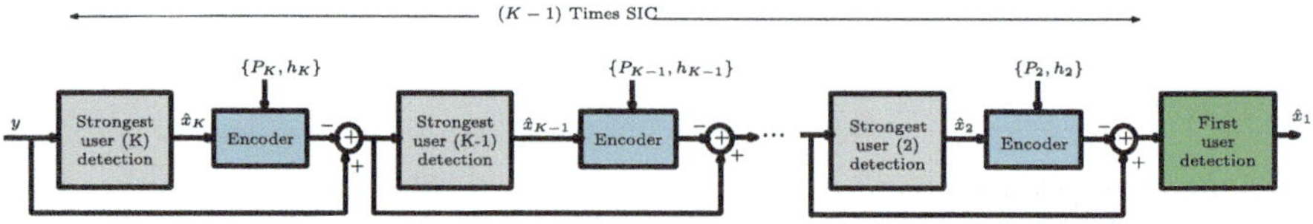

Fig. 2.7 SIC-based signal detection in uplink scenario

$$\hat{x}_K = \arg \min_{x_K \in \mathcal{A}_K} ||y - \sqrt{P_K} h_K x_K||^2, \tag{2.9}$$

where $\mathcal{A}_K$ is the Kth user constellation. Signal after Kth user SIC is written as

$$y_K = \sum_{i=1}^{K} \sqrt{P_i} h_i x_i + w - \sqrt{P_K} h_K \hat{x}_K = \sum_{i=1}^{K-1} \sqrt{P_i} h_i x_i + w + \underbrace{[\sqrt{P_K} h_K (x_K - \hat{x}_K)]}_{\text{imperfectSIC}}. \tag{2.10}$$

If SIC is perfect ($x_K = \hat{x}_K$), Kth user signal completely removed in the signal y_K. Similarly, $(K-1)$ user symbol is detected using MLD as

$$\hat{x}_{(K-1)} = \arg \min_{x_{(K-1)} \in \mathcal{A}_{(K-1)}} ||y_K - \sqrt{P_{(K-1)}} h_{(K-1)} x_{(K-1)}||^2. \tag{2.11}$$

Signal after $(K-1)$th user SIC is written as

$$\begin{aligned} y_{K-1} &= \sum_{i=1}^{K-1} \sqrt{P_i} h_i x_i + w - \sqrt{P_{K-1}} h_{K-1} \hat{x}_{K-1} \\ &= \sum_{i=1}^{K-2} \sqrt{P_i} h_i x_i + w + \underbrace{[\sqrt{P_{K-1}} h_{K-1} (x_{K-1} - \hat{x}_{K-1})]}_{\text{ISIC}}. \end{aligned} \tag{2.12}$$

If SIC is perfect ($x_{K-1} = \hat{x}_{K-1}$), $(K-1)$th user signal completely removed in the signal y_{K-1}. Similarly, we perform other users SIC, and signal of first user (assuming perfect SIC) is expressed as

$$y_1 = \sqrt{P_1} h_1 x_1 + w. \tag{2.13}$$

Thus, user-1 detects its signal without any interference, unlike other users in the PD-NOMA system. Kth user experiences interference from the $(K-1)$ users and interference level decreases as we remove successive users signal from the received signal after applying the SIC. Thus, the first user gets full diversity, and the Kth user has the least diversity in the NOMA system.

2.4.1 The Advantages of PD-NOMA

PD-NOMA offers several key advantages in modern wireless communication systems such as

Enhanced Spectral Efficiency: PD-NOMA enables multiple users to share the same resource block, significantly increasing the spectral efficiency compared to traditional orthogonal multiple access schemes. For example, if two users have their channel gain difference, NOMA can significantly enhance the spectral efficiency of the system [4].

Improved User Fairness: PD-NOMA ensures that all users can be accommodated in the system, even those with weaker channel conditions, promoting fairness in resource allocation.

Robustness to Channel Variability: PD-NOMA's ability to adapt power levels based on individual user channels makes it robust in handling variations in channel conditions, providing better performance in dynamic environments.

Low Complexity Implementation: PD-NOMA receivers require moderate complexity, making it feasible for practical implementations in real-world communication systems.

Low Latency: PD-NOMA supports low-latency communication, making it suitable for applications that require real-time data transmission, such as IoT devices and mission-critical services.

2.4.2 Downlink PD-NOMA

Downlink PD-NOMA is applied in the downlink transmission of wireless communication systems. In downlink communication, the base station (BS) or access point transmits data to multiple user devices simultaneously. In traditional downlink communications, such as in OMA schemes like TDMA or FDMA, each user is allocated exclusive time or frequency resources, resulting in separate transmissions for each user. Downlink PD-NOMA, on the other hand, addresses these challenges by exploiting the non-orthogonality principle, allowing multiple users to share the same time-frequency resource block simultaneously. By efficiently sharing the available resources among multiple users, downlink NOMA can significantly boost system capacity and throughput, making it a promising candidate for future 5G and beyond communication networks.

The downlink NOMA transmission process involves the three steps:

1. User Clustering: Users are grouped into different clusters based on their channel conditions and quality-of-service (QoS) requirements. Each cluster includes users with different channel strengths.
2. Power Allocation: Within each cluster, the users are allocated different power levels for their transmitted signals. The users with weaker channel conditions are allocated higher power, while users with better channel conditions are assigned lower power.
3. Successive Interference Cancellation (SIC): At the receiver side, the users with strong channels decode their signals first. Then, the decoded signals are subtracted from the received signal to cancel their contributions. This process continues iteratively until all users' signals are successfully recovered.

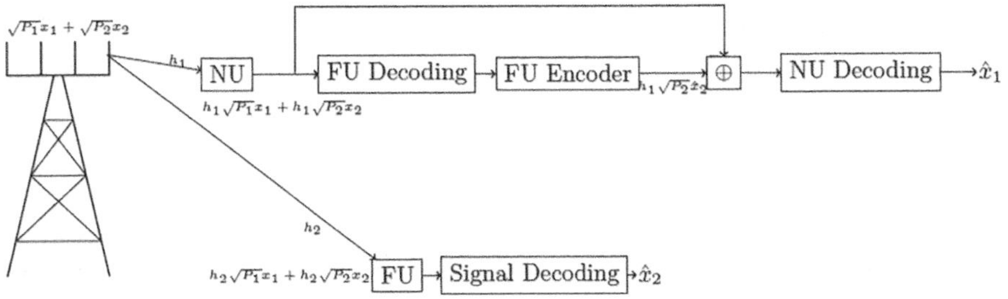

Fig. 2.8 Two user downlink NOMA system

For example, downlink PD-NOMA system by considering two users is shown in Fig. 2.8. NU and FU have channel response h_1 and h_2, respectively with $|h_1|^2 >> |h_2|^2$. Thus, power allocation at the BS for NU and FU is P_1 and P_2, respectively and $P_1 < P_2$. The composite transmitted signal x_{com} at BS is expressed as

$$x_{\text{com}} = \sqrt{P_1}x_1 + \sqrt{P_2}x_2. \tag{2.14}$$

Received signal y_1 at the NU is expressed as

$$y_1 = h_1[\sqrt{P_1}x_1 + \sqrt{P_2}x_2] + w_1, \tag{2.15}$$

where w_1 is AWGN of zero mean and σ^2 variance. Since, $P_1 < P_2$, NU applies the SIC for FU signal detection and y_1 is re-written as

$$y_1 = h_1\sqrt{P_2}x_2 + \underbrace{h_1\sqrt{P_1}x_1}_{\text{interference}} + w_1. \tag{2.16}$$

Using the MLD, decoded FU symbol at NU is $\hat{x}_2^{\text{nu}}$. After SIC, signal is expressed as

$$\tilde{y}_1 = h_1\sqrt{P_1}x_1 + \underbrace{h_1\sqrt{P_2}[x_2 - \hat{x}_2^{\text{nu}}]}_{I \quad \text{ISIC}} + w_1. \tag{2.17}$$

If SIC is perfect, $I = h_1\sqrt{P_2}[x_2 - \hat{x}_2^{\text{nu}}] = 0$, and $\tilde{y}_1 = h_1\sqrt{P_1}x_1 + w_1$. NU symbol is decoded using MLD as $\hat{x}_1$ from the $\tilde{y}_1$. The signal-to-interference plus noise ration (SINR) of the NU is expressed as

$$\text{SINR}_{\text{nu}} = \frac{P_1|h_1|^2}{\sigma_I^2 + \sigma^2}, \tag{2.18}$$

where σ_I^2 is the variance of interference I. Therefore, the sum-rate C_{nu} of the NU is expressed as

$$C_{\text{nu}} = B\log_2(1+\text{SINR}_{\text{nu}}) = B\log_2\left(1+\frac{P_1|h_1|^2}{\sigma_I^2+\sigma^2}\right). \tag{2.19}$$

In perfect SIC case, the sum-rate of NU is given as $C_{\text{nu}} = B\log_2\left(1+\frac{P_1|h_1|^2}{\sigma^2}\right)$, which is free form any interference and similar to point-to-point communication with transmitted signal power P_1.

Received signal y_2 at the FU is written as

$$y_2 = h_2[\sqrt{P_1}x_1 + \sqrt{P_2}x_2] + w_2 = h_2\sqrt{P_2}x_2 + \underbrace{h_2\sqrt{P_1}x_1}_{\text{interference}} + w_2, \tag{2.20}$$

where w_2 is the AWGN at the FU. $P_2 > P_1$, thus interference power is lower than the signal power, and FU signal is directly decoded from the received signal y_2 by considering NU as interference. The SINR of the FU is expressed as

$$\text{SINR}_{\text{fu}} = \frac{P_2|h_2|^2}{P_1|h_2|^2+\sigma^2}, \tag{2.21}$$

where $P_1|h_2|^2$ is the power of interference. Therefore, the sum-rate C_{fu} of the FU is expressed as

$$C_{\text{fu}} = B\log_2(1+\text{SINR}_{\text{fu}}) = B\log_2\left(1+\frac{P_2|h_2|^2}{P_1|h_2|^2+\sigma^2}\right). \tag{2.22}$$

Example 2.3 In Fig. 2.9, average BER performance of the NOMA system is plotted by considering channel gain of NU and FU as $|h_1|^2 = 32|h_2|^2$. Power allocation of NU and FU is considered as $P_1 = \{0.3, 0.1\}$ and $P_2 = \{0.7, 0.9\}$, respectively. NU first implement the SIC to remove the interference and subsequently decodes own signal using MLD. FU directly decodes it's signal in the presence of interference due to NU. Thus, FU's BER deteriorates as compared to NU as observed in Fig. 2.9. Both users use BPSK for data transmission in Fig. 2.9. For $P_1 = 0.3$ and $P_2 = 0.7$, FU experiences more deterioration in BER due to higher interference.

In the K user downlink PD-NOMA system, kth user sum-rate is expressed as

$$C_k = B\log_2\left(1+\frac{P_k|h_k|^2}{\sum_{i=1}^{k-1}P_i|h_k|^2+\sigma^2}\right), \quad k = 1, 2, \ldots, K. \tag{2.23}$$

Here, we assume that perfect SIC is implemented. If $\gamma_k = \frac{|h_k|^2}{\sigma^2}$, C_k is expressed as

$$C_k = B\log_2\left(1+\frac{P_k\gamma_k}{\sum_{i=1}^{k-1}P_i\gamma_k+1}\right). \tag{2.24}$$

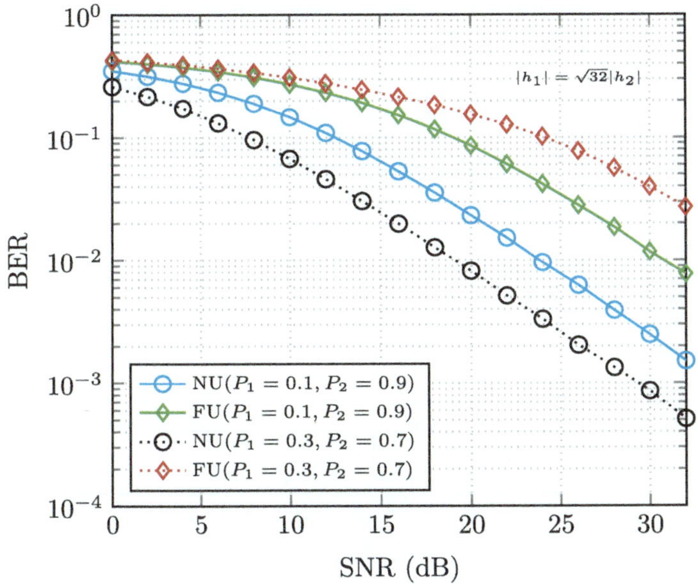

Fig. 2.9 BER versus SNR performance of two users NOMA system

Example 2.4 In Fig. 2.10, sum-rate of NOMA system is plotted by varying the number of users K. Channel gains of users are sorted as $|h_1|^2 < |h_2|^2 < \cdots < |h_K|^2$, and their corresponding power allocation coefficients are ordered as $\alpha_1 > \alpha_2 > \cdots \alpha_K$. When performing power allocation, it's imperative to ensure that α_1 is greater than the sum of $\alpha_2, \alpha_3, \ldots \alpha_K$, α_2 is greater than the sum of $\alpha_3, \alpha_4, \ldots \alpha_K$, α_3 is greater than the sum of $\alpha_4, \ldots \alpha_K$, and so forth. SIC is implemented at strong users to remove the interference. Sumrate of each users is calculated as

$$C_k = \log_2\left(1 + \frac{P_k\gamma_k}{\sum_{i=1}^{k-1} P_i\gamma_k + 1}\right). \tag{2.25}$$

where $P_k = \alpha_k P$ and P is the total power. Initially sum-rate of NOMA users increases and after particular value of K sum-rate starts decreasing, as observed in Fig. 2.10. Based on our simulation results, it is evident that continuously increasing the number of multiplexed users within the same carrier is not feasible.

Example 2.5 SNRs of user-1 and user-2 are given SNR$=\frac{P|h_1|^2}{N_0} = 0\,\text{dB}$ and SNR$=\frac{P|h_2|^2}{N_0} = 40\,\text{dB}$, respectively. Total power is one unit and power split is $P_1 = 0.7$ and $P_2 = 0.3$ between NOMA users. Calculate the sum-rate of NOMA users.
$P = 1$, User-1 SNR $\frac{P|h_1|^2}{N_0} = 0\,\text{dB}$, $\frac{|h_1|^2}{N_0} = 1$, $|h_1|^2 = N_0$ and user-2 SNR $\frac{P|h_2|^2}{N_0} = 40\,\text{dB}$, $\frac{|h_2|^2}{N_0} = 10^4$. User-1 (weak user) sum-rate in NOMA is $R_1 =$

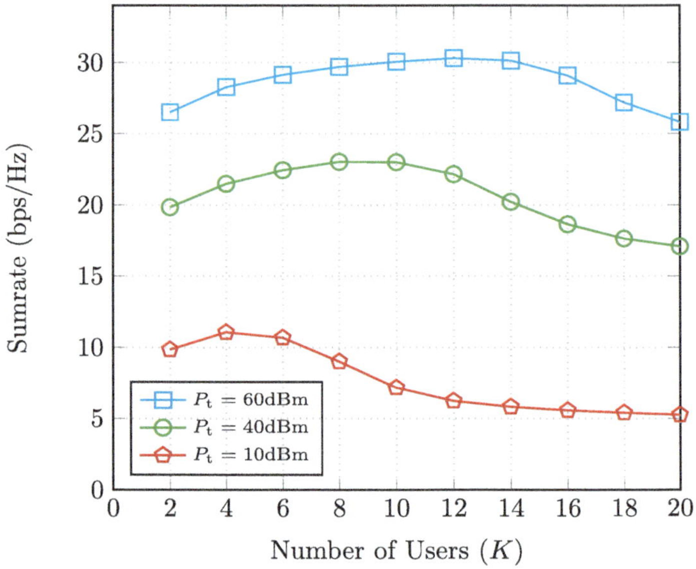

Fig. 2.10 Sumrate versus number of users

$\log_2\left(1 + \frac{P_1|h_1|^2}{P_2|h_1|^2+N_0}\right) = \log_2\left(1 + \frac{.7\times N_0}{.3\times N_0+N_0}\right) = 0.62149$ bps/Hz. User-2 (strong user) sum-rate is $R_2 \log_2\left(1 + \frac{P_2|h_2|^2}{N_0}\right) = \log_2(1 + 0.3 \times 10^4) =$ 11.5512 bps/Hz. When only single user is present, then corners points sum-rate is calculated as $\log_2\left(1 + \frac{P|h_2|^2}{N_0}\right) = 13.2879$ bps/Hz, and $\log_2\left(1 + \frac{P|h_1|^2}{N_0}\right) = 1$ bps/Hz. Sum rate points of users are shown in Fig. 2.11.

Example 2.6 The sum-rate of User-1 and User-2 is illustrated in Fig. 2.12, showcasing the variation in their received signal power levels. As the difference between power levels increases, NOMA demonstrates enhanced sum-rate performance compared to OMA. However, the sum-rate gain of NOMA diminishes as the power levels of the two users approach each other, as observed in Fig. 2.12.

Example 2.7 Sumrate of NOMA and OMA users is shown in Fig. 2.13. NOMA users have different sum-rate.

Despite its benefits, downlink NOMA also presents challenges related to power allocation, interference management, and complexity at the receiver side. Researchers and industry stakeholders continue to explore innovative solutions to overcome these challenges and fully unleash the potential of downlink NOMA for advanced wireless communication systems.

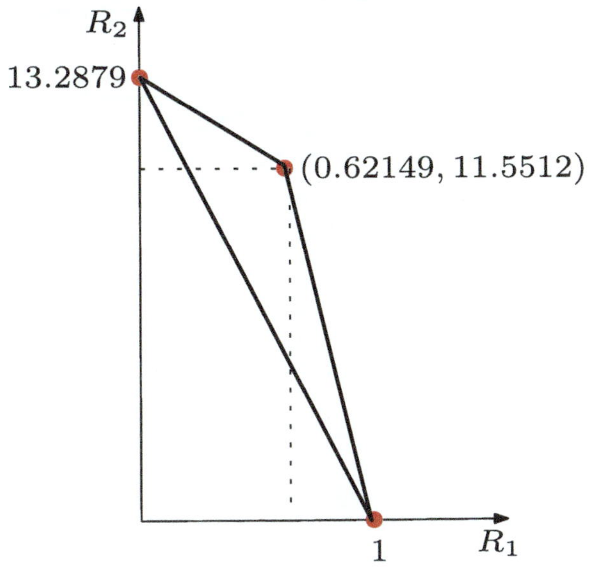

Fig. 2.11 Sumrate points

The fundamental concept of PD-NOMA revolves around the implementation of multiple access in the power domain. Consequently, efficient power allocation plays a pivotal role in realizing the maximum advantages of NOMA transmission. In the next subsection, we explore optimal power allocation strategies for various PD-NOMA schemes.

2.4.3 Optimal Power Allocation in PD-NOMA System

The NOMA scheme allows for the adaptable control of users' attainable data rates and offers an effective approach to improve user fairness. A commonly employed fairness measure is maximin fairness (MMF), which aims to maximize the lowest sum-rate of an user in the NOMA system. The power allocation required to attain MMF is determined through the resolution of the subsequent optimization problem in a two-user NOMS system:

$$\begin{aligned} \max_{P_1, P_2} \quad & \min \{C_{\text{nu}}(P_1, P_2), C_{\text{fu}}(P_1, P_2)\} \\ \text{s.t.} \quad & 0 \leq P_1 \leq P_2 \\ & P_1 + P_2 \leq P \end{aligned} \tag{2.26}$$

Here, P is the total available power at the BS. The optimal solution of the above optimization problem is given as [6]

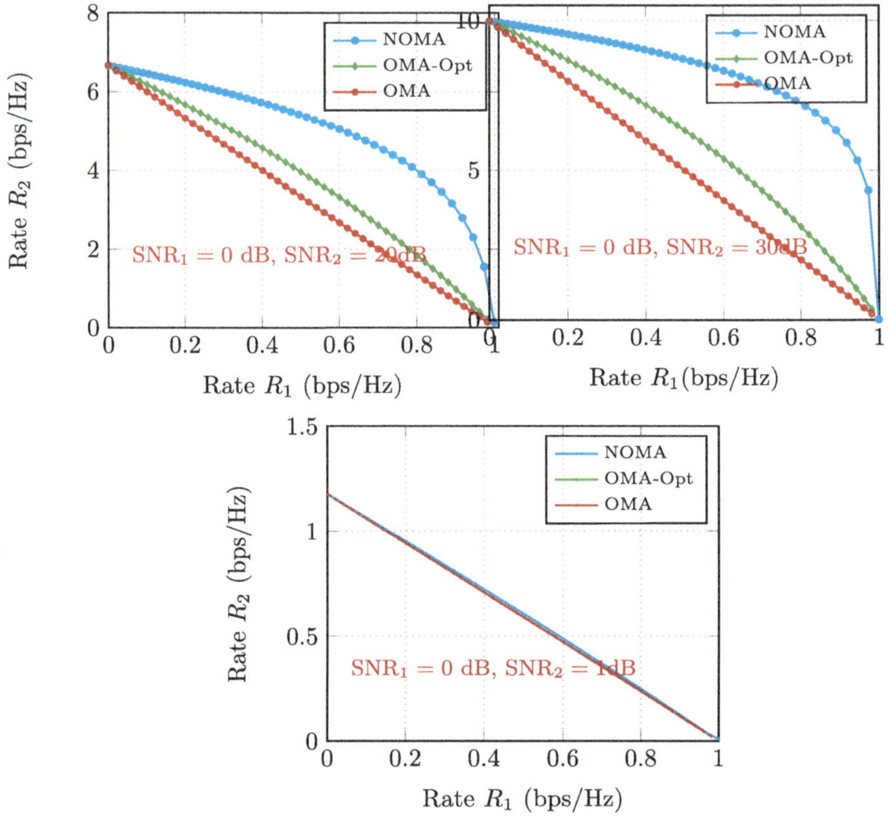

Fig. 2.12 Sum rate plot of two users

$$P_1^{\star} = \frac{-(\gamma_1 + \gamma_2) + \sqrt{(\gamma_1 + \gamma_2)^2 + 4\gamma_1\gamma_2^2 P}}{2\gamma_1\gamma_2} \quad \text{and} \quad P_2^{\star} = P - P_1^{\star}. \tag{2.27}$$

Here, we assume $\gamma_1 \geq \gamma_2$ and $\gamma_k = \frac{|h_k|^2}{\sigma^2}$, $k = 1, 2$. Both users attain an identical rate at the optimal point. This demonstrates that, according to the MMF criterion, the NOMA system ensures complete fairness between the two users sharing a single channel [6].

Example 2.8 Two FU and NU users have SNRs of 10 dB and 30 dB in a NOMA system, respectively. The total available power is one unit. Calculate the optimal power allocation for these two users.
$\gamma_1 = 10\,\text{dB} = 10$, $\gamma_2 = 20\,\text{dB} = 100$, and $P = 1$, Thus

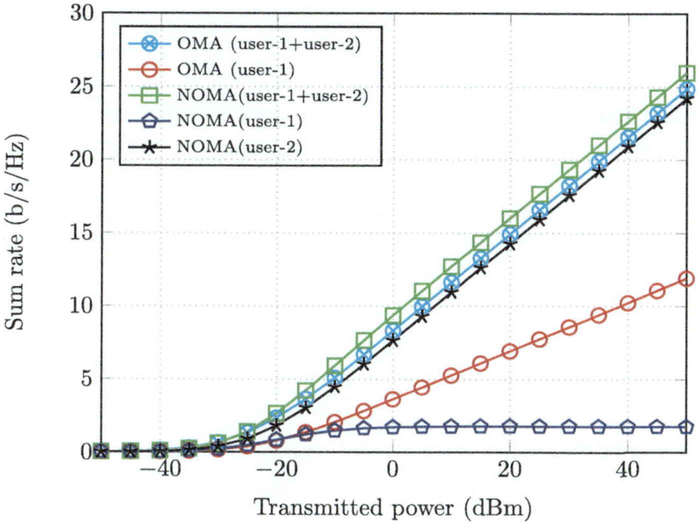

Fig. 2.13 Sumrate versus transmitted power plot

$$P_1^\star = \frac{-(\gamma_1+\gamma_2)+\sqrt{(\gamma_1+\gamma_2)^2+4\gamma_1\gamma_2^2 P}}{2\gamma_1\gamma_2} = \frac{-110+\sqrt{110^2+40.10000}}{20.100} = 0.2660 \tag{2.28}$$

$P_2 = 0.7340.$

Weighted SR Maximization: weighted sum-rate of the NOMA system expressed as

$$C_{\text{wsr}} = W_1 C_{\text{nu}}(P_1, P_2) + W_2 C_{\text{fu}}(P_1, P_2), \tag{2.29}$$

where W_1 and W_2 denote the weight of NU and FU sum-rate, respectively. The weighted sum-rate optimization is formulated as

$$\begin{aligned} \max_{P_1, P_2} \quad & \max\{W_1 C_{\text{nu}}(P_1, P_2) + W_2 C_{\text{fu}}(P_1, P_2)\} \\ \text{s.t.} \quad & 0 \le P_1 \le P_2 \\ & P_1 + P_2 \le P \end{aligned} \tag{2.30}$$

It is worth noting that the weighted sum-rate presents a non-convex problem because of the interference between NU and FU. However, its optimal solution can still be determined using the following approach.

$$P_1^\star = \frac{W_2\gamma_2 - W_1\gamma_1}{\gamma_1\gamma_2(W_1 - W_2)} \quad \text{and} \quad P_2^\star = P - P_1^\star. \tag{2.31}$$

Here, we assume that $1 < \frac{W_2}{W_1} < \frac{\gamma_1}{\gamma_2}$, $P > 2P_1^\star$ and $\gamma_1 \geq \gamma_2$. The weighted sum-rate is optimized by setting the derivative of $F = W_1 C_{\text{nu}}(P_1, P_2) + W_2 C_{\text{fu}}(P_1, P_2)$ with respect to P_1 is zero by substituting $P_1 \leq P_2$ and $P_1 \leq P/2$ [6].

SR Maximization with QoS: Now, we focus to maximize the sum-rate while adhering to QoS constraints, and in this scenario, the power allocation problem is formulated as

$$\begin{aligned} \max_{P_1, P_2} \quad & \max \{C_{\text{nu}}(P_1, P_2) + C_{\text{fu}}(P_1, P_2)\} \\ \text{s.t.} \quad & 0 \leq P_1 \leq P_2 \\ & P_1 + P_2 \leq P \\ & C_{\text{nu}}(P_1, P_2) \geq C_{\text{nu}}^{\min} \\ & C_{\text{fu}}(P_1, P_2) \geq C_{\text{fu}}^{\min}, \end{aligned} \tag{2.32}$$

where $C_{\text{nu}}^{\min}$ and $C_{\text{fu}}^{\min}$ are the minimum sum-rate of NU and FU, respectively to maintain QoS of users in the NOMA system. The optimization problem is re-expressed as

$$F(P_1) = B \log_2(1 + P_1\gamma_1) + B \log_2 \left(\frac{1 + P\gamma_2}{1 + P_1\gamma_2} \right), \tag{2.33}$$

Here, we consider $P_1 \leq P_2$, $P_1 \leq P/2$, and $P_2 = P - P_1$. By substituting $\frac{\partial F}{dP_1} = 0$, and $F(P_1)$ is monotonically nondecreasing, Thus, the maximum sum-rate is achieved at the upper bound of P_1. Using QoS sum-rate constraints, the optimal power allocation is given as [6]

$$P_1^\star = \frac{P\gamma_2 - A_2 + 1}{A_2\gamma_2} \quad \text{and} \quad P_2^\star = P - P_1^\star, \quad P \geq \Gamma \tag{2.34}$$

where $A_2 = 2^{\frac{C_{\text{fu}}^{\min}}{B}}$, $A_1 = 2^{\frac{C_{\text{nu}}^{\min}}{B}}$, $\Gamma = \frac{A_2(A_1-1)}{\gamma_1} + \frac{A_2-1}{\gamma_2}$, and $A_2 \geq 2$ with $\frac{A_1-1}{\gamma_1} \leq P_1 \leq \frac{P\gamma_2 - A_2+1}{A_2\gamma_2}$

Power Allocation to maximize Weighted EE Maximization: Energy efficiency (EE) in wireless systems refers to the ability of a wireless communication network to transfer data while minimizing the energy consumption required for data transmission and processing. It encompasses optimizing the utilization of resources, reducing power consumption in network components, and ensuring that energy is used effectively to transmit data. The optimal power allocation for maximizing EE, defined as the ratio between the data rate and the power consumption, and it is determined as

$$\begin{aligned} \max_{P_1, P_2} \quad & \max \left\{ \frac{W_1 C_{\text{nu}}(P_1, P_2) + W_2 C_{\text{fu}}(P_1, P_2)}{P_1 + P_2 + P_c} \right\} \\ \text{s.t.} \quad & 0 \leq P_1 \leq P_2 \\ & P_1 + P_2 \leq P \end{aligned} \tag{2.35}$$

where P_c denotes the power consumption of the circuit. The optimization problem is also expressed as $\max\left\{\frac{W_1 C_{\text{nu}}(P_1)+W_2 C_{\text{fu}}(P_1,P_q)}{P_q+P_c}\right\}$, where $P_2 = P_q - P_1$ and $P_1 \leq P_q/2$, $P_q \leq P$ with P_q is a auxiliary variable. To address this, we present the following objective function:

$$F(P_1, P_q, \alpha) = W_1 C_{\text{nu}}(P_1) + W_2 C_{\text{fu}}(P_1, P_q) - \alpha(P_q + P_c), \quad \alpha \geq 0 \tag{2.36}$$

Above optimization problem is convex problem with following conditions [7]

$$\begin{gathered} W_1 \geq W_2 \\ 1 < \frac{W_2}{W_1} \leq \frac{(\gamma_1 + \gamma_1\gamma_2 P_q)^2}{(\gamma_2 + \gamma_1\gamma_2 P_q)^2}. \end{gathered} \tag{2.37}$$

The optimal solution for power allocation can be determined by solving $F(P_1, P_q, \alpha)$, which is parameterized by α, and then iteratively updating α until $F^\star(P_1, P_q, \alpha) = 0$. To achieve this, we begin by solving $F(P_1, P_q, \alpha)$ with a specific α value [6]. Thus, optimal power allocation (considering convexity) is expressed as by considering $\gamma_1 \geq \gamma_2$ and $P/2 \geq P_1^\star$ [7]

$$P_1^\star = \frac{W_2\gamma_2 - W_1\gamma_1}{\gamma_1\gamma_2(W_1 - W_2)}, \quad P_2^\star = P_q^\star - P_1^\star. \tag{2.38}$$

Here

$$P_q^\star = \max\left\{2P_1^\star, \ \min\left(\frac{W_2 B}{\alpha \log 2} - \frac{1}{\gamma_2}, P\right)\right\}. \tag{2.39}$$

Example 2.9 – Downlink scenario two users SNR is expressed as strong user $\text{SNR}_1 = 20\,\text{dB}$ and weak user $\text{SNR}_2 = 0\,\text{dB}$, respectively. Total power P is one unit. Calculate sum-rate of NOMA users.

- Strong user $\frac{|h_1|^2}{N_0} = 100$, Weak user $\frac{|h_2|^2}{N_0} = 1$,
- OMA:-$R_1(\text{SU}) = \frac{1}{2}\log_2(1 + \frac{|h_1|^2}{N_0}) = \frac{1}{2}\log_2(1 + 101) = 3.33\,\text{bits/s/Hz}$, $R_2(\text{WU})$ $= \frac{1}{2}\log_2(1 + \frac{|h_2|^2}{N_0}) = \frac{1}{2}\log_2(1 + 1) = 1/2 = 0.5\text{bits/s/Hz}$
- NOMA: split power, Let $P_1 = 0.2$, $P_2 = 0.8$

$$R_2(FU) = \log\left(1 + \frac{P_2|h_2|^2}{(P_1|h_2|^2 + N_0)}\right) = \log\left(1 + \frac{0.8 \times N_0}{(0.2 \times N_0 + N_0)}\right)$$

$R_2(FU) = \log_2(1 + 0.8/1.2) = 0.7370$ bits/s/Hz

$$R_1(NU) = \log\left(1 + \frac{P_1|h_1|^2}{N_0}\right) \quad \text{: get full diversity}$$

$-R_1(NU) = \log_2(1 + \frac{0.2\times 100 N_0}{N_0}) = \log_2(1 + 20) = 4.3923$ bits/s/Hz

- Both users' data rate is increased and spectral efficiency improves.
- Corner points: $\log\left(1+\frac{P_2|h_2|^2}{N_0}\right)=1$bits/s/Hz and $\log\left(1+\frac{P_1|h_1|^2}{N_0}\right)=$ 6.6582 bits/s/Hz.

2.4.4 Uplink PD-NOMA

Uplink NOMA is used in the uplink transmission of wireless communication systems. In uplink communication, multiple user devices transmit data to the BS or access point. In conventional uplink communications, users transmit their data independently to the BS, and OMA schemes are commonly used. However, these OMA techniques may lead to inefficient utilization of uplink resources, especially in scenarios with a large number of users or varying traffic loads. The key idea is to employ PD-NOMA techniques to multiplex users' signals within the same resource block.

NU and FU transmit their symbols x_1 and x_2 to the BS, as shown in Fig. 2.14. The received signal y at the BS is expressed as

$$y=\sum_{i=1}^{K} h_i\sqrt{P_i}x_i+w, \tag{2.40}$$

where K denotes the total number of users. To separate the overlapped signals, an SIC receiver is employed at the BS. Before the BS detects the message from the kth UE, it first decodes the messages of the preceding ith UEs (where $i<k$) and removes them from its signal successively. The remaining symbols from the $(K-k)$ UEs are treated as interferences. Consequently, this process determines the achievable data rate for the kth UE. Thus, kth user sum-rate is expressed as

$$R_k=B\log_2\left(1+\frac{P_k|h_k|^2}{\sum_{i=k+1}^{K} P_i|h_i|^2+\sigma^2}\right) \tag{2.41}$$

For example two users ($K=2$) uplink PD-NOMA, sum-rate of both users is expressed as

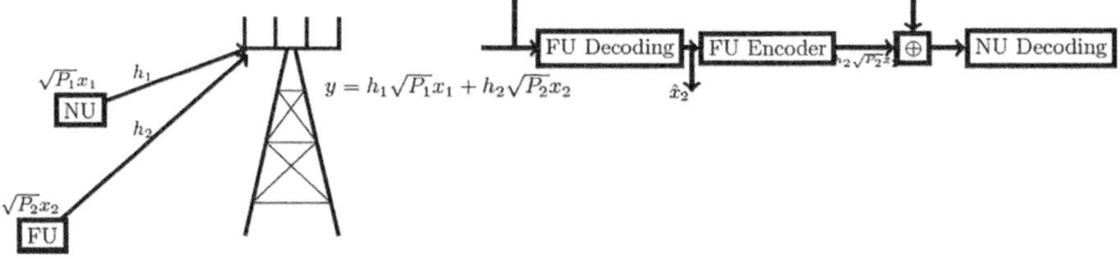

Fig. 2.14 Uplink NOMA system

$$R_1 = B\log_2\left(1+\frac{P_1|h_1|^2}{P_2|h_2|^2+\sigma^2}\right) \quad \text{if } P_1|h_1|^2 > P_2|h_2|^2$$
$$R_2 = B\log_2\left(1+\frac{P_2|h_2|^2}{\sigma^2}\right). \tag{2.42}$$

We should reverse the above signal detection order if $P_1|h_1|^2 < P_2|h_2|^2$. Further, if both users signals are added together, total sum-rate is expressed as

$$R_1 + R_2 = B\log_2\left(1+\frac{P_2|h_2|^2+P_1|h_1|^2}{\sigma^2}\right) \text{ bps/Hz}. \tag{2.43}$$

The above sum-rate asserts that the overall throughput must not surpass the capacity of a point-to-point channel with the sum of the received powers of the two users. This constraint is valid as the signals sent by the two users are independent, resulting in the aggregate received signal power being the sum of the powers of the individual received signals (Fig. 2.15).

Example 2.10 In Fig. 2.16, the sum-rate of individual users is depicted by considering a two-user NOMA uplink system. The distances from the BS are categorized as Set-1 $= \{d_1 = 800, d_2 = 600\}$ m and Set-2 $= \{d_1 = 600, d_2 = 300\}$ m, respectively. The pathloss factor is denoted as $\eta = 3$, and the system bandwidth is 1MHz, with a reference pathloss of 30 dBm at 1 m. The PSD is -174 dBm. Hence, the noise power is calculated as

$$N_0 = 10^{\frac{(-174-30)}{10}} \times 10^6 = 3.9811e^{-15}.$$

Additionally, the transmitted power of each user is expressed as $P_t = 10^{\frac{(-P-30-30)}{10}}$, where P is in dBm. In this scenario, it's evident that one user achieves interference-

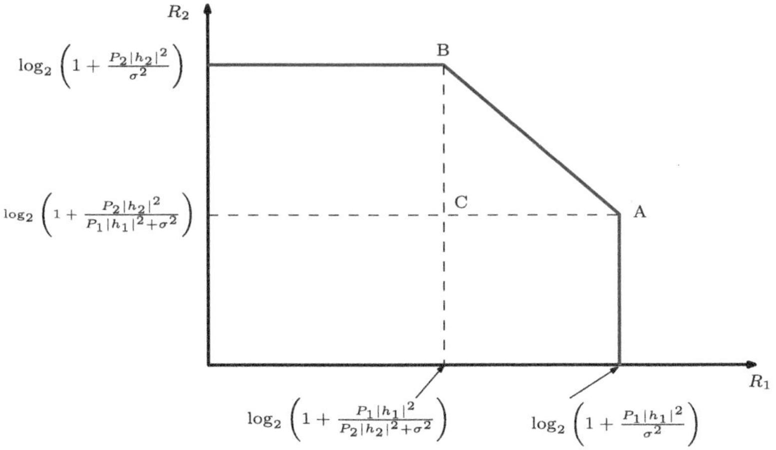

Fig. 2.15 Capacity region of uplink two user NOMA system

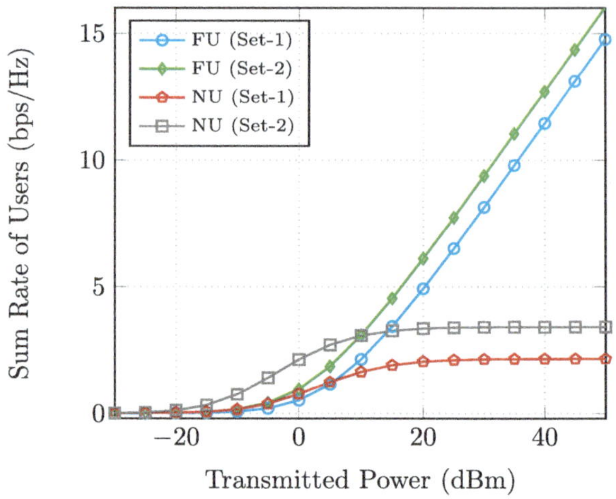

Fig. 2.16 Sumrate of the uplink NOMA system by considering two users

free sum-rate after SIC, while the sum-rate of the other user is affected by the interference from another user, as observed in Fig. 2.16.

2.4.5 Uplink Power Control for PD-NOMA

The BS necessitates equalizing the received power from various UEs to a specified target power level to prevent inter-user interferences. The fundamental transmit power P for each UE is calculated using the following expression [8]

$$P = \min\left\{P_{\max}, P_u + 10\log_{10}(M) + \omega P_{\text{loss}}\right\}, \tag{2.44}$$

where P_u is the target received power and it is informed to all UEs via control signaling or system information. $P_{\max}$ represents the maximum transmit power, and M denotes the number of allocated resource blocks for each UE communication. P_{loss} encompasses both distance-dependent path loss and lognormal shadowing. The parameter ω signifies the path loss attenuation factor associated with fractional transmission power control. In the context of uplink PD-NOMA, a diverse arrived power is essential for the SIC receiver to distinguish multiplexing UEs. To achieve this diversity, a power back-off scheme is introduced for uplink PD-NOMA transmission. Let ρ denote the power back-off step of the target received power. The transmit power of the kth UE in a NOMA set is then expressed as

$$P_k = \min\left\{P_{\max}, P_u - (k-1)\rho + 10\log_{10}(M_i) + \omega P_{k,\text{loss}}\right\}, k = 1, 2, \ldots, K \tag{2.45}$$

Above expression illustrates that the arrived power of NOMA UEs undergoes a gradual degradation with a step of ρ. For instance, the arrived power of the second strongest UE is ρ dB lower than that of the strongest UE, and the arrived power of the third strongest UE is ρ dB lower than that of the second strongest UE, and so forth. This design proves advantageous in systematically mitigating co-channel interferences. It is important to note that the scenario where P_u equals Pmax only arises in extreme cases. Since all UEs in a PD-NOMA set share the same resource blocks, the disparity between P_k and $P_{k'}$ (where $k' \geq k$) can be articulated as

$$P_k - P_{k'} = \omega(P_{k,\text{loss}} - P_{k^\star,\text{loss}}) + (k' - k)\rho \quad \text{dB} \tag{2.46}$$

$$\frac{P_k}{P_{k'}} = \frac{P_{k,\text{loss}}}{P_{k^\star,\text{loss}}} 10^{\left(\frac{(k'-k)\rho}{10}\right)} \quad \text{watt} \tag{2.47}$$

Before a PD-NOMA UE determines its transmit power and sends messages, the BS allocates multiple UEs to form a NOMA set and establishes the order of UEs based on the path loss between the UEs and the BS. UEs with larger path losses are assigned higher order values, indicating that their corresponding arrived power is smaller than that of other UEs. This approach proves beneficial in conserving UE power consumption, particularly for UEs situated farther away from the BS, such as those at the cell edge. Following this, the BS communicates the assigned order index to the respective UE via individual control channels. Assuming the order of the kth UE precedes that of the k'th UE, i.e., $P_k|h_k|^2 \geq P_{k'}|h_{k'}|^2$.

2.4.6 Issues in SIC

SIC is a key technique used in PD-NOMA to mitigate interference among users. However, SIC has its challenges and limitations in the context of PD-NOMA. Below we highlight some issues in SIC-based received design.

Error Propagation: If the decoding of an early user's signal is incorrect during the SIC process, subsequent users' signals may be affected, leading to error propagation. This can degrade the overall system performance. For example, FU signal is not decoded correctly at the NU, then FU signal residual affects the NU signal detection. In practice, error propagation is minimal at high SNR region

Complexity: SIC involves decoding multiple signals in succession, which increases the computational complexity, especially when dealing with a large number of users. The complexity grows linearly with the number of users in the NOMA set.

Impact of Channel Conditions: SIC's performance is sensitive to the channel conditions of different users. In scenarios where channels experience deep fades or high levels of interference, SIC may struggle to reliably decode signals. SIC may face challenges in dynamic and rapidly changing environments. Sudden variations in channel conditions or user dynamics may impact the effectiveness of successive interference cancellation.

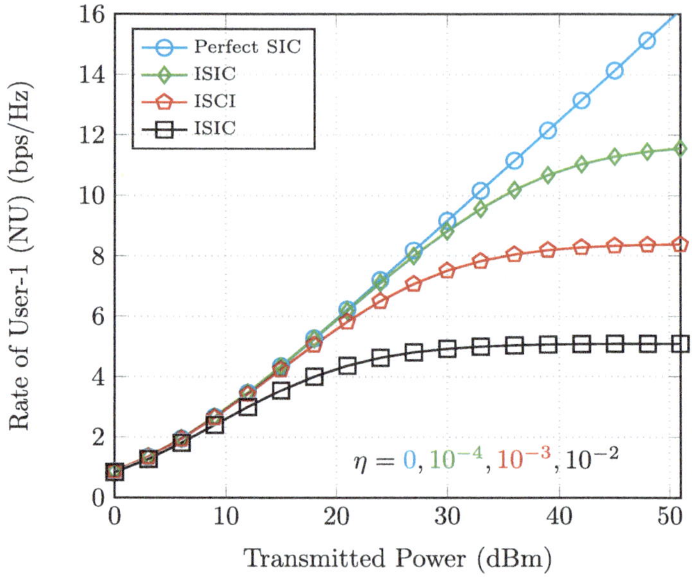

Fig. 2.17 Impact of ISIC on the sum-rate performance

Optimal Order Determination: The optimal order in which users should be decoded is a critical factor. Determining this order can be challenging, and selecting a suboptimal order may result in inefficient interference cancellation.
Finite Precision Effects: In practical implementations with finite precision, the cancellation process may introduce quantization errors, leading to performance degradation

Example 2.11 Impact of imperfect SIC (ISIC) on the sum-rate performance is shown in Fig. 2.17. NU and FU distances from the BS are 500 m and 1000 m, respectively. Transmitted signal BW is I MHz and noise PSD is −174 dBm. Pathloss exponent is 4 with power allocation coefficients 0.25 and 0.75 for the NU and FU, respectively. In ISIC case, sum-rate of NU and FU is expressed as

$$R_{\text{nu}} = B \log_2 \left(1 + \frac{P_1|h_1|^2}{\eta P_2|h_1|^2 + \sigma^2}\right)$$
$$R_{\text{fu}} = B \log_2 \left(1 + \frac{P_2|h_2|^2}{\sigma^2}\right). \tag{2.48}$$

where η denotes the residue of ISIC at the NU. Sumrate of NOMA system is saturate at high SNR region depending on the η value as shown in Fig. 2.17 due to both NU and FU experience interference.

2.4.7 PD-NOMA Resource Allocation and Power Control

Common algorithms in power domain NOMA for resource allocation are [4, 6]

User grouping: Users are grouped based on their channel conditions and quality-of-service (QoS) requirements. Two common NOMA user groups are typically formed: a strong user (or NU) with a better channel condition and a weak user (or FU) with a relatively worse channel condition.
SIC is a key algorithm in PD-NOMA that mitigates interference by successively decoding and canceling signals from stronger to weaker users.
Fractional transmit power control adjusts the transmit power of users to achieve a balance in received power levels, contributing to improved fairness and system performance.
Proportional fairness algorithms allocate resources in a way that balances fairness and efficiency. Users with better channel conditions receive more resources, while ensuring a fair distribution among all users.
Game theory models interactions among users, where users act strategically to maximize their own utility. Cooperative game theory, non-cooperative game theory, and coalition formation are applied to address resource allocation challenges.
Machine learning techniques, such as reinforcement learning, neural networks, and deep learning, are increasingly applied to optimize resource allocation in PD-NOMA systems, leveraging their ability to adapt to dynamic environments.
Particle swarm optimization (PSO) is a nature-inspired optimization algorithm that simulates the social behavior of particles. It is used for global optimization in resource allocation problems.
Genetic algorithms (GA) is an evolutionary algorithm that mimics the process of natural selection. It is employed for optimizing resource allocation parameters in PD-NOMA systems.
Maximizes the sum-rate or the sum of the logarithms of data rates while satisfying power constraints. Essentially allocates more power to users with better channel conditions using water filling algorithm.
Ensures fairness by providing lower data rates to the FU and higher rates to the NU by considering maximizes the minimum user data rate.

Optimal power allocation in NOMA is a complex optimization problem, and the choice of strategy may depend on system requirements, the specific application, and the channel models used. The key is to balance fairness, maximize system efficiency, and meet the QoS needs of all users. For example, QoS of FU is fixed, then we can allocate remaining power of frequency to the NU in the system.

2.5 Impact of Hardware Impairment (HI) in PD-NOMA

The impact of hardware impairment (HI) in PD-NOMA can significantly affect system performance. Hardware impairments introduce non-idealities in the transmission and reception processes, leading to distortions in the signals. Here are some key impacts of HI in PD-NOMA:

- Hardware impairments, such as non-linearities and distortions, can degrade the quality of transmitted signals. This degradation affects the accuracy of decoding at the receiver, leading to reduced reliability in signal detection.
- HI can amplify interference levels, especially in scenarios where successive interference cancellation (SIC) is employed. Interference from one user may be inaccurately decoded as a useful signal, leading to higher interference levels for subsequent users.
- The presence of hardware impairments can reduce the overall system capacity. This reduction is particularly significant in PD-NOMA, where efficient interference management is crucial for achieving high spectral efficiency.
- HI may introduce variations in the received power levels among users, impacting the fairness of resource allocation. Users with impaired hardware may experience suboptimal performance compared to others, leading to unfair distribution of resources.
- Hardware impairments contribute to higher error rates in signal transmission and reception. This, in turn, affects the reliability of communication links, impacting the achievable data rates and system performance.
- HI introduces additional challenges in designing algorithms for resource allocation and power control. Algorithms need to account for hardware impairments and adjust parameters accordingly to mitigate their effects.

Here, we highlight the effect of HIs and imperfect channel state information (ICSI) in the NOMA system. The composite signal at the BS in the presence of HI is expressed as

$$x_c = \sqrt{P_1}x_1 + \sqrt{P_2}x_2 + \eta_t, \tag{2.49}$$

Here, η_t denotes the HI effect in the transmitted signal [9]. We can model η_t as zero mean and $P\sigma^2_{\eta_t}$ variance. Variance depends on the transmitted power of the composite signal. ICSIs ($\hat{h}_1$ and $\hat{h}_2$) are available at the receiver and channel impulse response can be modeled as $h_1 = \hat{h}_1 + e_1$ and $h_2 = \hat{h}_2 + e_2$. Here, we assume that h_1 and h_2 are perfect channel of NU and FU respectively in the downlink PD-NOMA system. $e_i, i = 1, 2$ is the error in channel estimation and follows the zero mean and $\sigma^2_{e_i}$ variance. Consequently, the erroneously estimated channel coefficients $\hat{h}_1$ and $\hat{h}_2$ follow $\mathcal{CN}(0, \sigma_1^2 + \sigma^2_{e_1})$ and $\mathcal{CN}(0, \sigma_2^2 + \sigma^2_{e_2})$ respectively.

The received signal at the NU is expressed as

$$y_1 = h_1 x_c + \eta_r + w = h_1\sqrt{P_2}x_2 + h_1\sqrt{P_1}x_1 + \underbrace{h_1\eta_t + \eta_r}_{\text{HI}} + w, \tag{2.50}$$

where η_r is the HI at the receiver and modeled complex RV with zero mean and $\sigma_{\eta_r}^2$ variance. Let $w^\star = h_1\eta_t + \eta_r + w$, and $w^\star$ follows zero mean and $P\sigma_1^2\sigma_{\eta_t}^2 + \sigma_{\eta_r}^2 + \sigma^2$ variance. The y_1 is re-expressed as

$$y_1 = h_1\sqrt{P_2}x_2 + h_1\sqrt{P_1}x_1 + w^\star \tag{2.51}$$

SIC is applied to decode the FU information x_2 at the NU first and the signal $\tilde{y}_1$ after ISIC is written as

$$\tilde{y}_1 = h_1\sqrt{P_1}x_1 + w^\star + \eta_{\text{isic}}, \tag{2.52}$$

where η_{isic} denotes the residual signal of ISIC at the NU. η_{isic} can be modeled as $\mathcal{CN}(0, \sigma_{\text{isic}}^2)$. The NU signal is decoded form the signal $\tilde{y}_1$. The sum-rate of NU is expressed as

$$R_1 = B\log_2\left(1 + \frac{P_1|h_1|^2}{\sigma_{\text{isic}}^2 + P\sigma_1^2\sigma_{\eta_t}^2 + \sigma_{\eta_r}^2 + \sigma^2}\right). \tag{2.53}$$

If system is perfect, the above sum-rate is equal to the earlier derived sum-rate of the NU.

Similarly, the received signal of the FU is expressed as

$$y_2 = h_2\sqrt{P_2}x_2 + h_2\sqrt{P_1}x_1 + w^\star \tag{2.54}$$

where $w^\star$ jointly represents the noise and HI at the FU. The sum-rate of FU is expressed as

$$R_2 = B\log_2\left(1 + \frac{P_2|h_2|^2}{P_1|h_2|^2 + P\sigma_2^2\sigma_{\eta_t}^2 + \sigma_{\eta_r}^2 + \sigma^2}\right). \tag{2.55}$$

2.6 MIMO and Beamforming for NOMA Systems

MIMO-based NOMA is a system that combines the principles of MIMO technology with NOMA for data transmission in a wireless system [10, 11]. In MIMO systems, multiple antennas are used for both transmitting and receiving data. This technology takes advantage of spatial multiplexing to increase data throughput, enhance reliability, and improve spectral efficiency by exploiting the spatial diversity. NOMA, on the other hand, is a multiple access scheme where multiple users share the same time-frequency resources non-orthogonally. NOMA allows different users' signals to occupy the same time and frequency resources by utilizing power domain multiplexing, superposition coding, and SIC at the receiver to decode multiple user signals.

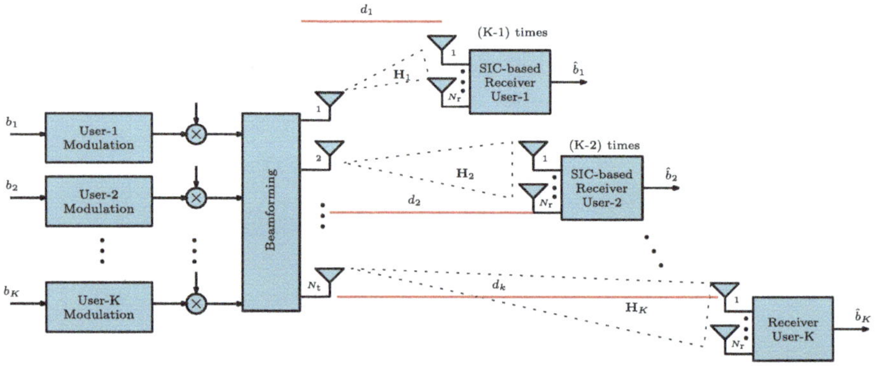

Fig. 2.18 MIMO-NOMA system

MIMO-based NOMA combines these two techniques [10]. It leverages the benefits of MIMO systems, such as spatial multiplexing gains and improved link reliability, with the advantages of NOMA in terms of efficient resource utilization and enhanced throughput. This integration aims to further increase the spectral efficiency and overall performance of wireless communication systems, especially in scenarios with high user density and limited resources. MIMO can be used to enhanced the diversity or multiplexing gain in the NOMA system. In Fig. 2.18, MIMO is used to enhance the diversity of the NOMA system.

The superimposed transmitted signal of K-users in Fig. 2.18 is written as

$$x_c = \sum_{k=1}^{K} \sqrt{P_k} x_k, \tag{2.56}$$

where P_k is the kth user assigned power. The received signal at NU is expressed as

$$\mathbf{y} = \mathbf{H}_{\text{nu}} \mathbf{w} x_c \in \mathcal{C}^{N_r \times 1}, \tag{2.57}$$

where $\mathbf{w} \in N_t \times 1$ is a beamforming vector at the transmitter. The $\mathbf{H}_{\text{nu}} \in N_t \times N_t$ is the channel matrix between BS and the NU.

Example 2.12 In Figs. 2.19 and 2.20, sum-rate of MIMO system by considering $N_t = 2, 4$ and $N_r = 1$ is shown. We consider distance $K = 2$ users and their distances from the BS are $d_1 = 500$ m and $d_2 = 200$ m, respectively. Pathloss factor $\eta = 3$ and system bandwidth is 1MHz with reference pathloss 30dBm at 1m. Noise PSD is -174 dBm. Power splitting coefficients among NOMA users are 0.75 (weak user) and 0.25 (strong user) and users use the TDMA approach for communication in OMA system. NOMA has better sum-rate performance as compared to OMA, as observed in Fig. 2.19. Further, individual users sum-rate in NOMA and OMA system is plotted in Fig. 2.20.

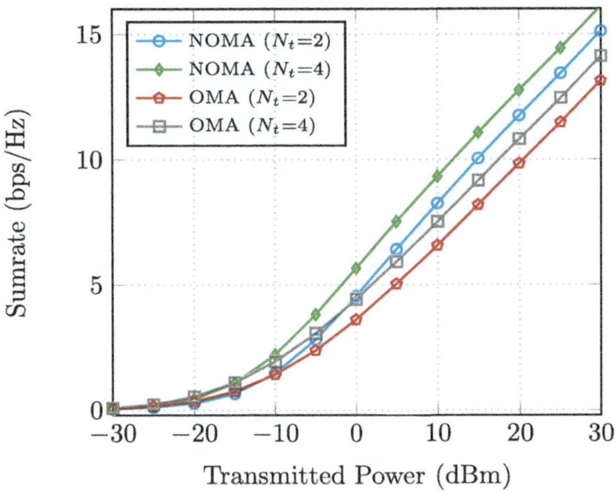

Fig. 2.19 Sumrate performance of NOMA and OMA system by selecting $N_t = 2$ and $N_t = 4$

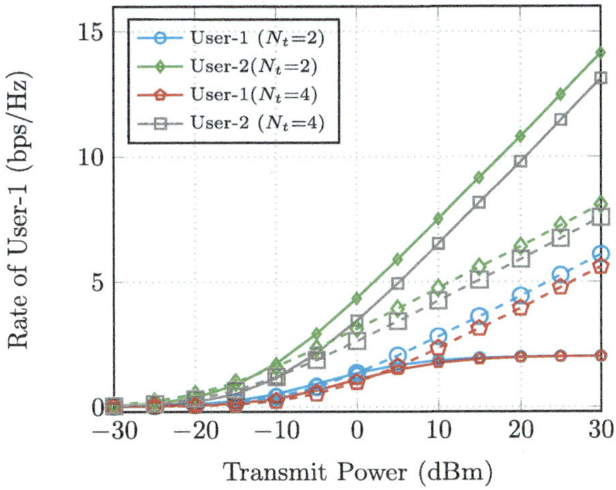

Fig. 2.20 Sumrate performance of individual users in the NOMA (solid lines) and OMA (dashed lines) system by selecting $N_t = 2$ and $N_t = 4$

2.7 Code Domain NOMA (CD-NOMA)

We summarize the CD-NOMA system mainly SCMA technique. SCMA is a key technique in CD-NOMA that facilitates simultaneous transmission of multiple users in the same time-frequency resource block by utilizing unique spreading codes for each user. Unlike PD-NOMA, which allocates different power levels to users, CD-NOMA allocates distinct spreading codes to multiplex users' signals. In CD-NOMA, the spreading codes play a crucial role in separating the signals of individual users at the receiver. Each user's data is spread using a unique code before transmission, enabling the signals to overlap in the same time-frequency resource block. At the receiver, sophisticated signal processing techniques, such as multi-user detection, are employed to successfully decode the individual signals, separating them from the overlapping transmissions.

The CD-NOMA transmission process involves the following steps:

1. Each user is assigned a unique spreading code called codebook that distinguishes their transmitted signal from other users in the same resource block.
2. Users transmit their spread signals simultaneously in the same time-frequency resource block, resulting in overlapping signals at the receiver.
3. The receiver employs advanced signal processing algorithms, such as multi-user detection or message passing algorithm (MPA), to separate and decode the individual user signals from the overlapping transmissions.

SCMA is an advanced multiple access technique that has gained significant attention in recent years for its potential to improve spectral efficiency and support massive connectivity in wireless communication systems. SCMA is designed for 5G and beyond networks to address the challenges posed by the increasing number of connected devices and diverse traffic demands. At its core, SCMA utilizes sparse codebooks to efficiently multiplex multiple users' data within the same time-frequency resource block. Unlike traditional OMA schemes that allocate exclusive resources to each user, SCMA allows users to share the same resource block simultaneously. This non-orthogonal sharing is made possible by exploiting the sparse nature of the codebooks.

The SCMA transmission process involves the following key steps:

Sparse Codebook Design: Each user is assigned a unique set of sparse codewords from the codebook. The sparse nature of the codebook ensures that only a few codewords are active for each user, making the transmitted signal sparse. For example, user-1 have only two non-zero values in each column in Fig. 2.21. SCMA reach user's data bits are directly mapped to a column vector of the predesigned codebook, as shown in Fig. 2.21.

Data Multiplexing: Users' data is mapped to the corresponding sparse codewords, creating sparse data streams for each user. SCMA reach user's data bits are directly mapped to a column vector of the predesigned codebook, as shown in Fig. 2.21.

Resource Allocation: Users transmit their sparse data streams simultaneously in the same time-frequency resource block, enabling concurrent transmissions. In Fig. 2.21, each user uses four resources for data transmission. Thus, size of the column vector is four.

Sparse Signal Detection: At the receiver, sophisticated signal processing techniques, such as compressed sensing or MPA, are employed to efficiently detect and separate the sparse data streams of individual users.

Example By considering six user ($J = 6$) and four resources ($K = 4$), we have shown SCMA system in Fig. 2.21, and in general system is referred to as $J \times K$ SCMA system. Each user has predefined codebooks ($\mathcal{X}_j, j = 1, \ldots, J$). Here, we consider that data is transmitted using two bits per symbol [12]. Thus, each codebook has four columns corresponding each incoming data bit combination and can be mapped as

$$b_1 = [00] \rightarrow \mathbf{x}_1, b_2 = [01] \rightarrow \mathbf{x}_2, b_1 = [10] \rightarrow \mathbf{x}_3, \text{ and } b_4 = [11] \rightarrow \mathbf{x}_4. \quad (2.58)$$

Therefore, each SCMA codebook in column format is expressed as

$$\mathcal{X}_j = \begin{bmatrix} \mathbf{x}_{j,1} & \mathbf{x}_{j,2} & \mathbf{x}_{j,3} & \mathbf{x}_{j,4} \end{bmatrix} \in \mathcal{C}^{4\times 4}, \quad j = 1, \ldots, J. \quad (2.59)$$

In Fig. 2.21, user-1 selects first column ($\mathbf{x}_1$) of the codebook depending upon the data bits b_1, and similarly user-2 chooses second column ($\mathbf{x}_2$) of the codebook depending upon the data bits b_2, and so on. The composite transmitted vector of all the six users is expressed as

$$\mathbf{x}_c = \mathbf{x}_{1,1} + \mathbf{x}_{2,2} + \mathbf{x}_{3,3} + \mathbf{x}_{4,4} + \mathbf{x}_{5,2} + \mathbf{x}_{6,1} = \sum_{j=1}^{J=6} \mathbf{x}_{j,i} \in \mathcal{C}^{4\times 1} \quad i = 1, 2, 3, 4. \quad (2.60)$$

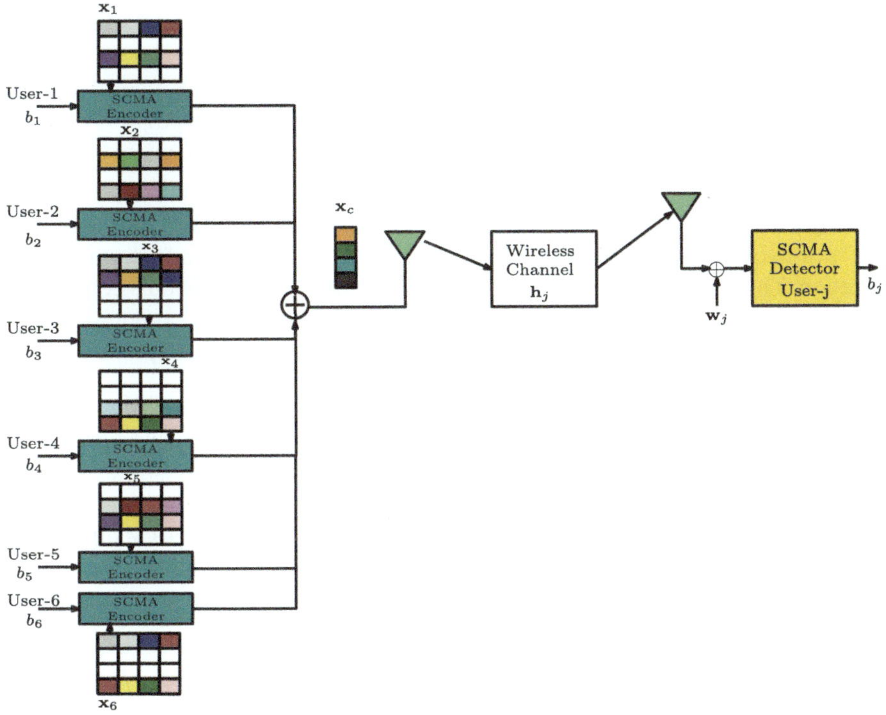

Fig. 2.21 6 × 4 SCMA system

At the jth user receiver, signal by considering a flat-fading channel $\mathbf{h}_j$ is written as

$$\mathbf{y}_j = \mathbf{h}_j\mathbf{x}_c + \mathbf{w}_j = \mathbf{h}_j\mathbf{x}_j + \underbrace{\mathbf{h}_j\left(\sum_{j'=1,j'\neq j}^{J=6} \mathbf{x}_{j'}\right)}_{\text{interference}} + \mathbf{w}_j \in \mathcal{C}^{4\times 1}. \quad (2.61)$$

In matrix form above expression is expressed as

$$\underbrace{\begin{bmatrix} y_{j,1} \\ y_{j,2} \\ y_{j,3} \\ y_{j,4} \end{bmatrix}}_{\mathbf{y}_j} = \underbrace{\begin{bmatrix} h_{j,1} \\ h_{j,2} \\ h_{j,3} \\ h_{j,4} \end{bmatrix}}_{\mathbf{h}_j} \odot \underbrace{\begin{bmatrix} x_{c,1} \\ x_{c,2} \\ x_{c,3} \\ x_{c,4} \end{bmatrix}}_{\mathbf{x}_c} + \underbrace{\begin{bmatrix} w_{j,1} \\ w_{j,2} \\ w_{j,3} \\ w_{j,4} \end{bmatrix}}_{\mathbf{w}_j} \in \mathcal{C}^{4\times 1}, \quad (2.62)$$

where $\odot$ denotes the element by element multiplication. Further, MLD is used for signal detection and is expressed as

$$\hat{\mathbf{x}}_j = \arg\min_{\mathbf{x}_j \in \boldsymbol{\mathcal{X}}_j} ||\mathbf{y}_j - \mathbf{h}_j\mathbf{x}_j||^2. \quad (2.63)$$

MLD detection has higher complexity, which further increases as the size of the codebook increases.

The MPA-based signal detection can be used in the SCMA system due to the sparse structure of the users' codebook [13]. MPA uses the factor graph (FG)-based signal detection approach, and 6×4 SCMA system FG is shown in Fig. 2.22.

MPA is used for signal detection in the SCMA system [14]. Let $U_{k\to j}^{(l)}$ denote the message sent from the kth resource node to the jth user node, and $V_{j\to k}^{(l)}$ represent the message sent from the jth user node to the kth resource node during the lth iteration. These messages are vectors with a length of M. The individual components $U_{k\to j}^{(l)}\left(\mathbf{x}_{jm}\right)$ and $V_{j\to k}^{(l)}\left(\mathbf{x}_{jm}\right)$ correspond to the mth codeword $\mathbf{x}_{jm}$ contained within the codebook $\mathcal{X}_j$. Five steps of MPA are summarized as follows [15].

Initialization: Initially, the messages start with equal probabilities. Hence, initialize $V_{j\to k}^{(0)}\left(\mathbf{x}_{jm}\right) = 1/M$. In SCMA, as user nodes do not directly receive signals from the channel, probability messages from user nodes are uniformly distributed among all potential symbols.

Resource node update: The resource nodes receive output signals $\mathbf{y}$ through the channel. The updating of messages from the resource nodes follows this procedure:

$$U_{k\to j}^{(l)}(\mathbf{x}_{jm}) = \sum_{\mathbf{c}\in\mathcal{C}_{kj}} \frac{1}{\pi N_0} \exp[-\frac{1}{N_0}|y_k - \sqrt{P_j}h_{kj}x_{jmk} - \sum_{j'\in\mathcal{M}_k^j} \sqrt{P_{j'}}h_{ki}c_{j'k}|^2] \prod_{j'\in\mathcal{M}_k^j} V_{j'\to k}^{(l-1)}\left(\mathbf{c}_{j'}\right) \quad (2.64)$$

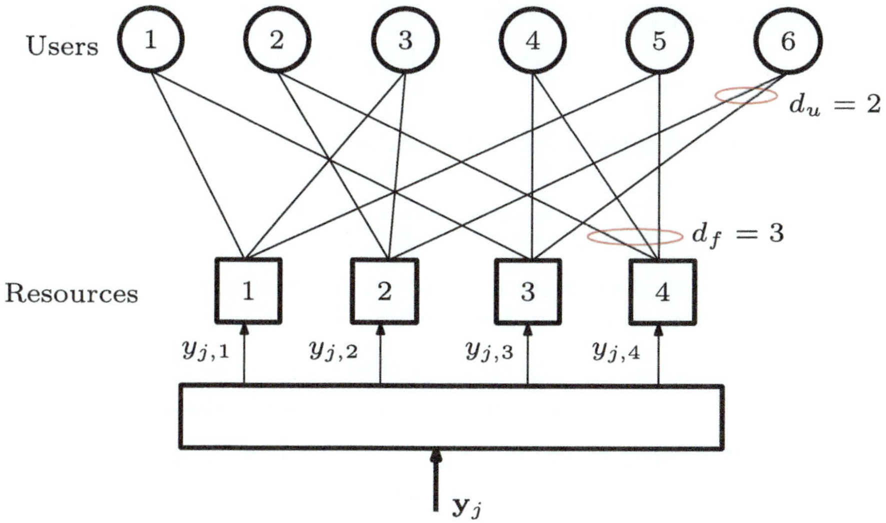

Fig. 2.22 Factor graph of 6 × 4 SCMA system

where,

- $\mathcal{M}_k$ is the set of the user nodes connected to the kth resource node and $\mathcal{M}_k^j = \mathcal{M}_k \setminus \{j\}$.
- Suppose $\mathcal{M}_k^j = \{j_1, j_2, \ldots, j_{d_f-1}\}$. Then, $\mathcal{C}_{kj}$ is the Cartesian product of the codebooks as defined below:

$$\mathcal{C}_{kj} = \mathcal{X}_{j_1} \times \mathcal{X}_{j_2} \times \cdots \times \mathcal{X}_{j_{d_f-1}}.$$

- $\mathbf{c} = \left(\mathbf{c}_{j_1}, \mathbf{c}_{j_2}, \ldots, \mathbf{c}_{j_{d_f-1}}\right)$ is a member of $\mathcal{C}_{kj}$ and $\mathbf{c}_{j'} = \left(c_{j'1}, c_{j'2}, \ldots, c_{j'K}\right)$.
- $\mathbf{x}_{jm} = \left(x_{jm1}, x_{jm2}, \ldots, x_{jmK}\right)$.

The step in (2.64) is known as the sum-product rule. The AWGN channel model is assumed.

User node update: The update process for messages originating from the user nodes occurs as follows:

$$V_{j\to k}^{(l)}\left(\mathbf{x}_{jm}\right) = \prod_{k' \in \mathcal{N}_j^k} U_{k'\to j}^{(l)}(\mathbf{x}_{jm}) \tag{2.65}$$

where $\mathcal{N}_j$ is the set of the resource nodes connected to the jth user node and $\mathcal{N}_j^k = \mathcal{N}_j \setminus \{k\}$.

Stopping rule: If the messages have converged or the maximum number of iterations T has been reached, halt the process and move forward to decision-making in Step (5). If not, increment l by 1 and proceed to Step (2).

Decision: Compute

$$V_j\left(\mathbf{x}_{jm}\right) = \prod_{k \in \mathcal{N}_j} U^{(l)}_{k \to j}(\mathbf{x}_{jm}). \tag{2.66}$$

The estimates of the transmitted codewords are found as follows:

$$\hat{\mathbf{x}}_j = \arg \max_{\mathbf{x}_{jm} \in \mathcal{X}_j} V_j\left(\mathbf{x}_{jm}\right).$$

Due to its numerical instability, it is essential to normalize the messages in each iteration within the probability-domain MPA.

For example, codebooks used for 6×4 SCMA system are shown in Table 2.1 [15]. Here, each user sends data using only two non-zero resources.

Example 2.13 In Fig. 2.23, SCMA symbol error rate (SER) performance is shown for different number (J) of users and resources (K). Overloading factor $\lambda = J/K$

Table 2.1 Codebooks of 6×4 SCMA system

$$\mathcal{X}_1 = \left\{ \begin{bmatrix} -1.2078 \\ 0 \\ -0.1339 - 0.3623i \\ 0 \end{bmatrix} \begin{bmatrix} -0.4022 \\ 0 \\ 0.7247 + 1.2078i \\ 0 \end{bmatrix} \begin{bmatrix} 0.4022 \\ 0 \\ -0.7247 - 1.2078i \\ 0 \end{bmatrix} \begin{bmatrix} 1.2078 \\ 0 \\ 0.1339 + 0.3623i \\ 0 \end{bmatrix} \right\}$$

$$\mathcal{X}_2 = \left\{ \begin{bmatrix} 0 \\ -0.1339 - 0.3623i \\ 0 \\ -1.2078 \end{bmatrix} \begin{bmatrix} 0 \\ 0.7247 + 1.2078i \\ 0 \\ -0.4022 \end{bmatrix} \begin{bmatrix} 0 \\ -0.7247 - 1.2078i \\ 0 \\ 0.4022 \end{bmatrix} \begin{bmatrix} 0 \\ 0.1339 + 0.3623i \\ 0 \\ 1.2078 \end{bmatrix} \right\}$$

$$\mathcal{X}_3 = \left\{ \begin{bmatrix} -0.7247 - 1.2078i \\ 0.3623 - 0.3623i \\ 0 \\ 0 \end{bmatrix} \begin{bmatrix} -0.1339 - 0.3623i \\ -0.7247 + 1.2078i \\ 0 \\ 0 \end{bmatrix} \begin{bmatrix} 0.1339 + 0.3623i \\ 0.7247 - 1.2078i \\ 0 \\ 0 \end{bmatrix} \begin{bmatrix} 0.7247 + 1.2078i \\ -0.3623 + 0.3623i \\ 0 \\ 0 \end{bmatrix} \right\}$$

$$\mathcal{X}_4 = \left\{ \begin{bmatrix} 0 \\ 0 \\ -1.2078 \\ 0.3623 - 0.3623i \end{bmatrix} \begin{bmatrix} 0 \\ 0 \\ -0.4022 \\ -0.7247 + 1.2078i \end{bmatrix} \begin{bmatrix} 0 \\ 0 \\ 0.4022 \\ 0.7247 - 1.2078i \end{bmatrix} \begin{bmatrix} 0 \\ 0 \\ 1.2078 \\ -0.3623 + 0.3623i \end{bmatrix} \right\}$$

$$\mathcal{X}_5 = \left\{ \begin{bmatrix} 0.3623 - 0.3623i \\ 0 \\ 0 \\ -0.7247 - 1.2078i \end{bmatrix} \begin{bmatrix} -0.7247 + 1.2078i \\ 0 \\ 0 \\ -0.1339 - 0.3623i \end{bmatrix} \begin{bmatrix} 0.7247 - 1.2078i \\ 0 \\ 0 \\ 0.1339 + 0.3623i \end{bmatrix} \begin{bmatrix} -0.3623 + 0.3623i \\ 0 \\ 0 \\ 0.7247 + 1.2078i \end{bmatrix} \right\}$$

$$\mathcal{X}_6 = \left\{ \begin{bmatrix} 0 \\ -1.2078 \\ 0.3623 - 0.3623i \\ 0 \end{bmatrix} \begin{bmatrix} 0 \\ -0.4022 \\ -0.7247 + 1.2078i \\ 0 \end{bmatrix} \begin{bmatrix} 0 \\ 0.4022 \\ 0.7247 - 1.2078i \\ 0 \end{bmatrix} \begin{bmatrix} 0 \\ 1.2078 \\ -0.3623 + 0.3623i \\ 0 \end{bmatrix} \right\}$$

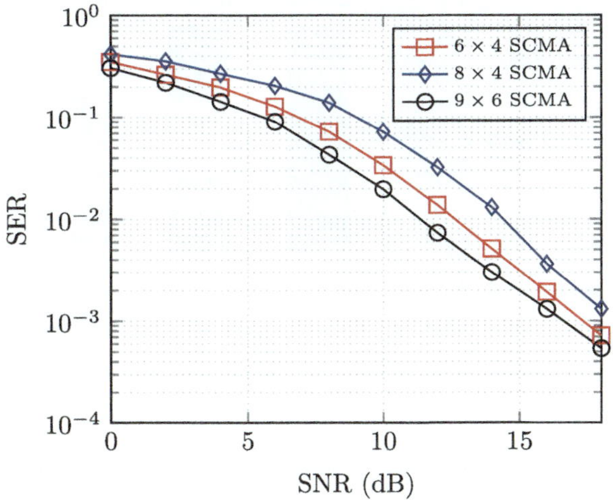

Fig. 2.23 BER performance of the SCMA system using various J and K in Rayleigh channel

200% [16] yields inferior performance as compared to 150% overloading. Moreover, the 9×6 SCMA system exhibits superior performance compared to the 6×4 system due to a reduced occurrence of symbol collisions among users. This observation holds true for the Rayleigh fading channel, as depicted in Fig. 2.23.

2.8 Pattern Division Multiple Access (PDMA)

PDMA is a multiple access technique used in communication systems, particularly in wireless networks. Unlike conventional techniques like TDMA or FDMA, which divide the communication resources based on time or frequency, PDMA divides resources based on patterns [17]. In PDMA, the available bandwidth is divided into various patterns or sequences. Each user is assigned a unique pattern, and communication occurs by using these distinct patterns. The simultaneous transmission and reception by different users rely on the orthogonality or distinguishability of these patterns.

The PDMA pattern establishes how transmitted data maps onto a resource group comprising time, frequency, spatial resources, or a combination thereof. This pattern serves to distinguish between signals of users utilizing shared resources. It's specifically designed with diverse orders of diversity and sparsity, allowing PDMA to leverage joint transmitter and receiver design to enhance system performance while keeping detection complexity at a manageable level.

PDMA employs a specific pattern to sparsely map data onto a resource group. This PDMA pattern is often represented as a binary vector, where the vector's dimension corresponds to the number of resources in a group. Each element in the vector denotes a resource within the resource group. A "1" in the vector indicates the mapping of

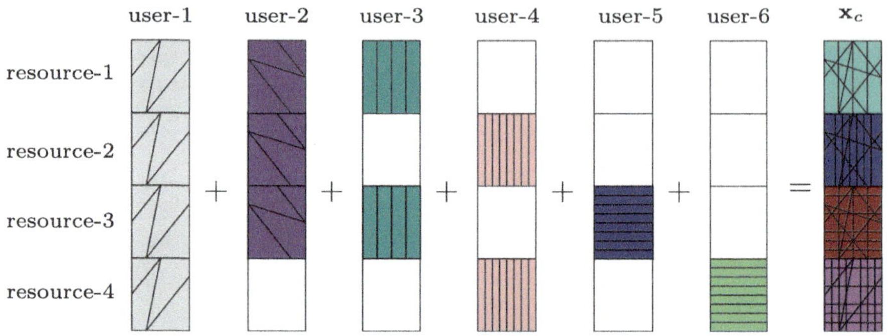

Fig. 2.24 PDMA pattern for six users on four resource nodes

data to the respective resource. The count of "1s" within the PDMA pattern defines its transmission diversity order. Figure 2.24 illustrates a resource mapping example following the PDMA pattern. In this scenario, four resource elements (REs) are used to multiplex data from six users. Each user is assigned a specific PDMA pattern. For instance, User 1's data is mapped to all four REs within the group, while User 2's data is mapped to the first three REs, and so forth. Consequently, the order of transmission diversity among the six users is: 4, 3, 2, 2, 1, and 1, respectively.

In non-orthogonal transmission employing an SIC receiver, the diversity order of each user changes according to the detection order. The user first detected has the lowest diversity order, while the last detected user possesses the highest diversity order. In a fading channel, diversity order significantly influences transmission reliability. Enhancing diversity order usually results in more reliable transmission. However, with the SIC receiver, the initial user's detection determines overall detection performance, yet regrettably, its diversity order is the lowest. To optimize system performance, achieving an identical pre-detection diversity order for each user is desirable.

PDMA utilizes a specific pattern to allocate transmitted data to a resource group, enabling diverse transmission orders. This pattern defines the mapping from transmitted data to a resource group comprising time, frequency, spatial resources, or a mix of these. The quantity of mapped resources in a group determines the transmission diversity order. Multiple users' data can be multiplexed onto a shared resource group using distinct PDMA patterns, facilitating non-orthogonal transmission. Assigning different PDMA patterns with varying diversity orders allows users to achieve disparate transmission diversity orders.

Assuming both the transmitter and receiver possess a single antenna. In this configuration, a PDMA encoder at the transmitter maps modulation symbols x_j onto resources, thereby generating a PDMA modulation vector $\mathbf{v}_j$. Conversely, at the receiver end, a PDMA multi-user detector is employed to detect the data of multiplexed users.

In an uplink PDMA system involving J users, their data gets mapped onto K resource elements (REs) using distinct PDMA patterns. The PDMA modulation vector $\mathbf{v}_j$ corresponding to user j, $j = 1, 2, \ldots, J$ is derived by distributing the user's modulation symbol x_j in accordance with the PDMA pattern $\mathbf{g}_j$. The PDMA modulation vector is expressed as

$$\mathbf{v}_j = \mathbf{g}_j x_j, \quad j = 1, 2, \ldots, J \tag{2.67}$$

The vector $\mathbf{g}_j$ consists of K elements in a binary format, each being either 0 or 1, where 1 indicates the mapping of the user's data to the respective RE, while 0 signifies otherwise. These J users' PDMA patterns across K REs collectively form a PDMA pattern matrix denoted as $\mathbf{G}_{\text{pdma}}$, structured with dimensions $K \times J$ as

$$\mathbf{G}_{\text{pdma}} = [\mathbf{g}_1, \mathbf{g}_2, \ldots, \mathbf{g}_J] \quad \text{of size} \quad K \times J$$

Uplink received signal is expressed as

$$\mathbf{y} = \sum_{j=1}^{J} \text{diag}[\mathbf{h}_j]\mathbf{v}_j + \mathbf{w} \tag{2.68}$$

In this context, $\mathbf{w}$ stands for the noise and interference experienced at the receiver; $\mathbf{h}_j$ denotes the uplink channel response pertaining to the jth user. The vectors $\mathbf{y}$, $\mathbf{w}$, and $\mathbf{h}_j = [k_{j1}, h_{j2}, \ldots, h_{jK}]^T$ possess a length of K. The expression diag$[\mathbf{h}_j]$ signifies the creation of a diagonal matrix incorporating elements from the vector $\mathbf{h}_j$. In matrix form received signal is expressed as

$$\mathbf{y} = \mathbf{H}\mathbf{x} + \mathbf{w}, \tag{2.69}$$

where $\mathbf{x} = [x_1, x_2, \ldots, x_J]^T$ represents a concatenated vector of symbols from J users. $\mathbf{H}$ refers to the PDMA equivalent channel response matrix, reflecting the multiplexing of J users over K REs, and is expressed as:

$$\mathbf{H} = \mathbf{H}_{\text{c}} \cdot \mathbf{G}_{\text{pdma}}, \quad \text{and} \quad \mathbf{H}_{\text{c}} = [\mathbf{h}_1, \mathbf{h}_2, \ldots, \mathbf{h}_J]$$

The (k, j), $\; k = 1, \ldots, K, j = 1, \ldots, J$ element of the matrix product $\mathbf{H}_{rmc}$ (size of $K \times J$) signifies the channel response from the kth user to the BS on the kth RE. The $\cdot$ symbol denotes the element-wise multiplication of two matrices.

Example 2.14 For instance, considering $K = 3$ and $J = 6$, the overload factor $\lambda = J/K$ is calculated as $J/K = 200\%$. This indicates that PDMA can accommodate twice the number of users compared to OMA. An illustration of the PDMA pattern matrix of size 3×6 is given by

$$\mathbf{G}_{\text{pdma}} = \begin{bmatrix} 1\,1\,0\,1\,0\,0 \\ 1\,0\,1\,0\,1\,0 \\ 0\,1\,1\,0\,0\,1 \end{bmatrix}. \tag{2.70}$$

Here, rows represent the RE and columns denote the number of user. First three users use 2 REs and remaining three users use only one RE, as observed in the above matrix $\mathbf{G}_{\text{pdma}}$. For $J = 6$ and $K = 3$, the received signal is given as

$$\underbrace{\begin{bmatrix} y_1 \\ y_2 \\ y_3 \end{bmatrix}}_{\mathbf{y}} = \underbrace{\begin{bmatrix} h_{1,1} & h_{1,2} & 0 & h_{1,4} & 0 & 0 \\ h_{2,1} & 0 & h_{2,3} & 0 & h_{2,5} & 0 \\ 0 & h_{3,2} & h_{3,3} & 0 & 0 & h_{3,6} \end{bmatrix}}_{\mathbf{H}} \underbrace{\begin{bmatrix} x_1 \\ x_2 \\ x_3 \\ x_4 \\ x_5 \\ x_6 \end{bmatrix}}_{\mathbf{x}} + \underbrace{\begin{bmatrix} w_1 \\ w_2 \\ w_3 \end{bmatrix}}_{\mathbf{w}} \tag{2.71}$$

To further expand, the process of constellation mapping and PDMA encoding can involve a direct mapping of the user's data bits to a PDMA modulation vector. This represents a collaborative action between PDMA encoding and constellation mapping, termed as joint PDMA modulation. This approach specifically involves the creation of a candidate set of PDMA modulation vectors, referred to as a codebook, tailored to the PDMA pattern of each user. The codebook, designed offline, is stored in a paired receiver and transmitter. Once the PDMA pattern for a user is established, the corresponding codebook is thereby determined and employed.

Pattern Matrix Design: Multiplexed PDMA patterns from users, allocated to the same resource group, collectively form a PDMA pattern matrix. Each column within this matrix signifies a distinct PDMA pattern. The matrix's properties, including its dimensions and sparsity level, impact both system performance and complexity. When dealing with a specific overload factor, there exist multiple pattern matrices that can be employed, provided J and K are appropriately chosen. For example three users over two RE can be expressed as

$$v_1 = x_1 + x_2 \tag{2.72}$$

$$v_2 = x_1 + x_3. \tag{2.73}$$

For instance, when all three users utilize BPSK modulation, the modulation symbols for user-1, user-2, and user-3 take values of either $+1$ or -1. Consequently, the composite transmission signal falls within the set $\{-2, 0, +2\}$. In a noiseless channel scenario, if the receiver detects -2 or $+2$ on a RE, it can deduce that the transmitted symbols on the RE are either $[-1, -1]$ or $[+1, +1]$. However, if the receiver receives 0, it becomes impossible to discern the transmitted symbols since both $[+1, -1]$ and $[-1, +1]$ result in the same output. Moreover, if each user opts for QPSK (quadrature phase shift keying) or 16QAM (quadrature amplitude modulation) modulation, the combined constellation comprises either nine or 49 constellation points. Thus, it's apparent that the combined constellation exhibits a nonuniform distribution, deviating from a one-to-one correspondence between a constellation point and an input user data. This amalgamation results in ambiguity within the system.

To address the ambiguity, the PDMA pattern matrix can incorporate power scaling and phase shifting. Particularly, before the amalgamation of symbols from two users, adjustments involving a power scaling factor (β) and a phase shifting factor (ϕ) are applied as follows:

$$v = \sqrt{\beta} x_1 e^{j\phi} + \sqrt{1-\beta} x_2. \tag{2.74}$$

In PDMA, the integration of power scaling and phase shifting can be seamlessly embedded within the PDMA pattern matrix. Specifically, the "1" values within the PDMA pattern matrix can be replaced by complex values that encompass both power scaling and phase shifting. This adaptation results in the creation of an extended PDMA pattern matrix. For example $J = 3$ and $K = 2$, PDMA matrix is written as

$$\mathbf{G} = \begin{bmatrix} \alpha_{11} e^{-j\phi_{11}} & \alpha_{21} e^{-j\phi_{21}} & 0 \\ \alpha_{12} e^{-j\phi_{12}} & 0 & \alpha_{32} e^{-j\phi_{32}} \end{bmatrix}, \tag{2.75}$$

where α_{jk} and ϕ_{jk} are the power scaling and phase shifting factors of the kth user on the jth RE. The ideal settings for power scaling and phase shifting are contingent upon the number of users and the configuration of the input constellation.

2.9 Rate Splitting Multiple Access System

Rate splitting multiple access (RSMA) is derived from the principle of rate-splitting (RS), has emerged as a promising transmission framework at the physical layer for non-orthogonal transmission, interference control, and multiple access strategies within the context of next-generation technology. RSMA's fundamental concept involves the division of user messages into common and private segments, facilitating partial decoding of interference while also treating interference as noise to a certain extent. This approach contrasts with the more extreme interference management methods employed in SDMA and NOMA. RSMA's adaptable nature enables efficient performance across various interference levels. Whether encountering weak or strong interference, RSMA can dynamically adjust by fine-tuning the powers and content of both the common and private streams, seamlessly transitioning between SDMA and NOMA [18] (Fig. 2.25).

Next-generation multiple access schemes are anticipated to exhibit enhanced interference management capabilities, dynamically adjusting the handling of interference based on its intensity [18]. Specifically, these schemes are designed to fully treat interference as noise (or completely decode it) in instances of low (or high) interference levels. Conversely, when the interference level is moderate, these schemes aim to partially decode the interference while treating the remainder as noise. Hence,

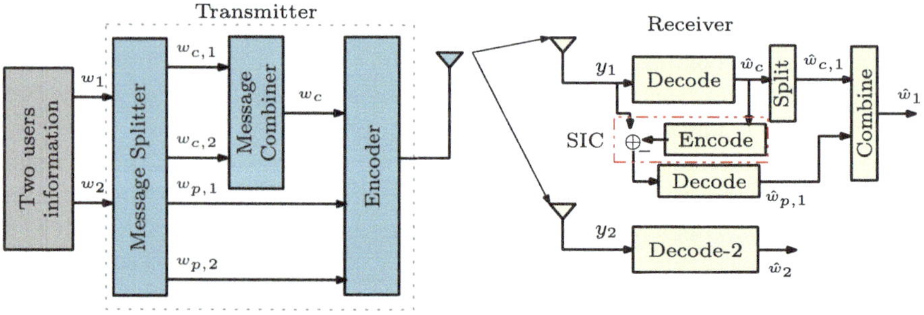

Fig. 2.25 RSMA for two users system

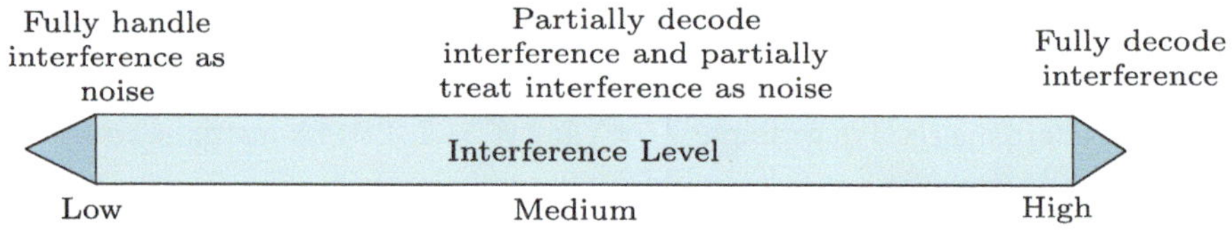

Fig. 2.26 Amount of interference to be decoded and treated as noise should be adaptively changed with the interference level [18]

RSMA stands as a novel paradigm shaping the landscape of multiple access and non-orthogonal transmission within upcoming wireless networks. Additionally, it introduces a fresh approach to interference management, offering the ability to reconcile existing interference challenges (Fig. 2.26).

2.10 Hybrid NOMA Techniques: Combining Orthogonal and Non-orthogonal Access

Hybrid NOMA (HNOMA) uses both PD-NOMA and CD-NOMA concepts to enhance spectral efficiency compared to individual NOMA systems [13, 19]. In HNOMA, users are grouped into group-1 (NU group) and group-2 (FU group), as shown in Fig. 2.27. The composite transmit signal of group-1 and group-2 at the BS is expressed as

$$\mathbf{x}_c = \sqrt{P}_1 \sum_{j=1}^{J} \mathbf{x}_{j1} + \sqrt{P}_2 \sum_{j=1}^{J} \mathbf{x}_{j2} = \sqrt{P}_1 \mathbf{x}_{g1} + \sqrt{P}_2 \mathbf{x}_{g2} \in \mathbb{C}^{K\times 1}$$

Here, J is the total number of users in each group, and K denotes the number of resources in the HNOMA system. P_1 and P_2 are the allocated powers for group-1 and group-2 users, respectively, with $P_1 < P_2$ and $||\mathbf{h}_1||^2 \gg ||\mathbf{h}_2||^2$. Additionally, we can consider that each user in the group experiences independent channel fading.

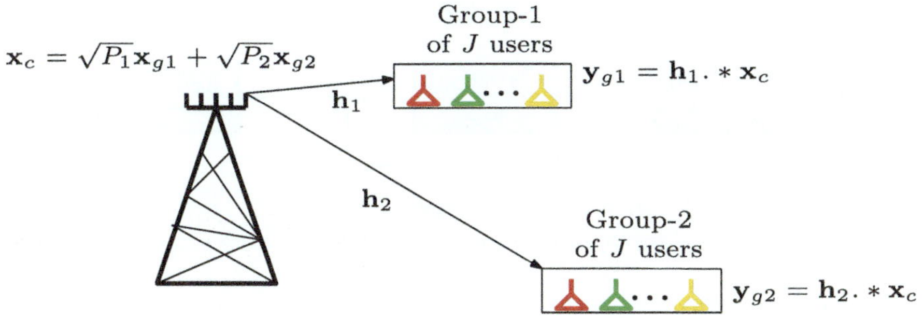

Fig. 2.27 Downlink HNOMA system block diagram

The received signal for group-1 is expressed as

$$\begin{aligned}\mathbf{y}_{g1} &= \underbrace{\text{diag}(\mathbf{h}_1)}_{\mathbf{H}_1}\mathbf{x}_c + \mathbf{n}_1 = \mathbf{h}_1 .* \mathbf{x}_c + \mathbf{n}_1 \in \mathbb{C}^{K\times 1}\\ &= \sqrt{P}_1\mathbf{H}_1\sum_{j=1}^{J}\mathbf{x}_{j1} + \sqrt{P}_2\mathbf{H}_1\sum_{j=1}^{J}\mathbf{x}_{j2} + \mathbf{n}_1\end{aligned} \tag{2.76}$$

First decode the group-2 users' signal ($\tilde{\mathbf{x}}_{j2}$) from the $\mathbf{y}_{g1}$ using MPA by considering group-1 users as interference. Apply the SIC at the group-1 to decode group-1 users' signal as

$$\tilde{\mathbf{y}}_{g1} = \mathbf{y}_{g1} - \sqrt{P}_2\mathbf{H}_1\sum_{j=1}^{J}\tilde{\mathbf{x}}_{j2}$$

Finally, decode the group-1 users' signal (($\tilde{\mathbf{x}}_{j1}$)) from $\tilde{\mathbf{y}}_{g1}$. SIC-based signal detection steps are highlighted in Fig. 2.28.

Group-2 received signal is expressed as

$$\begin{aligned}\mathbf{y}_{g2} &= \underbrace{\text{diag}(\mathbf{h}_2)}_{\mathbf{H}_2}\mathbf{x}_c + \mathbf{n}_2 = \mathbf{h}_2 .* \mathbf{x}_c + \mathbf{n}_2 \in \mathbb{C}^{K\times 1}\\ &= \sqrt{P}_2\mathbf{H}_2\sum_{j=1}^{J}\mathbf{x}_{j2} + \underbrace{\sqrt{P}_1\mathbf{H}_2\sum_{j=1}^{J}\mathbf{x}_{j1} + \mathbf{n}_1}_{\text{interference}}\end{aligned} \tag{2.77}$$

Group-2 users signal is decoded from the $\mathbf{y}_{g2}$.

HNOMA has more complexity in the signal detection and channel estimation compared to a single NOMA system due to multiple MPA and SIC. Allocating resources such as power, subcarriers, and time slots optimally among NOMA and OMA users is complex. Hybrid NOMA systems require intelligent user pairing and grouping

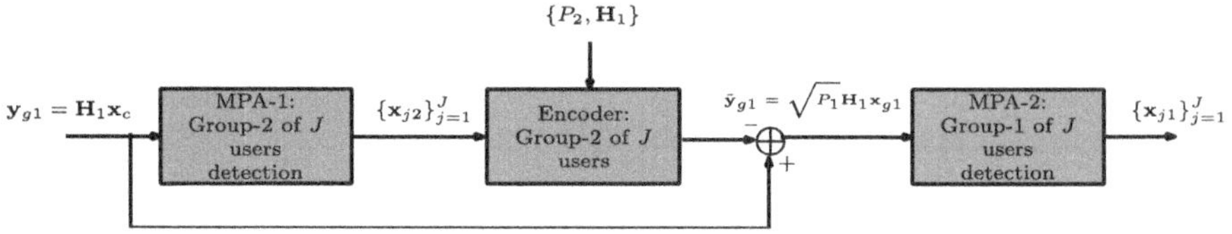

Fig. 2.28 SIC at group-1 users in the HNOMA system

strategies to maximize the benefits of NOMA while ensuring fairness among users. Performance improvement in HNOMA depends on the channel conditions and overall optimization of the system design.

References

1. Chen Y, Bayesteh A, Wu Y, Ren B, Kang S, Sun S, Xiong Q, Qian C, Yu B, Ding Z et al (2018) Toward the standardization of non-orthogonal multiple access for next generation wireless networks. IEEE Commun Mag 56(3):19–27
2. Saito Y, Kishiyama Y, Benjebbour A, Nakamura T, Li A, Higuchi K (2013) Non-orthogonal multiple access (NOMA) for cellular future radio access. In: 2013 IEEE 77th vehicular technology conference (VTC Spring). IEEE, pp 1–5
3. Zeng M, Yadav A, Dobre OA, Poor HV (2019) Energy-efficient joint user-RB association and power allocation for uplink hybrid NOMA-OMA. IEEE Internet Things J 6(3):5119–5131
4. Dai L, Wang B, Ding Z, Wang Z, Chen S, Hanzo L (2018) A survey of non-orthogonal multiple access for 5G. IEEE Commun Surv & Tutor 20(3):2294–2323
5. Proakis JG (2008) Digital communications. McGraw-Hill, Higher Education
6. Zhu J, Wang J, Huang Y, He S, You X, Yang L (2017) On optimal power allocation for downlink non-orthogonal multiple access systems. IEEE J Sel Areas Commun 35(12):2744–2757
7. Vaezi M, Ding Z, Poor HV (2019) Multiple access techniques for 5G wireless networks and beyond, vol 159. Springer
8. Zhang N, Wang J, Kang G, Liu Y (2016) Uplink nonorthogonal multiple access in 5G systems. IEEE Commun Lett 20(3):458–461
9. Can M, Altunbas I, Basar E (2021) MBM-aided uplink cooperative NOMA with hardware impairments and imperfect CSI. IEEE Commun Lett 25(6):1830–1834
10. Ding Z, Adachi F, Poor HV (2015) Performance of MIMO-NOMA downlink transmissions. In: 2015 IEEE global communications conference (GLOBECOM). IEEE, pp 1–6
11. Sharma S, Deka K, Beferull-Lozano B (2021) Low-complexity detection for uplink massive MIMO SCMA systems. IET Commun 15(1):51–59
12. Vikas V, Rajesh A, Deka K, Sharma S (2022) A comprehensive technique to design SCMA codebooks. IEEE Commun Lett 26(8):1735–1739
13. Sharma S, Deka K, Bhatia V, Gupta A (2019) Joint power-domain and SCMA-based NOMA system for downlink in 5G and beyond. IEEE Commun Lett 23(6):971–974
14. Thomas A, Deka K, Raviteja P, Sharma S (2022) Convolutional sparse coding based channel estimation for OTFS-SCMA in uplink. IEEE Trans Commun 70(8):5241–5257
15. Sharma S, Deka K (2023) Sparse code multiple access (SCMA) technique. arXiv:2309.09127
16. Deka K, Priyadarsini M, Sharma S, Beferull-Lozano B (2020) Design of SCMA codebooks using differential evolution. In: 2020 IEEE international conference on communications workshops (ICC Workshops). IEEE, pp 1–7

17. Chen S, Ren B, Gao Q, Kang S, Sun S, Niu K (2016) Pattern division multiple access-a novel nonorthogonal multiple access for fifth-generation radio networks. IEEE Trans Veh Technol 66(4):3185–3196
18. Mao Y, Dizdar O, Clerckx B, Schober R, Popovski P, Poor HV (2022) Rate-splitting multiple access: fundamentals, survey, and future research trends. IEEE Commun Surv & Tutor 24(4):2073–2126
19. Deka K, Sharma S (2022) Hybrid NOMA for future radio access: design, potentials and limitations. Wireless Personal Commun 1–16

Chapter 3
Co-located and Cell-Free Massive MIMO Based Underlay Spectrum Sharing Networks

Enukonda Venkata Pothan and Salil Kashyap

Abstract In this chapter, we develop modeling and analysis to compare performance of co-located and cell-free massive MIMO (mMIMO) based underlay spectrum sharing networks in terms of interference outage probability (IOP) and spectral efficiency (SE). We consider that the secondary co-located and cell-free mMIMO systems operate based on estimates acquired via shared pilots and under an IOP constraint. This constraint controls the fraction of time interference threshold is exceeded at any of the primary users (PUs). Based on this study, we conclude that cell-free mMIMO based underlay network gives lower IOP at PUs and improves SE of PUs and weaker secondary users when compared to users in co-located mMIMO networks.

Keywords Co-located mMIMO · Cell-free · mMIMO · Interference outage probability · Spectral efficiency · Cognitive radio

3.1 Introduction

To meet the ever-increasing requirement for higher spectral efficiency (SE) and the necessity to support diverse wireless services and applications, the design of wireless communication networks continues to evolve. It is expected that in the next few years, billions of IoT devices, autonomous cars, and unmanned aerial vehicles (UAVs) will be connected to the network [1]. However, the SE and the number of user equipments that can be supported depend largely on the available wireless spectrum. We wish to emphasize that most of the bandwidth below 6 GHz is already assigned to diverse wireless technologies such as 2G–5G and Wi-Fi services such as IEEE 802.11 ac and IEEE 802.11 ax. This is the reason why spectrum regulatory authorities across the world are now deliberating on permitting sharing of spectrum by unlicensed cognitive users (CUs) to enhance spectrum utilization efficiency. For example, the

E. V. Pothan · S. Kashyap (✉)
Department of Electronics and Electrical Engineering, Indian Institute of Technology Guwahati, Guwahati, India
e-mail: salilkashyap@iitg.ac.in

S. Sharma et al. (eds.), *Next-Generation Wireless Systems*,
Transactions on Computer Systems and Networks,
https://doi.org/10.1007/978-981-96-3185-8_3

Federal Communications Commission (FCC) in the USA has opened up 150 MHz bandwidth in the 3.5 GHz band [2] for shared use and 1.2 GHz bandwidth in the 6 GHz band [3] which is currently being used by the primary users (PUs) such as satellite services or public safety services for providing service to unlicensed CUs such as 5G new radio unlicensed (NR-U) and IEEE 802.11 ax/be [4]. Thus, these CUs must coexist with the PUs who are the licensed users and who continue to use the spectrum in a cognitive radio (CR) network [5].

CR networks usually operate in three different paradigms, namely, overlay, interweave, and underlay [6, 7]. While in the overlay mode of operation, the CUs are expected to have knowledge of the codebook and messages being transmitted by the PUs to enable interference cancellation at their end using sophisticated signal processing techniques, like dirty paper coding. In the interweave mode, the CUs transmit only when the PUs are silent. To achieve this, currently, a listen-before-talk (LBT) strategy is being utilized in long term evolution (LTE) based license assisted access (LAA), MulteFire, and Wi-Fi standards. In LBT, the secondary transmitter first senses the spectrum and transmits only if it finds the spectrum idle. The main challenge with such an approach is to monitor the activity of PUs and correctly detect the voids in space, time, and frequency commonly referred to as the spectrum holes to avoid interference to PUs. In other words, the PUs and the CUs in such networks can transmit simultaneously only when the spectrum hole is detected incorrectly.

However, to improve the utilization of the scarce spectrum, it makes sense to allow concurrent spectrum access in which the CUs can transmit even when the PUs are using the spectrum provided the interference caused to the primary due to secondary transmissions is below a certain threshold. For example, in [8], the investigations related to a citizens broadband radio service (CBRS) based spectrum sharing illustrate that concurrent spectrum sharing is indeed viable and does not harm the reception by the primary naval radar systems. This implementation-friendly mode of concurrent spectrum access that we focus on in this chapter is commonly referred to as underlay spectrum sharing. To prevent adverse effects of spectrum sharing and ensure that the interference constraint is obeyed, the CUs must control their transmit power based on the nature of the interference constraint[1] and the quality of channel estimates that is available at their end. This gives rise to a limit on the maximum SE that the CUs can achieve in underlay networks.

Massive MIMO (mMIMO) has already been identified as a core component of 5G wireless systems, in which hundreds of antennas at the base station (BS) can simultaneously serve tens of users through spatial multiplexing at the same time and over the same frequency [9–11]. Since it provides unprecedented multiplexing gain, huge array gain, high spatial resolution, favorable propagation, and reduced inter-user interference, the integration of mMIMO in underlay networks can significantly

[1] The cognitive transmissions may be subject to an interference outage probability (IOP) constraint or an average (peak) interference constraint. While the IOP constraint ensures that interference power at the primary does not exceed a threshold more than a certain fraction of time, an average (peak) interference constraint mandates that the average (instantaneous) power of interference at primary is below a threshold.

boost the SE of users in unlicensed cognitive network, where interference generated at the PUs, and the quality and the availability of the CSI at the cognitive BS are of major concern and restrict the power with which a cognitive BS can transmit [12]. The co-located mMIMO architecture has low backhaul requirements. However, these systems offer lower probability of coverage and relatively poorer quality of service to cell edge users when compared to a distributed mMIMO system commonly referred to as cell-free massive MIMO.

Cell-free mMIMO systems in which a large number of geographically distributed access points (APs) coordinate to serve users over the same time-frequency resource without cell boundaries, is a promising 6G technology [13–15]. These systems offer significant gains in SE, since these geographically distributed APs offer uncorrelated fading to different users and can help mitigate shadowing due to large obstacles. As these APs can coordinate to perform joint beamforming, these systems offer macro-diversity gains and unprecedented degrees-of-freedom [7, 16]. Furthermore, the short effective transmission distance between the distributed APs and the users reduces path loss, enables transmission at lower transmit powers, and simplifies interference management [16]. In cell-free mMIMO systems, the APs are connected to the central processing unit (CPU) via fronthaul links, which are of limited capacity. These links are generally used for (1) transportation of physical layer signals that need to be transmitted on the downlink (DL), (2) exchange of CSI, (3) transmission of the received uplink (UL) data signals from APs to the CPUs, (4) phase synchronization [17].

3.1.1 Organization of Chapter

We discuss system modeling, channel estimation, data transmission, and performance analysis in terms of IOP and SE for underlay co-located mMIMO networks in Sect. 3.2. The corresponding modeling, channel estimation, and analysis for underlay cell-free mMIMO networks are discussed in Sect. 3.3. This is followed by a numerical results in Sect.3.4, where we compare the performance of these two networks and make new inferences in Sect. 3.5.

3.2 Co-located mMIMO Based Underlay Spectrum Sharing

In this section, we discuss the system model shown in Fig. 3.1, channel estimation procedure, data transmission model, analysis of IOP and SE for co-located underlay spectrum sharing networks. The notations are summarized in Table 3.1.

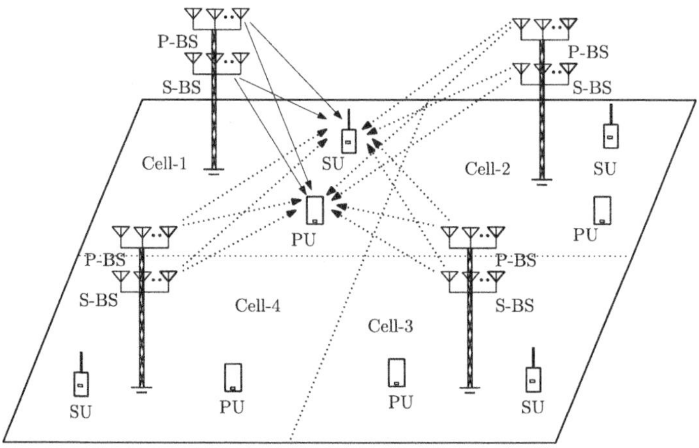

Fig. 3.1 Underlay co-located mMIMO network ($L_{cl} = 4$)

Table 3.1 Table of notations (Underlay co-located mMIMO network)

Parameter	Definition
L_{cl}	No. of cells
M_{cl}, N_{cl}	No. of antennas at PBS, SBS
K_p, K_s	No. of PUs, SUs
β_{bc}^{xa} $x \in \{p, s\}$	Large-scale fading coefficient between cth PU/SU in bth cell and PBS/SBS in ath cell
ϕ_p, ϕ_s	Pilot matrix of PUs, SUs in each cell
τ_p, τ_c	Length of pilot, coherence interval
ζ_{ac}^{xa} $x \in \{p, s\}$	Variance of channel estimate between cth PU/SU and PBS/SBS in ath cell
I_{ab}^{pa}	Interference power at bth PU in ath cell due to SBS in ath cell
p_t^p, p_t^s	Pilot power of PU, SU
P_d^{pa}, P_d^{sa}	DL transmission power of PBS, SBS in ath cell
η_{ab}^p	Power control coefficient between PBS and bth PU in ath cell
P_o	IOP
λ_c	Carrier wavelength
σ_w^2	Variance of noise
η_a^s	Power back-off at SBS in ath cell
$\overline{I}_p$	Interference threshold
z_{bc}^{xa} $x \in \{p, s\}$	Shadow fading coefficient between cth PU/SU in bth cell and PBS/SBS in ath cell

3.2.1 Mathematical Modeling

We consider a co-located mMIMO network with L_{cl} cells. Each cell consists of a primary BS (PBS) and a secondary BS (SBS). Both PBS and SBS are equipped with M_{cl} and N_{cl} antennas each, respectively. We consider no coordination among the BSs in different cells in co-located mMIMO network [18]. This is unlike cell-free mMIMO network studied in Sect. 3.3, where the APs coordinate to serve users [14]. We consider that there are K_p PUs and K_s secondary users (SUs) in each cell. The channel between the k_p^{th} PU in cell l and serving PBS is represented by $\boldsymbol{h}_{lk_p}^{pl} \in \mathbb{C}^{M_{cl}\times 1}$. We consider Rayleigh fading. Thus, $\boldsymbol{h}_{lk_p}^{pl} \sim \mathcal{CN}(\boldsymbol{0}, \beta_{lk_p}^{pl}\boldsymbol{I}_{M_{cl}})$. The channel between the k_p^{th} PU in the ith cell and the PBS in the lth cell is $\boldsymbol{h}_{ik_p}^{pl} \sim \mathcal{CN}(\boldsymbol{0}, \beta_{ik_p}^{pl}\boldsymbol{I}_{M_{cl}})$. The channel between the k_s^{th} SU in cell l and the serving SBS is represented as $\boldsymbol{h}_{lk_s}^{sl} \sim \mathcal{CN}(0, \beta_{lk_s}^{sl}\boldsymbol{I}_{N_{cl}})$. The channel between the k_s^{th} SU in cell i and the SBS in cell l is denoted as $\boldsymbol{h}_{ik_s}^{sl} \sim \mathcal{CN}(0, \beta_{ik_s}^{sl}\boldsymbol{I}_{N_{cl}})$ for $i = 1$ to L_{cl} and $l = 1$ to L_{cl}.

The interference channel between the k_p^{th} PU in cell l and the SBS in cell l is represented as $\boldsymbol{g}_{lk_p}^{sl} \sim \mathcal{CN}(\boldsymbol{0}, \beta_{lk_p}^{sl}\boldsymbol{I}_{N_{cl}})$. The interference channel between the k_p^{th} PU in cell i and the SBS in cell l is represented as $\boldsymbol{g}_{ik_p}^{sl} \sim \mathcal{CN}(\boldsymbol{0}, \beta_{ik_p}^{sl}\boldsymbol{I}_{N_{cl}})$. The interference channel between the k_s^{th} SU in cell l and the PBS in cell l is represented as $\boldsymbol{g}_{lk_s}^{pl} \sim \mathcal{CN}(0, \beta_{lk_s}^{pl}\boldsymbol{I}_{M_{cl}})$. The interference channel between the k_s^{th} SU in cell i and the PBS in cell l is represented as $\boldsymbol{g}_{ik_s}^{pl} \sim \mathcal{CN}(0, \beta_{ik_s}^{pl}\boldsymbol{I}_{M_{cl}})$.

UL Training: Let $\boldsymbol{\phi}^p \in \mathbb{C}^{\tau_p\times K_p}$ denote the pilot transmitted by the PUs in each cell and $\boldsymbol{\phi}^s \in \mathbb{C}^{\tau_p\times K_s}$ represents the pilot transmitted by the SUs in each cell. The length of the pilot sequences τ_p needs to be at least equal to $K_p + K_s$ to maintain orthogonality among pilots of users within a cell. The pilot matrices $\boldsymbol{\phi}^p$ and $\boldsymbol{\phi}^s$ are mutually orthogonal to each other. We assume that the length of the coherence interval τ_c is greater than τ_p. Both primary and secondary networks are also considered to operate in synchronism. The baseband signal observed at mth antenna of the lth cell PBS during UL pilot transmission by PUs and SUs is modeled as

$$[\boldsymbol{y}^{pl}]_m = \sum_{i=1}^{L_{cl}}\sum_{k_p=1}^{K_p}\sqrt{p_t^p\tau_p}[h_{ik_p}^{pl}]_m\boldsymbol{\phi}_{k_p}^p + \sum_{i=1}^{L_{cl}}\sum_{k_s=1}^{K_s}\sqrt{p_t^s\tau_p}[g_{ik_s}^{pl}]_m\boldsymbol{\phi}_{k_s}^s + \boldsymbol{n}_m^{pl}, \tag{3.1}$$

where $[\boldsymbol{y}^{pl}]_m \in \mathbb{C}^{\tau_p}$, $\boldsymbol{\phi}_{k_p}^p \in \mathbb{C}^{\tau_p}$ denotes the k_p^{th} column of $\boldsymbol{\phi}^p$ such that $||\boldsymbol{\phi}_{k_p}^p||^2 = 1$, $\boldsymbol{\phi}_{k_s}^s \in \mathbb{C}^{\tau_p}$ denotes the k_s^{th} column of $\boldsymbol{\phi}^s$ such that $||\boldsymbol{\phi}_{k_s}^s||^2 = 1$, p_t^p and p_t^s denote the power with which pilots are sent by PUs and SUs, and $\boldsymbol{n}_m^{pl} \sim \mathcal{CN}(\boldsymbol{0}, \sigma_w^2\boldsymbol{I}_{\tau_p})$. Here, we assume that the pilot sequences in one cell are being reused in other cells. To estimate the channel between the mth antenna of the PBS and k_p^{th} PU of the lth cell, $[\boldsymbol{y}^{pl}]_m$ is projected on $\boldsymbol{\phi}_{k_p}^{p\,H}$ to obtain

$$[y^{pl}_{lk_p}]_m = \boldsymbol{\phi}^{p\ H}_{k_p}[\boldsymbol{y}^{pl}]_m = \sum_{i=1}^{L_{cl}} \sqrt{p^p_t \tau_p}[h^{pl}_{ik_p}]_m + \boldsymbol{\phi}^{p\ H}_{k_p}\boldsymbol{n}^{pl}_m, \text{ for } k_p = 1, \ldots, K_p. \tag{3.2}$$

where we assume that k_p^{th} PU in every cell uses the same sequence and $\boldsymbol{\phi}^{p\ H}_{k_p}\boldsymbol{\phi}^{p}_{k_p'} = \delta_{k_p k_p'}$ which equals 1 if $k_p = k_p'$ and 0 for $k_p \neq k_p'$. Based on $[y^{pl}_{lk_p}]_m$ the MMSE estimate of $[h^{pl}_{lk_p}]_m$ equals

$$[\widehat{h}^{pl}_{lk_p}]_m = \left(\frac{\sqrt{p^p_t \tau_p}\beta^{pl}_{lk_p}}{\sum_{i=1}^{L_{cl}} p^p_t \tau_p \beta^{pl}_{ik_p} + \sigma^2_w}\right)[y^{pl}_{lk_p}]_m. \tag{3.3}$$

where it can be shown that $[\widehat{h}^{pl}_{lk_p}]_m \sim \mathcal{CN}(0, \zeta^{pl}_{lk_p})$ and $\zeta^{pl}_{lk_p} = \frac{p^p_t \tau_p (\beta^{pl}_{lk_p})^2}{\sum_{i=1}^{L_{cl}} p^p_t \tau_p \beta^{pl}_{ik_p} + \sigma^2_w}$. The MMSE estimate $[\widehat{h}^{pl}_{lk_p}]_m$ can be written as

$$\begin{aligned}[\widehat{h}^{pl}_{lk_p}]_m =& [h^{pl}_{lk_p}]_m - \left(\frac{\sigma^2_w [h^{pl}_{lk_p}]_m - \sqrt{p^p_t \tau_p}\beta^{pl}_{lk_p}\boldsymbol{\phi}^{p\ H}_{k_p}\boldsymbol{n}^p_l}{\sum_{i=1}^{L_{cl}} p^p_t \tau_p \beta^{pl}_{ik_p} + \sigma^2_w}\right), \\ =& [h^{pl}_{lk_p}]_m - [\widetilde{h}^{pl}_{lk_p}]_m, \end{aligned} \tag{3.4}$$

where the channel estimation error $[\widetilde{h}^{pl}_{lk_p}]_m \sim \mathcal{CN}(0, \beta^{pl}_{lk_p} - \zeta^{pl}_{lk_p})$ and is independent of $[\widehat{h}^{pl}_{lk_p}]_m$.

The signal observed at the nth antenna of the SBS in the lth cell during the UL training is modeled as

$$[\boldsymbol{y}^{sl}]_n = \sum_{i=1}^{L_{cl}}\sum_{k_s=1}^{K_s} \sqrt{p^s_t \tau_p}[h^{sl}_{ik}]_n \boldsymbol{\phi}^s_{k_s} + \sum_{i=1}^{L_{cl}}\sum_{k_p=1}^{K_p} \sqrt{p^p_t \tau_p}[g^{sl}_{ik}]_m \boldsymbol{\phi}^p_{k_p} + \boldsymbol{n}^{sl}_n, \tag{3.5}$$

where $[\boldsymbol{y}^{sl}]_n \in \mathbb{C}^{\tau_p}$ and $\boldsymbol{n}^{sl}_n \sim \mathcal{CN}(\boldsymbol{0}, \sigma^2_w \boldsymbol{I}_{\tau_p})$. To estimate the channel between the nth antenna of the SBS and k_s^{th} SU of the lth cell, $[\boldsymbol{y}^{sl}]_n$ is projected on $\boldsymbol{\phi}^{s\ H}_{k_s}$ to obtain

$$[y^{sl}_{lk_s}]_n = \boldsymbol{\phi}^{s\ H}_{k_s}[\boldsymbol{y}^{sl}]_n = \sum_{i=1}^{L_{cl}} \sqrt{p^p_t \tau_p}[h^{sl}_{ik_s}]_n + \boldsymbol{\phi}^{s\ H}_{k_s}\boldsymbol{n}^{sl}_n, \text{ for } k_s = 1, \ldots, K_s. \tag{3.6}$$

Based on $[y^{sl}_{lk_s}]_n$, the MMSE estimate of $[h^{sl}_{lk_s}]_n$ is

$$[\widehat{h}^{sl}_{lk_s}]_n = \left(\frac{\sqrt{p^s_t \tau_p}\beta^{sl}_{lk_s}}{\sum_{i=1}^{L_{cl}} p^s_t \tau_p \beta^{sl}_{ik_s} + \sigma^2_w}\right)[y^{sl}_{lk_s}]_n, \tag{3.7}$$

where it can be shown that $[\widehat{h}^{sl}_{lk_s}]_n \sim \mathcal{CN}(0, \zeta^{sl}_{lk_s})$ and $\zeta^{sl}_{lk_s} = \frac{p^s_l \tau_p (\beta^{sl}_{lk_s})^2}{\sum_{i=1}^{L_{cl}} p^s_i \tau_p \beta^{sl}_{ik_s} + \sigma^2_w}$. The MMSE estimate $[\widehat{h}^{sl}_{lk_s}]_n$ can be written as

$$
\begin{aligned}
[\widehat{h}^{sl}_{lk_s}]_n =& [h^{sl}_{lk_s}]_n - \left(\frac{\sigma^2_w [h^{sl}_{lk_s}]_n - \sqrt{p^p_i \tau_p \beta^{sl}_{lk_s}} \boldsymbol{\phi}^{s\ H}_{k_s} \boldsymbol{n}^s_l}{\sum_{i=1}^{L_{cl}} p^s_i \tau_p \beta^{sl}_{ik_s} + \sigma^2_w} \right), \\
=& [h^{sl}_{lk_s}]_n - [\widetilde{h}^{sl}_{lk_s}]_n,
\end{aligned} \tag{3.8}
$$

where the estimation error $[\widetilde{h}^{sl}_{lk_s}]_n \sim \mathcal{CN}(0, \beta^{sl}_{lk_s} - \zeta^{sl}_{lk_s})$ and is uncorrelated to $[\widehat{h}^{sl}_{lk_s}]_n$.

We assume that the pilots of PUs are exchanged periodically with SBS through a backhaul link [11, 16, 19, 20]. Therefore, to estimate the interference channel between the nth antenna of the SBS and $k'_p{}^{\text{th}}$ PU of the lth cell, $[y^{sl}_{lk'_p}]_n$ is projected on $\boldsymbol{\phi}^{p\ H}_{k'_p}$ to obtain

$$
[y^{sl}_{lk'_p}]_n = \boldsymbol{\phi}^{p\ H}_{k'_p} [\boldsymbol{y}^{sl}]_n = \sum_{i=1}^{L_{cl}} \sqrt{p^p_i \tau_p} [g^{sl}_{ik'_p}]_n + \boldsymbol{\phi}^{p\ H}_{k'_p} \boldsymbol{n}^{sl}_n, \text{ for } k'_p = 1, \ldots, K_p. \tag{3.9}
$$

Thus, based on $[y^{sl}_{lk'_p}]_n$, the MMSE estimate of $[g^{sl}_{lk'_p}]_n$ is

$$
[\widehat{g}^{sl}_{lk'_p}]_n = \left(\frac{\sqrt{p^s_l \tau_p} \beta^{sl}_{lk'_p}}{\sum_{i=1}^{L_{cl}} p^s_i \tau_p \beta^{sl}_{ik'_p} + \sigma^2_w} \right) [y^{sl}_{lk'_p}]_n. \tag{3.10}
$$

where it can be shown that $[\widehat{g}^{sl}_{lk'_p}]_n \sim \mathcal{CN}(0, \zeta^{sl}_{lk'_p})$ and $\zeta^{sl}_{lk'_p} = \frac{p^s_l \tau_p (\beta^{sl}_{lk'_p})^2}{\sum_{i=1}^{L_{cl}} p^s_i \tau_p \beta^{sl}_{ik'_p} + \sigma^2_w}$. The MMSE estimate $[\widehat{g}^{sl}_{lk'_p}]_n$ can be written as

$$
\begin{aligned}
[\widehat{g}^{sl}_{lk'_p}]_n =& [h^{sl}_{lk'_p}]_n - \left(\frac{\sigma^2_w [g^{sl}_{lk'_p}]_n - \sqrt{p^p_i \tau_p \beta^{sl}_{lk'_p}} \boldsymbol{\phi}^{p\ H}_{k'_p} \boldsymbol{n}^{sl}_n}{\sum_{i=1}^{L_{cl}} p^s_i \tau_p \beta^{sl}_{ik'_p} + \sigma^2_w} \right), \\
=& [g^{sl}_{lk'_p}]_n - [\widetilde{g}^{sl}_{lk'_p}]_n,
\end{aligned} \tag{3.11}
$$

where the channel estimation error $[\widetilde{g}^{sl}_{lk'_p}]_n \sim \mathcal{CN}(0, \beta^{sl}_{lk'_p} - \zeta^{sl}_{lk'_p})$ and is independent of $[\widehat{g}^{sl}_{lk'_p}]_n$.

Downlink Transmission: The PBS employs maximum ratio (MR) precoding designed based on estimated channels to transmit information to PUs on the DL. The signal sent by the lth PBS is given by

$$\boldsymbol{x}_l^p = \sum_{k_p=1}^{K_p} \sqrt{\eta_{lk_p}^p} \widehat{\boldsymbol{h}}_{lk_p}^{pl\ *} q_{k_p}, \tag{3.12}$$

where $q_{k_p} \sim \mathcal{CN}(0, 1)$ represents the information symbol corresponding to the k_p^{th} PU, $\mathbb{E}(q_{k_p} q_{k_p'}) = 0$ for $k_p \neq k_p'$ and $\eta_{lk_p}^p$ is the power control coefficient given by

$$\eta_{lk_p}^p = \frac{P_d^{pl}}{\sum_{j=1}^{K_p} \mathbb{E}||\widehat{\boldsymbol{h}}_{lj}^{pl}||^2} = \frac{P_d^{pl}}{\sum_{j=1}^{K_p} M_{cl} \zeta_{lj}^{pl}}, \text{ for } k_p = 1, 2, \ldots, K_p. \tag{3.13}$$

This power allocation ensures that $\mathbb{E}||\boldsymbol{x}_l^p||^2 \leq P_d^{pl}$, where P_d^{pl} is the DL transmission power of PBS in the lth cell. The SBS in the lth cell performs normalized MR precoding based on the estimates obtained and the signal transmitted by the SBS in the lth cell equals

$$\boldsymbol{x}_l^s = \sum_{k_s=1}^{K_s} \sqrt{\frac{P_d^{sl}}{K_s}} \frac{\widehat{\boldsymbol{h}}_{lk_s}^{sl\ *}}{||\widehat{\boldsymbol{h}}_{lk_s}^{sl}||} s_{k_s}, \tag{3.14}$$

where $s_{k_s} \sim \mathcal{CN}(0, 1)$ is the data symbol of the k_s^{th} SU, $\mathbb{E}(s_{k_s} s_{k_s'}) = 0$ for $k_s \neq k_s'$. The signal $\boldsymbol{x}_l^s$ obeys the power constraint $\mathbb{E}||\boldsymbol{x}_l^s||^2 \leq P_d^{sl}$, where P_d^{sl} is the data transmission power of SBS of the lth cell.

The data signal obtained at the k_p^{th} PU of the lth cell when all the PBSs and SBSs transmit simultaneously is given by

$$y_{lk_p}^{pl} = \underbrace{\sum_{i=1}^{L_{cl}} \boldsymbol{h}_{ik_p}^{pl\ T} \boldsymbol{x}_i^p}_{①} + \underbrace{\sum_{j=1}^{Lcl} \boldsymbol{g}_{jk_p}^{sl\ T} \boldsymbol{x}_j^s}_{②} + \underbrace{n_{k_p}^{pl}}_{③}. \tag{3.15}$$

In Eq. (3.15), term ① denotes signal received by the k_p^{th} PU of lth cell from L_{cl} PBSs, term ② represents the interference at the k_p^{th} PU due to transmission by L_{cl} SBSs and term ③ denotes noise at the k_p^{th} PU and $n_{k_p}^{pl} \sim \mathcal{CN}(0, \sigma_w^2)$. The interference $I_{lk_p}^{pl}$ generated at the k_p^{th} PU of the lth cell due to spectrum access by the SBSs is given by

$$I_{lk_p}^{pl} = \mathbb{E}\left|\sum_{j=1}^{L_{cl}} \sum_{k_s'=1}^{K_s} \sqrt{\frac{P_d^{sj}}{K_s}} \frac{\boldsymbol{g}_{jk_p}^{sl\ T} \widehat{\boldsymbol{h}}_{jk_s'}^{sj\ *} s_{k_s'}}{||\widehat{\boldsymbol{h}}_{jk_s'}^{sj}||}\right|^2 = \sum_{k_s'=1}^{K_s} \left|\sum_{j=1}^{L_{cl}} \sqrt{\frac{P_d^{sj}}{K_s}} \frac{\boldsymbol{g}_{jk_p}^{sl\ T} \widehat{\boldsymbol{h}}_{jk_s'}^{sj\ *}}{||\widehat{\boldsymbol{h}}_{jk_s'}^{sj}||}\right|^2. \tag{3.16}$$

The signal received at the k_s^{th} SU of the lth cell when all the SBSs and PBSs transmit simultaneously is given by

$$y_{lk_s}^{sl} = \underbrace{\sum_{i=1}^{L_{cl}} \boldsymbol{h}_{ik_s}^{sl\ T} \boldsymbol{x}_i^s}_{①} + \underbrace{\sum_{j=1}^{L_{cl}} \boldsymbol{g}_{jk_s}^{pl\ T} \boldsymbol{x}_j^p}_{②} + \underbrace{n_{k_s}^{sl}}_{③}. \tag{3.17}$$

In (3.17), term ① captures the signal received by the k_s^{th} SU of lth cell from L_{cl} SBSs, term ② represents the interference at the k_s^{th} SU due to transmission from L_{cl} PBSs and term ③ denotes the noise at the k_s^{th} SU and $n_{k_s}^{sl} \sim \mathcal{CN}(0, \sigma_w^2)$.

3.2.2 *Performance Analysis*

IOP Analysis and Power Policy: We consider an underlay spectrum access system where the SBS transmission power is adapted based on an IOP constraint. This probabilistic constraint requires that the power of interference due to transmission by SBS of the lth cell does not exceed an interference threshold $\overline{I}_p$ at any of the K_p PUs in the lth cell more than P_o fraction of the time [11, 16]. Mathematically, this constraint mandates that

$$\Pr(I_{l1}^{pl} \le \overline{I}_p, I_{l2}^{pl} \le \overline{I}_p, \ldots, I_{lK_p}^{pl} \le \overline{I}_p) \ge (1 - P_o), \tag{3.18}$$

where P_o denotes IOP.

We design a simple power control rule that helps us decide the power back-off that must be adopted to satisfy the constraint in (3.18). Based on this rule, as a function of interference channel estimates obtained during UL training, the SBS of the lth cell transmits with power

$$P_d^{sl} = \frac{\overline{I}_p}{\eta_l^s \max\limits_{1\le k_p \le K_p} \left\{ \frac{1}{K_s} \sum\limits_{k_s'=1}^{K_s} \left| \frac{\widehat{\boldsymbol{g}}_{lk_p}^{sl\ T} \widehat{\boldsymbol{h}}_{lk_s'}^{sl\ *}}{\|\widehat{\boldsymbol{h}}_{lk_s'}^{sl}\|} \right|^2 \right\}}, \tag{3.19}$$

where the power back-off factor $\eta_l^s (> 1)$ is chosen such that the IOP constraint in (3.18) is satisfied with equality for a given P_o. The SBS in every cell would compute the back-off factor based on this rule depending on the channel estimate that is available to them. For the power policy in (3.19), the back-off η_l^s is taken so that

$$\Pr(I_{l1}^{pl} \le \overline{I}_p, I_{l2}^{pl} \le \overline{I}_p, \ldots, I_{lK_p}^{pl} \le \overline{I}_p) = (1 - P_o), \tag{3.20}$$

where

$$I_{lk_p}^{pl} = \sum_{k_s'=1}^{K_s} \left| \sum_{j=1}^{L_{cl}} \sqrt{\frac{\overline{I}_p}{K_s \, \eta_j^s \max\limits_{1 \le k_p''' \le K_p} \left\{ \frac{1}{K_s} \sum\limits_{k_s''=1}^{K_s} \left| \frac{\boldsymbol{g}_{jk_p'''}^{sj\ T} \widehat{\boldsymbol{h}}_{jk_s''}^{sj\ *}}{\|\widehat{\boldsymbol{h}}_{lk_s''}^{sj}\|} \right|^2 \right\}}} \frac{\boldsymbol{g}_{jk_p}^{sl\ T} \widehat{\boldsymbol{h}}_{jk_s'}^{sj\ *}}{\|\widehat{\boldsymbol{h}}_{jk_s'}^{sj}\|} \right|^2 . \tag{3.21}$$

DL SE of PUs and SUs: We now analyze the DL SE of the k_p^{th} PU. Based on (3.4), we express the true channel $\boldsymbol{h}_{lk_p}^{pl} = \widehat{\boldsymbol{h}}_{lk_p}^{pl} + \widetilde{\boldsymbol{h}}_{lk_p}^{pl}$ and re-write the signal acquired at the k_p^{th} PU in (3.15) as

$$\begin{aligned}
y_{lk_p}^{pl} &= \sum_{i=1}^{L_{cl}} \boldsymbol{h}_{ik_p}^{pl\ T} \boldsymbol{x}_i^p + \sum_{j=1}^{L_{cl}} \boldsymbol{g}_{jk_p}^{sl\ T} \boldsymbol{x}_j^s + n_{k_p}^{pl}, \\
&= \boldsymbol{h}_{lk_p}^{pl\ T} \boldsymbol{x}_l^p + \sum_{i \neq l}^{L_{cl}} \boldsymbol{h}_{ik_p}^{pl\ T} \boldsymbol{x}_i^p + \sum_{j=1}^{L_{cl}} \boldsymbol{g}_{jk_p}^{sl\ T} \boldsymbol{x}_j^s + n_{k_p}^{pl}, \\
&= \boldsymbol{h}_{lk_p}^{pl\ T} \left(\sum_{k_p'=1}^{K_p} \sqrt{\eta_{lk_p'}^p} \widehat{\boldsymbol{h}}_{lk_p'}^{pl\ *} q_{k_p'} \right) + \sum_{i \neq l}^{L_{cl}} \boldsymbol{h}_{ik_p}^{pl\ T} \boldsymbol{x}_i^p + \sum_{j=1}^{L_{cl}} \boldsymbol{g}_{jk_p}^{sl\ T} \boldsymbol{x}_j^s + n_{k_p}^{pl}, \\
&= \sqrt{\eta_{lk_p}^p} \|\widehat{\boldsymbol{h}}_{lk_p}^{pl}\|^2 q_{k_p} + \sqrt{\eta_{lk_p}^p} \widetilde{\boldsymbol{h}}_{lk_p}^{pl\ T} \widehat{\boldsymbol{h}}_{lk_p}^{pl\ *} q_{k_p} + \sum_{k_p' \neq k_p}^{K_p} \sqrt{\eta_{lk_p'}^p} \boldsymbol{h}_{lk_p}^{pl\ T} \widehat{\boldsymbol{h}}_{lk_p'}^{pl\ *} q_{k_p'} \\
&\quad + \sum_{i \neq l}^{L_{cl}} \boldsymbol{h}_{ik_p}^{pl\ T} \boldsymbol{x}_i^p + \sum_{j=1}^{L_{cl}} \boldsymbol{g}_{jk_p}^{sl\ T} \boldsymbol{x}_j^s + n_{k_p}^{pl}.
\end{aligned} \tag{3.22}$$

By using use and then forget bound [21], we can write $y_{lk_p}^{pl}$ as

$$\begin{aligned}
y_{lk_p}^{pl} &= \sqrt{\eta_{lk_p}^p} \mathbb{E}\|\widehat{\boldsymbol{h}}_{lk_p}^{pl}\|^2 q_{k_p} + \sqrt{\eta_{lk_p}^p} \|\widehat{\boldsymbol{h}}_{lk_p}^{pl}\|^2 q_{k_p} - \sqrt{\eta_{lk_p}^p} \mathbb{E}\|\widehat{\boldsymbol{h}}_{lk_p}^{pl}\|^2 q_{k_p} + \sqrt{\eta_{lk_p}^p} \widetilde{\boldsymbol{h}}_{lk_p}^{pl\ T} \widehat{\boldsymbol{h}}_{lk_p}^{pl\ *} q_{k_p} \\
&\quad + \sum_{k_p' \neq k_p}^{K_p} \sqrt{\eta_{lk_p'}^p} \boldsymbol{h}_{lk_p}^{pl\ T} \widehat{\boldsymbol{h}}_{lk_p'}^{pl\ *} q_{k_p'} + \sum_{i \neq l}^{L_{cl}} \boldsymbol{h}_{ik_p}^{pl\ T} \boldsymbol{x}_i^p + \sum_{j=1}^{L_{cl}} \boldsymbol{g}_{jk_p}^{sl\ T} \boldsymbol{x}_j^s + n_{k_p}^{pl}.
\end{aligned} \tag{3.23}$$

We note that the terms in (3.23) are uncorrelated to each other. Using (3.23), the signal to interference plus noise ratio (SINR) corresponding to the k_p^{th} PU can be expressed as

$$\mathrm{SINR}_{k_p}^{pl} = \frac{\left|\sqrt{\eta_{lk_p}^{p}}\mathbb{E}||\widehat{\boldsymbol{h}}_{lk_p}^{pl}||^2\right|^2}{\left(\begin{array}{l}\eta_{lk_p}^{p}\mathbb{E}\left|(||\widehat{\boldsymbol{h}}_{lk_p}^{pl}||^2-\mathbb{E}||\widehat{\boldsymbol{h}}_{lk_p}^{pl}||^2)\right|^2+\eta_{lk_p}^{p}\mathbb{E}\left|\widetilde{\boldsymbol{h}}_{lk_p}^{pl\ T}\widehat{\boldsymbol{h}}_{lk_p}^{pl\ *}\right|^2+\sum\limits_{k_p'\neq k_p}^{K_p}\eta_{lk_p'}^{p}\mathbb{E}\left|\boldsymbol{h}_{lk_p}^{pl\ T}\widehat{\boldsymbol{h}}_{lk_p'}^{pl*}\right|^2\\ +\sum\limits_{k_p'=1}^{K_p}\mathbb{E}\left|\sum\limits_{i\neq l}^{L_{cl}}\sqrt{\eta_{lk_p'}^{p}}\boldsymbol{h}_{ik_p}^{pl\ T}\widehat{\boldsymbol{h}}_{ik_p'}^{pi*}\right|^2+\sum\limits_{k_s'=1}^{K_s}\mathbb{E}\left|\sum\limits_{j=1}^{L_{cl}}\sqrt{\frac{P_d^{sj}}{K_s}}\frac{\boldsymbol{g}_{lk_p}^{sj\ T}\widehat{\boldsymbol{h}}_{jk_s'}^{sj\ *}}{||\widehat{\boldsymbol{h}}_{jk_s'}^{sj}||}\right|^2+\mathbb{E}\left|n_{k_p}^{pl}\right|^2\end{array}\right)}. \tag{3.24}$$

The lower bound on the DL SE of the k_p^{th} PU in a concurrent spectrum access network based on estimates acquired using shared pilot is given by

$$\mathrm{SE}_{k_p}^{pl} = \left(1 - \frac{\tau_p}{\tau_c}\right)\log\left(1 + \mathrm{SINR}_{k_p}^{pl}\right). \tag{3.25}$$

We next analyze the DL SE of the k_s^{th} SU. From (3.8), we express the channel $\boldsymbol{h}_{lk_s}^{sl} = \widehat{\boldsymbol{h}}_{lk_s}^{sl} + \widetilde{\boldsymbol{h}}_{lk_s}^{sl}$ and re-arrange the signal obtained at the k_s^{th} SU in (3.17) as

$$\begin{aligned}
y_{lk_s}^{sl} &= \sum_{i=1}^{L_{cl}} \boldsymbol{h}_{ik_s}^{sl\ T}\boldsymbol{x}_i^s + \sum_{j=1}^{L_{cl}} \boldsymbol{g}_{jk_s}^{pl\ T}\boldsymbol{x}_j^p + n_{k_s}^{sl},\\
&= \boldsymbol{h}_{lk_s}^{sl\ T}\boldsymbol{x}_l^s + \sum_{i\neq l}^{L_{cl}} \boldsymbol{h}_{ik_s}^{sl\ T}\boldsymbol{x}_i^s + \sum_{j=1}^{L_{cl}} \boldsymbol{g}_{jk_s}^{pl\ T}\boldsymbol{x}_j^p + n_{k_s}^{sl},\\
&= \boldsymbol{h}_{lk_s}^{sl\ T}\left(\sum_{k_s'=1}^{K_s}\sqrt{\frac{P_d^{sl}}{K_s}}\frac{\widehat{\boldsymbol{h}}_{lk_s'}^{sl\ *}}{||\widehat{\boldsymbol{h}}_{lk_s'}^{sl}||}s_{k_s'}\right) + \sum_{i\neq l}^{L_{cl}} \boldsymbol{h}_{ik_s}^{sl\ T}\boldsymbol{x}_i^s + \sum_{j=1}^{L_{cl}} \boldsymbol{g}_{jk_s}^{pl\ T}\boldsymbol{x}_j^p + n_{k_s}^{sl},\\
&= \sqrt{\frac{P_d^{sl}}{K_s}}||\widehat{\boldsymbol{h}}_{lk_s}^{sl}||s_{k_s} + \sqrt{\frac{P_d^{sl}}{K_s}}\widetilde{\boldsymbol{h}}_{lk_s}^{sl\ T}\frac{\widehat{\boldsymbol{h}}_{lk_s}^{sl\ *}}{||\widehat{\boldsymbol{h}}_{lk_s}^{sl}||}s_{k_s} + \sum_{k_s'\neq k_s}^{K_s}\sqrt{\frac{P_d^{sl}}{K_s}}\boldsymbol{h}_{lk_s}^{sl\ T}\frac{\widehat{\boldsymbol{h}}_{lk_s'}^{sl\ *}}{||\widehat{\boldsymbol{h}}_{lk_s'}^{sl}||}s_{k_s'}\\
&\quad + \sum_{i\neq l}^{L_{cl}} \boldsymbol{h}_{ik_s}^{sl\ T}\boldsymbol{x}_i^s + \sum_{j=1}^{L_{cl}} \boldsymbol{g}_{jk_s}^{pl\ T}\boldsymbol{x}_j^p + n_{k_s}^{sl}.
\end{aligned} \tag{3.26}$$

By using use and then forget bound [21], we can write $y_{lk_s}^{sl}$ as

$$\begin{aligned}
y_{lk_s}^{sl} &= \mathbb{E}\left(\sqrt{\frac{P_d^{sl}}{K_s}}||\widehat{\boldsymbol{h}}_{lk_s}^{sl}||\right)s_{k_s} + \sqrt{\frac{P_d^{sl}}{K_s}}||\widehat{\boldsymbol{h}}_{lk_s}^{sl}||s_{k_s} - \mathbb{E}\left(\sqrt{\frac{P_d^{sl}}{K_s}}||\widehat{\boldsymbol{h}}_{lk_s}^{sl}||\right)s_k + \sqrt{\frac{P_d^{sl}}{K_s}}\frac{\widetilde{\boldsymbol{h}}_{lk_s}^{sl\ T}\widehat{\boldsymbol{h}}_{lk_s}^{sl\ *}s_{k_s}}{||\widehat{\boldsymbol{h}}_{lk_s}^{sl}||}\\
&\quad + \sum_{k_s'\neq k_s}^{K_s}\sqrt{\frac{P_d^{sl}}{K_s}}\frac{\boldsymbol{h}_{lk_s}^{sl\ T}\widehat{\boldsymbol{h}}_{lk_s'}^{sl\ *}s_{k_s'}}{||\widehat{\boldsymbol{h}}_{lk_s'}^{sl}||} + \sum_{i\neq l}^{L_{cl}} \boldsymbol{h}_{ik_s}^{sl\ T}\boldsymbol{x}_i^s + \sum_{j=1}^{L_{cl}} \boldsymbol{g}_{jk_s}^{pl\ T}\boldsymbol{x}_j^p + n_{k_s}^{sl}.
\end{aligned} \tag{3.27}$$

We note that all terms in (3.26) are uncorrelated to each other. Using (3.26), the SINR corresponding to the k_s^{th} SU can be expressed as

$$\text{SINR}_{k_s}^{sl} = \frac{\left|\mathbb{E}\left(\sqrt{\frac{P_d^{sl}}{K_s}}||\widehat{\boldsymbol{h}}_{lk_s}^{sl}||\right)\right|^2}{\left(\begin{array}{l}\mathbb{E}\left|\sqrt{\frac{P_d^{sl}}{K_s}}||\widehat{\boldsymbol{h}}_{lk_s}^{sl}|| - \mathbb{E}\left(\sqrt{\frac{P_d^{sl}}{K_s}}||\widehat{\boldsymbol{h}}_{lk_s}^{sl}||\right)\right|^2 + \mathbb{E}\left|\sqrt{\frac{P_d^{sl}}{K_s}}\frac{\widetilde{\boldsymbol{h}}_{lk_s}^{sl\ T}\widehat{\boldsymbol{h}}_{lk_s}^{sl\ *}}{||\widehat{\boldsymbol{h}}_{lk_s}^{sl}||}\right|^2 + \sum_{k_s' \neq k_s}^{K_s}\mathbb{E}\left|\sqrt{\frac{P_d^{sl}}{K_s}}\frac{\boldsymbol{h}_{lk_s}^{sl\ T}\widehat{\boldsymbol{h}}_{lk_s'}^{sl\ *}}{||\widehat{\boldsymbol{h}}_{lk_s'}^{sl}||}\right|^2 \\ + \sum_{k_s'=1}^{K_s}\mathbb{E}\left|\sum_{i \neq l}^{L_{cl}}\sqrt{\frac{P_d^{si}}{K_s}}\frac{\boldsymbol{h}_{ik_s}^{sl\ T}\widehat{\boldsymbol{h}}_{ik_s'}^{si\ *}}{||\widehat{\boldsymbol{h}}_{ik_s'}^{si}||}\right|^2 + \sum_{k_p'=1}^{K_p}\mathbb{E}\left|\sum_{j=1}^{L_{cl}}\sqrt{\eta_{jk_p'}^{p}}\boldsymbol{g}_{jk_s}^{pl\ T}\widehat{\boldsymbol{h}}_{jk_p'}^{pj*}\right|^2 + \mathbb{E}\left|n_{k_s}^{sl}\right|^2\end{array}\right)}. \tag{3.28}$$

The lower bound on the DL SE of the k_s^{th} SU in a concurrent spectrum access network under pilot contamination is given by

$$\text{SE}_{k_s}^{sl} = \left(1 - \frac{\tau_p}{\tau_c}\right)\log\left(1 + \text{SINR}_{k_s}^{sl}\right). \tag{3.29}$$

3.3 Cell-Free mMIMO Based Underlay Spectrum Sharing

In this section, we describe the cell-free mMIMO aided underlay CR set-up shown in Fig. Fig. 3.2, UL training, IOP at PUs, power policy at secondary APs (SAPs), and SE of PUs and SUs. The notations are summarized in Table 3.2.

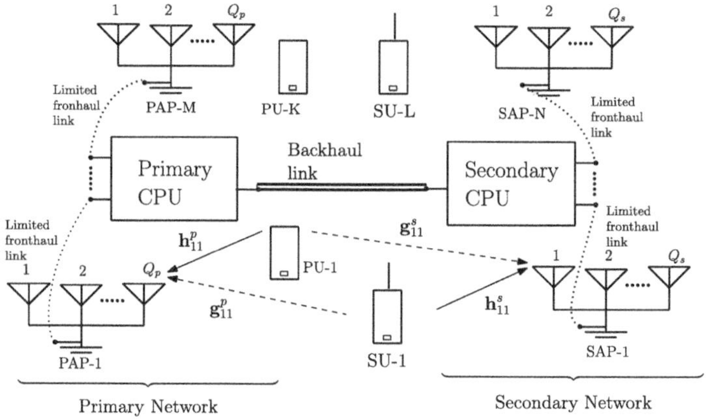

Fig. 3.2 Underlay cell-free mMIMO network

Table 3.2 Table of notations (Underlay cell-Free mMIMO network)

Parameter	Definition
M, N	no. of PAPs, SAPs
Q_p, Q_s	no. of antennas at PAP, SAP
K, L	no. of PUs, SUs
β^x_{ab} $x \in \{p, s\}$	Large-scale fading coefficient between ath PAP/SAP and bth PU/SU
ζ^x_{ab} $x \in \{p, s\}$	Variance of channel estimate between ath PAP/SAP and bth PU/SU
τ_p, τ_c	Length of pilot, coherence interval
I^p_k	Interference power at kth PU due to SAPs
$P^p_d, \widehat{P}^s_d$	DL transmission power of PAPs, SAPs
ϕ_p, ϕ_s	Pilot matrix of PUs, SUs
P_o	IOP
λ_c	Carrier wavelength
p^p_t, p^s_t	Pilot power at PUs, SUs
η	Back-off factor
z^x_{ab} $x \in \{p, s\}$	Shadow fading coefficient between ath PAP/SAP and bth PU/SU
$\overline{I}_p$	Interference threshold
η_{ab}	Power control coefficient between ath PAP and bth PU
σ^2_w	Variance of noise

3.3.1 Mathematical Modeling

We study DL of a cell-free mMIMO based spectrum sharing network. The network comprises a cell-free based primary system and a cell-free based secondary system that accesses the spectrum under an IOP constraint. The primary network consists of M multi-antenna primary APs (PAPs) with Q_p antennas each and serving K PUs, and also consists of N multi-antenna SAPs with Q_s antennas each and serving L SUs. The channel response between the mth PAP and kth PU is denoted by $\boldsymbol{h}^p_{mk} \sim \mathcal{CN}(\mathbf{0}, \beta^p_{mk}\boldsymbol{I}_{Q_p})$, where β^p_{mk} represents large scale fading coefficient. The interfering channel between the nth SAP and kth PU is denoted by $\boldsymbol{g}^s_{nk} \sim \mathcal{CN}(\mathbf{0}, \beta^s_{nk}\boldsymbol{I}_{Q_s})$. Furthermore, we denote the channel from the nth SAP to lth SU by $\boldsymbol{h}^s_{nl} \sim \mathcal{CN}(\mathbf{0}, \beta^s_{nl}\boldsymbol{I}_{Q_s})$. The interfering channel between the mth PAP and lth SU is denoted by $\boldsymbol{g}^p_{ml} \sim \mathcal{CN}(\mathbf{0}, \beta^p_{ml}\boldsymbol{I}_{Q_p})$.

UL Training : Let $\boldsymbol{\phi}_p$ denote the pilots sent out by the PUs and $\boldsymbol{\phi}_s$ represent the pilots transmitted by the SUs, where the pilot length $\tau_p \leq \tau_c$ with τ_c representing the length of the coherence block. We consider pilot sharing among SUs and among

PUs. The signal obtained at q_p^{th} antenna of the mth PAP during pilot transmission of PUs and SUs is given by

$$\left[\boldsymbol{y}_m\right]_{q_p} = \sum_{k'=1}^{K} \sqrt{p_t^p \tau_p} [h_{mk'}^p]_{q_p} \boldsymbol{\phi}_p^{k'} + \sum_{l=1}^{L} \sqrt{p_t^s \tau_p} [g_{ml}^p]_{q_p} \boldsymbol{\phi}_s^l + [\boldsymbol{w}_m]_{q_p}, \tag{3.30}$$

where $\left[\boldsymbol{y}_m\right]_{q_p} \in \mathbb{C}^{\tau_p \times 1}$, $\boldsymbol{\phi}_p^{k'} \in \mathbb{C}^{\tau_p \times 1}$ represents the k'^{th} column of $\boldsymbol{\phi}_p$, $\boldsymbol{\phi}_s^l \in \mathbb{C}^{\tau_p \times 1}$ denotes the lth column of $\boldsymbol{\phi}_s$, $[\boldsymbol{w}_m]_{q_p} \in \mathbb{C}^{\tau_p \times 1}$ and $[\boldsymbol{w}_m]_{q_p} \sim \mathcal{CN}(\boldsymbol{0}, \sigma_w^2 \boldsymbol{I}_{\tau_p})$, p_t^p and p_t^s represent the pilot powers of PUs and SUs, respectively. To obtain the channel estimate of the kth PU, $\left[\boldsymbol{y}_m\right]_{q_p}$ is spread over $\boldsymbol{\phi}_p^{k\,\mathrm{H}}$ to obtain

$$[y_{mk}]_{q_p} = \boldsymbol{\phi}_p^{k\,\mathrm{H}} \left[\boldsymbol{y}_m\right]_{q_p} = \sqrt{p_t^p \tau_p} \left[h_{mk}^p\right]_{q_p} + \boldsymbol{\phi}_p^{k\,\mathrm{H}} [\boldsymbol{w}_m]_{q_p}, \tag{3.31}$$

where we exploit the fact that $\boldsymbol{\phi}_p^{k\,\mathrm{H}} \boldsymbol{\phi}_p^{k'} = \delta_{kk'}$ which equals 1 if $k = k'$ and 0 for k not equal to k' and $\boldsymbol{\phi}_p^{k\,\mathrm{H}} \boldsymbol{\phi}_s^l = 0$. From $[y_{mk}]_{q_p}$, the MMSE estimate of $\left[h_{mk}^p\right]_{q_p}$ equals [22]

$$\begin{aligned}
[\widehat{h}_{mk}^p]_{q_p} &= \left[h_{mk}^p\right]_{q_p} - \left(\frac{\sigma_w^2 \left[h_{mk}^p\right]_{q_p} - \sqrt{p_t^p \tau_p} \beta_{mk}^p \boldsymbol{\phi}_p^{k\,\mathrm{H}} [\boldsymbol{w}_m]_{q_p}}{\sum_{i \in \mathcal{P}_k} \tau_p p_t^p \beta_{mi}^p + \sigma_w^2} \right), \\
&= \left[h_{mk}^p\right]_{q_p} - [\widetilde{h}_{mk}^p]_{q_p},
\end{aligned} \tag{3.32}$$

where it can be shown that $[\widehat{h}_{mk}^p]_{q_p} \sim \mathcal{CN}\left(0, \zeta_{mk}^p\right)$, $\zeta_{mk}^p = \frac{p_t^p \tau_p (\beta_{mk}^p)^2}{\sum_{i \in \mathcal{P}_k} \tau_p p_t^p \beta_{mi}^p + \sigma_w^2}$ and the error incurred in estimation $[\widetilde{h}_{mk}^p]_{q_p} \sim \mathcal{CN}\left(0, \beta_{mk}^p - \zeta_{mk}^p\right)$ and it is uncorrelated to $[\widehat{h}_{mk}^p]_{q_p}$, and $\mathcal{P}_k$ denotes subset of PUs that use the same pilot as kth PU. The signal obtained at antenna q_s of the nth SAP during UL training is written as

$$\left[\boldsymbol{y}_n\right]_{q_s} = \sum_{l'=1}^{L} \sqrt{p_t^s \tau_p} [h_{nl'}^s]_{q_s} \boldsymbol{\phi}_s^{l'} + \sum_{k'=1}^{K} \sqrt{p_t^p \tau_p} \left[g_{nk'}^s\right]_{q_s} \boldsymbol{\phi}_p^{k'} + [\boldsymbol{w}_n]_{q_s}, \tag{3.33}$$

where $[\boldsymbol{w}_n]_{q_s} \in \mathbb{C}^{\tau_p \times 1}$, $\left[\boldsymbol{y}_n\right]_{q_s} \in \mathbb{C}^{\tau_p \times 1}$, $\boldsymbol{\phi}_s^{l'} \in \mathbb{C}^{\tau_p \times 1}$, $\boldsymbol{\phi}_p^{k'} \in \mathbb{C}^{\tau_p \times 1}$ and $[\boldsymbol{w}_n]_{q_s} \sim \mathcal{CN}(\boldsymbol{0}, \sigma_w^2 \boldsymbol{I}_{\tau_p})$. To acquire channel estimate of the lth SU, $\left[\boldsymbol{y}_n\right]_{q_s}$ is spread over $\boldsymbol{\phi}_s^{l\,H}$ to yield

$$[y_{nl}]_{q_s} = \boldsymbol{\phi}_s^{l\,\mathrm{H}} \left[\boldsymbol{y}_n\right]_{q_s} = \sqrt{p_t^s \tau_p} \left[h_{nl}^s\right]_{q_s} + \boldsymbol{\phi}_s^{l\,\mathrm{H}} [\boldsymbol{w}_n]_{q_s}. \tag{3.34}$$

From $[y_{nl}]_{q_s}$, the MMSE estimate of $\left[h_{nl}^s\right]_{q_s}$ is [22]

$$[\widehat{h}^s_{nl}]_{q_s} = [h^s_{nl}]_{q_s} - \left(\frac{\sigma_w^2 [h^s_{nl}]_{q_s} - \sqrt{p_t^p \tau_p \beta^s_{nl}} \boldsymbol{\phi}_s^{l^{\mathrm{H}}} [\boldsymbol{w}_n]_{q_s}}{\sum\limits_{i \in \mathcal{P}_l} \tau_p p_t^p \beta^p_{mi} + \sigma_w^2} \right),$$
$$= [h^s_{nl}]_{q_s} - [\widetilde{h}^s_{nl}]_{q_s}, \tag{3.35}$$

where it can be shown that $[\widehat{h}^s_{nl}]_{q_s} \sim \mathcal{CN}\left(0, \zeta^s_{nl}\right)$, $\zeta^s_{nl} = \frac{p_t^s \tau_p (\beta^s_{nl})^2}{\sum_{i \in \mathcal{P}_l} \tau_p p_t^p \beta^p_{mi} + \sigma_w^2}$ and the estimation error $[\widetilde{h}^s_{nl}]_{q_s} \sim \mathcal{CN}\left(0, \beta^s_{nl} - \zeta^s_{nl}\right)$ and it is uncorrelated to $[\widehat{h}^s_{nl}]_{q_s}$, and $\mathcal{P}_l$ denotes subset of SUs that share the same pilot as lth SU.

We assume that the pilots of PUs are exchanged periodically with SAPs over a backhaul link that connects the primary CPU to the secondary CPU [19, 20, 23, 24]. To find channel estimate between the nth SAP and the kth PU, $[\boldsymbol{y}_n]_{q_s}$ can be spread over $\boldsymbol{\phi}_p^{k^{\mathrm{H}}}$ to obtain

$$[y_{nk}]_{q_s} = \boldsymbol{\phi}_p^{k^{\mathrm{H}}} [\boldsymbol{y}_n]_{q_s} = \sqrt{p_t^p \tau_p} [g^s_{nk}]_{q_s} + \boldsymbol{\phi}_p^{k^{\mathrm{H}}} [\boldsymbol{w}_n]_{q_s}. \tag{3.36}$$

From $[y_{nk}]_{q_s}$, the MMSE estimate of $[g^s_{nk}]_{q_s}$ is given by [22]

$$[\widehat{g}^s_{nk}]_{q_s} = [g^s_{nk}]_{q_s} - \left(\frac{\sigma_w^2 [g^s_{nk}]_{q_s} - \sqrt{p_t^p \tau_p \beta^s_{nk}} \boldsymbol{\phi}_p^{k^{\mathrm{H}}} [\boldsymbol{w}_n]_{q_s}}{\sum\limits_{i \in \mathcal{P}_k} \tau_p p_t^p \beta^p_{mi} + \sigma_w^2} \right),$$
$$= [g^s_{nk}]_{q_s} - [\widetilde{g}^s_{nk}]_{q_s}, \tag{3.37}$$

where $[\widehat{g}^s_{nk}]_{q_s} \sim \mathcal{CN}\left(0, \zeta^s_{nk}\right)$, $\zeta^s_{nk} = \frac{p_t^p \tau_p (\beta^s_{nk})^2}{\sum_{i \in \mathcal{P}_k} \tau_p p_t^p \beta^p_{mi} + \sigma_w^2}$ and the estimation error $[\widetilde{g}^s_{nk}]_{q_s}$ $\sim \mathcal{CN}\left(0, \beta^s_{nk} - \zeta^s_{nk}\right)$ and is independent of $[\widehat{g}^s_{nk}]_{q_s}$.

DL Transmission: The PAPs employ MR precoding based on channel estimates to transfer information to PUs on the DL. The signal transmitted by the mth PAP is given by

$$\boldsymbol{x}_m = \sum_{k=1}^{K} \sqrt{\eta_{mk}}\, \widehat{\boldsymbol{h}}^{p*}_{mk}\, q_k, \tag{3.38}$$

where $q_k \sim \mathcal{CN}(0, 1)$ represents information corresponding to the kth PU, $\mathbb{E}\left(q_{k'} q_k\right) = 0$ for k not equal to k' and η_{mk} is the power allocation coefficient written as

$$\eta_{mk} = \frac{P_d^p}{\sum\limits_{j=1}^{K} \mathbb{E}\|\widehat{\boldsymbol{h}}^p_{mj}\|^2} = \frac{P_d^p}{\sum\limits_{j=1}^{K} Q_p \zeta^p_{mj}}, \quad \text{for } k = 1, \ldots, K. \tag{3.39}$$

With this choice, $\mathbb{E}\|\boldsymbol{x}_m\|^2 \leq P_d^p$, where P_d^p is the data power of PAPs. Based on the estimates obtained, the SAPs perform conjugate normalized beamforming. The signal sent out by nth SAP is given by

$$\boldsymbol{x}_n = \sum_{l=1}^{L} \sqrt{\frac{P_d^s}{L}} \frac{\widehat{\boldsymbol{h}}_{nl}^{s*}}{||\widehat{\boldsymbol{h}}_{nl}^{s}||} s_l, \tag{3.40}$$

where $s_l \sim \mathcal{CN}(0, 1)$ denotes the information symbol for the lth SU and $\mathbb{E}(s_l s_{l'}) = 0$ for l not equal to l'. Note that $\mathbb{E}\|\boldsymbol{x}_n\|^2 \leq P_d^s$, where P_d^s is the power with which data is transmitted by SAPs.

The signal obtained at the kth PU when both PAPs and SAPs transmit concurrently equals

$$y_k = \underbrace{\sum_{k'=1}^{K}\sum_{m=1}^{M} \sqrt{\eta_{mk'}} \boldsymbol{h}_{mk}^{p\mathrm{T}} \widehat{\boldsymbol{h}}_{mk'}^{p*} q_{k'}}_{①} + \underbrace{\sum_{l=1}^{L}\sum_{n=1}^{N} \sqrt{\frac{P_d^s}{L}} \frac{\boldsymbol{g}_{nk}^{s\mathrm{T}} \widehat{\boldsymbol{h}}_{nl}^{s*} s_l}{||\widehat{\boldsymbol{h}}_{nl}^{s}||}}_{②} + \underbrace{n_k}_{③}. \tag{3.41}$$

In the equation above, term ① represents the signal obtained at the kth PU from M PAPs, term ② denotes the interference at the kth PU due to simultaneous transmission by N SAPs and term ③ captures noise at the kth PU and $n_k \sim \mathcal{CN}(0, \sigma_w^2)$. The interference power I_k^p at the kth PU due to sharing of spectrum by the SAPs is given by

$$I_k^p = \mathbb{E}\left(\left|\sum_{l=1}^{L}\sum_{n=1}^{N} \sqrt{\frac{P_d^s}{L}} \frac{\boldsymbol{g}_{nk}^{s\mathrm{T}} \widehat{\boldsymbol{h}}_{nl}^{s*}}{||\widehat{\boldsymbol{h}}_{nl}^{s}||} s_l\right|^2\right) = \sum_{l=1}^{L} \frac{P_d^s}{L} \left|\sum_{n=1}^{N} \frac{\boldsymbol{g}_{nk}^{s\mathrm{T}} \widehat{\boldsymbol{h}}_{nl}^{s*}}{||\widehat{\boldsymbol{h}}_{nl}^{s}||}\right|^2. \tag{3.42}$$

The signal received at the lth SU equals

$$y_l = \underbrace{\sum_{l'=1}^{L}\sum_{n=1}^{N} \frac{\sqrt{P_d^s} \boldsymbol{h}_{nl}^{s\mathrm{T}} \widehat{\boldsymbol{h}}_{nl'}^{s*} s_{l'}}{\sqrt{L} ||\widehat{\boldsymbol{h}}_{nl'}^{s}||}}_{①} + \underbrace{\sum_{k=1}^{K}\sum_{m=1}^{M} \sqrt{\eta_{mk}} \boldsymbol{g}_{ml}^{p\mathrm{T}} \widehat{\boldsymbol{h}}_{mk}^{p*} q_k}_{②} + \underbrace{n_l}_{③}. \tag{3.43}$$

In the above equation term ① represents the signal received by the lth SU from N SAPs, term ② denotes the interference generated due to transmission by M PAPs and term ③ denotes the noise at lth SU and $n_l \sim \mathcal{CN}(0, \sigma_w^2)$.

3.3.2 Performance Analysis

IOP Analysis and Power Policy: We study cell-free mMIMO underlay network architecture where the SAPs adjust their transmission power based on IOP constraint, in which, as mentioned before, the interference power arising due to spectrum access by the SAPs must not exceed a threshold $\overline{I}_p$ at any of the K PUs more than P_o fraction of times. This can be written as

$$\Pr(I_1^p \le \overline{I}_p, I_2^p \le \overline{I}_p, \ldots, I_K^p \le \overline{I}_p) \ge (1 - P_o), \tag{3.44}$$

where P_o refers to IOP.

As before, we design a simple power control rule that facilitates us to determine the power back-off that need to be taken in order to satisfy the condition in (3.44) with equality. Based on this rule, the transmission power of SAPs is written as

$$P_d^s = \frac{\overline{I}_p}{\eta \max\limits_{1\le k\le K}\left\{\frac{1}{L}\sum\limits_{l=1}^{L}\left|\sum\limits_{n=1}^{N}\frac{\widehat{\boldsymbol{g}}_{nk}^{s\,\mathrm{T}}\widehat{\boldsymbol{h}}_{nl}^{s\,*}}{||\widehat{\boldsymbol{h}}_{nl}^{s}||}\right|^2\right\}}, \tag{3.45}$$

where the power back-off η (> 1) is chosen such that the IOP constraint in (3.44) is met with equality given a target P_o. In other words, the power back-off η is taken such that

$$\Pr(I_1^p \le \overline{I}_p, I_2^p \le \overline{I}_p, \ldots, I_K^p \le \overline{I}_p) = (1 - P_o), \tag{3.46}$$

where

$$I_k^p = \frac{\frac{\overline{I}_p}{L}\sum\limits_{l=1}^{L}\left|\sum\limits_{n=1}^{N}\frac{\widehat{\boldsymbol{g}}_{nk}^{s\,\mathrm{T}}\widehat{\boldsymbol{h}}_{nl}^{s\,*}}{||\widehat{\boldsymbol{h}}_{nl}^{s}||}\right|^2}{\eta \max\limits_{1\le k\le K}\left\{\frac{1}{L}\sum\limits_{l=1}^{L}\left|\sum\limits_{n=1}^{N}\frac{\widehat{\boldsymbol{g}}_{nk}^{s\,\mathrm{T}}\widehat{\boldsymbol{h}}_{nl}^{s\,*}}{||\widehat{\boldsymbol{h}}_{nl}^{s}||}\right|^2\right\}}. \tag{3.47}$$

DL SE of PUs and SUs: We now analyze DL SE of the kth PU. From (3.32), we represent the channel as $\boldsymbol{h}_{mk}^p = \widehat{\boldsymbol{h}}_{mk}^p + \widetilde{\boldsymbol{h}}_{mk}^p$ and using use and then forget bound [21], the signal obtained at the kth PU in (3.41) can be re-written as

$$\begin{aligned} y_k = &\sum_{m=1}^{M}\sqrt{\eta_{mk}}\mathbb{E}||\widehat{\boldsymbol{h}}_{mk}^p||^2 q_k + \sum_{m=1}^{M}\sqrt{\eta_{mk}}\left(||\widehat{\boldsymbol{h}}_{mk}^p||^2 - \mathbb{E}||\widehat{\boldsymbol{h}}_{mk}^p||^2\right)q_k + \sum_{m=1}^{M}\sqrt{\eta_{mk}}\widetilde{\boldsymbol{h}}_{mk}^{p\,\mathrm{T}}\widehat{\boldsymbol{h}}_{mk}^{p\,*}q_k \\ &+ \sum_{k'\ne k}^{K}\sum_{m=1}^{M}\sqrt{\eta_{mk'}}\boldsymbol{h}_{mk}^{p\,\mathrm{T}}\widehat{\boldsymbol{h}}_{mk'}^{p\,*}q_{k'} + \sum_{l=1}^{L}\sum_{n=1}^{N}\sqrt{\frac{P_d^s}{L}}\frac{\boldsymbol{g}_{nk}^{s\,\mathrm{T}}\widehat{\boldsymbol{h}}_{nl}^{s\,*}}{||\widehat{\boldsymbol{h}}_{nl}^{s}||}s_l + n_k. \end{aligned} \tag{3.48}$$

We note that the terms in (3.48) are uncorrelated to each other. Using (3.48), the SINR corresponding to the kth PU can be written as

$$\mathrm{SINR}_k = \frac{\left|\sum_{m=1}^{M}\sqrt{\eta_{mk}}\mathbb{E}||\widehat{\boldsymbol{h}}^{p}_{mk}||^2\right|^2}{\left(\begin{array}{c}\mathbb{E}\left|\sum_{m=1}^{M}\sqrt{\eta_{mk}}\left(||\widehat{\boldsymbol{h}}^{p}_{mk}||^2-\mathbb{E}||\widehat{\boldsymbol{h}}^{p}_{mk}||^2\right)\right|^2+\mathbb{E}\left|\sum_{m=1}^{M}\sqrt{\eta_{mk}}\widetilde{\boldsymbol{h}}^{p\mathrm{T}}_{mk}\widehat{\boldsymbol{h}}^{p*}_{mk}\right|^2\\ +\sum_{k'\neq k}^{K}\mathbb{E}\left|\sum_{m=1}^{M}\sqrt{\eta_{mk}}\boldsymbol{h}^{p\mathrm{T}}_{mk}\widehat{\boldsymbol{h}}^{p*}_{mk'}\right|^2+\sum_{l=1}^{L}\mathbb{E}\left|\sum_{n=1}^{N}\sqrt{\frac{P^s_d}{L}}\frac{\boldsymbol{g}^{s\mathrm{T}}_{nk}\widehat{\boldsymbol{h}}^{s*}_{nl}}{||\widehat{\boldsymbol{h}}^{s}_{nl}||}\right|^2+\mathbb{E}\left|n_k\right|^2\end{array}\right)}. \tag{3.49}$$

The SE of the kth PU in a network where the SAPs control their transmit power based on the IOP constraint under pilot contamination is written as

$$\mathrm{SE}_k = \left(1-\frac{\tau_p}{\tau_c}\right)\log\left(1+\mathrm{SINR}_k\right). \tag{3.50}$$

We now analyze the DL SE of the lth SU. From (3.35), we express true channel as $\boldsymbol{h}^s_{nl} = \widehat{\boldsymbol{h}}^s_{nl} + \widetilde{\boldsymbol{h}}^s_{nl}$ and the signal obtained at the lth SU in (3.43) is written as

$$\begin{aligned} y_l =& \sum_{n=1}^{N}\frac{\mathbb{E}\left(\sqrt{P^s_d}||\widehat{\boldsymbol{h}}^s_{nl}||\right)s_l}{\sqrt{L}} + \sum_{n=1}^{N}\left(\frac{\sqrt{P^s_d}||\widehat{\boldsymbol{h}}^s_{nl}||}{\sqrt{L}} - \frac{\mathbb{E}(\sqrt{P^s_d}||\widehat{\boldsymbol{h}}^s_{nl}||)}{\sqrt{L}}\right)s_l + \sum_{n=1}^{N}\sqrt{\frac{P^s_d}{L}}\frac{\widetilde{\boldsymbol{h}}^{s\mathrm{T}}_{nl}\widehat{\boldsymbol{h}}^{s*}_{nl}}{||\widehat{\boldsymbol{h}}^s_{nl}||}s_l \\ &+ \sum_{l'\neq l}^{L}\sum_{n=1}^{N}\frac{\sqrt{P^s_d}\boldsymbol{h}^{s\mathrm{T}}_{nl}\widehat{\boldsymbol{h}}^{s*}_{nl'}s_{l'}}{\sqrt{L}||\widehat{\boldsymbol{h}}^s_{nl'}||} + \sum_{k=1}^{K}\sum_{m=1}^{M}\sqrt{\eta_{mk}}\boldsymbol{g}^{p\mathrm{T}}_{ml}\widehat{\boldsymbol{h}}^{p*}_{mk}q_k + n_l. \end{aligned} \tag{3.51}$$

Again, the terms in (3.51) are uncorrelated to each other. Using (3.51), the SINR corresponding to the lth SU can be expressed as

$$\mathrm{SINR}_l = \frac{\left|\sum_{n=1}^{N}\frac{\mathbb{E}\left(\sqrt{P^s_d}||\widehat{\boldsymbol{h}}^s_{nl}||\right)}{\sqrt{L}}\right|^2}{\left(\begin{array}{c}\mathbb{E}\left|\sum_{n=1}^{N}\left(\frac{\sqrt{P^s_d}||\widehat{\boldsymbol{h}}^s_{nl}||}{\sqrt{L}}-\frac{\mathbb{E}\left(\sqrt{P^s_d}||\widehat{\boldsymbol{h}}^s_{nl}||\right)}{\sqrt{L}}\right)\right|^2+\mathbb{E}\left|\sum_{n=1}^{N}\frac{\sqrt{P^s_d}\widetilde{\boldsymbol{h}}^{s\mathrm{T}}_{nl}\widehat{\boldsymbol{h}}^{s*}_{nl}}{\sqrt{L}||\widehat{\boldsymbol{h}}^s_{nl}||}\right|^2+\sum_{l'\neq l}^{L}\mathbb{E}\left|\sum_{n=1}^{N}\frac{\sqrt{P^s_d}\boldsymbol{h}^{s\mathrm{T}}_{nl}\widehat{\boldsymbol{h}}^{s*}_{nl'}}{\sqrt{L}||\widehat{\boldsymbol{h}}^s_{nl'}||}\right|^2\\ +\sum_{k'\neq k}^{K}\mathbb{E}\left|\sum_{m=1}^{M}\sqrt{\eta_{mk}}\boldsymbol{h}^{p\mathrm{T}}_{mk}\widehat{\boldsymbol{h}}^{p*}_{mk'}\right|^2+\sum_{l=1}^{L}\mathbb{E}\left|\sum_{n=1}^{N}\sqrt{\frac{P^s_d}{L}}\frac{\boldsymbol{g}^{s\mathrm{T}}_{nk}\widehat{\boldsymbol{h}}^{s*}_{nl}}{||\widehat{\boldsymbol{h}}^s_{nl}||}\right|^2+\mathbb{E}\left|n_k\right|^2\end{array}\right)}. \tag{3.52}$$

The lth SU SE when the SAP transmission power is controlled based on an IOP constraint is given by

$$\mathrm{SE}_l = \left(1-\frac{\tau_p}{\tau_c}\right)\log\left(1+\mathrm{SINR}_l\right). \tag{3.53}$$

3.4 Results and Discussions

We first describe the simulation set-up for both co-located mMIMO and cell-free mMIMO based underlay spectrum sharing networks. We then present simulation results to compare the performance of the two networks and elucidate effect of parameters on system performance.

3.4.1 Simulation Set-Up

Co-located mMIMO Underlay Network: We consider $L_{cl} = 4$ square cells in 1 Km × 1 Km area. Each cell consists of a PBS at a height of 20 meters and SBS at a height of 15 meters located at the center of the cell. The height of PUs and SUs are taken to be 1.65 m. Both PBS and SBS are equipped with $M_{cl} = N_{cl} = 100$ antennas each. We consider that there are $K_p = 10$ PUs and $K_s = 10$ SUs in each cell. We take $\tau_c = 200$ samples, $P_d^{pl} = 16.02$ dB, $\tau_p = 20$, $p_t^p = p_t^s = 0\,\text{dBm}$ and choose bandwidth (BW) to be 20 MHz and pilots are shared across cells. The noise power spectral density (PSD) is taken as −174 dBm/Hz. The large scale fading coefficient β_{ik}^{pl} is modeled as [14],

$$\beta_{ik}^{xl} = 10^{\frac{PL_{ik}^{xl}}{10}}.10^{\frac{\sigma_{sh} z_{ik}^{xl}}{10}}, \tag{3.54}$$

where k belongs to the set $\{k_p, k_s\}$ and equals k_p for PUs and equals k_s for SUs, such that $1 \leq k_p \leq K_p$ and $1 \leq k_s \leq K_s$, $x \in \{p, s\}$ depending on whether x corresponds to PBS or SBS, $1 \leq l \leq 4$, $1 \leq i \leq 4$ where PL_{ik}^{xl} denotes path loss and is obtained based on the three slope model [14], and $10^{\frac{\sigma_{sh} z_{ik}^{xl}}{10}}$ represents shadow fading with standard deviation $\sigma_{sh} = 8$ dB, and $z_{ik}^{xl} \sim \mathcal{N}(0, 1)$.

Cell-Free mMIMO: We consider that the APs in cell-free mMIMO network and the users are distributed randomly in a 1 Km × 1 Km area. The height of both PAPs and SAPs is considered to be 15 m, the height of PUs and SUs is taken to be of 1.65 m. We consider $\tau_p = 20$, $\tau_c = 200$, $P_d^p = 0\,\text{dBm}$, $f_c = 1.9\,\text{GHz}$, BW $= 20\,\text{MHz}$, $p_t^p = p_t^s = 0\,\text{dBm}$ (one pilot shared by four users) and noise PSD of −174 dBm/Hz. We model large-scale fading β_{ab}^x between ath AP and bth user as [14],

$$\beta_{ab}^{x} = 10^{\frac{PL_{ab}^{x}}{10}}.10^{\frac{\sigma_{sh} z_{ab}^{x}}{10}}, \tag{3.55}$$

where $x \in \{p, s\}$ based on whether x corresponds to PAP or SAP, $1 \leq a \leq M$ or $1 \leq a \leq N$ depending on whether it corresponds to PAP or SAP, $1 \leq b \leq K$ or $1 \leq b \leq L$ depending on whether it corresponds to PUs or SUs. In addition to that, the path loss PL_{ab}^x corresponding to the link between the ath AP and bth user is calculated using the three slope model [14], and $10^{\frac{\sigma_{sh} z_{ab}^{x}}{10}}$ represents shadow fading

$\sigma_{sh} = 8$ dB, and $z^x_{ab} \sim \mathcal{N}(0, 1)$. To make a fair comparison between co-located and cell-free mMIMO we consider $M = N = 100$ APs with four antennas each, $K = 40$ PUs, $L = 40$ SUs and 4 users (PUs or SUs) share the same pilot.

3.4.2 Numerical Results

Figure 3.3 plots $(1 - P_o)$ vs η for cell-free and co-located mMIMO networks. We observe that the SBS in co-located mMIMO underlay network must reduce its transmit power by a larger amount (higher back-off) in comparison to the reduction in transmit power at SAPs in cell-free mMIMO underlay network in order to maintain P_o at a fixed level. This is because SAPs transmit at relatively lower power when compared to a SBS in the data phase. This can be observed form (3.19) and (3.45) and based on the fact that the effective transmission distance is shorter between the SAPs and PUs in cell-free mMIMO underlay network when compared to distance between SBS and PUs in co-located mMIMO underlay network.

Figure 3.4 compares $(1 - P_o)$ vs η for co-located and cell-free mMIMO networks when the power with which the users transmit pilots is increased to 20 dBm. We observe that for higher pilot power, lower back-off in transmit power is required at SBS when compared to Fig. 3.3. This is, because as the pilot power increases, the channel estimation accuracy increases, which results in relatively lower violation of target P_o at the PUs. When compared to the result in Fig. 3.3, a marginally lower back-off is required at SAPs in cell-free mMIMO underlay networks when the pilot power is increased to 20 dBm.

Figure 3.5 plots $(1 - P_o)$ vs η for different number of SBS antennas in co-located mMIMO and different number of SAPs in cell-free mMIMO networks. We observe

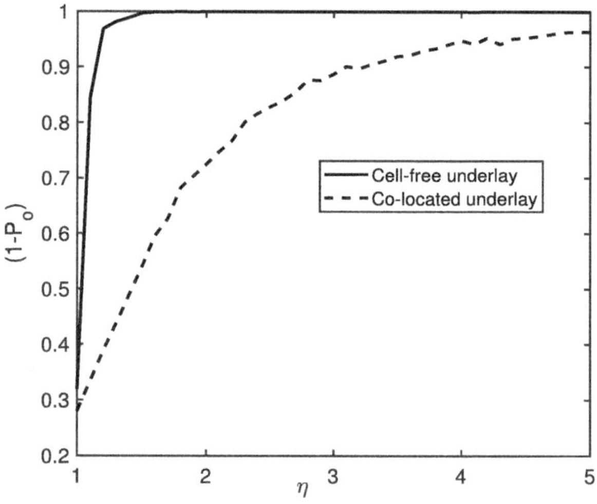

Fig. 3.3 $(1 - P_o)$ vs. η for **a** co-located mMIMO network ($L_{cl} = 4$, $M_{cl} = N_{cl} = 100$, $K_p = K_s = 10$, $\tau_p = 20$, $p^p_t = 0$ dBm) **b** cell-free mMIMO network ($M = N = 100$, $Q_p = Q_s = 4$, $K = L = 40$, $\tau_p = 20$, $p^p_t = 0$ dBm)

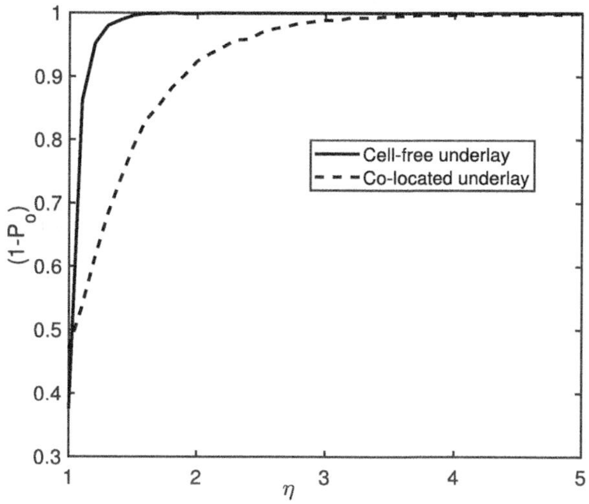

Fig. 3.4 Impact of pilot power: $(1 - P_o)$ vs. η for **a** co-located mMIMO network ($L_{cl} = 4$, $M_{cl} = N_{cl} = 100$, $K_p = K_s = 10$, $\tau_p = 20$, $p_t^p = 20\,\text{dBm}$) **b** cell-free mMIMO network ($M = N = 100$, $Q_p = Q_s = 4$, $K = L = 40$, $\tau_p = 20$, $p_t^p = 20\,\text{dBm}$)

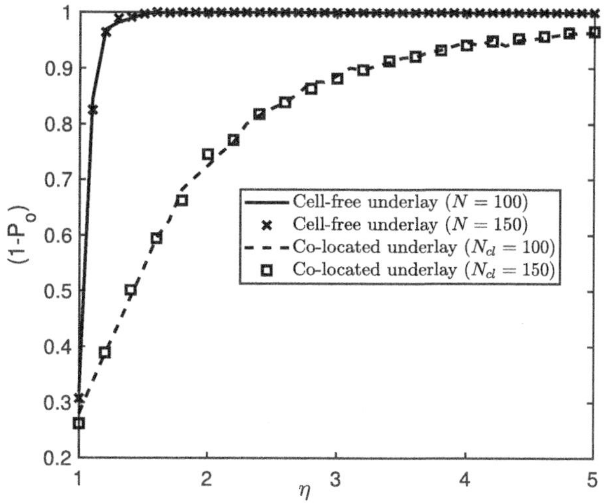

Fig. 3.5 Effect of number of SBS antennas and number of SAPs: $(1 - P_o)$ vs. η for **a** co-located mMIMO network ($L_{cl} = 4$, $M_{cl} = 100$, $K_p = K_s = 10$, $\tau_p = 20$, $p_t^p = 0\,\text{dBm}$) **b** cell-free mMIMO network ($M = 100$, $Q_p = Q_s = 4$, $K = L = 40$, $\tau_p = 20$, $p_t^p = 0\,\text{dBm}$)

that P_o is independent of number of SBS antennas and number of SAPs. This is due to the use of normalized MR precoding at the SBS and SAPs during DL data transmission. Thus, we obtain a crucial design intuition that P_o at PUs in co-located mMIMO and cell-free mMIMO underlay networks remains constant regardless of the changes in the number of SBS antennas and the number of SAPs.

Figure 3.6 plots the CDF of DL SE of PUs and SUs for two different values of the interference threshold $\bar{I}_p$ namely -80 dB and -90 dB. We observe that as $\bar{I}_p$ increases, the IOP constraint gets loose, as a result, the probability for SUs to obtain higher UL SE increases and the probability for PUs to obtain higher UL SE decreases, since higher $\bar{I}_p$ results in more interference at PUs for both co-located and

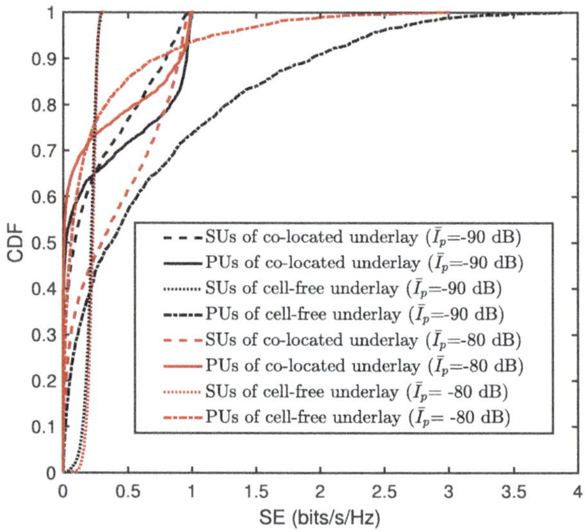

Fig. 3.6 Effect of $\overline{I}_p$: CDF of DL SE for **a** co-located mMIMO network ($L_{cl} = 4$, $M_{cl} = N_{cl} = 100$, $K_p = K_s = 10$, $\tau_p = 20$, $p_t^p = p_t^s = 0\,\text{dBm}$, $P_d^{pl} = 16.020$ dB) **b** cell-free mMIMO network ($M = N = 100$, $Q_p = Q_s = 4$, $K = L = 40$, $\tau_p = 20$, $p_t^p = p_t^s = 0\,\text{dBm}$, $P_d^p = 0\,\text{dBm}$)

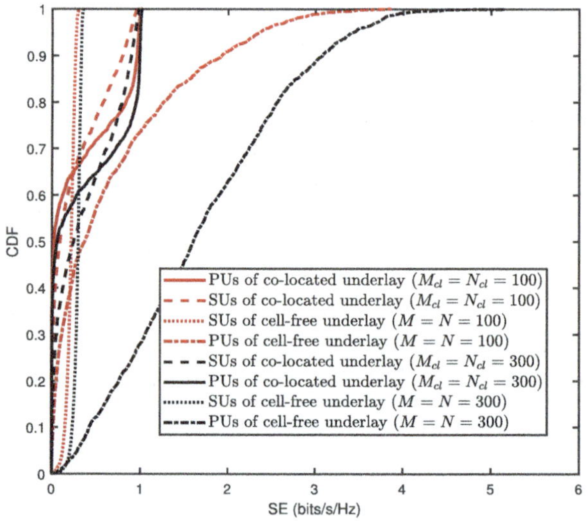

Fig. 3.7 Impact of number of SBS antennas and number of SAPs: CDF of SE for **a** co-located mMIMO network ($L_{cl} = 4$, $K_p = K_s = 10$, $\tau_p = 20$, $P_d^{pl} = 16.020$ dB, $\overline{I}_p = -90$ dB, $p_t^p = p_t^s = 0\,\text{dBm}$) **b** cell-free mMIMO network ($Q_p = Q_s = 4$, $K = L = 40$, $\tau_p = 20$, $\overline{I}_p = -90$ dB, $P_d^p = 0\,\text{dBm}$, $p_t^p = p_t^s = 0\,\text{dBm}$)

cell-free mMIMO underlay networks. We also observe that cell-free mMIMO based underlay network gives higher SE for PUs and boosts the performance of weaker SUs in comparison to users in co-located mMIMO underlay networks.

Figure 3.7 plots the CDF of DL SE of PUs and SUs for different values of M_{cl}, N_{cl}, M and N. As we increase the number of antennas in co-located mMIMO BSs (SBS and PBS) or as we increase the number of SAPs and PAPs, the probability of obtaining a higher SE increases for both PUs and SUs. We also observe that cell-free mMIMO based underlay network gives higher SE for PUs when compared against co-located mMIMO based underlay networks, while the performance of weaker SUs is boosted in cell-free mMIMO underlay networks.

3.5 Conclusions

In this chapter, we presented a study to compare performance of co-located and cell-free mMIMO based underlay networks. To this end, we first developed the model and presented a detailed performance analysis for co-located mMIMO based underlay network under pilot sharing induced contamination and when the SBSs transmit under an IOP constraint. We then developed the corresponding model and analysis for cell-free mMIMO based underlay networks.

We inferred that the SAPs in cell-free mMIMO underlay networks need to back-off their transmit power by a lower amount when compared against SBS in co-located mMIMO underlay networks for a given IOP. We also observed that cell-free mMIMO based underlay network boosts the performance of PUs and the weaker SUs when compared against co-located mMIMO based underlay networks under pilot contamination and when these networks operate under an IOP constraint.

References

1. Cisco Annual Internet Report (2018–2023) White Paper. Ericsson, Stockholm, Sweden (2020)
2. FCC (2015) Amendment of the commission's rules with regard to commercial operations in the 3550-3650 MHz band. FCC-15-47
3. FCC (2018) Amendment of the commission's rules with regard to commercial operations in the 3.7-24 GHz band. FCC-18-147
4. Sarvendranath R, Mehta NB (2020) Exploiting power adaptation with transmit antenna selection for interference-outage constrained underlay spectrum sharing. IEEE Trans Commun 68(1):480–492
5. FCC (2020) Unlicensed use of the 6 GHz band; expanding flexible use in mid-band spectrum between 3.7 and 24 GHz. FCC-20-51
6. Goldsmith A, Jafar SA, Maric I, Srinivasa S (2009) Breaking spectrum gridlock with cognitive radios: an information theoretic perspective. Proc IEEE 97(5):894–914
7. Galappaththige DL, Baduge GAA (2021) Exploiting underlay spectrum sharing in cell-free massive MIMO systems. IEEE Trans Commun 69(11):7470–7488
8. Krishnan NN, Kumbhkar R, Mandayam NB, Seskar I, Kompella S (2017) Coexistence of radar and communication systems in CBRS bands through downlink power control. In: Proceedings IEEE Military Communications Conference, pp 713–718
9. Larsson EG, Tufvesson F, Edfors O, Marzetta TL (2014) Massive MIMO for next generation wireless systems. IEEE Commun Mag 52(2):186–195
10. Björnson E, Larsson EG, Marzetta TL (2016) Massive MIMO: ten myths and one critical question. IEEE Commun Mag 54(2):114–123
11. Pothan EV, Kashyap S (2022) Massive MIMO based underlay spectrum access under incomplete and/or imperfect channel state information. IEEE Trans Cognitive Commun Netw 8(3):1482–1496

12. Geraci G, Rodriguez AG, Perez DL, Bonfante A, Giordano G, Claussen H (2017) Operating massive MIMO in unlicensed bands for enhanced coexistence and spatial reuse. IEEE J Sel Areas Commun 35(6):1282–1293
13. Giordani M, Polese M, Mezzavilla M, Rangan S, Zorzi M (2020) Toward 6G networks: use cases and technologies. IEEE Commun Mag 58(3):55–61
14. Ngo HQ, Ashikhmin A, Yang H, Larsson EG, Marzetta TL (2017) Cell-free massive MIMO versus small cells. IEEE Trans Wireless Commun 16(3):1834–1850
15. Elhoushy S, Ibrahim M, Hamouda W (2022) Cell-free massive MIMO: a survey. IEEE Commun Surveys Tuts 24(1):492–523
16. Pothan EV, Kashyap S (2023) Cell-free massive MIMO enabled wireless communication with UAVs in underlay spectrum access networks. IEEE Trans Commun 71(12):7363–7377
17. Demir Ö, Björnson E, Sanguinetti L (2021) Foundations of user-centric cell-free massive MIMO. Now Publishers
18. Björnson E, Sanguinetti L (2019) Cell-free versus cellular massive MIMO: what processing is needed for cell-free to win? In: Proceeding of the SPAWC, pp 1–5
19. Hao W, Muta O, Gacanin H, Furukawa H (2018) Power allocation for massive MIMO cognitive radio networks with pilot sharing under SINR requirements of primary users. IEEE Trans Veh Technol 67(2):1174–1186
20. Wang W, Yu G, Huang A (2013) Cognitive radio enhanced interference coordination for femtocell networks. IEEE Commun Mag 51(6):37–43
21. Marzetta TL, Larsson EG, Yang H, Ngo HQ (2016) Fundamentals of massive MIMO. Cambridge University Press, Cambridge, U.K
22. Kay SM (1993) Fundamentals of statistical signal processing: estimation theory, 1st edn. Prentice-Hall Inc, USA
23. Khan FA, Masouros C, Ratnarajah T (2012) Interference-driven linear precoding in multiuser MISO downlink cognitive radio network. IEEE Trans Veh Technol 61(6):2531–2543
24. Noh J-H, Oh S-J (2013) Beamforming in a multi-user cognitive radio system with partial channel state information. IEEE Trans Wireless Commun 12(2):616–625

Chapter 4
Intelligent Reflecting Surface Assisted Wireless Communications: Design and Performance Analysis

Dharmendra Dixit, Sanjeev Sharma, and Kuntal Deka

Abstract This chapter presents an exploration of intelligent reflecting surface (IRS) assisted wireless communications, delving into both its design principles and an in-depth analysis of its performance. The chapter begins with an introduction to IRS, detailing its fundamental principles and role in enhancing wireless communication systems. Subsequently, a section on "Design, Implementation, and Real-world Considerations" delves into the architecture of a typical IRS setup, various methods for IRS reconfiguration, practical reflection models employed by IRS, and considerations regarding the power consumption of IRS. Following this, the chapter presents a comprehensive analysis of the performance of a single-IRS-aided single-input single-output system in Rayleigh fading channels. This analysis includes the derivation of approximate outage probability (OP) expressions, asymptotic OP analysis, and diversity order analysis based on asymptotic OP. Additionally, approximate average symbol error rate (ASER) expressions are derived for both coherent and non-coherent modulation schemes, along with asymptotic ASER analysis and diversity order analysis based on asymptotic ASER. In addition, the chapter also includes a presentation of numerical and simulation results, accompanied by a detailed discussion of their implications.

Keywords Intelligent reflecting surfaces · Outage probability · Average symbol error rate

---

D. Dixit (✉)
MNNIT Allahabad, Prayagraj, India
e-mail: dharmendradixit@mnnit.ac.in

S. Sharma
IIT (BHU) Varanasi, Varanasi, India
e-mail: sanjeev.ece@iitbhu.ac.in

K. Deka
IIT Guwahati, Guwahati, India
e-mail: kuntaldeka@iitg.ac.in

S. Sharma et al. (eds.), *Next-Generation Wireless Systems*,
Transactions on Computer Systems and Networks,
https://doi.org/10.1007/978-981-96-3185-8_4

4.1 Introduction

Amidst the widespread implementation of fifth-generation (5G) mobile communication networks, researchers from diverse backgrounds, including both industry and academia, have initiated exploration beyond 5G (B5G) wireless networks. In the pursuit of significantly elevating key performance metrics like data rate, latency, and connectivity density, a wave of innovative technologies has surfaced. Examples include the expansive realm of extremely large-scale multiple-input multiple-output (XL-MIMO), the frontier of TeraHertz (THz) communication, and the ingenuity embedded in intelligent reflecting surfaces (IRS). Figure 4.1 lists three main B5G technologies. Especially noteworthy, the utilization of IRS stands out as a technology with great promise, aiming to realize wireless communication that is both cost-effective and energy-efficient through the proactive manipulation of the radio propagation environment [1, 2].

4.1.1 What is IRS?

The IRS, as man-made electromagnetic surfaces, possesses electronic control through integrated electronics, endowing it with distinctive capabilities in the realm of wireless communication. Present applications encompass traditional reflect-arrays, surfaces incorporating liquid crystal technology, and meta-surfaces defined through software. Unlike other technologies employed in wireless networks and the prevailing design principles of wireless communications, the IRS stands out due to its unique feature. It enables telecommunication operators to control the environment, offering the capability to shape and fully control the electromagnetic response of distributed objects within the network. Thus, the IRS is specifically designed to deliberately and predictably manipulate the propagation environment, with the objective of enhancing

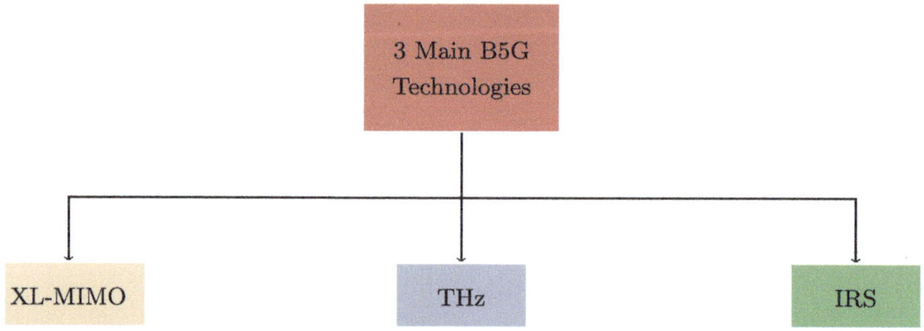

Fig. 4.1 B5G technologies

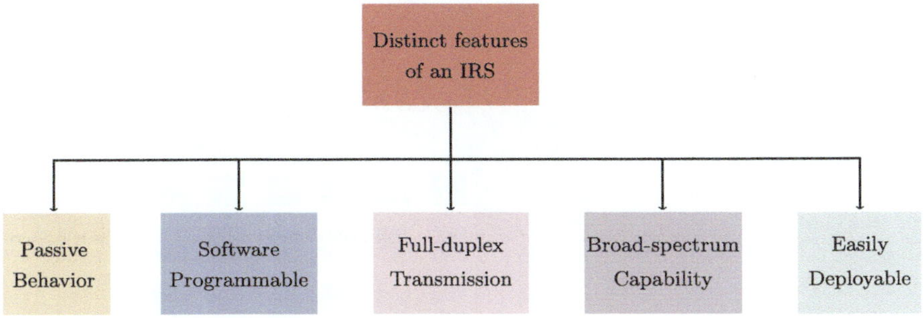

Fig. 4.2 Distinct features of an IRS

signal quality at the receiver. In other words, the advent of the IRS has sparked the emergence of the innovative concept known as "smart radio environments".

It is noteworthy that the IRS distinguishes itself from various technologies in wireless networks, including relaying, MIMO beamforming, and backscatter communications, which may initially seem related. The IRS stands apart with unique characteristics that set it apart from these other technologies. The IRS exhibits the following distinct features [1]:

1. It exhibits near-passive behavior and, ideally, requires no dedicated energy source.
2. Perceived as a seamless surface, each point ideally possesses the capability to shape the impinging wave, embodying the concept of soft programming.
3. Unaffected by receiver noise, the ideal IRS doesn't require analog-to-digital/digital-to-analog converters (ADCs and DACs) or power amplifiers, eliminating noise amplification during signal reflection and enabling inherently full-duplex transmission.
4. Boasting a broad-spectrum capability, they ideally exhibit a full-band response, ensuring effective operation across diverse operating frequencies.
5. Its deployment versatility extends to various settings, including building facades, factory ceilings, indoor spaces, and even human clothing.

4.2 Design, Implementation, and Real-World Considerations

Achieving practical control over reflections with IRS is made feasible by harnessing the capabilities of digitally reconfigurable or programmable meta-surfaces already in place [2]. As we know, a meta-surface is a planar array comprised of meticulously crafted reflecting elements, or meta-atoms, each with an electrical thickness typically on the subwavelength scale relative to the signal of the interest [3]. Tailoring

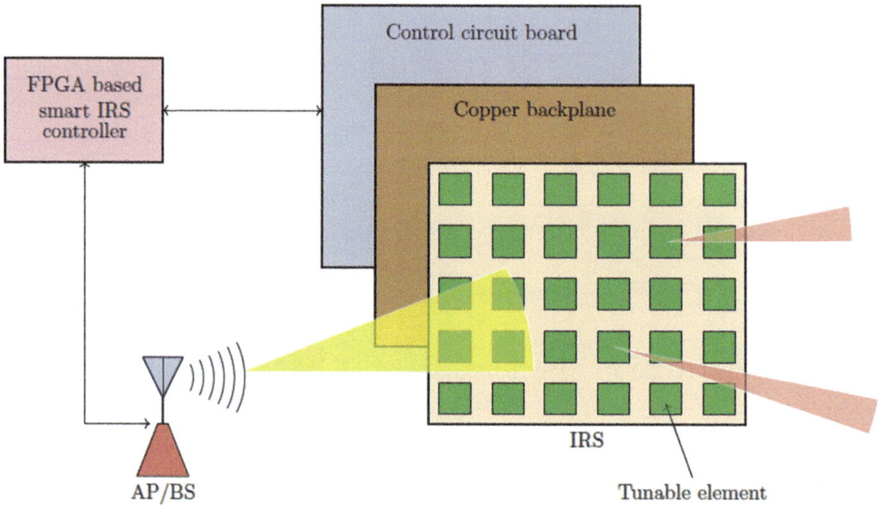

Fig. 4.3 A typical IRS architecture

their geometric shape (e.g., square or split-ring), dimensions, orientation, arrangement, and other parameters enables the realization of specific signal responses (e.g., amplitude modulation or phase shift) for each individual element or atom. Since, within wireless communications, the channel fluctuates over time, affected by the movement of the transmitter/receiver and surrounding objects. This dynamic nature requires real-time adaptability in response to IRS to match channel variations. For this purpose, IRS elements must be manufactured with reflection coefficients that can be adjusted dynamically. In addition, the IRS system needs to interface with wireless networks to gather information about the surrounding communication environment, thereby facilitating its real-time adaptive reflection capabilities.

Figure 4.3 shows a typical IRS architecture consisting of 3 layers with a smart controller. The first layer consists of multiple tunable or reconfigurable metal patches made on a dielectric substrate, facilitating direct manipulation of incident signals. The second layer is made of copper plate which reduces the signal energy leakage during the reflection process by IRS. Next, the third layer contains a control circuit board that serves to activate the reflective elements and dynamically adjust their reflection properties, including amplitude and phase shift, in real time. Furthermore, the adjustment of the reflections is initiated and managed by an smart controller associated with each IRS, with implementations often using field-programmable gate arrays (FPGAs). In addition, the smart controller serves as a gateway for communication with various network components, such as base stations (BS), access points (AP), and user terminals, through wired or wireless backhaul/control links. In practical applications, specialized sensors can be strategically deployed within the first layer to enhance the ability of the IRS to learn from its environment. These sensors, integrated with the reflective elements of the IRS, monitor relevant surrounding radio

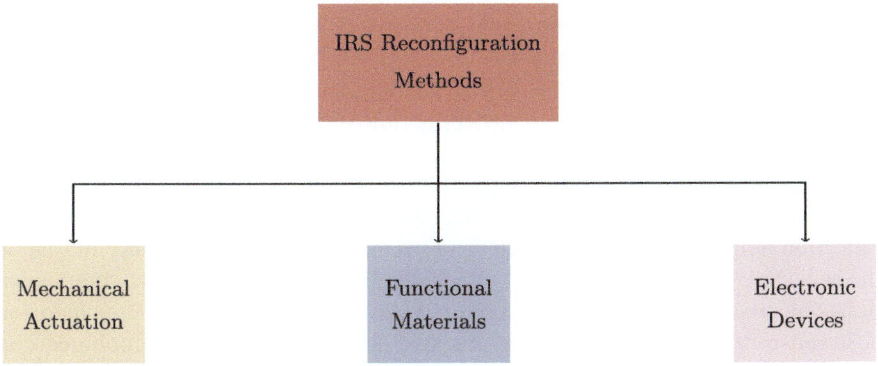

Fig. 4.4 IRS reconfiguration methods

signals. This data helps the smart controller to effectively optimize the reflection coefficient.

Figure 4.4 shows existing IRS reconfiguration methods. To achieve highly controllable reflection from IRS elements, there are three primary methods [4]:

1. mechanical actuation (such as mechanical rotation and translation)
2. using functional materials (such as liquid crystals and graphene), and
3. using electronic devices. (such as PIN diode, field-effect transistor (FET), or MEMS switch)

Out of these three approaches, the third one has garnered considerable attention in real-world application owing to its rapid response time, minimal reflection loss, and comparatively lower energy usage and hardware expenses.

Figure 4.5 shows a sample design of a reflective element with its equivalent circuit, which includes a PIN diode located in the center of the element. Applying varied

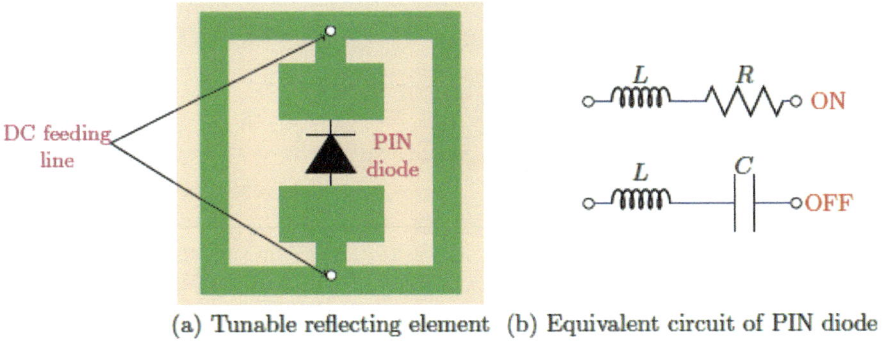

Fig. 4.5 Illustration of a PIN diode-based tunable reflecting element

biasing voltages to the PIN diode through a direct-current (DC) feeding line facilitates switching the PIN diode between its "ON" and "OFF" states. This capability allows the element to induce π phase-shift difference in the incident signal. As mentioned in [5], a PIN diode can achieve a switching time of less than 1 microsecond (1 μs). This significantly shorter switching time, in contrast to typical channel coherence times of milliseconds (ms), makes it particularly suitable for mobile applications with time-varying channels. In addition to adjusting phase shift, the ability to control the reflection amplitude of each IRS element provides additional flexibility in shaping the reflected signal, increasing the effectiveness of achieving diverse communications objectives. This approach also offers a versatile method of balancing hardware cost and reflection performance in practical scenarios. Typically, amplitude control involves lower implementation costs than phase control, providing an efficient trade-off solution. There are multiple methods available to accomplish amplitude adjustment for IRS. One prevalent approach involves modifying the load resistance or impedance in each element [6]. For instance, altering the resistance of individual elements allows for the dissipation of a specific portion of the incident signal energy as heat, thereby enabling a dynamic range of reflection amplitudes within the range of [0, 1]. In practical applications, it is better to achieve independent control over both the amplitude and phase shift of each IRS element to optimize reflection designs. However, this necessitates more sophisticated hardware designs, such as the multilayer surface design, compared to configurations where only one parameter is controlled separately [4].

Ideally, each element of the IRS could independently and continuously adjust both reflection amplitude and phase shift, resulting in their respective possible sets:

$$\mathcal{S}_a = \{a | a \in [0, 1]\} \tag{4.1}$$

$$\mathcal{S}_\theta = \{\theta | \theta \in [0, 2\pi]\} \tag{4.2}$$

Keep in mind that the aforementioned ideal possible sets for IRS reflecting elements typically establish the theoretical performance benchmarks of IRS-enhanced

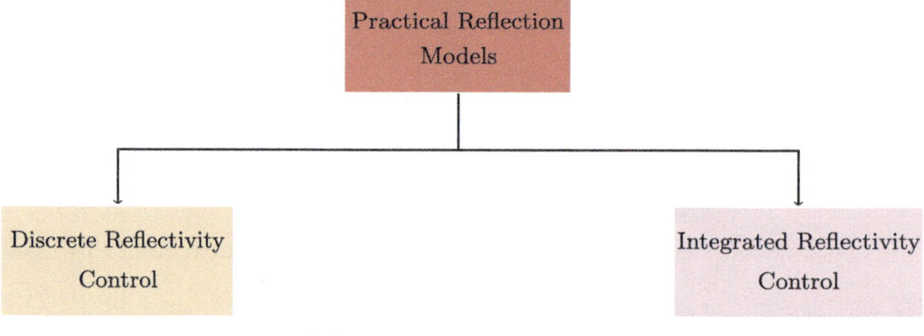

Fig. 4.6 Practical reflection models

wireless communication systems within practical reflection models. Figure 4.6 shows practical reflection models. Two practical reflection models are elaborated as below.

1. **Discrete Reflectivity Control:**
   Continuous tuning of the reflection coefficient offers advantages for optimizing communication performance. However, its practical implementation poses challenges because of the heightened cost and complexity of hardware design required for higher-resolution reflecting elements. Consider this: To achieve 16 levels of phase shifts per IRS element, a minimum of $\log_2(16) = 4$ PIN diodes are necessary. However, this presents a challenge in element design due to size constraints and requires additional control pins at the IRS controller to manage the necessary PIN diodes. Using a single varactor diode to accomplish multi-level phase shifts can be an option, but it demands a broad range of biasing voltages, making it a costlier choice to implement. Therefore, for practical IRS configurations, which typically contain multiple reflecting elements, it is more economically viable to employ discrete and finite amplitude/phase-shift levels. This approach requires just a few control bits for each element, such as implementing binary amplitude control (reflected or absorbed) and/or binary phase-shift control (0 or π) [7].
   Generally, let b_a and b_θ represent the number of bits needed to control the respective levels of reflection amplitude and phase shift, denoted as N_a and N_θ respectively, where $N_a = 2^{b_a}$ and $N_\theta = 2^{b_\theta}$. Consequently, the sets of possible discrete reflection amplitudes and phase shifts at each element are given as:

$$\mathcal{S}_a^{'} = \{a_1, a_2, \ldots a_m \ldots, a_{N_a}\} \tag{4.3}$$

$$\mathcal{S}_\theta^{'} = \{\theta_1, \theta_2 \ldots, \theta_l \ldots, \theta_{N_\theta}\} \tag{4.4}$$

   where $0 \leq a_m \leq 1$ for $1 \leq m \leq N_a$ and $0 \leq \theta_l \leq 2\pi$ for $1 \leq l \leq N_\theta$. For example, if we assume that discrete amplitude and phase-shift values are acquired through uniform quantization of the intervals $[0, 1]$ and $[0, 2\pi]$ respectively, then we can represent these values as

$$\mathcal{S}_a^{'} = \{0, \Delta_a, \ldots, (N_a - 1)\Delta_a\} \tag{4.5}$$

$$\mathcal{S}_\theta^{'} = \{0, \Delta_\theta, \ldots, (N_\theta - 1)\Delta_\theta\} \tag{4.6}$$

   where $\Delta_a = 1/(N_a - 1)$ and $\Delta_\theta = 2\pi/(N_\theta - 1)$. In contrast to the ideal continuous reflection amplitude/phase-shift models outlined in Eqs. (4.1) and (4.2), their quantized versions as depicted in Eqs. (4.5) and (4.6) invariably lead to a less refined control over reflected signal amplitude/phase, consequently leading to degraded communication performance. Moreover, they introduce fresh challenges in optimizing the amplitudes and phase-shifts of IRS reflecting elements in practical applications, as the discrete optimization variables they produce are generally more complex to manage than their continuous counterparts. In practical scenarios, to minimize hardware costs and design intricacies, one might opt for either discrete phase-shift control or discrete amplitude control exclusively. In general, phase-shift control, also known as phase beamforming, tends to incur higher implementation costs for IRS compared to amplitude control, or amplitude beamforming. However, despite its higher cost, phase-shift control typically offers superior passive beamforming performance compared to amplitude control when the same number of control bits or discrete levels per reflecting element is employed.

2. **Integrated Reflectivity Control:**
   While individual control over reflection amplitude and phase shift provides unparalleled design flexibility, it presents substantial hurdles in element design. In [8], an IRS reflection model is presented where each reflecting element is represented as a resonant circuit defined by distinct inductance, capacitance, and resistance parameters. This model reveals that the amplitude response of the reflecting element is typically intertwined with its phase shift in a non-linear manner, indicating that they cannot be adjusted independently. In particular, the reflection amplitude generally reaches its minimum at zero phase shift, then gradually rises, approaching its maximum value of one as the phase shift approaches either π or $-\pi$. This occurs because in each reflecting element, achieving any phase shift involves tuning its effective capacitance or resistance, which unavoidably alters the reflection amplitude. Put simply, when the phase shift approaches zero, the reflective currents align with the element currents, boosting both the electric field and current flow within the element. This leads to maximum energy dissipation and the lowest reflection amplitude. Conversely, when the phase shift is approximately at π or $-\pi$, the reflective currents oppose the element currents, causing a decrease in the electric field and current flow within the element. This results in minimal energy loss and the highest reflection amplitude. This analytical IRS reflection model aligns with experimental findings documented in [9]. The relationship between reflection amplitude and phase shift significantly influences the design of optimal reflection coefficients for IRS elements. It's crucial to strike a balance between signal amplitude and phase in each element to ensure that the combined signals at the receiver maximize power or achieve the highest signal-to-noise ratio (SNR).

4.2.1 Power Consumption of IRS

Apart from the comparison between ideal and practical IRS reflection models, it's worth noting that while IRS is theoretically passive, it does require power in practice. Even though IRS merely reflects signals without amplification, it still necessitates a power supply to maintain the process of adjusting its reflecting elements and smart controller. For example, when using PIN diodes to adjust phase shifts in IRS reflecting elements, the element consumes no energy when the diode is OFF. However, when the diode is ON, it consumes a specific amount of power, such as approximately

0.33 milliwatts (mW) according to [10], and 50 microwatts (μW) according to [11]. However, in comparison to the power usage of active antennas like XL-MIMO and multi-antenna relay systems, the power consumed by IRS is significantly lower and can typically be disregarded for comparison purposes. Conversely, the power consumption of the smart controller relies on the specific circuit implementation, such as FPGA, and the communication module utilized. For example, the power usage of an IRS controller with 256 reflecting elements is approximately 0.72 W, as reported in [10], considerably reduce than that of active BSs or relays typically encountered in practice. Hence, IRS typically requires much less energy than the active relays currently used in wireless networks.

4.3 Performance Analysis: Single IRS Aided Single-Input Single-Output System

We consider an IRS-aided single-input single-output (SISO) scheme which is shown in Fig. 4.7. A source node S with a single-antenna establishes communication with a single-antenna node D using an IRS comprising L elements. A direct path between nodes S and D is absent due to man-made or natural obstructions. The function of the IRS is to provide adjustable phase shifts that are controlled by and programmed

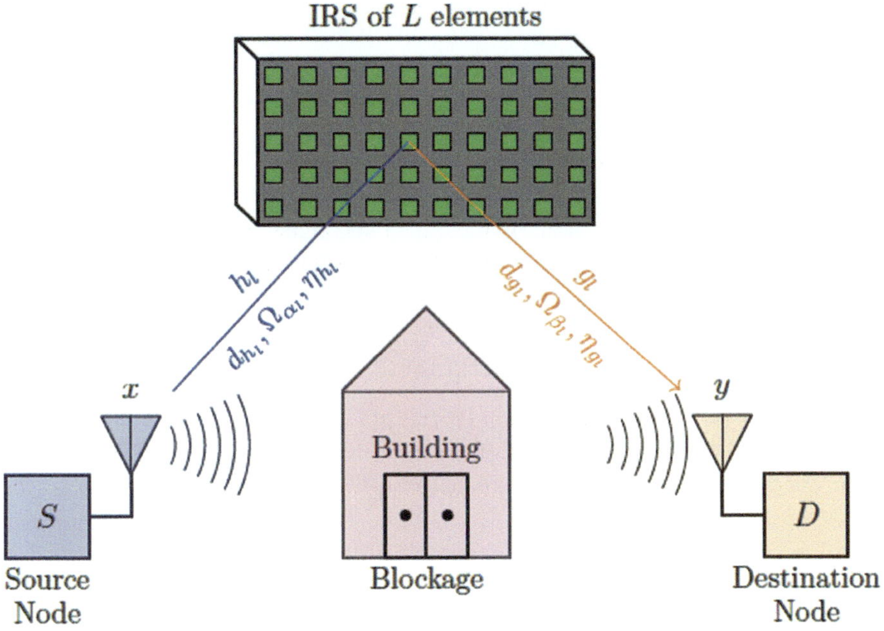

Fig. 4.7 An IRS-enabled dual-hop SISO communication scenario

through communication-oriented software, assuming perfect knowledge of the channel phases for both hops is known by the IRS.

We assume a channel model characterized by slow variations and flat fading for our analysis. Hence, the expression for the base-band equivalent of the received signal at the destination is given by

$$y = \left[\sum_{l=1}^{L} h_l e^{j\phi_l} g_l \right] x + n, \tag{4.7}$$

where h_l and g_l are the complex channel coefficients between node S and the lth reflecting element and between the lth reflecting element and node D, respectively, ϕ_l denotes the adjustable phase introduced by the lth reflecting element within the IRS, x represents the data symbol of average energy E_t chosen from an M-ary digital modulation constellation, and n denotes the additive white Gaussian noise (AWGN) sample characterized by zero mean and variance N_0. The complex channel coefficients can also be expressed as $h_l = \alpha_l e^{j\theta_l}$ and $g_l = \beta_l e^{j\psi_l}$. Alternatively, (4.7) can be expressed in matrix form as [1, 12]

$$y = \mathbf{g}^T \Phi \mathbf{h} x + n, \tag{4.8}$$

where $\mathbf{h} = [h_1, \ldots, h_L]^T$ and $\mathbf{g} = [g_1, \ldots, g_L]^T$ denote the vectors of channel coefficients connecting the two nodes and the IRS, and $\Phi = \text{diag}\left[e^{j\phi_1}, \ldots, e^{j\phi_L}\right]$ is a diagonal matrix incorporating the phase shifts from the IRS reflecting elements. We make the assumption that Rayleigh fading impacts both hops of the channels. Therefore, the formula for the probability density function (PDF) of a Rayleigh-distributed random variable (RV) $Z \in \{\alpha_l, \beta_l, 1 \leq l \leq L\}$ is given by [13, (2.6)]

$$f_Z(z) = \frac{2z}{\Omega_z} \exp\left(-\frac{z^2}{\Omega_z}\right), \tag{4.9}$$

where $\Omega_z = \mathbf{E}[Z^2]$ represents the average power, with $\mathbf{E}[\cdot]$ denoting the expectation operator. Utilizing [14, (8.310.1)], one can derive the nth moment of the RV Z, i.e.,

$$\mathbf{E}[Z^n] = \int_0^\infty z^n f_Z(z) dz = \Omega_z^{0.5n} \Gamma(0.5n + 1). \tag{4.10}$$

Deriving from (4.7), the formulation of the instantaneous SNR at the node D can be expressed as

$$\gamma = \frac{\left| \sum_{l=1}^{L} \alpha_l \beta_l e^{j(\phi_l - \theta_l - \psi_l)} \right|^2 E_t}{N_0}, \tag{4.11}$$

where $E_t = \mathbf{E}\left[|x|^2\right]$. For the maximization of γ, the optimal selection for ϕ_l is given by $\phi_l = \theta_l + \psi_l$ for $l = 1, 2, \ldots, L$. Hence, the formulation of the maximized SNR is expressed by [15]

$$\gamma = \frac{\left|\sum_{l=1}^{L} \alpha_l \beta_l\right|^2 E_t}{N_0}. \tag{4.12}$$

In order to conduct performance analysis, we need the PDF and cumulative distribution function (CDF) expressions for γ, respectively.

4.3.1 Approximating PDF and CDF for SNR γ

To analyze performance, we require the expressions for the PDF $f_\gamma(\cdot)$ and CDF $F_\gamma(\cdot)$ of the SNR γ. It is important to highlight that, as far as the authors are aware, deriving an exact tractable closed-form expression for either $F_\gamma(\cdot)$ or $f_\gamma(\cdot)$ is presently not viable. Furthermore, our goal is to develop precise approximations for the PDF and CDF of γ. In this context, we will utilize two approximation techniques: the central limit theorem (CLT) and Laguerre series expansion (LSE) [16–18].

The instantaneous SNR at the node D due to the IRS can be re-expressed as

$$\gamma = \mathcal{B}^2 \frac{E_t}{N_0} \tag{4.13}$$

where $\mathcal{B} = \sum_{i=1}^{L} \mathcal{B}_l$ and $\mathcal{B}_l = \alpha_l \beta_l$. Given that α_l and β_l are independent RVs following the Rayleigh distribution with parameters Ω_{α_l} and Ω_{β_l}, respectively. Hence, using (4.10), the mean and the square of the standard deviation of RV $\mathcal{B}_l$ can be given as

$$\mu_{\mathcal{B}_l} = \mathbf{E}[\mathcal{B}_l] = \mathbf{E}[|\alpha_l|]\mathbf{E}[|\beta_l|] = \frac{\pi}{4}\sqrt{\Omega_{\alpha_l}\Omega_{\beta_l}}, \tag{4.14}$$

and

$$\begin{aligned}\sigma_{\mathcal{B}_l}^2 &= \mathbf{E}[\mathcal{B}_l^2] - (\mathbf{E}[\mathcal{B}_l])^2 = \mathbf{E}[|\alpha_l|^2]\mathbf{E}[|\beta_l|^2] - (\mathbf{E}[|\alpha_l|]\mathbf{E}[|\beta_l|])^2 \\ &= \left(1 - \frac{\pi^2}{16}\right)\Omega_{\alpha_l}\Omega_{\beta_l},\end{aligned} \tag{4.15}$$

respectively. Observe that $\mathbf{E}[|\alpha_l|]$ and $\mathbf{E}[|\beta_l|]$ are determined by substituting $n = 1$ into (4.10), while $\mathbf{E}[|\alpha_l|^2]$ and $\mathbf{E}[|\beta_l|^2]$ are obtained by substituting $n = 2$ into (4.10). Therefore, the average and the square of the standard deviation of the RV $\mathcal{B}$ can be expressed as

$$\mu_{\mathcal{B}} = \sum_{l=1}^{L} \mu_{\mathcal{B}_l} = \frac{\pi}{4} \sum_{l=1}^{L} \sqrt{\Omega_{\alpha_l} \Omega_{\beta_l}}, \tag{4.16}$$

and

$$\sigma_{\mathcal{B}}^2 = \sum_{l=1}^{L} \sigma_{\mathcal{B}_l}^2 = \left(1 - \frac{\pi^2}{16}\right) \sum_{l=1}^{L} \Omega_{\alpha_l} \Omega_{\beta_l}, \tag{4.17}$$

respectively. Furthermore, we can write $\mu_{\mathcal{B}}$ and $\sigma_{\mathcal{B}}^2$ in non-identical and identical scenarios as

$$\mu_{\mathcal{B}} = \begin{cases} \frac{\pi}{4} \sum_{l=1}^{L} \sqrt{\Omega_{\alpha_l} \Omega_{\beta_l}}, & \text{Non-identical} \left(\Omega_{\alpha_l} \neq \Omega_{\beta_l}\right) \\ \frac{\pi}{4} L\Omega, & \text{Identical} \left(\Omega_{\alpha_l} = \Omega_{\beta_l} = \Omega\right) \end{cases} \tag{4.18}$$

$$\sigma_{\mathcal{B}}^2 = \begin{cases} \left(1 - \frac{\pi^2}{16}\right) \sum_{l=1}^{L} \Omega_{\alpha_l} \Omega_{\beta_l}, & \text{Non-identical} \left(\Omega_{\alpha_l} \neq \Omega_{\beta_l}\right) \\ \left(1 - \frac{\pi^2}{16}\right) L\Omega^2, & \text{Identical} \left(\Omega_{\alpha_l} = \Omega_{\beta_l} = \Omega\right) \end{cases} \tag{4.19}$$

CLT Method for Approximation: When the number of reflecting surfaces L, is sufficiently large (i.e., $L >> 1$), the CLT theorem can be employed. This allows modeling the RV $\mathcal{B}$ as a normally distributed RV with variance $\sigma_{\mathcal{B}}^2$ and mean $\mu_{\mathcal{B}}$ [19, 20]. Hence, the PDF of $\mathcal{B}$ can be given by [21]

$$f_{\mathcal{B}}(b) = \frac{\Delta_{\mathcal{B}}}{\sqrt{2\pi\sigma_{\mathcal{B}}^2}} \exp\left(-\frac{(b - \mu_{\mathcal{B}})^2}{2\sigma_{\mathcal{B}}^2}\right), \tag{4.20}$$

where $\Delta_{\mathcal{B}} = \left(0.5 + 0.5\,\mathrm{erf}\left(\sqrt{\frac{\mu_{\mathcal{B}}^2}{2\sigma_{\mathcal{B}}^2}}\right)\right)^{-1}$ and $\mathrm{erf}(\cdot)$ is the error function [21]. The corresponding CDF of $\mathcal{B}$ is expressed as

$$F_{\mathcal{B}}(b) = \int_{-\infty}^{b} f_{\mathcal{B}}(t)dt = 1 - \Delta_{\mathcal{B}}\, Q\left(\frac{b - \mu_{\mathcal{B}}}{\sigma_{\mathcal{B}}}\right), \tag{4.21}$$

where $Q(\cdot)$ denotes the Gaussian Q-function [21]. Now, the CDF of γ can be obtained as follows

$$F_{\gamma}(\gamma_{th}) = \Pr(\gamma \leq \gamma_{th}) = \Pr\left(\mathcal{B}^2 \frac{E_t}{N_0} \leq \gamma_{th}\right) = \Pr\left(\mathcal{B} \leq \sqrt{\frac{\gamma_{th} N_0}{E_t}}\right)$$
$$= F_{\mathcal{B}}\left(\sqrt{\frac{\gamma_{th}}{\bar{\gamma}}}\right) = 1 - \Delta_{\mathcal{B}}\, Q\left(\frac{\sqrt{\frac{\gamma_{th}}{\bar{\gamma}}} - \mu_{\mathcal{B}}}{\sigma_{\mathcal{B}}}\right) \tag{4.22}$$

where $\bar{\gamma} = \frac{E_t}{N_0}$ denotes the average SNR. Furthermore, by employing the transformation of RVs, we can get the PDF of γ as

$$f_{\gamma}(\gamma) = \frac{\Delta_{\mathcal{B}}}{\sqrt{8\gamma\bar{\gamma}\pi\sigma_{\mathcal{B}}^2}} \exp\left(-\frac{\left(\sqrt{\frac{\gamma}{\bar{\gamma}}} - \mu_{\mathcal{B}}\right)^2}{2\sigma_{\mathcal{B}}^2}\right). \tag{4.23}$$

LSE Method for Approximation: Unlike the CLT, the LSE method offers a better approximation even for a small number of IRS elements. The LSE method involves the first-term expansion of $\mathcal{B}$ [18]. This approach provides a highly accurate approximation of its PDF as [22]

$$f_{\mathcal{B}}(b) = \frac{b^{\tau} \exp\left(-\frac{b}{\Lambda}\right)}{\Lambda^{\tau+1}\Gamma(\tau+1)}, \tag{4.24}$$

where $\tau = \frac{\mu_{\mathcal{B}}^2}{\sigma_{\mathcal{B}}^2} - 1$ and $\Lambda = \frac{\sigma_{\mathcal{B}}^2}{\mu_{\mathcal{B}}}$. Furthermore, the CDF expression of $\mathcal{B}$ can be formulated as:

$$F_{\mathcal{B}}(b) = \int_0^b f_{\mathcal{B}}(t)dt = \frac{\Upsilon\left(\tau+1, \frac{b}{\Lambda}\right)}{\Gamma(\tau+1)}, \tag{4.25}$$

where $\Upsilon(\cdot, \cdot)$ is the lower incomplete gamma function [14, (8.350.1)]. Hence, we can now derive another CDF expression for γ as follows:

$$F_{\gamma}(\gamma_{th}) = \Pr(\gamma \leq \gamma_{th}) = \Pr\left(\mathcal{B}^2 \frac{E_t}{N_0} \leq \gamma_{th}\right) = \frac{\Upsilon\left(\tau+1, \frac{\sqrt{\gamma_{th}}}{\Lambda\sqrt{\bar{\gamma}}}\right)}{\Gamma(\tau+1)}. \tag{4.26}$$

Additionally, through the utilization of RV transformations, we can obtain the PDF of γ as

$$f_{\gamma}(\gamma) = \frac{\gamma^{(\tau-1)/2} \exp\left(-\frac{\sqrt{\gamma}}{\Lambda\sqrt{\bar{\gamma}}}\right)}{2(\bar{\gamma}\Lambda)^{\tau+1}\,\Gamma(\tau+1)}. \tag{4.27}$$

4.3.2 Asymptotic PDF and CDF for SNR γ

Asymptotic PDF and CDF for SNR γ provides valuable insights in a very high SNR scenario. In the asymptotic analysis, we make the assumption that $\bar{\gamma} \to \infty$ or $\Omega_{\alpha_l} \to \infty$ and $\Omega_{\beta_l} \to \infty$. The asymptotic OP into two different fading conditions is derived next.

Non-Identical Fading Conditions: We assume that the fading conditions are non-identical nature. For ease of representation, the PDF of RV $Z \in \{\alpha_l, \beta_l, 1 \le l \le L\}$ is re-expressed as [13, (2.6)]

$$f_Z(z) = 2\lambda_z\, z\, \exp\left(-\lambda_z z^2\right), \tag{4.28}$$

where $\lambda_z = \frac{1}{\Omega_z}$. Now, using [23, (2.3.16.1)], the PDF expression of $\mathcal{B}_l$ can be obtain as

$$f_{\mathcal{B}_l}(b_l) = \int_0^\infty \frac{1}{\alpha_l} f_{\alpha_l}(\alpha_l)\, f_{\beta_l}\left(\frac{b_l}{\alpha_l}\right) d\alpha_l = 4\lambda_{\alpha_l}\lambda_{\beta_l}\, b_l K_0\left(2b_l\sqrt{\lambda_{\alpha_l}\lambda_{\beta_l}}\right). \tag{4.29}$$

A relevant relation for the modified zero-order second kind Bessel function $K_0(\cdot)$ in the case of small arguments can be found in [24, (9.6.8)]. This relation states that $\lim_{x\to 0} K_0(x) \approx -\ln x$. By utilizing this approximation, we can approximate the asymptotic PDF expression of $\mathcal{B}_l$ as[1]

$$f_{\mathcal{B}_l}^{\infty}(b_l) \approx 4\lambda_{\alpha_l}\lambda_{\beta_l}\Theta_l\, b_l, \tag{4.30}$$

where $\Theta_l = -\ln\left(2\sqrt{\lambda_{\alpha_l}\lambda_{\beta_l}}\right)$ and $\ln(\cdot)$ representing the natural logarithm.[2] Furthermore, we can express the asymptotic moment generating function (MGF) of $\mathcal{B}_l$ as

$$\mathcal{M}_{\mathcal{B}_l}^{\infty}(s) = \mathcal{L}\left\{f_{\mathcal{B}_l}^{\infty}(b_l)\right\} \approx 4\lambda_{\alpha_l}\lambda_{\beta_l}\Theta_l\, s^{-2}, \tag{4.31}$$

where $\mathcal{L}\{\cdot\}$ and s denote the Laplace transform (LT) operator and variable, respectively. Since $\mathcal{B}_l$s are independent RVs, the expression for the asymptotic MGF of $\mathcal{B}$ is as follows.

$$\mathcal{M}_{\mathcal{B}}^{\infty}(s) = \prod_{i=1}^{L} \mathcal{M}_{\mathcal{B}_l}^{\infty}(s) \approx \left(4^L \prod_{l=1}^{L} \lambda_{\alpha_l}\lambda_{\beta_l}\Theta_l\right) s^{-2L}. \tag{4.32}$$

By performing the inverse LT on (4.32), we can derive the asymptotic PDF expression for $\mathcal{B}$ as

---

[1] Note that, in this approximation, we have omitted the $-\ln b_l$ term.

[2] Note that Θ_l can be computed numerically by MATHEMATICA/MATLAB software assuming large values of Ω_{α_l} and Ω_{β_l}.

$$f_{\mathcal{B}}^{\infty}(b) = \mathcal{L}^{-1}\left\{\mathcal{M}_{\mathcal{B}}^{\infty}(s)\right\} \approx \frac{\left(4^L \prod_{l=1}^{L} \lambda_{\alpha_l}\lambda_{\beta_l}\Theta_l\right) b^{2L-1}}{\Gamma\,(2L)}, \tag{4.33}$$

where $\mathcal{L}^{-1}\{\cdot\}$ denotes the inverse LT operator. Applying the transformation of RVs yields the asymptotic PDF expression for γ as

$$f_{\gamma}^{\infty}(\gamma) \approx \frac{\left(4^L \prod_{l=1}^{L} \lambda_{\alpha_l}\lambda_{\beta_l}\Theta_l\right) (\gamma/\bar{\gamma})^{L-1}}{2\Gamma\,(2L)\,\bar{\gamma}}. \tag{4.34}$$

By applying the LT to (4.34), we can derive the asymptotic MGF expression for γ as

$$\mathcal{M}_{\gamma}^{\infty}(s) = \mathcal{L}\left\{f_{\gamma}^{\infty}(\gamma)\right\} \approx \frac{\Gamma\,(L)\left(4^L \prod_{l=1}^{L} \lambda_{\alpha_l}\lambda_{\beta_l}\Theta_l\right) s^{-L}}{2\Gamma\,(2L)\,\bar{\gamma}^L}. \tag{4.35}$$

Hence, the CDF expression of RV γ can be obtained as

$$F_{\gamma}^{\infty}(\gamma) = \mathcal{L}^{-1}\left\{\frac{\mathcal{M}_{\gamma}^{\infty}(s)}{s}\right\} \approx \mathcal{G}\frac{\gamma^L}{\bar{\gamma}^L}, \tag{4.36}$$

where $\mathcal{G} = \left(4^L \prod_{l=1}^{L} \lambda_{\alpha_l}\lambda_{\beta_l}\Theta_l\right)/\Gamma\,(2Ł + 1)$.

Identical Fading Conditions: For identical fading conditions, the PDF of RV $Z \in \{\alpha_l, \beta_l, 1 \leq l \leq L\}$ is re-written as [13, (2.6)]

$$f_Z(z) = 2\lambda\, z \, \exp\left(-\lambda z^2\right), \tag{4.37}$$

where $\lambda = \frac{1}{\Omega}$. Now, using [23, (2.3.16.1)], the PDF expression of $\mathcal{B}_l$ can be obtained as

$$f_{\mathcal{B}_l}(b_l) = \int_0^{\infty} \frac{1}{\alpha_l} f_{\alpha_l}(\alpha_l)\, f_{\beta_l}\left(\frac{b_l}{\alpha_l}\right) d\alpha_l = 4\lambda^2\, b_l\, K_0\,(2b_l\lambda)\,. \tag{4.38}$$

A pertinent relation for the modified zero-order second kind Bessel function $K_0(\cdot)$ when dealing with small arguments is given in [24, (9.6.8)]. This relation approximates $\lim_{x\to 0} K_0(x) \approx -\ln x$. By leveraging this relationship, the asymptotic PDF expression of $\mathcal{B}_l$ can be approximate as[3]

$$f_{\mathcal{B}_l}(b_l) \approx 4\,\theta\,\lambda^2\, b_l; \qquad b_l \geq 0 \tag{4.39}$$

where $\theta = -\ln\,(2\lambda)$.[4] Consequently, we can express the asymptotic MGF of $\mathcal{B}_l$ as follows:

---

[3] Note that, in this approximation, we have omitted the $-\ln b_l$ term.

[4] It is worth noting that θ can be numerically computed using MATHEMATICA/MATLAB software, under the assumption of a large value for Ω.

$$\mathcal{M}_{\mathcal{B}_l}^{\infty}(s) = \mathcal{L}\left\{f_{\mathcal{B}_l}^{\infty}(b_l)\right\} \approx 4\,\theta\,\lambda^2\,s^{-2}. \tag{4.40}$$

Since $\mathcal{B}_l$s are independent and identically RVs, we can represent the asymptotic MGF of $\mathcal{B}$ as

$$\mathcal{M}_{\mathcal{B}}^{\infty}(s) = \prod_{l=1}^{L} \mathcal{M}_{\mathcal{B}_l}^{\infty}(s) \approx \mathcal{U}\,s^{-2L}, \tag{4.41}$$

where $\mathcal{U} = \left(4\theta\,\lambda^2\right)^L$. Applying the inverse LT to (4.41) yields the asymptotic PDF expression for $\mathcal{B}$ as

$$f_{\mathcal{B}}^{\infty}(b) = \mathcal{L}^{-1}\left\{\mathcal{M}_{\mathcal{B}}^{\infty}(s)\right\} \approx \frac{\mathcal{U}b^{2L-1}}{\Gamma(2L)}. \tag{4.42}$$

By applying an RV transformation, we derive the asymptotic PDF expression for γ as

$$f_{\gamma}^{\infty}(\gamma) \approx \frac{\mathcal{U}\gamma^{L-1}}{2\Gamma(2L)\bar{\gamma}^L}. \tag{4.43}$$

Utilizing the LT on (4.43), we can derive the asymptotic MGF expression for γ as

$$\mathcal{M}_{\gamma}^{\infty}(s) = \mathcal{L}\left\{f_{\gamma}^{\infty}(\gamma)\right\} \approx \frac{\mathcal{U}\Gamma(L)}{2\Gamma(2L)\bar{\gamma}^L s^L}. \tag{4.44}$$

Hence, the CDF expression of RV γ can be obtained as

$$F_{\gamma}^{\infty}(\gamma) = \mathcal{L}^{-1}\left\{\frac{\mathcal{M}_{\gamma}^{\infty}(s)}{s}\right\} \approx \frac{\mathcal{U}\gamma^L}{\Gamma(2L+1)\bar{\gamma}^L}. \tag{4.45}$$

4.3.3 Analyzing Outage Probability: Evaluation and Insights

For wireless systems subject to fading, an essential performance metric is the outage probability (OP), represented as P_{out}. It represents the likelihood that the probability that the output SNR, γ, drops below a specified threshold, γ_{th}. In mathematical terms, we have

$$P_{\text{out}} = \Pr\left(\gamma \leq \gamma_{th}\right) = F_{\gamma}\left(\gamma_{th}\right) = \int_0^{\gamma_{th}} f_{\gamma}\left(\gamma\right) d\gamma, \tag{4.46}$$

where $F_{\gamma}(\cdot)$ represents the CDF, while $f_{\gamma}(\cdot)$ corresponds to the PDF of γ. To compute the OP, it is necessary to derive an expression for either $F_{\gamma}(\cdot)$ or $f_{\gamma}(\cdot)$.

Approximate OP Expressions via CLT and LSE: By substituting either (4.22) or (4.26) into (4.46), we can obtain two approximate OP for the SISO wireless system under consideration, i.e.,

$$P_{\text{out}} = \begin{cases} 1 - \Delta_{\mathcal{B}}\, Q\left(\dfrac{\sqrt{\frac{\gamma_{th}}{\bar{\gamma}}} - \mu_{\mathcal{B}}}{\sigma_{\mathcal{B}}}\right), & \text{CLT} \\ \dfrac{\Upsilon\left(\tau+1, \frac{\sqrt{\frac{\gamma_{th}}{\bar{\gamma}}}}{\Lambda}\right)}{\Gamma(\tau+1)}, & \text{LSE} \end{cases} \tag{4.47}$$

Asymptotic OP Expressions: Using (4.36) and (4.45) into (4.46), the asymptotic OP of the SISO wireless system under consideration can be written as

$$P_{\text{out}}^{\infty} = \begin{cases} \mathcal{G}\frac{\gamma^L}{\bar{\gamma}^L}, & \text{Non-Identical } (\Omega_{\alpha_l} \neq \Omega_{\beta_l}) \\ \frac{\mathcal{U}\gamma^L}{\Gamma(2L+1)\bar{\gamma}^L}, & \text{Identical } (\Omega_{\alpha_l} = \Omega_{\beta_l} = \Omega) \end{cases} \tag{4.48}$$

Diversity Order Analysis: The diversity order of the SISO wireless system under consideration can be obtained as

$$d = \lim_{\bar{\gamma}\to\infty} \frac{-\log P_{\text{out}}^{\infty}}{\log \bar{\gamma}}. \tag{4.49}$$

After substituting (4.48) into (4.49) and performing simplification, the result is

$$d = \begin{cases} L, \text{ Non-Identical } (\Omega_{\alpha_l} \neq \Omega_{\beta_l}) \\ L, \text{ Identical } (\Omega_{\alpha_l} = \Omega_{\beta_l} = \Omega) \end{cases} \tag{4.50}$$

The diversity order of the SISO wireless system under consideration depends on only the number of IRS elements, as expected.

4.3.4 Analyzing Average Symbol Error Rate: Evaluation and Insights

The derivation of the average symbol error rate (ASER) expression for a wireless system operating over fading channels involves the statistical averaging of the conditional symbol error rate (CSER) expression over the probability density function (PDF) of the output SNR γ, i.e.,[13, 25, 26]

$$P(e) = \int_0^{\infty} P_{e|\gamma}(\gamma) f_{\gamma}(\gamma)\, d\gamma, \tag{4.51}$$

where $P_{e|\gamma}(\gamma)$ and $f_{\gamma}(\gamma)$ denote CSER expression and the PDF expression of SNR γ, respectively [27]. In this context, for analytical ASER performance analysis, we employ the computationally efficient method based on MGF. Consequently, it is necessary to derive an expression for the MGF of the SNR γ, which is outlined in the subsequent subsection.

Approximate MGF Expression of SNR γ via CLT: Deriving a computationally efficient closed-form PDF for $\mathcal{B}$ is, to the best of our knowledge, unfeasible for every value of L. Nevertheless, for sufficiently large values of L, an accurate approximation of the RV $\mathcal{B}$ as a Gaussian-distributed RV with a mean $\mu_{\mathcal{B}}$ and a square of standard deviation $\sigma_{\mathcal{B}}^2$-can be achieved by applying the CLT. Therefore, the RV $\gamma = \mathcal{B}^2 \frac{E_t}{N_0}$ will be a non-central chi-square RV with one degree of freedom. Thus, MGF of SNR γ can be given as [1, 21]

$$\mathcal{M}_{\gamma}(s) = \mathcal{L}\{f_{\gamma}(\gamma)\} \simeq \left(\frac{\Delta_{\mathcal{B}}^2 \delta_1}{\delta_1 + s\bar{\gamma}}\right)^{\frac{1}{2}} \exp\left(-\frac{s\,\delta_2\,\bar{\gamma}}{\delta_1 + s\bar{\gamma}}\right), \qquad L >> 1 \tag{4.52}$$

where $\delta_1 = \frac{1}{2\sigma_{\mathcal{B}}^2}$, and $\delta_2 = \frac{\mu_{\mathcal{B}}^2}{2\sigma_{\mathcal{B}}^2}$.

Asymptotic MGF Expressions of SNR γ: Using (4.35) and (4.44) the asymptotic MGF expressions of SNR γ can be written as

$$\mathcal{M}_{\gamma}^{\infty}(s) = \begin{cases} \frac{\Gamma(L)\left(4^L \prod_{l=1}^{L} \lambda_{\alpha_l} \lambda_{\beta_l} \Theta_l\right)}{2\Gamma(2L)\bar{\gamma}^L s^L}, & \text{Non-indentical } \left(\Omega_{\alpha_l} \neq \Omega_{\beta_l}\right) \\ \frac{\mathcal{U}\Gamma(L)}{2\Gamma(2L)\bar{\gamma}^L s^L}, & \text{Indentical } \left(\Omega_{\alpha_l} = \Omega_{\beta_l} = \Omega\right) \end{cases} \tag{4.53}$$

Approximate ASER Expression for Coherent Modulation Schemes: The CSER for various coherent modulation schemes can be represented as [28]

$$P_{e|\gamma}(\gamma) = c_1\, \text{erfc}\left(\sqrt{c_2 \gamma}\right), \tag{4.54}$$

where the values of c_1 and c_2 are determined by the selected modulation scheme, as mentioned in Table 4.1, and $\text{erfc}(x) = (2/\sqrt{\pi}) \int_x^{\infty} e^{-t^2} dt$ denotes complementary error function. An alternative expression of $\text{erfc}(\cdot)$ can be given as [13, (4.2)]

$$\text{erfc}(x) = \frac{2}{\pi} \int_0^{\frac{\pi}{2}} \exp\left(-\frac{x^2}{\sin^2 \phi}\right) d\phi, \qquad x \geq 0 \tag{4.55}$$

Substituting (4.54) into (4.51) and utilizing the alternative representation provided in (4.55), followed by additional algebraic manipulations, allows us to reformulate the expression for ASER, $P(e)$ as

Table 4.1 Modulation parameters c_1 and c_2

Modulation schemes		c_1	c_2	M (Modulation order)
Coherent schemes	BPSK	0.5	1	2
	BFSK	0.5	0.5	2
	M-DEPSK	2	$\sin^2\left(\frac{\pi}{M}\right)$	≥ 2
	QPSK	1	0.5	4
	M-PSK	1	$\sin^2\left(\frac{\pi}{M}\right)$	>4
	M-FSK	$\frac{M-1}{2}$	0.5	>2
	Square M-QAM	$2-2/\sqrt{M}$	$\frac{1.5}{M-1}$	≥ 4
Non-coherent Schemes	DBPSK	0.5	1	2
	M-NFSK	$\sum_{l=1}^{M-1}\frac{(-1)^{l+1}}{l+1}$	$\frac{l}{l+1}$	≥ 2

$$
\begin{aligned}
P(e) &= \int_0^{\infty} c_1 \operatorname{erfc}\left(\sqrt{c_2\gamma}\right) f_\gamma(\gamma)\, d\gamma \\
&= \frac{2c_1}{\pi}\int_0^{\frac{\pi}{2}}\int_0^{\infty} \exp\left(\frac{-c_2\gamma}{\sin^2\phi}\right) f_\gamma(\gamma)\, d\gamma\, d\phi \\
&= \frac{2c_1}{\pi}\int_0^{\frac{\pi}{2}} \mathcal{M}_\gamma\left(\frac{c_2}{\sin^2\phi}\right) d\phi.
\end{aligned} \tag{4.56}
$$

Now, by substituting (4.52) into (4.56) and subsequently employing two successive changes of variables, $t = \frac{\delta_1\delta_2\sin^2\phi}{\delta_1\sin^2\phi + c_2\bar{\gamma}}$ and $v = \left(\frac{\delta_1 + c_2\bar{\gamma}}{\delta_1\delta_2}\right)t$, respectively, for the ASER expression, we can derive a closed-form expression as

$$
P(e) \simeq \frac{2c_1\Delta_{\mathcal{B}}\sqrt{c_2\delta_1\bar{\gamma}}\exp(-\delta_2)}{\pi(c_2\bar{\gamma}+\delta_1)}\Phi_1^{(2)}\left(1;\, 1;\, 1.5;\, \frac{\delta_1}{c_2\bar{\gamma}+\delta_1}, \frac{\delta_1\delta_2}{c_2\bar{\gamma}+\delta_1}\right), \qquad L >> 1 \tag{4.57}
$$

where $\Phi_1^{(2)}(\cdot;\cdot;\cdot;\cdot,\cdot)$ denotes the bi-variate confluent Lauricella's hypergeometric function (BCHF), which is given as [25, 29, 30]

$$
\begin{aligned}
\Phi_1^{(2)}(m;\, p;\, q;\, w_1, w_2) &= \frac{\Gamma(q)}{\Gamma(m)\Gamma(q-m)} \\
&\times \int_0^1 \nu^{m-1}\exp(w_2\,\nu)(1-\nu)^{q-m-1}(1-w_1\,\nu)^{-p}\, d\nu.
\end{aligned} \tag{4.58}
$$

Approximate ASER Expression for Non-Coherent Modulation Schemes: The CSER for various non-coherent modulation schemes can be represented as [13, 31]

$$P_{e|\gamma}(\gamma) = c_1 \exp\left(-c_2\gamma\right), \tag{4.59}$$

where c_1 and c_2 are determined by the selected modulation scheme, as given in Table 4.1. By substituting (4.59) into (4.51) and performing the necessary integration manipulations, we can express the ASER for non-coherent schemes in terms of the MGF of γ as

$$P(e) \simeq c_1\, \mathcal{M}_{\gamma}(c_2), \qquad L >> 1. \tag{4.60}$$

Asymptotic ASER Expression for Coherent Modulation Scheme: Using (4.53) and (4.56), the asymptotic ASEP expression for coherent modulation schemes can be written as

$$P^{\infty}(e) = \frac{2c_1}{\pi}\int_0^{\frac{\pi}{2}} \mathcal{M}_{\gamma}^{\infty}\left(\frac{c_2}{\sin^2\phi}\right) d\phi, \qquad \forall L. \tag{4.61}$$

Utilizing [14, (8.380.2)], we can provide a closed-form solution for $P^{\infty}(e)$, expressed as

$$P^{\infty}(e) = \begin{cases} \frac{c_1\Gamma(L)\left(4^L\prod_{l=1}^{L}\lambda_{\alpha_l}\lambda_{\beta_l}\Theta_l\right)}{2\pi\,\Gamma(2L)\bar{\gamma}^L c_2^L} B(L+0.5,0.5), & \text{Non-indentical } \left(\Omega_{\alpha_l} \neq \Omega_{\beta_l}\right) \\ \frac{c_1\mathcal{U}\Gamma(L)}{2\pi\,\Gamma(2L)\bar{\gamma}^L c_2^L} B(L+0.5,0.5), & \text{Indentical } \left(\Omega_{\alpha_l} = \Omega_{\beta_l} = \Omega\right) \end{cases} \tag{4.62}$$

where $B(\cdot,\cdot)$ denotes the beta function [14, (8.380.1)].

Asymptotic ASER Expression for Non-Coherent Modulation Schemes: Using (4.53) and (4.60), we can obtain asymptotic ASER expression for non-coherent modulation schemes as

$$\begin{aligned} P^{\infty}(e) &= c_1\, \mathcal{M}_{\gamma}^{\infty}(c_2) \\ &\approx \begin{cases} \frac{c_1\Gamma(L)\left(4^L\prod_{l=1}^{L}\lambda_{\alpha_l}\lambda_{\beta_l}\Theta_l\right)}{2\Gamma(2L)\bar{\gamma}^L c_2^L}, & \text{Non-indentical } \left(\Omega_{\alpha_l} \neq \Omega_{\beta_l}\right) \\ \frac{c_1\mathcal{U}\Gamma(L)}{2\Gamma(2L)\bar{\gamma}^L c_2^L}, & \text{Indentical } \left(\Omega_{\alpha_l} = \Omega_{\beta_l} = \Omega\right) \end{cases} \end{aligned} \tag{4.63}$$

Diversity Order Analysis: The diversity order of the analyzed SISO wireless communication system aided by IRS is obtained as

$$d = \lim_{\bar{\gamma}\to\infty} \frac{-\log P^{\infty}(e)}{\log \bar{\gamma}}. \tag{4.64}$$

By substituting either (4.62) or (4.63) into (4.64) and simplifying, the resulting value is $d = L$. It is important to highlight that the diversity order of the analyzed SISO wireless system with IRS assistance is solely dependent on L.

4.3.5 Numerical and Simulation Results and Discussion

Here, we investigate the analyzed SISO communication system using our derived analytical expressions. In addition, we present Monte-Carlo simulations to validate these expressions. To ensure clarity, we make several parameter assumptions: regarding distances as indicated in Fig. 4.7, we set $d_{h_l} = d_1$ and $d_{g_l} = d_2$ for all l; for path loss exponents (PLEs) as indicated in Fig. 4.7, we assume $\eta_{h_l} = \eta_1$ and $\eta_{g_l} = \eta_2$ for all l. We also define $\Omega_{\alpha_l} = d_{h_l}^{-\eta_{h_l}} = d_1^{-\eta_1}$ and $\Omega_{\beta_l} = d_{g_l}^{-\eta_{g_l}} = d_2^{-\eta_2}$ for all l. Our scenario aligns with a sub-6G setup similar to [32], featuring a system bandwidth of 180 KHz and a power spectral density of noise at -173 dBm/Hz. Distances are fixed at $d_1 = d_2 = 150$ m, and PLEs are set to $\eta_1 = \eta_2 = 2$. The reference path loss at a distance of $d_0 = 1$ m is -30 dB, with a threshold SNR of $\gamma_{th} = 10$ dB. Each figure illustrates the performance measures plotted against the average energy of data symbol E_t (dBm).

Figure 4.8 presents a graphical representation of the OP experienced by the single-IRS-assisted SISO communications across a range of values for the parameter L. Upon examination, it becomes apparent that the OP derived through the LSE method closely corresponds to the exact simulation outcomes when L is greater than or equal to 2, showcasing a notable consistency not observed with the CLT method. Furthermore, it is worth highlighting that the OP calculated using the CLT method exhibits a similar level of alignment with the exact simulation results for L greater than or equal to 64. Additionally, an intuitive observation emerges from the results, revealing a clear and consistent upward trend in the OP as L increases.

Figure 4.9 provides compelling evidence about the accuracy of the asymptotic OP expression within the high SNR range at different values of L. It is clearly evident from the graph that the higher SNR approximation closely mirrors the actual results, confirming the validity of the asymptotic OP expression. Furthermore, an interesting observation emerges, indicating the diversity order of L for the system.

Figure 4.10 shows the performance of ASER for single-IRS-assisted SISO communication. This figure shows a comparison between the approximated ASER obtained using the CLT method and the exact ASER obtained through simulation. Interestingly, it is observed that ASER results from the CLT method exhibit inconsistencies for low values of L, but gradually yield accurate ASER results when L reaches or exceeds 64. This is consistent with the expected behavior, as ASER results obtained from the CLT method are expected to increase their accuracy with increasing value of L.

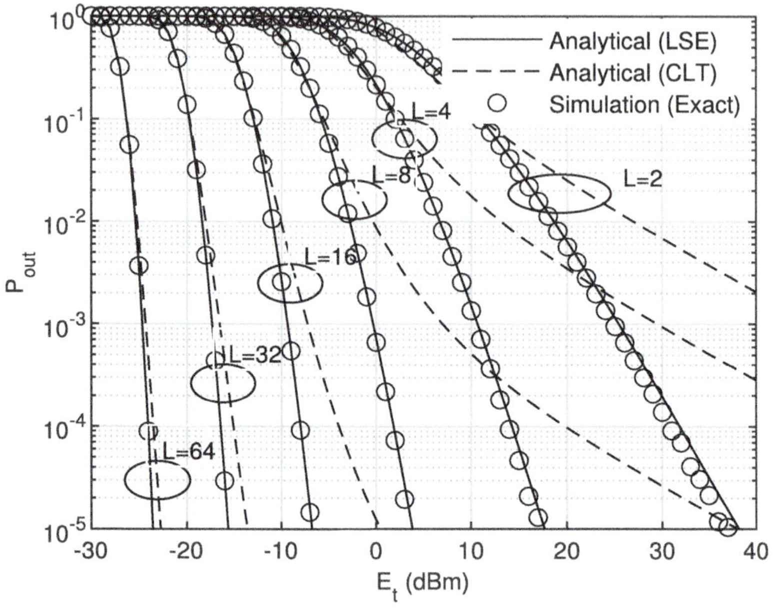

Fig. 4.8 OP performance for different L

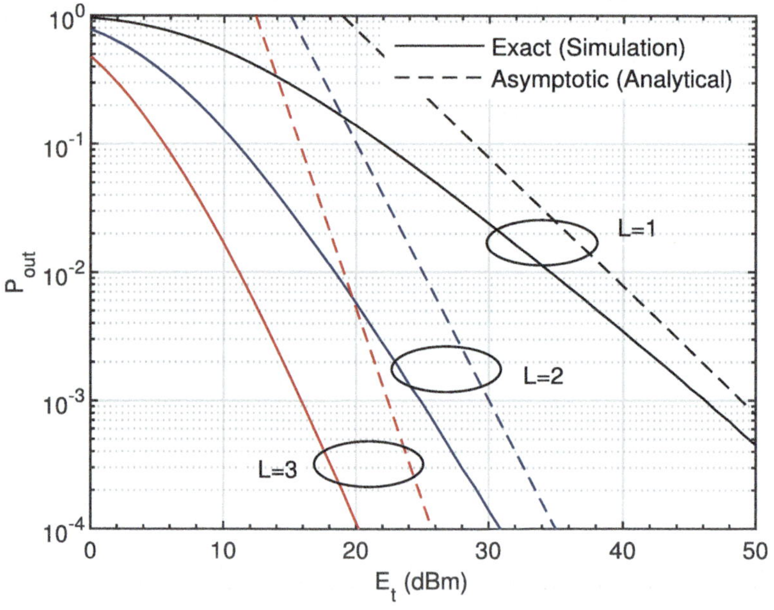

Fig. 4.9 OP performance for different L

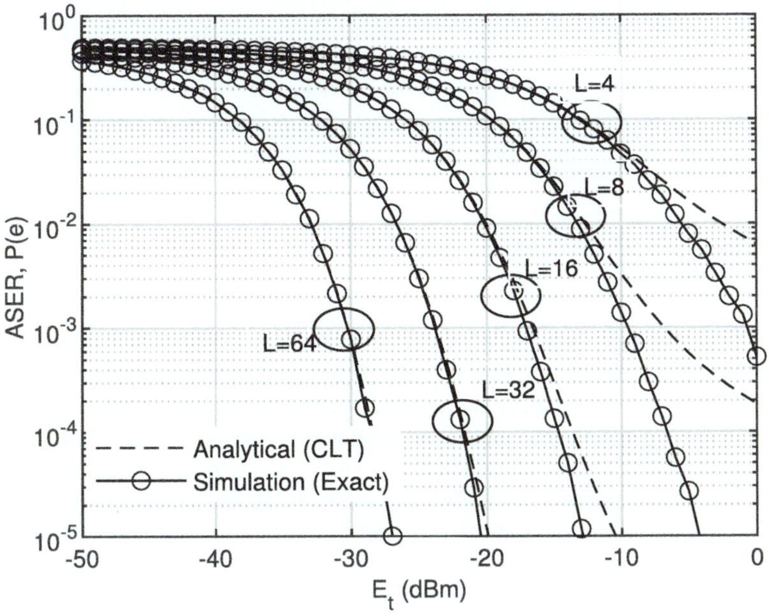

Fig. 4.10 ASER performance of BPSK modulation schemes for different L

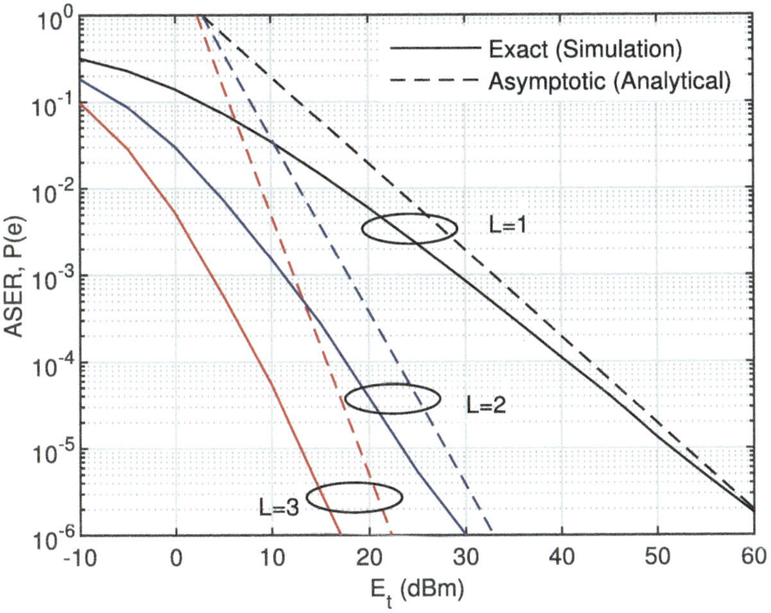

Fig. 4.11 ASER performance of BPSK modulation schemes for different L

Figure 4.11 provides strong evidence about the accuracy of the asymptotic ASER expression within the high SNR range at different values of L. The graph clearly shows that the higher SNR approximation tracks the exact ASER results closely, confirming the accuracy of the asymptotic ASER expression. Additionally, an interesting insight emerges, which reveals the diversity order of L for the system.

4.4 Conclusion

In conclusion, this chapter has provided a comprehensive overview of IRS-assisted wireless communications, covering both design principles and performance analysis. We began by introducing the concept of IRS and discussing its fundamental role in enhancing wireless communication systems. Subsequently, we explored various aspects of IRS design, implementation, and real-world considerations, including typical IRS architectures, reconfiguration methods, practical reflection models, and power consumption considerations. Moving forward, we conducted a thorough performance analysis of a single-IRS-aided SISO system in Rayleigh fading channels. This analysis involved the derivation of approximate OP expressions, asymptotic OP analysis, diversity order analysis based on asymptotic OP, as well as the derivation of approximate ASER expressions for both coherent and non-coherent modulation schemes, along with asymptotic ASER analysis and diversity order analysis based on asymptotic ASER. The diversity order of the considered system in Rayleigh fading channels reveals that it depends solely on the number of IRS elements. Finally, numerical and simulation results were presented and discussed, providing valuable insights into the effectiveness and limitations of IRS-assisted wireless communications. Overall, this chapter contributes to a deeper understanding of the design and performance aspects of IRS technology and its potential impact on future wireless communication systems.

Acknowledgements The authors would like to thank REC Sonbhadra, MNNIT Allahabad, IIT (BHU) Varanasi, and IIT Guwahati for all their support. This research is supported by the Science and Engineering Research Board (SERB), Department of Science and Technology, Government of India, under State University Research Excellence (SURE) Grant: SUR/2022/004497.

References

1. Basar E, Di Renzo M, De Rosny J, Debbah M, Alouini M-S, Zhang R (2019) Wireless communication through reconfigurable intelligent surfaces. IEEE Access 7:116 753–116 773
2. Wu Q, Zhang S, Zheng B, You C, Zhang R (2021) Intelligent reflecting surface-aided wireless communications: a tutorial. IEEE Trans Commun 69(5):3313–3351
3. Cui TJ, Qi MQ, Wan X, Zhao J, Cheng Q (2014) Coding metamaterials, digital metamaterials and programmable metamaterials. Light: Sci & Appl 3(10):e218
4. Nayeri P, Yang F, Elsherbeni AZ (2018) Reflectarray antennas: theory, designs, and applications

5. Zhang L, Chen XQ, Liu S, Zhang Q, Zhao J, Dai JY, Bai GD, Wan X, Cheng Q, Castaldi G et al (2018) Space-time-coding digital metasurfaces. Nat Commun 9(1):4334
6. Yang H, Chen X, Yang F, Xu S, Cao X, Li M, Gao J (2016) Design of resistor-loaded reflectarray elements for both amplitude and phase control. IEEE Antennas Wirel Propag Lett 16:1159–1162
7. Wu Q, Zhang R (2019) Beamforming optimization for intelligent reflecting surface with discrete phase shifts. In: ICASSP 2019-2019 IEEE international conference on acoustics, speech and signal processing (ICASSP). IEEE, pp 7830–7833
8. Abeywickrama S, Zhang R, Wu Q, Yuen C (2020) Intelligent reflecting surface: practical phase shift model and beamforming optimization. IEEE Trans Commun 68(9):5849–5863
9. Zhu BO, Zhao J, Feng Y (2013) Active impedance metasurface with full 360 reflection phase tuning. Sci Rep 3(1):3059
10. Tang W, Chen MZ, Chen X, Dai JY, Han Y, Di Renzo M, Zeng Y, Jin S, Cheng Q, Cui TJ (2020) Wireless communications with reconfigurable intelligent surface: path loss modeling and experimental measurement. IEEE Trans Wireless Commun 20(1):421–439
11. Kaina N, Dupré M, Lerosey G, Fink M (2014) Shaping complex microwave fields in reverberating media with binary tunable metasurfaces. Sci Rep 4(1):6693
12. Kumar Hindustani R, Dixit D, Sharma S (2023) Outage probability analysis of multiple intelligent reflecting surface-assisted single-input single-output system with switched diversity. Int J Commun Syst 36(14):e5550
13. Simon MK, Alouini M-S (2005) Digital communication over fading channels. Wiley
14. Gradshteyn IS, Ryzhik IM (2014) Table of integrals, series, and products. Academic
15. Hindustani R, Dixit D, Sharma S, Bhatia V (2023) Outage probability of multiple-IRS-assisted SISO wireless communications over Rician fading. Phys Commun 59:102102
16. Cramér H (1999) Mathematical methods of statistics. Princeton University Press, vol 26
17. Lebedev NN, Silverman RA, Livhtenberg D (1965) Special functions and their applications
18. Amarakone U, Ekanayake N (2009) Laguerre series expansion for the density function of sum of random variables and its application to analysis of EGC diversity systems. In: 2009 international conference on industrial and information systems (ICIIS). IEEE, pp 143–147
19. Dixit D, Joshi KC, Sharma S (2023) Performance of intelligent reflecting surface-based wireless communications using QAM signaling. In: 2023 IEEE global conference on artificial intelligence and Internet of Things (GCAIoT). IEEE, pp 168–173
20. Tao Q, Wang J, Zhong C (2020) Performance analysis of intelligent reflecting surface aided communication systems. IEEE Commun Lett 24(11):2464–2468
21. Proakis JG, Salehi M (2001) Digital communications, 4th edn. McGraw-Hill Companies Inc, New York, NY
22. Salhab AM, Samuh MH (2021) Accurate performance analysis of reconfigurable intelligent surfaces over Rician fading channels. IEEE Wireless Commun Lett 10(5):1051–1055
23. Prudnikov A, Brychkov YA, Marichev OI (1986) Integrals and series. Volume. 1: elementary functions. Cordon and Beach SP, New York
24. Abramowitz M, Stegun IA (1964) Handbook of mathematical functions with formulas, graphs, and mathematical tables, vol 55. US Government Printing Office
25. Kumar Bhagat S, Dixit D, Ranjan Sahu P (2023) Average bit error probability performance of intelligent reflecting surface-aided wireless communications over Hoyt fading. Int J Commun Syst 36(16):e5580
26. Chatterjee S, Dixit D, Kumar N, Verma P, Pandey P (2023) Exact analysis of two-branch SC receiver over α -κ-μ fading. Internet Technol Lett 6(6):e479
27. Dixit D, Pandey P, Verma P, Sigroha D, Tripathi D (2023) New exact ABER for MRC receiver in κ-μ faded channels. In: 2023 international conference on IoT, communication and automation technology (ICICAT). IEEE, pp 1–5
28. Sagias NC, Zogas DA, Karagiannidis GK (2005) Selection diversity receivers over nonidentical Weibull fading channels. IEEE Trans Veh Technol 54(6):2146–2151
29. Exton H (1976) Multiple hypergeometric functions and applications. Ellis Horwood

30. Lopez-Martinez FJ, Pawula RF, Martos-Naya E, Paris JF (2014) A clarification of the proper-integral form for the Gaussian Q-function and some new results involving the F-function. IEEE Commun Lett 18(9):1495–1498
31. Dixit D, Kumar N, Mandpura AK (2021) On the ASER performance of SC receiver with RQAM and HQAM over κ-μ fading. AEU-Int J Electr Commun 138:153883
32. Huang C, Zappone A, Alexandropoulos GC, Debbah M, Yuen C (2019) Reconfigurable intelligent surfaces for energy efficiency in wireless communication. IEEE Trans Wireless Commun 18(8):4157–4170

Chapter 5
SFFT-Based Two-Stage OTFS: Fundamentals and Applications

Anna Thomas, Kuntal Deka, and Sanjeev Sharma

Abstract Next-generation wireless networks are designed to address high delay and Doppler (DD) scenarios for achieving ubiquitous connectivity. However, the performance of current waveforms, such as orthogonal frequency division multiplexing (OFDM), degrades in high-mobility scenarios due to significant Doppler effects. In this chapter, we highlight the fundamentals of the orthogonal time-frequency space modulation (OTFS) system, including its basic principles and block-level implementation. We also discuss the relationship between various communication systems, such as time domain, frequency domain and DD domain. Additionally, this chapter focuses on the OFDM-compatible OTFS system, referred to as the two-stage OTFS system.

Keywords 6G network · OFDM · OTFS · Delay-Doppler communication

5.1 Introduction

5.1.1 Evolution of OTFS

As the wireless communication has evolved through generations, several modulation techniques were implemented in each stage. A brief summary of different modulation techniques is given in Fig. 5.1. 2G started with TDMA (Time Division Multiple Access), when it is moved on to 3G the modulation technique implemented was CDMA (Code Division Multiple Access). By the evolution of 4G, OFDM (Orthog-

A. Thomas
Yenepoya Institute of Technology, Moodabidri, Mangalore, India
e-mail: annathomas@yit.edu.in

K. Deka (✉)
IIT Guwahati, Guwahati, India
e-mail: kuntaldeka@iitg.ac.in

S. Sharma
IIT (BHU) Varanasi, Varanasi, India

S. Sharma et al. (eds.), *Next-Generation Wireless Systems*,
Transactions on Computer Systems and Networks,
https://doi.org/10.1007/978-981-96-3185-8_5

Fig. 5.1 Modulation techniques in wireless communication standard

onal Frequency Division Multiple Access) was the standard modulation technique. And OFDM continues to be the modulation technique for 5G and beyond generations. There are several challenges faced by OFDM while rendering the facilities and applications promised by 5G and beyond network. The performance degradation in high Doppler environments like V2X communication and mmWave communication. The conventional signal representation of time and frequency domain has limitations in the case of high Doppler scenario. Hence, an efficient modulation technique that can overcome the drawbacks of existing modulation techniques needs to be designed, which is Orthogonal Time-Frequency Space (OTFS) modulation [1].

In the following section, the features of TF domain analysis in wireless communication channel are explained. The multipath propagation and Doppler effect in a wireless channel will affect the TF signal analysis, which limits the performance of OFDM.

5.1.2 *Time-Frequency (TF) Domain Representation*

Time and frequency are considered as the two conventional domains of signal representation in wireless communication. The existing modulation techniques are also based on these two fundamental domains of time and frequency, namely, TDMA and FDMA, respectively. However, the features of wireless channel are highly random and a point-to-point transmission and reception is not feasible. The reflectors present in the channel play a vital role in wireless communication. Hence, there are multiple paths present in a channel. In addition, the presence of dynamic reflectors in the channel also has its impact on data transmission and reception. These features make wireless channel analogous to a linear time-variant (LTV) channel [2], which is explained in the following section.

Time Dispersive Channels and Frequency Selectivity Consider a wireless channel with multiple scatterers resulting in P number of multipaths, each of which will have a time delay of $\tau_1, \tau_2, \ldots, \tau_P$. Impulse responses of the paths are $h_1, h_2, \ldots, h_P$ and for any path p, the $h_p = |h_p|e^{j\phi_p}$. These channels are termed as time-dispersive channels in general. Let us analyse the signal transmission and reception along time-dispersive wireless channel. For any transmitted signal $s(t)$, the received signal is given by

$$r(t) = h_1 s(t - \tau_1) + h_2 s(t - \tau_2) + \cdots + h_P s(t - \tau_P)$$

Let $R(f)$ denote the Fourier transform of $r(t)$, given by

$$R(f) = \int_{-\infty}^{\infty} r(t)e^{-j2\pi f t}dt$$

$$= S(f)h_1e^{-j2\pi f\tau_1} + S(f)h_2e^{-j2\pi f\tau_2} + \cdots + S(f)h_Pe^{-j2\pi f\tau_P}.$$

Consider the case of two multipaths where $P = 2$, and the magnitude of $R(f)$ is obtained as

$$|R(f)| = |S(f)|\sqrt{|h_1|^2 + |h_2|^2 + 2|h_1||h_2|Cos(2\pi(\tau_1 - \tau_2)f - (\phi_1 - \phi_2))} \tag{5.1}$$

From (5.1), we can observe that the attenuation factor of the received signal depends on the transmit signal frequency. Hence, time-dispersive channels are frequency selective in nature.

Frequency-Dispersive Channels and Time Selectivity Next, consider the case of wireless channel with dynamic reflectors which result in Doppler effect. Let us consider a channel where there are P different Doppler frequencies $\nu_1, \nu_2, \ldots, \nu_P$ where all have same delay of τ_0. The received signal in frequency domain is given by

$$R(f) = [h_1S(f - \nu_1) + h_2S(f - \nu_2) + \cdots + h_PS(f - \nu_P)]e^{-j2\pi\tau_0 f}.$$

Let $P = 2$ and the magnitude of the received signal is given by

$$|r(t)| = |s(t - \tau_0)|\sqrt{|h_1|^2 + |h_2|^2 + 2|h_1||h_2|Cos(2\pi(\nu_1 - \nu_2)(t - \tau_0) + (\phi_1 - \phi_2))} \tag{5.2}$$

From (5.2), we can see that the attenuation factor of the received signal depends on the time delay of the corresponding reflectors. Hence, we can infer that frequency-dispersive channels are time selective by default. Hence, wireless channels are analogous to LTV systems which is a challenge to design effective wireless communication system.

TF Domain and Heisenberg's Uncertainity Principle Now let us consider the major limitations of TF domain representation, which results in drawbacks of existing modulation techniques especially in high Doppler environments. Consider the basis pulses shown in Fig. 5.2. The basis pulse of time domain and frequency domain is 1D. As per Heisenberg's uncertainity principle, a pulse can be localised either in time or frequency domain simultaneously but not in both domains. The conventional modulation techniques like TDMA and FDMA use these 1D localised basis pulse. This results in performance degradation of TDMA and FDMA in wireless channels, which are both time-selective and frequency-selective simultaneously. If we follow TDMA modulation, then it will be a spread spectrum in frequency domain. And the frequency selectivity of wireless channel will result in poor performance. Similarly, if we follow FDMA modulation, in time domain the spectrum is spread. The time

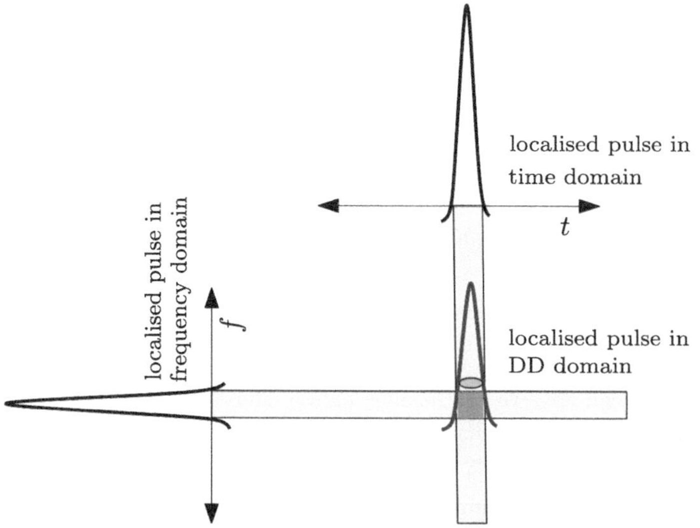

Fig. 5.2 Basis pulse in time domain, frequency domain and DD domain

selectivity of wireless channel will result in performance loss. There exists another domain of signal representation called delay-Doppler (DD) domain which can overcome this drawback of TDMA and FDMA [3]. The concept of 2D localisation is the major feature of DD domain which is explained in the following section.

5.1.3 Delay-Doppler Domain Representation

Delay-Doppler domain is a domain of signal representation which is already existing, but less explored. OTFS is a newly designed modulation technique, which explores the features of the existing DD domain. The significance of DD domain in wireless communication is that delay and Doppler are two parameters that represent the reflectors in a more accurate way. These parameters of a channel vary slowly compared to the time and frequency parameters. Therefore, eventually there is advantage in the intervals in which channel estimation is to be performed.

Two-Dimensional Localisation in DD Domain The major advantage of DD domain over TF domain is that there exists 2D localised basis pulse in DD domain. Observe in Fig. 5.2, where the basis pulse in DD domain is shown. We can obtain a basis pulse that is localised in both delay and Doppler domains. Compared to the basis pulses of TDMA and FDMA, the basis pulse in DD domain can deal with the time selectivity and frequency selectivity simultaneously. This infers that the DD basis pulse has the features of localised pulse in both time and frequency domains simultaneously. This 2D localisation is the basic property explored in OTFS modulation technique.

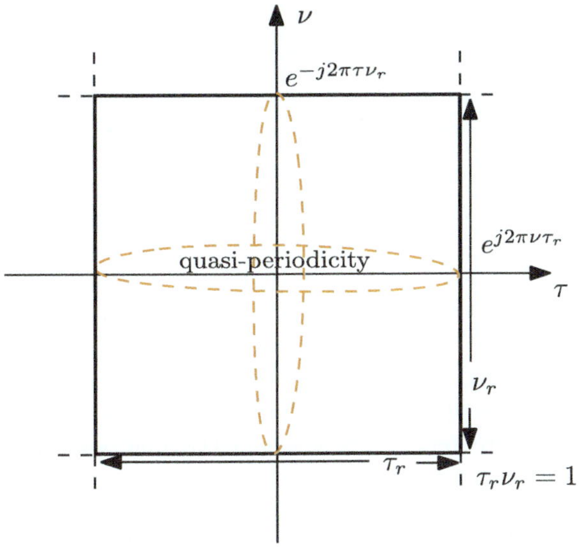

Fig. 5.3 Quasi-periodic pulse

Quasi-periodicity in DD Domain Quasi-periodicity refers to a partial periodicity condition where a signal in DD domain is periodic with respect to magnitude and not with respect to phase. In DD domain, a signal follows this quasi-periodicity, which is explained as follows [1]. Let τ_r and ν_r be the periodicity in delay and Doppler domains, respectively, such that $\tau_r.\nu_r = 1$. As per the quasi-periodicity property, for a signal in DD domain $\phi(\tau, \nu)$ and for any integers m and n

$$\phi(\tau + m\tau_r, \nu + n\nu_r) = e^{j2\pi(n\nu\tau_r - m\tau\nu_r)}\phi(\tau, \nu)$$

From the above condition of quasi-periodicity, the signal will be periodic with a phase shift provided the signal is shifted at integer multiples of ν_r or τ_r along the corresponding dimensions. Figure 5.3 describes the concept of quasi-periodicity in DD domain. As shown in the figure, if the signal is periodically shifted along the delay/Doppler axis, it results only in phase shift of the signal. The only condition to be satisfied is that the periodic window should be chosen such that $\nu_r\tau_r = 1$. This property plays a key role in the evolution of OTFS modulation technique.

5.1.4 DD ⇔ TF Conversion

Fourier transform (FT) is a well-known transform applied in time to frequency conversion and vice versa. The basic transform used in TF to DD conversion is also Fourier transform, applied in a step-by-step pattern. There are two types of approaches for this transformation, single-stage and two-stage transformation.

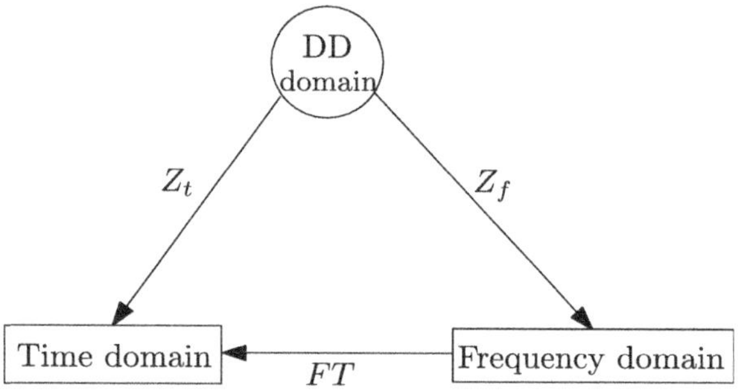

Fig. 5.4 Zak transform: delay-Doppler to Time/Frequency conversion

Single-Stage Conversion Zak transform is the mathematical tool for transforming between DD and time/frequency domain in a single stage [4]. Mathematically, Zak transform for conversion from DD to time domain and DD to frequency domain can be expressed as

$$\phi(t) = Z_t(\phi) = \int_0^{\nu_r} e^{j2\pi t\nu}\phi(t, \nu)d\nu$$

$$\phi(f) = Z_f(\phi) = \int_0^{\tau_r} e^{-j2\pi f\tau}\phi(\tau, f)d\tau$$

The Zak transform conversions are realised as the periodic Fourier integration along the delay or Doppler dimension [5]. The transformation cycle for the conversions among time, frequency and delay-Doppler domain can be represented as shown in Fig. 5.4.

An interesting property of the transformation triangle given in Fig. 5.4 is that any transformation can be represented as the composition of any other two transformations. In general, we can represent the relation as

$$FT = Z_t \circ Z_f^{-1}$$

Therefore, any conversion from one domain to another can be interpreted using a composition of Fourier transform and Zak transform. The periodicities of DD window, ν_r and τ_r, are the important parameters that decide the Zak transform coefficients.

Two-Stage Conversion The DD to time/frequency conversion can be done in a two-stage process also. This is implemented using Symplectic finite Fourier transform (SFFT) and Heisenberg's transform. In this two-stage process, initially the DD signal is transformed to TF domain by applying SFFT. In the second stage of transform, the TF domain is converted to time using Heisenberg's transform. A detailed discussion regarding the two-stage transformation is included in Sect. 5.2.

DD Response and TF Spectrum In Sect. 5.1.3, it is mentioned that in the DD domain 2D localised basis pulse exists. Let's analyse this concept in detail explaining the advantages when transformed to TF domain. In Fig. 5.5, we have shown the DD

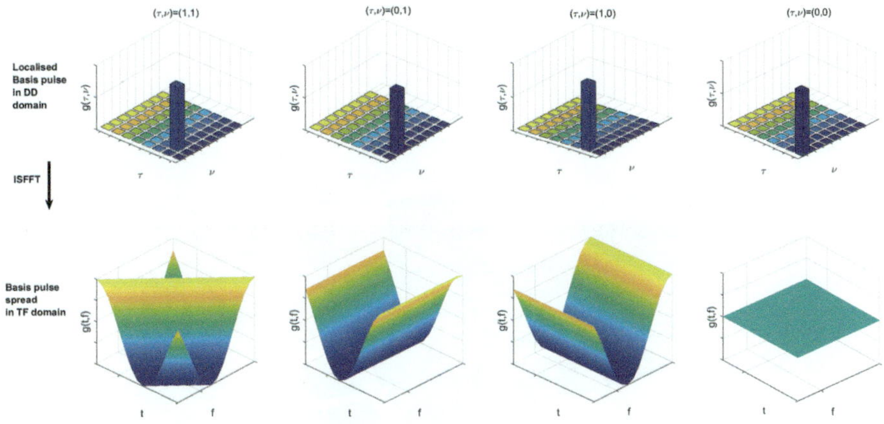

Fig. 5.5 2D localised pulse in DD domain and corresponding TF spread function

responses ($g(\tau, \nu)$) of four different sets of delay and Doppler parameters (τ, ν). Also, the spread spectrum in TF domain corresponding to each of these set of parameters. The spread spectrum of all the sample responses is distinguishable from one another. To be more specific, if there are two paths with same delay tap and different Doppler taps, then the channel response of both the paths can be identified. Similarly, in the case of paths with same Doppler and different delay also, both the paths are distinguishable. The DD response can be recovered correctly from the TF spectrum, which aids to the basic principle of OTFS [6].

5.1.5 Channel Signal Coupling

Wireless channel can be represented most judiciously in DD domain since all the reflectors can be identified by distinct delay and Doppler parameters. Figure 5.6 shows a sample DD channel response with four paths of distinct delay and Doppler parameters [6]. We can see that the corresponding bins in DD domain directly give the delay, Doppler and channel strength parameters, h_1, h_2, h_3 and h_4.

One of the major properties of OTFS modulation is that in addition to the channel, data is also represented in DD domain. Therefore, the channel signal coupling in OTFS modulation is highly effective compared to the existing modulation techniques. Figure 5.7 clearly explains the advantages of OTFS modulation from the perspective of channel signal coupling [1].

A wireless channel with five reflectors R0, R1, R2, R3 and R4 is demonstrated in Fig. 5.7. Consider the case of TDMA modulation, where the resolution can be achieved only with respect to time. Although R1 and R2 are two different reflectors, they will result only in a single faded response. Similarly consider the case of FDMA modulation where the resolution is achievable only in frequency domain. Hence the

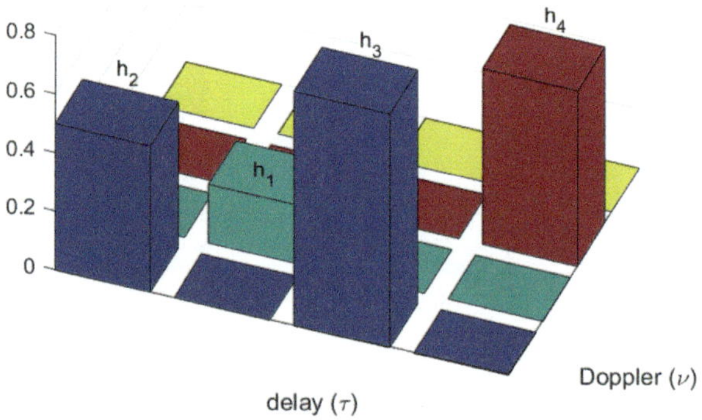

Fig. 5.6 Channel response in DD domain

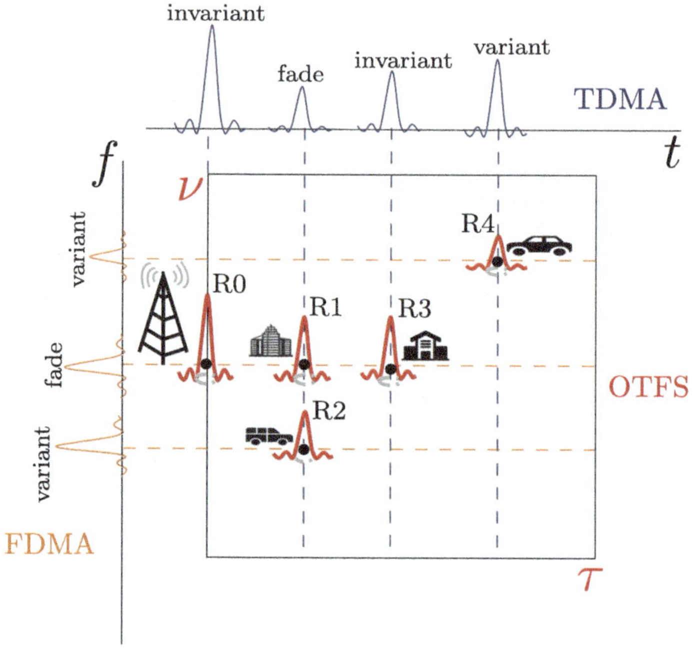

Fig. 5.7 Channel signal coupling in TDMA, FDMA and OTFS

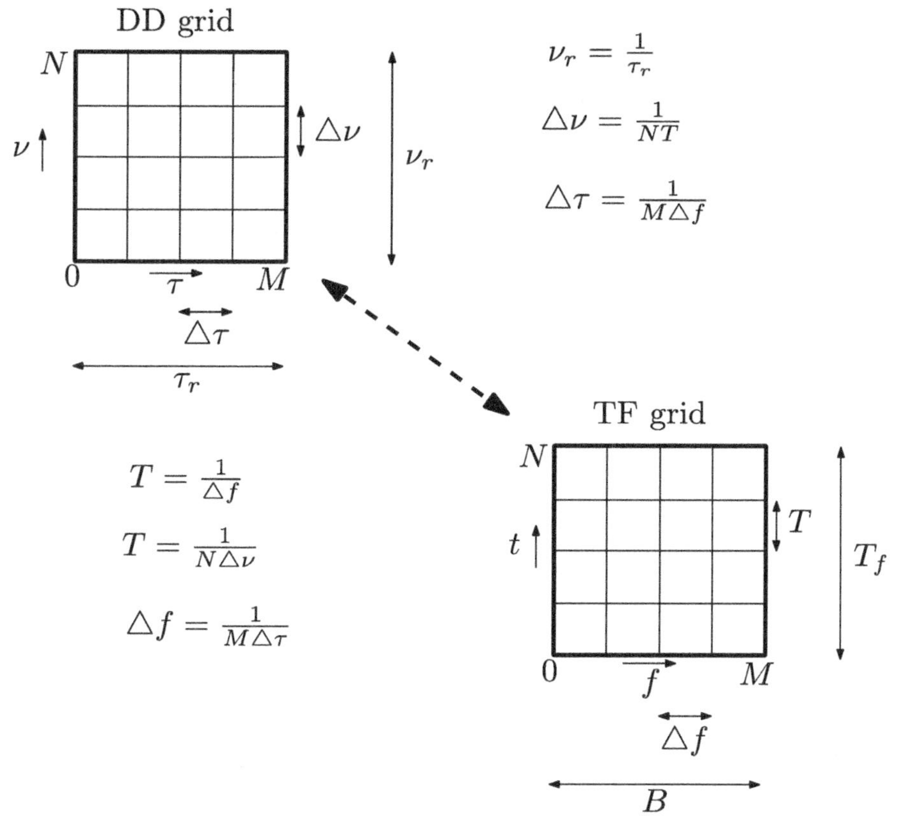

Fig. 5.8 DD grid and TF grid

reflectors R0, R1 and R3 are not distinguishable. Now, consider the case of OTFS modulation where all the four reflectors are distinguishable from each other based on their delay and Doppler. The channel response will include all the five channel gains specifying their delay and Doppler parameters, rather than their time or frequency.

5.2 OTFS Modulation–Demodulation Principle

5.2.1 OTFS Grid Design

In OTFS modulation, the data is represented in a grid of delay and Doppler bins. Let M and N be the number of delay bins and Doppler bins, respectively, and $\Delta\tau$ and $\Delta\nu$ be their corresponding resolutions in the DD grid. For designing the parameters, mainly the selection of M and N has to be done based on the channel conditions. The

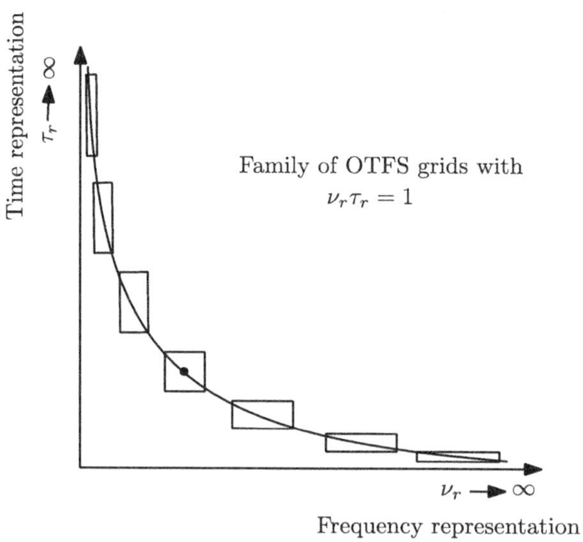

Fig. 5.9 TDMA and FDMA as limiting cases of OTFS

maximum delay of the channel, $\tau_{max} < M\triangle\tau = \tau_r$ and maximum Doppler present in the channel, $\nu_{max} < N\triangle\nu = \nu_r$. Let $\triangle f$ be the subcarrier frequency and the sampling interval $T = \frac{1}{\triangle f}$. We have $T = \frac{1}{N\triangle\nu}$ and $\triangle f = \frac{1}{M\triangle\tau}$, then M and N should be selected such that $\nu_{max} < \frac{1}{T}$ and $\tau_{max} < \frac{1}{\triangle f}$. A total of NM data symbols can be transmitted over a single OTFS frame. The mapping from DD grid to TF grid is shown in Fig. 5.8.

TDMA and FDMA: Limiting Cases of OTFS As seen in the previous section, for transmitting NM symbols over a DD grid, it can be designed using any values of N and M such that $\nu_r\tau_r = 1$. From Fig. 5.9, we can observe that as $\nu_r \rightarrow \infty$ the OTFS modulation performs similar to FDMA and as $\tau_r \rightarrow \infty$ it is similar to a TDMA system. Hence, OTFS modulation is considered as a limiting case of TDMA and FDMA depending on the design of OTFS grid.

The principle of OTFS modulation and demodulation is shown in Fig. 5.10. Here we discuss on the two-stage OTFS modulation and demodulation, which is initially proposed [1, 7]. The important transforms used at the transmitter and receiver sections of OTFS are explained in the figure.

5.2.2 *OTFS Transmitter*

Symplectic Finite Fourier Transform (SFFT): Let $(x[k, l])$ be the data symbols in DD domain over a grid of $N \times M$ dimension. These symbols are initially converted to TF domain using SFFT transform. SFFT can be interpreted as FT along the delay

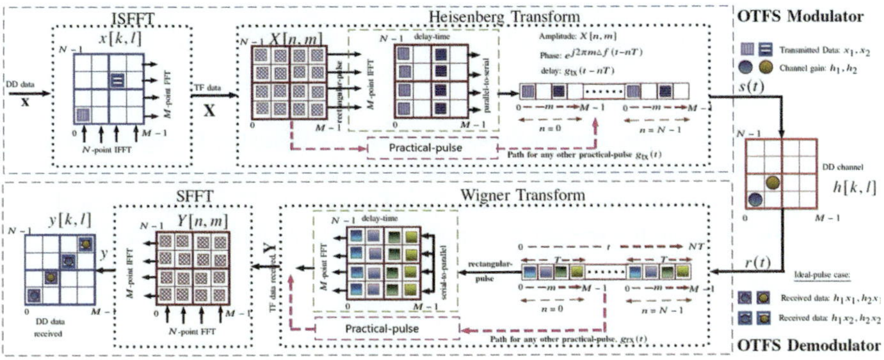

Fig. 5.10 OTFS modulation and demodulation

dimension and inverse FT along the Doppler dimension. This is mathematically represented as

$$X[n, m] = \frac{1}{MN} \sum_{k=0}^{N-1} \sum_{l=0}^{M-1} x[k, l] e^{j2\pi(\frac{nk}{N} - \frac{ml}{M})}. \tag{5.3}$$

Heisenberg's Transform: Now the data symbols are in the conventional TF domain, and by applying Heisenberg's transform to obtain the dats in delay-time domain. This is same as applying inverse FT along the frequency domain for the data symbols in TF domain. The data is then serialised to obtain the time domain signal and using a basis pulse $g_{tx}(t)$, the signal is transmitted as $s(t)$. The entire operation is represented as

$$s(t) = \sum_{n=0}^{N-1} \sum_{m=0}^{M-1} X[n, m] e^{j2\pi m \triangle f (t-nT)} g_{\text{tx}}(t - nT). \tag{5.4}$$

5.2.3 OTFS Receiver

Wigner's Transform: Let $h(\nu, \tau)$ be the channel response, and then the received signal $r(t)$ is given as

$$r(t) = \iint h(\nu, \tau) e^{j2\pi \nu (t-\tau)} s(t - \tau)\, d\tau d\nu + z(t) \tag{5.5}$$

where $z(t)$ is the AWGN present in the channel. At the receiver side, using the basis pulse $g_{rx}(t)$, the received signal is converted to TF domain using Wigner transform given by

$$Y[n,m] = \int_{-\infty}^{\infty} g_{\text{rx}}^{*}(t-nT)r(t)e^{-j2\pi m\Delta f(t-nT)}dt \tag{5.6}$$

Inverse Symplectic Finite Fourier Transform (ISFFT): Once the TF signal is obtained, then we have to finally convert that to the basic DD domain. This is achieved by applying ISFFT on the TF signal, where ISFFT is same as FT along the time domain and inverse FT along the frequency domain. This can be represented as

$$y[k,l] = \sum_{n=0}^{N-1}\sum_{m=0}^{M-1} Y[n,m]e^{-j2\pi(\frac{nk}{N}-\frac{ml}{M})}. \tag{5.7}$$

Input–Output Relation: Considering a channel with DD impulse response, $h(\tau, \nu)$, which is sparse in nature. For a channel with P multipaths, the sparse channel representation in DD domain is given as

$$h(\tau,\nu) = \sum_{i=1}^{P} h_i\delta(\tau-\tau_i)\delta(\nu-\nu_i) \tag{5.8}$$

where h_i, τ_i and ν_i correspond to the channel gain, delay and Doppler of ith path. Considering the transformations at the transmitter and receiver sides, the complete input–output relation of OTFS modulation is expressed as

$$y[k,l] = \sum_{k_i=0}^{N-1}\sum_{l_i=0}^{M-1} h[k_i,l_i]x[[k-k_i]_N,[l-l_i]_M] + z[k_i,l_i] \tag{5.9}$$

where $[\cdot]_N$ denotes the modulo operation with respect to N, $\lceil k_i = \frac{\nu_i}{\Delta\nu}\rceil$, $\lceil l_i = \frac{\tau_i}{\Delta\tau}\rceil$ and $z[k_i, l_i]$ is the noise signal present in that DD bin. The relation given in (5.9) clearly denotes that OTFS is a 2D circular convolution operation.

5.2.4 OTFS Compatibility with OFDM

One of the major advantages of two-stage OTFS modulator is that it is compatible with the existing OFDM modulator. Consider Fig. 5.11 which shows the simplified block diagram of Fig. 5.10. We can observe that OTFS is similar to OFDM with additional pre-processing and post-processing operations. Once the signals are in TF domain, the operations in OTFS modulation are same as that of OFDM modulation. In other words, SFFT and ISFFT in OTFS are similar to a pre-processing and post-processing steps for OFDM. In TF domain, for the ideal case the input–output relation is given by

$$Y[n,m] = H[n,m]X[n,m] + Z[n,m] \tag{5.10}$$

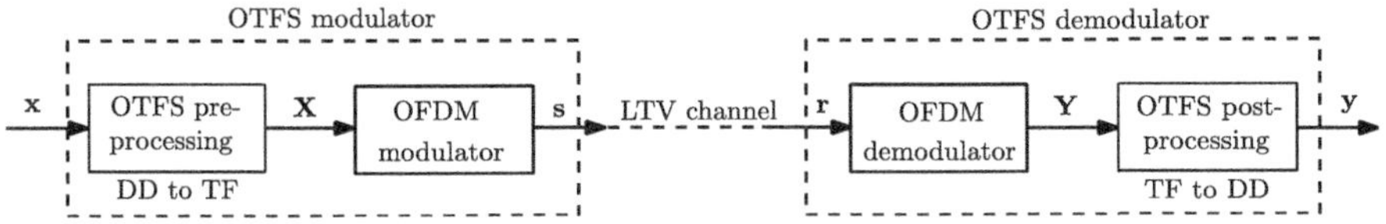

Fig. 5.11 OTFS as OFDM compatible

5.2.5 Pulse-Shaping and Interferences in OTFS Modulation

The performance of OTFS modulation is dependent on the pulse-shaping at the transmitter and receiver sides. The selection of $g_{tx}(t)$ and $g_{rx}(t)$ is significant in exploring the benefits of OTFS fully [7]. The pulses that satisfy bi-orthogonality property can be represented as

$$\int e^{-j2\pi m\Delta f(t-nT)} g^*_{rx}(t-nT) g_{tx}(t)\, dt = \delta(m)\delta(n) \tag{5.11}$$

The use of practical basis pulse instead of ideal basis pulse will result in inter-symbol interference (ISI) and inter-channel interference (ICI). In this case, the input–output relation in TF domain is given by

$$\begin{aligned} Y[n,m] &= \sum_{n'=n-1}^{n} \sum_{m'=0}^{M-1} H[n',m']X[n',m'] \\ &= H[n,m]X[n,m] + \sum_{m'=0,m'\neq m}^{M-1} H[n,m']X[n,m'] \\ &\quad + \sum_{m'=0}^{M-1} H[n-1,m']X[n-1,m'] \end{aligned} \tag{5.12}$$

where the additional terms compared to (5.10) are the ICI and ISI, respectively.

5.2.6 Fractional Doppler and Inter-Doppler Interference (IDI)

In a high Doppler environment, the Doppler frequency can be more accurately denoted as $\nu_i = (k_i + \kappa_i)\Delta\nu$, where k_i is an integer and $-\frac{1}{2} < \kappa_i < \frac{1}{2}$ and κ_i is called fractional Doppler. If we consider the fractional Doppler, then it will result in inter-Doppler interference (IDI).

$$y[k,l] = \sum_{i=1}^{P} \sum_{q=-N_i}^{N_i} (\mathbf{D}(q,\kappa_i) h_i e^{-j2\pi \nu_i \tau_i} \times x[[k - k_i + q]_N, [l - l_i]_M] \tag{5.13}$$

where the interference signal is given by

$$\mathbf{D}(q,\kappa_{\nu_i}) = \frac{e^{-j2\pi(-q-\kappa_{\nu_i})} - 1}{Ne^{-j\frac{2\pi}{N}(-q-\kappa_{\nu_i})} - N}.$$

As observed from (5.13), each of the paths has interference from the nearby $2N_i$ Doppler taps also, which results in IDI.

5.2.7 *OTFS Channel Representation*

Consider the input–output relation given in (5.9). This can be expressed in matrix form as

$$\mathbf{y} = \mathbf{H}\mathbf{x} + \mathbf{z}$$

where, $\mathbf{y}, \mathbf{x}, \mathbf{z} \in \mathbb{C}^{NM \times 1}$ and $\mathbf{H} \in \mathbb{C}^{NM \times NM}$. The convolution property of OTFS modulation results that $\mathbf{H}$ is a block-circulant matrix. Consider an OTFS grid with $N = M = 2$ and the wireless channel has two paths, $P = 2$. If $h_{k,l}$ represents the channel gain of the path with Doppler tap as k and delay tap as l, let's take the two paths as $h_{0,0}$ and $h_{1,1}$. And the remaining channel coefficients are zero. In this case, the channel matrix can be represented as,

$$\mathbf{H} = \left[\begin{array}{cc|cc} h_{0,0} & 0 & 0 & h_{1,1} \\ 0 & h_{0,0} & h_{1,1} & 0 \\ \hline 0 & h_{1,1} & h_{0,0} & 0 \\ h_{1,1} & 0 & 0 & h_{0,0} \end{array}\right] \tag{5.14}$$

Due to the block-circulant nature, all the rows and columns of $\mathbf{H}$ matrix will have P non-zero values. In OTFS modulation, the channel matrix $\mathbf{H}$ is always considered as sparse matrix. The wireless channel is underspread and we have $k_{max} \ll N$ and $l_{max} \ll M$.

5.3 Multiple Access Techniques in OTFS

The principle of OTFS modulation and the initial studies of OTFS was based on the single-user scenario. The multiple access techniques of OTFS gained attention in later stages. As we know multiple access techniques are broadly classified into two types, Orthogonal Multiple Access (OMA) and Non-Orthogonal Multiple Access (NOMA) techniques.

5.3.1 Orthogonal Multiple Access Techniques in OTFS

The resource allocation for multiple users can be performed in different ways in the case of OTFS-OMA. The resources can be multiplexed either in contiguous pattern or interleaved pattern. A single resource is allocated only to a single user, and same resource is not shared among users.

Contiguous DD Resource Allocation In this method of orthogonal multiple access, the DD resources are assigned to each user in contiguous fashion [8]. The resource allocation in the case of four users is shown in Fig. 5.12.

There are two schemes of contiguous DD resource allocation, either along Doppler (Scheme-1) or along the delay (Scheme-2) dimensions. The resources for a particular user are allocated continuously along the Doppler dimension, while different users are allocated the resources along the delay dimension. In Scheme-2, same user is allocated with continuous resources along the delay dimension, and along the Doppler dimension all users have equal number of resources. However, both Scheme-1 and Scheme-2 will perform equally in the case of OTFS-OMA scheme.

Interleaved DD Resource Allocation Another scheme of OTFS-OMA resource allocation is in interleaved fashion, we name it as Scheme-3. In the case of four users, the resource allocation in this scheme is as shown in Fig. 5.13. The resources are allocated not contiguously either along the delay or along the Doppler dimension. They are allocated in an interleaved fashion, along both the delay and Doppler dimension. Figure a shows the resource allocation in DD domain for Scheme-3. It is observable that none of the users occupy adjacent resources. Figure b shows the corresponding TF mapping of Scheme-3. It can be observed that each user's data has a block of resources which is in a non-overlapping fashion.

TF Resource Allocation Schemes The schemes described in the above section are with respect to resource allocation in DD domain. However, the resource allocation can be done in such a way that the resources are multiplexed in TF domain in specific patterns [9]. All the resource allocation patterns which are described in DD domain

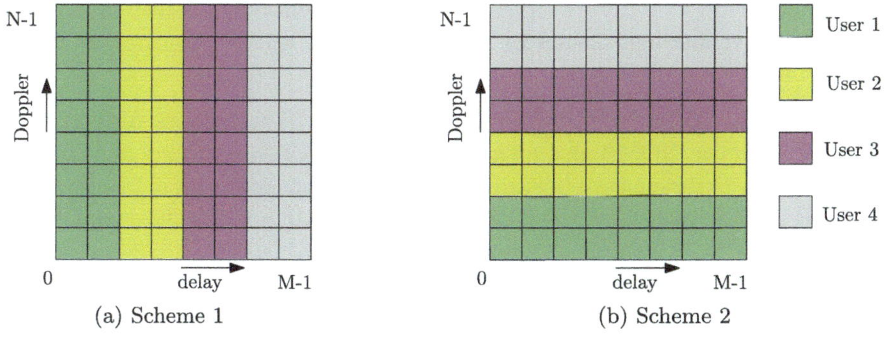

Fig. 5.12 Contiguous DDRB allocation schemes for $K_u = 4$

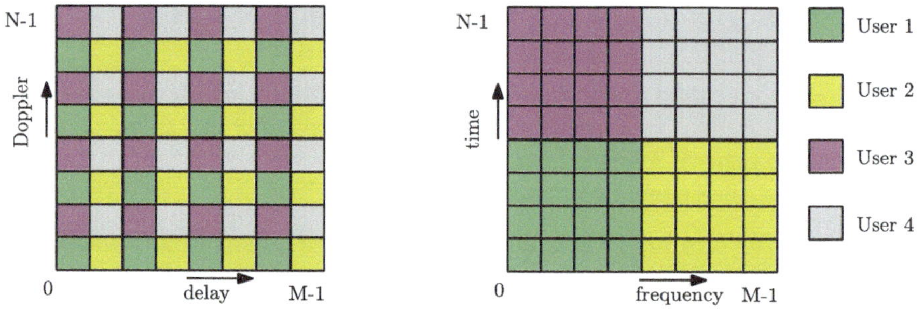

(a) Scheme 3 DDRB allocation (b) TF mapping of Scheme 3

Fig. 5.13 Interleaved DDRB allocation scheme and TF mapping

can be implemented in TF domain also. The resource sharing among users can be performed in TF domain in both contiguous and interleaved fashions.

5.3.2 Non-Orthogonal Multiple Access Techniques in OTFS

The multiple access techniques described in Sect. 5.3.1 describes OTFS-OMA techniques. As we all know, non-orthogonal multiple access is also an efficient way of multiple access scheme, where same resources are shared by different users. OTFS-NOMA schemes are broadly classified into power-domain NOMA and code-domain NOMA.

Power-Domain OTFS-NOMA Power-domain OTFS-NOMA schemes are based on the user velocity of different users served on an OTFS frame [10]. One of the major schemes of resource allocation in OTFS-PD-NOMA scheme is as shown in Fig. 5.14. It is an heterogeneous method of resource allocation where the users are served in both DD domain and TF domain depending on the velocity. As observed from Fig. 5.14, the DD domain is reserved explicitly for high-mobility users while the users with low mobility are served in TF domain itself. Hence, in the TF mapping, we can observe that same resources are shared among users of higher mobility and low mobility.

Code-Domain OTFS-NOMA Considering the NOMA techniques for OTFS, we have code-domain NOMA also for OTFS. OTFS-SCMA is a code-domain NOMA technique where the data symbols are coded as SCMA codewords [11]. Sparse Code Multiple Access (SCMA) is a code-domain NOMA technique which is implemented along with OTFS modulation to design OTFS-SCMA scheme. OTFS-SCMA scheme has the same overloading factor as of the underlying SCMA system. Since OTFS is a 2D grid, we have two schemes of arrangements of codewords either along the Doppler dimension (Scheme-1) or along the delay dimension (Scheme-2).

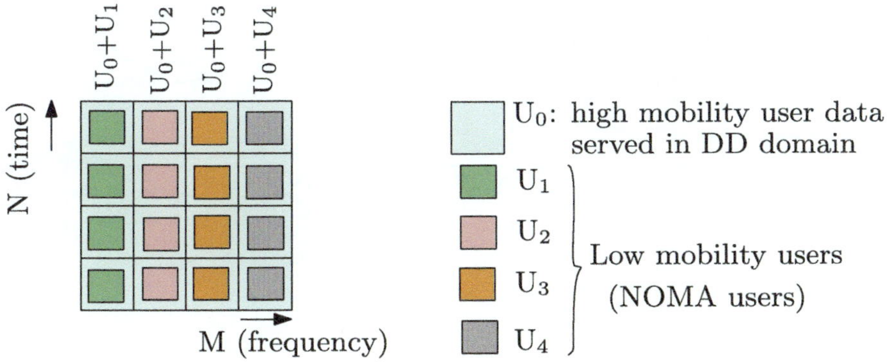

Fig. 5.14 TF mapping of OTFS power-domain NOMA

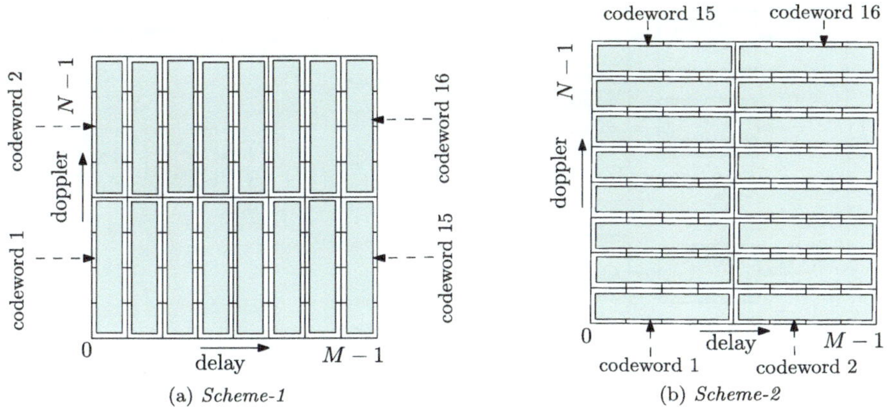

Fig. 5.15 Codeword allocation schemes of OTFS-SCMA (K=4)

The codeword arrangement of OTFS-SCMA is as shown in Fig. 5.15. Since a data symbol is represented by $K \times 1$ codeword for Scheme-1 and Scheme-2, we should have $[N]_K = 0$ and $[M]_K = 0$, respectively. The total number of data symbols that can be transmitted over an $N \times M$ DD grid is $J\frac{NM}{K}$. Hence, we have an overloading factor of $\frac{J}{K}$ which is same as that of the underlying SCMA scheme. The detailed explanation of OTFS-SCMA in downlink and uplink is given in the following section.

OTFS-SCMA in downlink: The schematic of OTFS-SCMA in downlink is as shown in Fig. 5.16. The SCMA codewords of each user are mapped to the OTFS grid and then superimposed. Once we have the superimposed data plane then OTFS modulation is applied. The input–output relation in the case of OTFS-SCMA downlink can be expressed as given in (5.15).

$$\mathbf{y}_{j,vec} = \mathbf{H}_j \mathbf{x}_{sum,vec} + \mathbf{z}_j \tag{5.15}$$

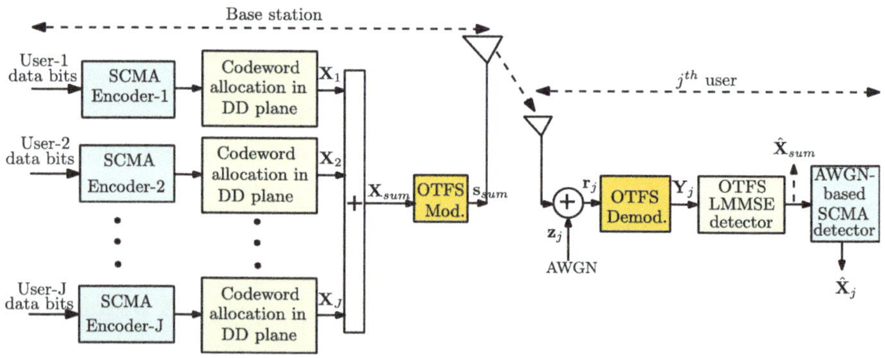

Fig. 5.16 Block diagram of OTFS-SCMA in downlink

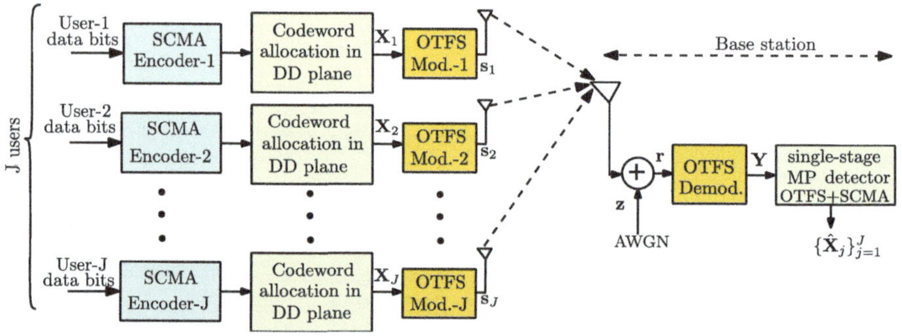

Fig. 5.17 Block diagram of OTFS-SCMA in uplink

where $\hat{\mathbf{x}}_{sum,vec}$ is the superimposed data over OTFS grid in vectorised form. As seen from Fig. 5.16, at the receiver side the data detection has to be done in two stages. First OTFS data detector is applied to retrieve the superimposed SCMA codewords, using any of the OTFS detection methods like LMMSE or message passing (MP) algorithm. Using LMMSE detection, we have

$$\hat{\mathbf{x}}_{sum,vec} = \mathbf{H}^{\dagger}{}_{j}[\mathbf{H}_{j}\mathbf{H}^{\dagger}{}_{j} + \sigma^{2}{}_{n}\mathbf{I}_{MN}]^{-1}\mathbf{y}_{j,vec} \tag{5.16}$$

Once OTFS data detection is performed then the data is similar to SCMA codewords in AWGN noise. Hence directly SCMA data detection can be applied to the superimposed codewords such that individual user data is retrieved back.

OTFS-SCMA in uplink The basic schematic of OTFS-SCMA in uplink is given in Fig. 5.17. OTFS modulation is applied to individual used data frame and is transmitted, where the OTFS frame of each user will have the codewords corresponding to the data to be transmitted.

The input–output relation for uplink is obtained as

$$\mathbf{y}_{vec} = \sum_{j=1}^{J} \mathbf{H}_j \mathbf{x}_{j,vec} + \mathbf{z} \tag{5.17}$$

Similar to OTFS-OMA uplink equation, this can be simplified to a linear matrix equation given by

$$\begin{aligned} \mathbf{y_{vec}} &= \mathbf{H}_{all}\mathbf{x}_{all} + \mathbf{z} \\ \text{where } \mathbf{H}_{all} &= [\mathbf{H_1 H_2 ... H_J}] \\ \text{and } \mathbf{x}_{all} &= [\mathbf{x}_{\mathbf{1,vec}}^{\mathbf{T}} \mathbf{x}_{\mathbf{2,vec}}^{\mathbf{T}} \ldots \mathbf{x}_{\mathbf{J,vec}}^{\mathbf{T}}]^T \end{aligned} \tag{5.18}$$

where $\mathbf{H}_{all}$ is an $MN \times JMN$ matrix and $\mathbf{x}_{all}$ is a JMN dimension vector. But we know that for an SCMA system all the MN resources will not be used by a user, instead only d_v resources are used by each user. Therefore we need to consider the linear equation with reduced dimension. The effective relation is represented as

$$\mathbf{y_{vec}} = \mathbf{H}_{all,compr}\mathbf{x}_{all,compr} + \mathbf{z} \tag{5.19}$$

where the compressed matrix $\mathbf{H}_{all,compr}$ is of dimension $MN \times \frac{JMN}{K} d_v$ and $\mathbf{x}_{all,compr}$ a $\frac{JMN}{K} d_v \times 1$ vector. Here $\mathbf{H}_{all,compr}$ is formed by selecting only the d_v channel coefficients in each row of $\mathbf{H}_{all}$, corresponding to non-zero symbol positions of $\mathbf{x}_{all}$. Since adjacent d_v elements in $\mathbf{x}_{all,compr}$ collectively represent a symbol, they are clubbed together and form a single variable node. Correspondingly, the adjacent d_v elements of each row of $\mathbf{H}_{all,compr}$ are also clubbed together to form a subvector h_{dc} of dimension $1 \times d_v$. In the corresponding factor graph, an edge appears if at least one of the coefficients of $\mathbf{h_{dc}}$ vector is non-zero. The expanded form of (5.19) is given by

$$\mathbf{y}_{\text{vec}} = \begin{bmatrix} y_1 \\ \vdots \\ y_{MN} \end{bmatrix} = \begin{bmatrix} \mathbf{h}_{1,1} & \cdots & \mathbf{h}_{1,\frac{JMN}{K}} \\ \vdots & \cdots & \vdots \\ \mathbf{h}_{MN,1} & \cdots & \mathbf{h}_{MN,\frac{JMN}{K}} \end{bmatrix} \begin{bmatrix} \mathbf{x}_1 \\ \vdots \\ \mathbf{x}_{\frac{JMN}{K}} \end{bmatrix} + \begin{bmatrix} z_1 \\ \vdots \\ z_{MN} \end{bmatrix}. \tag{5.20}$$

Now we have a single equation where $\mathbf{H}_{all,compr}$ has the combined effect of both SCMA and OTFS modulations. Since $\mathbf{H}_{all,compr}$ is still a sparse matrix, a more efficient combined MPA detector is used. The resultant MPA detector increases the complexity, but gives superior performance for OTFS-SCMA system in uplink.

5.3.3 Diversity Analysis of OTFS

The diversity gain of a system is a key parameter that decides the performance of communication system. For QAM data symbols, the asymptotic diversity order of OTFS is 1, in the case of both single-user and multi-user scenarios. However, in the case of OTFS-SCMA, the diversity order is dependent on the channel condition and the SCMA codeword length, K. Consider a channel with k' distinct Doppler taps and l' distinct delay taps. For Scheme-1 codeword allocation $[k']_K$ and for Scheme-2 codeword allocation $[l']_K$ will be the asymptotic diversity orders. Hence, depending on the channel parameters we have to decide the codeword allocation scheme to achieve maximum diversity order.

5.4 Channel Estimation Techniques in OTFS

In the case of OTFS modulation, both the data and channel are represented in DD domain. A total of NM resources is available in an OTFS frame. For a channel with P multipaths, the delay (l_i) and Doppler taps (k_i) can take any value such that $k_i < N$ and $l_i < M$. In the case of underspread channel, $k_i < \frac{N}{2}$ and $l_i < \frac{M}{2}$. The total number of paths in the channel, $P \ll NM$, which clearly point out that DD channel response is highly sparse. There are a few channel estimation techniques that are specifically designed for OTFS channel estimation. The 2D convolution process involved in OTFS modulation process makes the channel estimation a complex procedure. Some of the major channel estimation techniques of OTFS are discussed in the following section.

5.4.1 QAM-Pilot- and Guard Band-Based Channel Estimation

The basic concept of QAM-pilot-based channel estimation is to transmit a QAM pilot symbol embedded in the OTFS data frame. The QAM pilot is embedded such that enough guard band is provided along the delay and Doppler dimension, such that in the received OTFS frame there is no interference between the data and pilot symbol [12]. Figure 5.18 shows the OTFS frame embedded with pilot and data, where $l_{max} = k_{max} = 1$. The guard band is provided in Fig. a, such that the negative Doppler taps are also accommodated. From the received grid shown in Fig. b, we can observe that the DD paths present in the channel are $(0, 0)$ and $(1, 1)$. Mathematically, this can be interpreted by applying thresholding to each of the DD resources in the observation region which is given by $|y_p| > \mathcal{T}$.

In this method of channel estimation, the guard band allocation will result in large overhead. As the channel condition varies, the maximum delay and Doppler

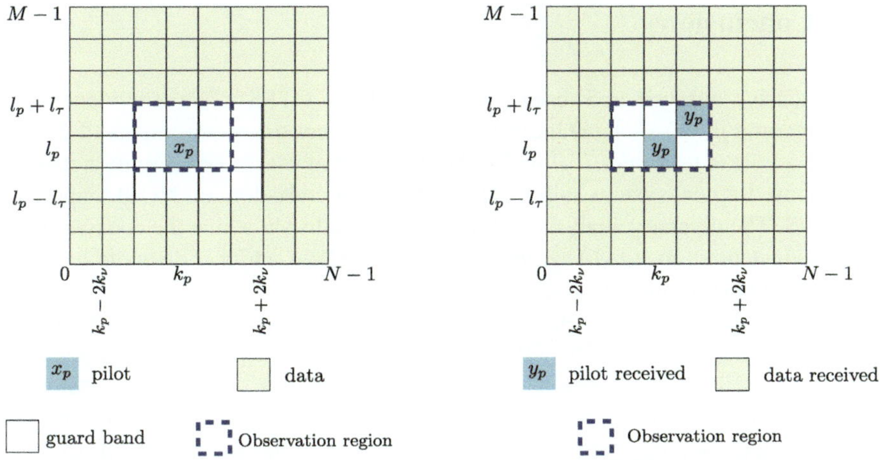

(a) *OTFS frame transmitted* (b) *OTFS frame received*

Fig. 5.18 QAM pilot embedded channel estimation for single user, $(k_i, l_i) = \{(0, 0), (1, 1)\}$

tap also increase. This demands an increased guard band allocation and the overhead also increases accordingly. In the case of multi-user scenario in uplink, all the users have to allocate independent resources for the pilot symbol and guard band. Hence, the overhead for channel estimation increases proportionally as the number of users increases.

5.4.2 Sparse Signal Recovery-Based Channel Estimation

The DD channel response is highly sparse in nature. Hence, it is possible to formulate the channel estimation problem as a sparse signal recovery problem. The input–output relation of OTFS can be expressed as $\mathbf{y} = \mathbf{X}\mathbf{h}$, where $\mathbf{X}$ is an $NM \times NM$ matrix formed from the $N \times M$ OTFS grid. Here $\mathbf{X}$ denotes the block-circulant matrix formed from the data grid, while $\mathbf{h}$ denotes the $MN \times 1$ channel coefficient vector. For a channel with P paths, the sparse signal recovery algorithm aims to estimate P coefficients from the unknown sparse vector. There are several approaches for the sparse signal recovery problem like orthogonal matching pursuit (OMP), subspace pursuit (SP) and so on [13, 14].

5.5 Conclusions

The extensive analysis and research in the field of OTFS made it evident that it can overcome the challenges faced by OFDM to meet the requirements of 5G and beyond wireless networks. To render the demands of mass-connectivity networks, we proposed a new system called OTFS-SCMA, a code-domain NOMA approach for OTFS. The diversity analysis and simulation results show that the codeword data symbols instead of the QAM symbol give OTFS-SCMA a higher diversity gain over the existing OTFS multi-user schemes.

To ensure a more practically feasible OTFS-SCMA system, a novel method of channel estimation for OTFS-SCMA based on CSC is proposed in the uplink. An impressive BER performance and spectral efficiency are obtained, maintaining a reduced guard band overhead and complexity. These results corroborate the proposed method's suitability for a high Doppler uplink scenario. The benefits of OTFS can be enhanced more effectively by designing an IRS-assisted OTFS system. The transmission protocol for the IRS-OTFS system and the channel estimation technique provide an overall scheme of practical implementation. We can infer from the simulation results that the IRS-OTFS system is a promising system model even in a multi-user scenario.

The complexity of data detection in OTFS is a primary concern regardless of its advantages over OFDM. Designing a low-complexity detector for OTFS is left as one of the future works. In addition to SCMA, there are other multiple access techniques. The analysis of OTFS from the perspective of these multiple access techniques will be taken up in the future. Additionally, the channel estimation, beamforming and detection of multi-user massive MIMO-OTFS systems are essential tasks. We plan to carry out these works specifically using machine learning techniques.

References

1. Hadani R, Rakib S, Tsatsanis M, Monk A, Goldsmith AJ, Molisch AF, Calderbank R (2017) Orthogonal time frequency space modulation. In: IEEE wireless communications and networking conference WCNC, pp 1–13
2. Franz Hlawatsch GM (2011) Wireless communications over rapidly time-varying channels. Elsevier Ltd., pp 0–99
3. Hadani R, Monk A (2018) OTFS: a new generation of modulation addressing the challenges of 5G. arXiv:1802.02623
4. Mohammed SK (2021) Derivation of OTFS modulation from first principles. IEEE Trans Veh Technol 70(8):7619–7636
5. Janssen A (1988) The zak transform: a signal transform for sampled time-continuous signals. Philips J Res 43:01
6. Kollengode Ramachandran M, Chockalingam A (2018) Mimo-otfs in high-doppler fading channels: signal detection and channel estimation. In: 2018 IEEE global communications conference (GLOBECOM), pp 206–212
7. Raviteja P, Phan KT, Hong Y, Viterbo E (2018) Interference cancellation and iterative detection for orthogonal time frequency space modulation. IEEE Trans Wirel Commun 17(10):6501–6515

8. Surabhi GD, Augustine RM, Chockalingam A (2019) Multiple access in the delay-doppler domain using OTFS modulation. arXiv:1902.03415
9. Khammammetti V, Mohammed SK (2019) OTFS-based multiple-access in high doppler and delay spread wireless channels. IEEE Wirel Commun Lett 8(2):528–531
10. Ding Z, Schober R, Fan P, Vincent Poor H (2019) OTFS-NOMA: an efficient approach for exploiting heterogenous user mobility profiles. IEEE Trans Commun 67(11):7950–7965
11. Deka K, Thomas A, Sharma S (2021) OTFS-SCMA: a code-domain NOMA approach for orthogonal time frequency space modulation. IEEE Trans Commun 69(8):5043–5058
12. Raviteja P, Phan KT, Hong Y (2019) Embedded pilot-aided channel estimation for OTFS in delay-Doppler channels. IEEE Trans Veh Technol 68(5):4906–4917
13. Rasheed OK, Surabhi GD, Chockalingam A (2020) Sparse delay-Doppler channel estimation in rapidly time-varying channels for multiuser OTFS on the uplink. In: Proceeding of the IEEE VTC2020-Spring, Antwerp, Belgium, pp 4–8
14. Thomas A, Deka K, Raviteja P, Sharma S (2022) Convolutional sparse coding based channel estimation for otfs-scma in uplink. IEEE Trans Commun 70(8):5241–5257

Chapter 6
Zak-OTFS: Concepts of Delay-Doppler Communication

Sangeeta Nalluru, Sanjeev Sharma, and Kuntal Deka

Abstract In this chapter, we will discuss the concepts and principles of the discrete Zak transform (DZT)-based orthogonal time-frequency space (OTFS) for doubly dispersive channels. OTFS is a two-dimensional modulation technique that translates a rapidly time-varying channel into a quasi-static, time-invariant channel by modulating the information symbols in the delay-Doppler (DD) domain. The DD domain originates from radar communication literature, where each target is identified by its delay- and velocity-induced Doppler. Therefore, we will first introduce some radar concepts and how they address the doubly dispersive nature of real-time channels. The core of the DZT lies in the DD domain. Hence, before formally introducing the DZT, we will explore features of the DD domain, including the concept of non-fading, crystallization, etc. Then, we will proceed with the definition of the DZT, its properties, and Zak basis functions. Following that, we will discuss how OTFS is constructed based on the DZT, its matrix representation, and also a comparison between pre-coded OTFS and DZT-based OTFS. Finally, we will conclude this chapter with some applications.

Keywords OTFS · DZT · Crystallization · Quasi-periodicity · DD domain

6.1 Introduction

With the increase of high-speed trains, unmanned aerial vehicles (UAVs), augmented reality (AR), and virtual reality (VR), there is a huge demand for massive data rate with reliable communication under high-mobility environments. Orthogonal frequency-division multiplexing (OFDM) [1, 2], which translates time-invariant frequency-selective channels into parallel frequency flat fading channels by keep-

S. Nalluru · K. Deka (✉)
IIT Guwahati, Guwahati, India
e-mail: kuntaldeka@iitg.ac.in

S. Sharma
IIT (BHU) Varanasi, Varanasi, India

S. Sharma et al. (eds.), *Next-Generation Wireless Systems*,
Transactions on Computer Systems and Networks,
https://doi.org/10.1007/978-981-96-3185-8_6

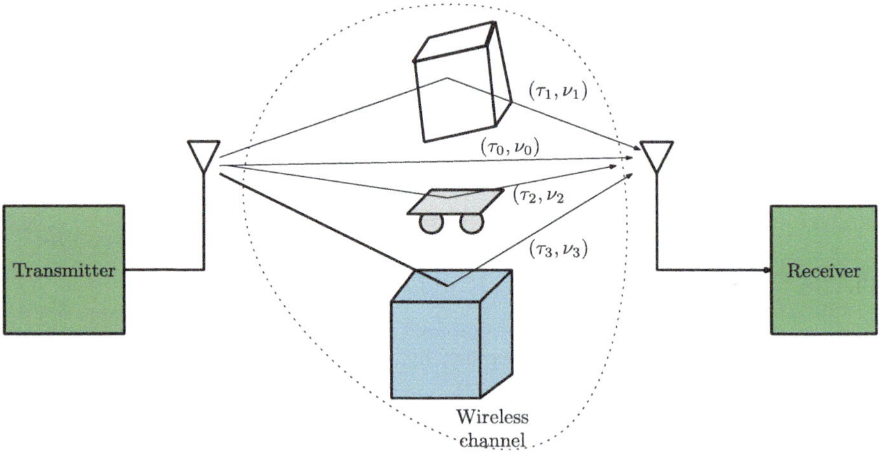

Fig. 6.1 Communication system scenario. Multiple paths between the transmitter and receiver exist. Each path has some delay (τ) and Doppler (ν) values

ing the information symbols on the orthogonal sub-carriers. Unfortunately, these sub-carriers are sensitive to Doppler shifts, which leads to the loss of orthogonality, making OFDM inefficient in rapidly varying channels. Hence, a new modulation technique, namely, orthogonal time-frequency space (OTFS), is introduced [3–5]. In OTFS, each reflector is identified by its delay and Doppler with respect to the transmitter. A typical communication system was shown in Fig. 6.1.

6.1.1 Review of Radar Concepts

The concepts of Zak transform play an important role in the practical realization of radar systems [6, 7]. Recently, this has been adopted in communication systems with Discrete Zak Transform (DZT) as a core processing element [8], utilizing the same communication and detection waveform. The entire signal processing of radar technology is equivalent to Zak processing, relying completely on detection and estimation. This is achieved by characterizing a target with its delay (τ) and Doppler (ν), which result from the relative distance and relative velocity with respect to the transmitter. Zak transform maps a one-dimensional periodic signal onto a two-dimensional signal [9]. Here, we map a discrete-time periodic sequence onto a 2D delay-Doppler (DD) plane through a series of operations. A detailed study of Zak transform is provided in Sect. 6.3. One such system is a *Pulse Radar*, as shown in Fig. 6.2.

Consider a pulse radar system with only one target as shown in Fig. 6.2. A pulse radar is a system which detects a target by transmitting a finite number of rectangular pulses. The job of the transmitter is to find the target's relative distance, i.e., range

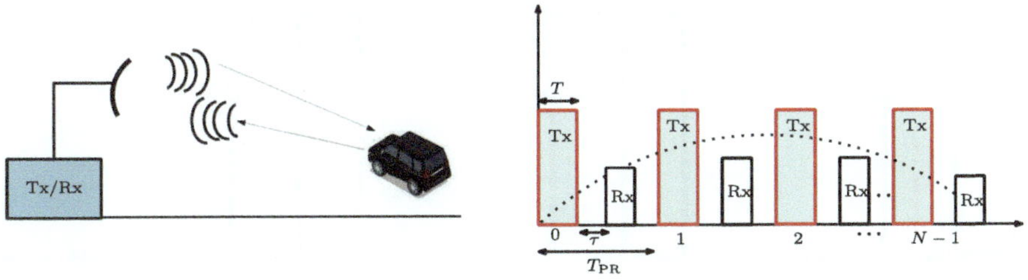

Fig. 6.2 Pulse radar to identified an object in the delay and Doppler domain

and the relative velocity, i.e., Doppler. This is done by transmitting a waveform which will be reflected by the target surface and hence the range and the Doppler will be determined by this reflected signal. Due to this reflection, the transmit signal will undergo delay and phase rotation. In this system, let us assume that the transmitter is sending a train of N rectangular pulses each with a duration of T and separated by a repetition interval of T_{PR}. This is illustrated on the left side of Fig. 6.2, highlighted in red color. Transmitted signal can be given by

$$s(t) = \sum_{k=0}^{N-1} p(t - kT_{\text{PR}})$$

Now, due to the reflection, this signal will undergo a delay of τ and phase rotation of ν. The received signal is shown in the black color in Fig. 6.2 given as

$$r(t) = \sum_{k=0}^{N-1} p(t - kT_{\text{PR}} - \tau)\, e^{j2\pi \nu k T_{\text{PR}}}$$

At the receiver, the processing is done by converting this analog signal into discrete one. At the receiver, the processing involves converting the analog signal into a discrete one. This is achieved by sampling the signal, resulting in $r(n) = r(t)|_{t=nT_s}$, where T_s is the sampling interval. Depending on the location of the non-zero element and its phase, the range and Doppler are estimated by arranging it in a range-Doppler matrix. Collect M consecutive samples and process in the discrete domain, as shown in Fig. 6.3.

6.1.2 Radar Systems Using Zak Transform

In this section, we will be discussing how the radars use Zak transform. Let the reflected signal from the target be a finite length periodic sequence $r(q)$ which is of length MN as one period of a periodic sequence with a period MN where $M, N \in$

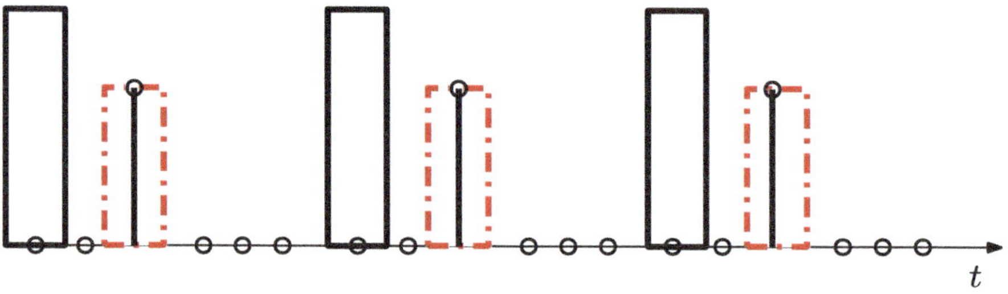

Fig. 6.3 Discrete-time signal processing using the sampling of the received analog signal

ℕ. Now the task is to find the range and the velocity of the target from this received signal. Consider 2D grid Z_r as the DZT on DD axis of size MxN with M bins along delay and N bins along the Doppler. $Z_r\ (k, l)$ corresponds to the element on the kth Doppler bin and lth delay bin. Basically DZT will be implemented in three steps as

Step-1: Collect first M consecutive samples from $x\ (q)$ and arrange them in the rows of the DD matrix. If we observe column-wise, it will be a vector formed by taking the samples separated by M with the stating index equal to l.
Step-2: Take DFT along the columns to get a delay-Doppler (range-Doppler) matrix.
Step-3: Depending the values of range-Doppler matrix, estimate the range and velocity of the target.

Mathematically this can be shown as

$$Z_x\ (l, k) = \frac{1}{\sqrt{N}} \sum_{n=0}^{N-1} x\ (l + nM)\, e^{-j2\pi \frac{n}{N}k} \quad \text{where } 0 \leq l \leq M-1 \text{ and } 0 \leq k \leq N-1.$$

At the transmitter, the inverse discrete Zak transform (IDZT) is performed, which reverses these operations.

Example 6.1 Consider a radar system which can capture a target within the range of four units along the delay, i.e., $M = 5$ and three units along the Doppler, i.e., $N = 4$, as shown in Fig. 6.4. Let us assume a target at zero delay and zero Doppler. It means that except for (0, 0), all remaining elements will be zero. Now, let us examine how the received sequence will be.

Let the entry in (0, 0) be 5, as observed in Fig. 6.5. Then the inverse discrete Fourier transform (IDFT) of first column will be $(1.25 + 0j, 1.25 + 0j, 1.25 + 0j, 1.25 + 0j)$ and for the remaining columns it will be (0, 0, 0, 0). Now these vectors has to be up-sampled with M and staggered to get a MN length sequence. Here as $M = 5$ so we will up-sampled by appending 4 zeros between every samples of each symbol and then stagger all these up-sampled sequences. So the discrete signal will be $(1.25 + 0j, 0, 0, 0, 0, 1.25 + 0j, 0, 0, 0, 0, 1.25 + 0j, 0, 0, 0, 0, 1.25 + 0j, 0, 0, 0, 0)$ as shown in Fig. 6.5.

It means if $r\ (q)$, $q = 0, 1, ..., MN - 1$ is of the abovementioned waveform, then it corresponds to zero delay and zero Doppler. For a non-zero delay and non-zero Doppler, the received wave form is shown Fig. 6.6. There we are considering a target

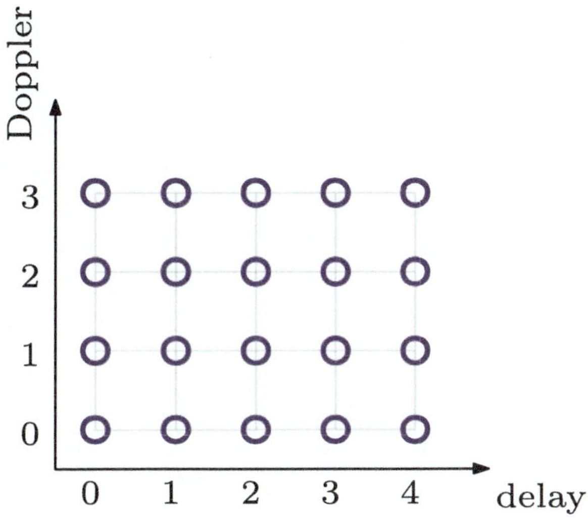

Fig. 6.4 Delay and Doppler domain grid formulation for $M = 5$ and $N = 4$

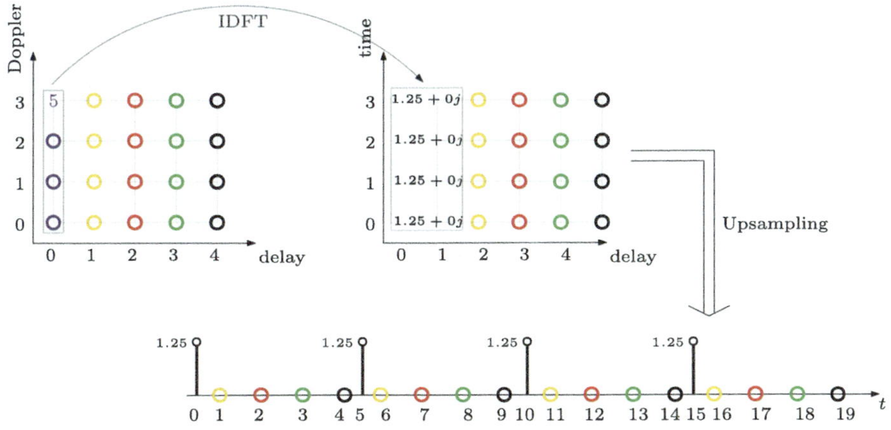

Fig. 6.5 Delay and Doppler domain signal processing for $M = 5$ and $N = 4$

with delay of 2 and Doppler of 0 and the corresponding real and imaginary parts of the $r\ (q)$ are given. Then non-zero element will start from $q = 2$ which is the target's delay and Doppler will be estimated from the rotations.

The range is estimated by the starting index of the non-zero element in the received reflected signal and Doppler from the rotation of the samples. So, a radar system is a single non-zero element at corresponding range and the Doppler in the delay-Doppler (range-Doppler) grid. If we consider MN non-zero elements, i.e., utilizing the entire grid, then it forms an OTFS frame, which will be discussed later. In summary, we can say that a radar system with MN targets is an OTFS system.

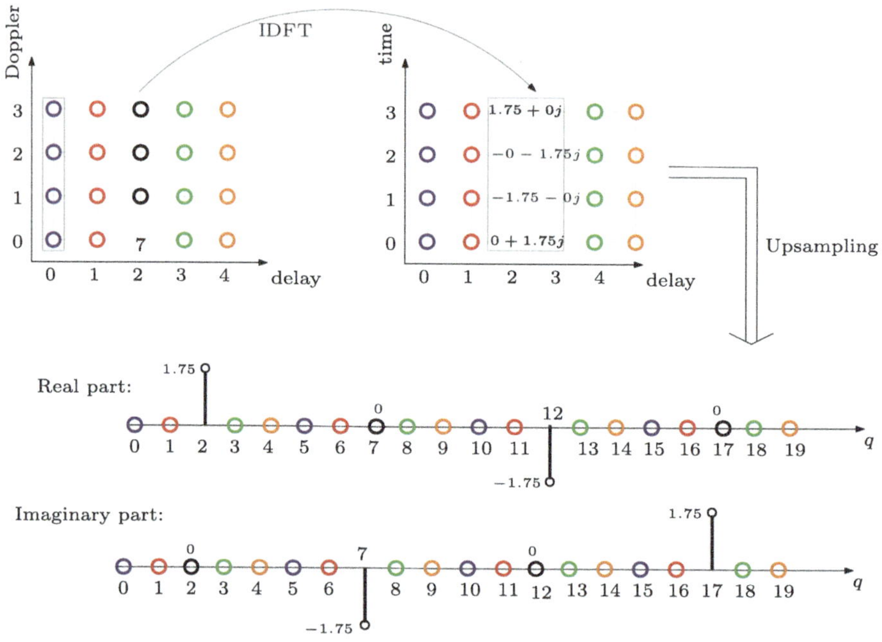

Fig. 6.6 Delay and Doppler domain signal processing for $M = 5$ and $N = 4$ for non-zero delay

6.2 Delay-Doppler Communication

Previous wireless technologies mostly depended on transmitting information using either orthogonal time-division multiplexing (TDM) or frequency-division multiplexing (FDM) slots. They all fall under one-dimensional modulation. Thus, they can handle either delay-only channels or Doppler-only channels. To address scenarios with both delay and Doppler, we turn to 2D modulation techniques, where information is transmitted on a 2D orthogonal resource. Before delving into the concept of delay-Doppler, let us take a look at TDM and FDM.

6.2.1 *Review of TDM and FDM*

A wireless environment is modeled by the number of the multipaths (reflectors) it has. Each reflector is characterized by its attenuation, delay and Doppler. Due to delay d, the information transmitted at time t will be received at $t + d$ and interferes with the symbols transmitted at $t + d$ location. This is called time dispersion which leads to a phenomena called inter-symbol interference (ISI). The effect of this in the frequency domain will be frequency selectivity. Due to Doppler ν, the information transmitted on the frequency f_c will be received at $f_c + \nu$ and interferes with the symbols transmitted on $f_c + \nu$ location. This is called frequency dispersion which leads to a phenomena called inter-carrier interference (ICI). The effect of this in the time domain will be time selectivity. When we have both delay and Doppler, it is called a doubly dispersive channel. Wireless channels are in general doubly dispersive in nature. The delay-Doppler spreading function for such systems is

$$h(\tau, \nu) = \sum_{i=1}^{P} h_i \delta(\tau - \tau_i)\delta(\nu - \nu_i), \tag{6.1}$$

where P is the total paths in the spreading function. Parameters $h_i, \tau_i, and \nu_i$ are the gain, delay, and Doppler of the ith path, respectively. The signal received at ith path will be $h_i x(t - \tau_i)e^{j2\pi\nu_i(t-\tau_i)}$, where $x(t)$ is the transmit signal.

In TDM, reflectors will be identified with delays only. So any two reflectors with same delay and different Dopplers will be overlapped. Channels with such distinct delays and no Dopplers are called delay-only channels. Similarly in FDM, reflectors will be identified with Doppler shifts only. So any two reflectors with same Doppler and different delays will be overlapped. Channels with such distinct Dopplers and no delays are called Doppler-only channels. But for the doubly dispersive channels, the pulses will be sent on a 2D delay-Doppler grid that can uniquely capture the effect of delay and Doppler. Irrespective of the modulation domain, we can transmit only time-domain signal in the physical channel. For TDM, as the pulse is already in time domain, so no need to change the signal (identity). In case of FDM, pulses are designed in the frequency domain. Hence in order to convert to time domain we use Inverse Fourier Transform (IFT). Similarly, when a pulse is in DD domain, we use a Inverse Zak transform to get its time equivalent signal that needs to be transmitted as shown in Fig. 6.7. A detailed study of DD domain is given in Sect. 6.2.2.

If we observe the transition of the applications of wireless techniques from low mobility scenarios, where the channel is time invariant, to high-mobility scenarios with time-variant channels, in low mobility cases, the channel will be simply represented as $h(\tau)$. However, for high mobility, it will be represented as $h(\tau, t)$, indicating that the delay taps change with respect to time. Another way to express this is that as the velocity of the reflector increases, the coherence time in the time-frequency (TF) domain will decrease. Consequently, for every small interval of time, we need to perform channel estimation, ultimately increasing the channel estimation overhead. Researchers have sought ways to capture such dynamics with less overhead. Before exploring those propositions, let us consider a small example of how the modulation domain impacts the system.

Example 6.2 Let us assume a transmitter wants to convey a sine signal of frequency ω_0. If it tries to convey it in the time domain, then it should send the values of $\sin(\omega_0 t)$ for an infinite duration. Now, if the same transmitter tries to send it in the frequency domain, it is enough to convey ω_0, which is a sine function with a frequency of ω_0 in the time domain. In this way, there will be a significant impact on the system due to the modulation domain.

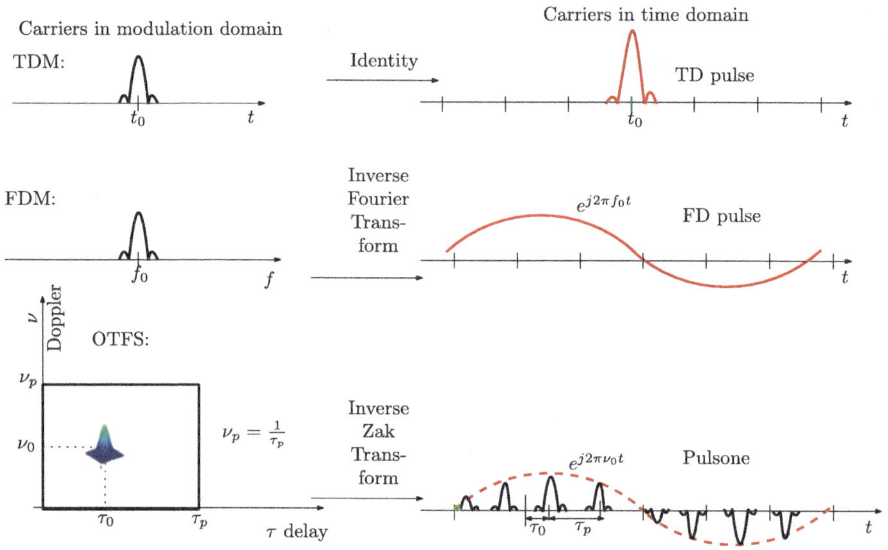

Fig. 6.7 TDM versus FDM versus DDM (delay-Doppler modulation)

Keeping this in mind, instead of representing the channel in the time-frequency domain, we represent it in its delay-Doppler equivalent. For doubly dispersive channels, the DD domain representation is very sparse. It has only P non-zero elements, where each one corresponds to one reflector (multipath). Another interesting aspect is that the stationary time of the DD representation is greater than the coherence time of the TF one. Therefore, we will represent both the information and the channel in the DD domain.

6.2.2 DD Domain and Crystallization

In this section, we will present a deep analysis of delay-Doppler communications. A delay-Doppler domain (τ, ν) is an infinitely extended 2D domain. But this can be completely characterized by a fundamental region defined in $\tau \in [0, \tau_p]$ and $\nu \in [0, \nu_p]$, where τ_p and ν_p are the delay and Doppler periods with which they repeat as shown in Fig. 6.8.

DD domain signal processing is governed by the Zak transform. It determines how a DD point should be mapped to a time-domain signal. A point in the DD plane can have an infinite number of time-domain realizations according to the Zak transform. To make such mapping unique, a property called quasi-periodicity is invoked. If we slide the fundamental rectangle along the Doppler axis, it is periodic, whereas along the delay axis, it is quasi-periodic. Due to this, it is enough to observe what is happening in the fundamental rectangle rather than the entire DD plane. Another

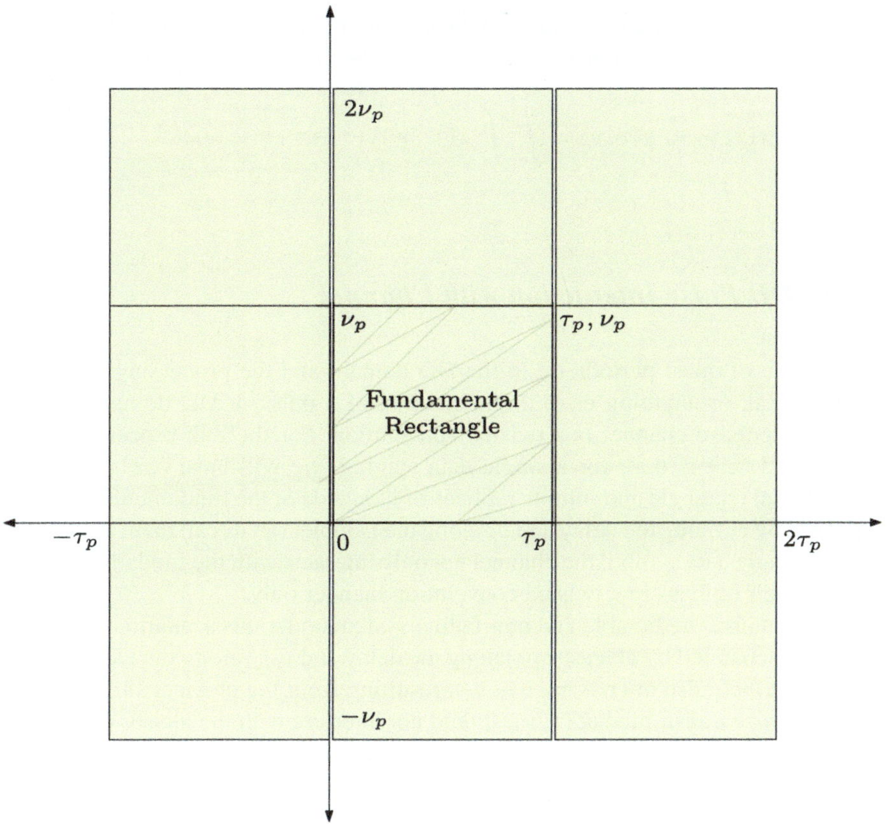

Fig. 6.8 DD domain grid with delay period τ_p and Doppler period ν_p

constraint imposed by these properties is that only the fundamental rectangle will be our resource region because the rest are replicas of this. So, we need to find a strategy to accommodate MN information symbols within the fundamental rectangle.

Additionally, if we place a point at some specific location in the fundamental rectangle, it implies that it is globally extended with replicas in the entire DD domain. This indicates that we can't find a set of basis functions that are globally orthogonal; instead, we should look for local orthogonality within the fundamental rectangle.

When a pulse is sent in time or frequency, the interaction of that pulse with the doubly dispersive channel will be a linear convolution, i.e.:

$$y(t) = h(t) \star x(t)$$

for time-domain pulse, where $\star$ is the convolution operator between two signals $x(t)$ and $y(t)$. $Y(f) = H(f) \star X(f)$ for frequency domain pulse.

Similarly when a pulse is sent in DD domain, the interaction of that pulse with the doubly dispersive channel will be "*twisted convolution*" denoted by $\star_\sigma$ given by

$$r(\tau, \nu) = x(\tau, \nu) \star_\sigma y(\tau, \nu) = \int_{\tau'} \int_{\nu'} x(\tau', \nu') y(\tau - \tau', \nu - \nu') e^{j2\pi\nu'(\tau-\tau')} d\nu' d\tau'. \tag{6.2}$$

6.2.3 DD Pulse Interaction with Channel

Irrespective of quasi-periodicity in the DD domain and the processing techniques (whether Zak or anything else) the interaction of a pulse in DD domain with the doubly dispersive channel is a twisted convolution. For the Zak processing, point to be noted here is that, for a single data symbol, we will have one pulse in the fundamental rectangle and infinite replicas of it outside of the fundamental rectangle separated by τ_p along the delay and ν_p along the Doppler (let us call them as replicas). So for a single data symbol the channel not only interacts with the fundamental pulse but also their replicas in a twisted convolution manner only.

To maintain a predictable and non-fading system under this scenario, the replicas should be separated by at least $\tau_{\max}$ along the delay and $\nu_{\max}$ along the Doppler. This means that the pulse at $(\tau + \tau_0, \nu + \nu_0)$, resulting from the channel shift of τ_0 and ν_0 of the pulse transmitted at (τ, ν), should not interfere with the aliases of the pulse at (τ, ν). In Fig. 6.9, the blue pattern should not overlap with the red pattern. This condition is referred to as the **Crystallization** condition, i.e.:

$$f(x) = \begin{cases} \tau_p > \max \tau_i - \min \tau_i \\ \\ \nu_p > \max \nu_i - \min \nu_i \end{cases} \quad \text{Crystallization condition.}$$

If we maintain the above condition, the aliases will not overlap. The only transform that allows twisted convolution is the Zak transform.

Example 6.3 Let us consider a single path channel with delay τ_1 and Doppler ν_1. For the above system, as shown in Fig. 6.10, the input–output relation is given by

$$y_{dd}(\tau, \nu) = [\delta(\tau - \tau_1)\delta(\nu - \nu_1)] \star_\sigma x_{dd}(\tau, \nu) \tag{6.3}$$

$$= \underbrace{e^{j2\pi\nu_1(\tau-\tau_1)}}_{\substack{\text{twisting} \\ \text{operation}}} \underbrace{x_{dd}(\tau - \tau_1, \nu - \nu_1)}_{\substack{\text{geometric shift} \\ \text{in DD domain}}} \tag{6.4}$$

The above equation can be interpreted in one more way. It is equivalent to shifting a pulse located at $x_{dd}(\tau, \nu)$ by τ_1 units along the delay and ν_1 units along the Doppler. So the resultant pulse will be a simple geometric shift and with a twist.

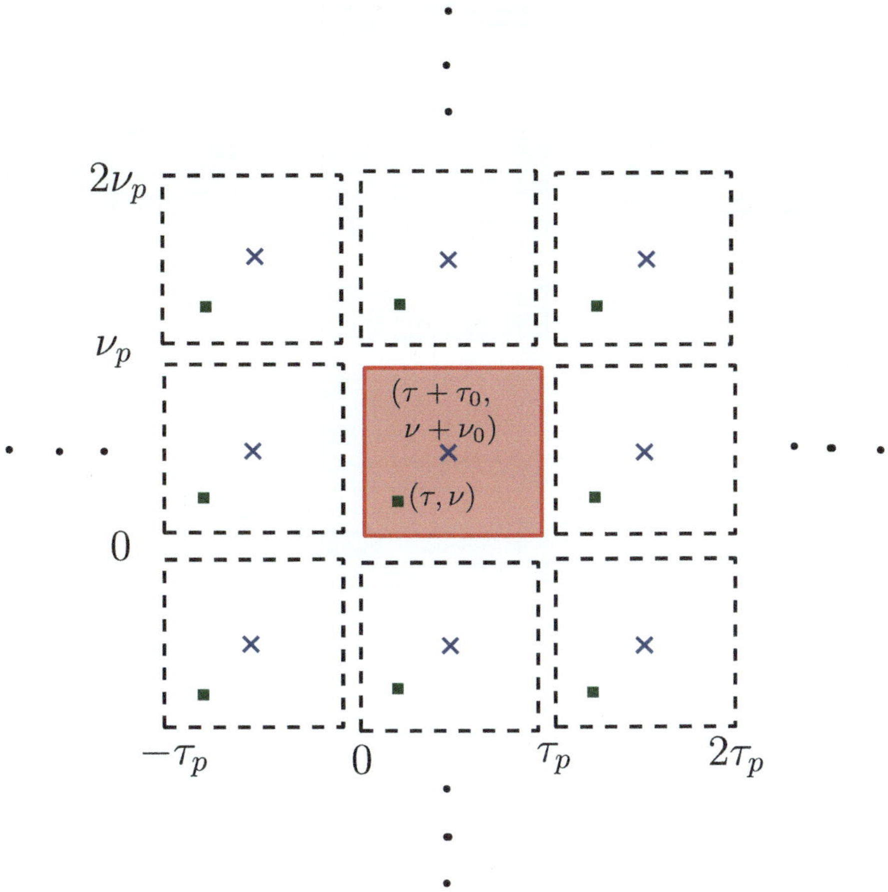

Fig. 6.9 Interaction of the DD pulse which is a single path channel

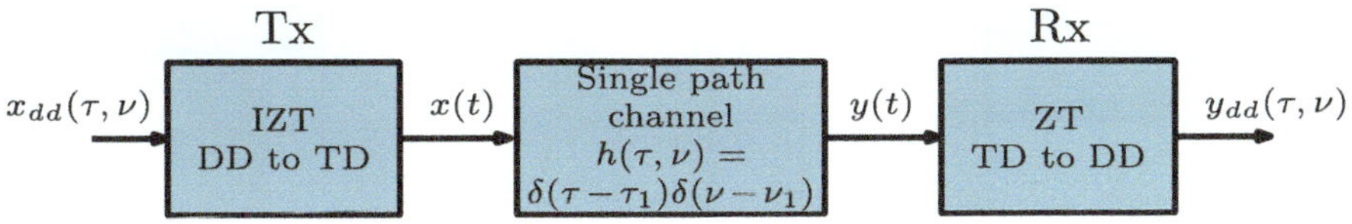

Fig. 6.10 Single path model

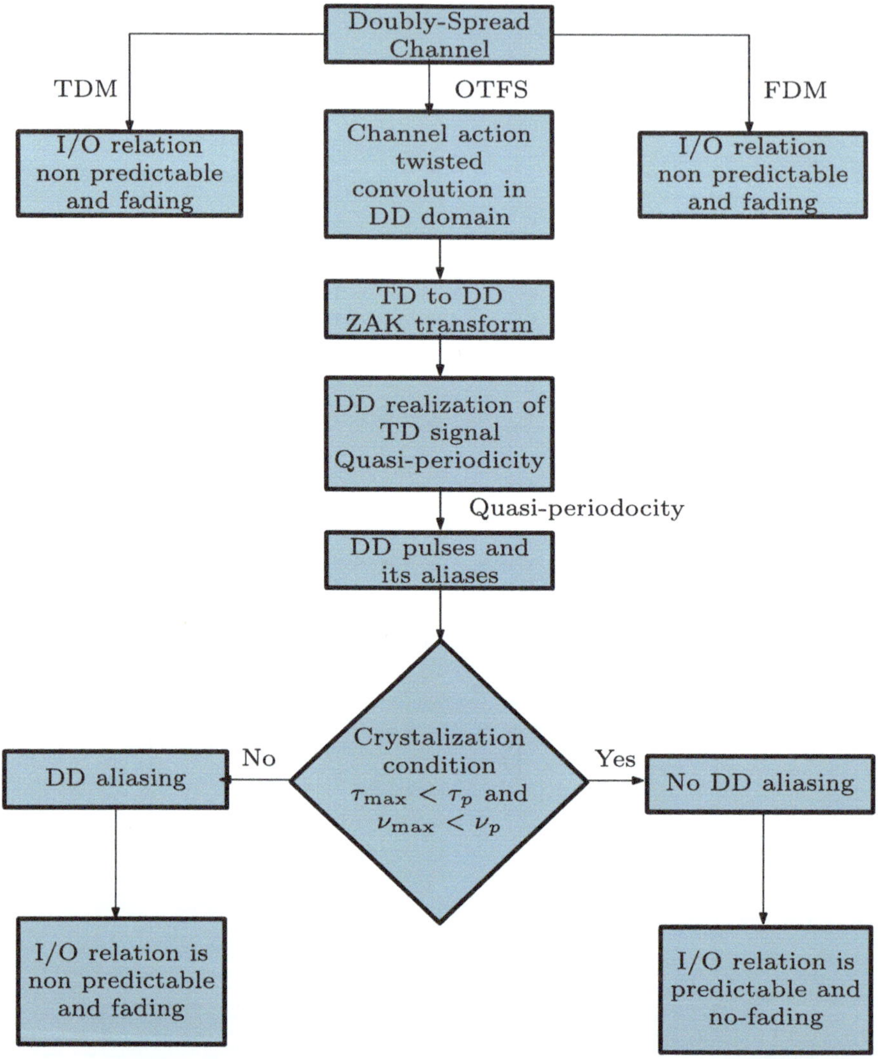

Fig. 6.11 Doubly spread channel characterization

The above discussion of TDM, FDM, and OTFS waveform can be summarized in a flow chart in Fig. 6.11.

Quasi-periodically extended impulses in the DD domain will lead to a periodic signal that is extended with infinite duration and bandwidth, which is practically unrealizable. Hence, we perform windowing in the time and frequency domains, resulting in a practical impulse, i.e., not necessarily zero occupancy in the DD domain. If we choose a pulse with a duration of T^{-1} along the Doppler and B^{-1} along the

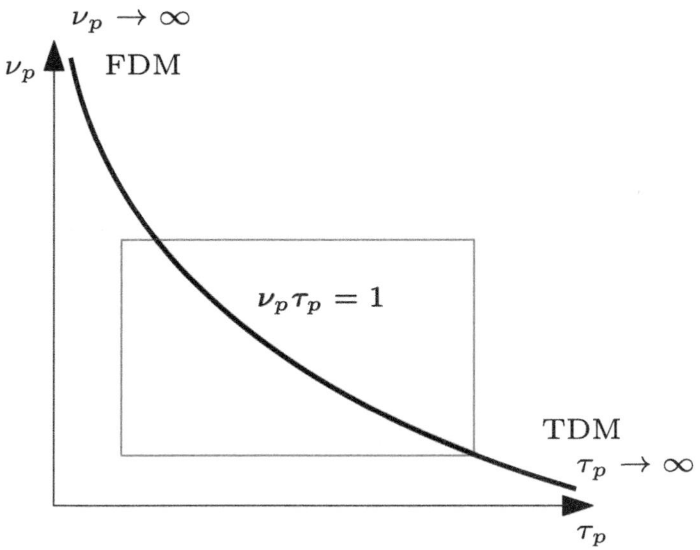

Fig. 6.12 Impact of τ_p and ν_p on OTFS system

delay, then its corresponding time-domain signal will have a duration of T and a bandwidth of B.

Impact of τ_p and ν_p: The values of τ_p and ν_p play an important role in designing a system. They are the determining factors for fading and predictability in the DD domain. Interestingly, it is observed that Zak-based OTFS is a combination of TDM and FDM. For $\tau_p = \infty$, we will get only one impulse in the time domain with a delay offset rather than an impulse train. So, in essence, we will have only one pulse per symbol in the time domain located at its delay. This is nothing but a TDM system. Additionally, $\tau_p = \infty$ leads to $\nu_p = 0$, i.e., the Doppler will vanish. Similarly, for $\nu_p = \infty$, we will get only one impulse in the frequency domain with a Doppler offset rather than an impulse train. So, in essence, we will have only one pulse per symbol in the frequency domain located at its Doppler. This is nothing but an FDM system. Additionally, $\nu_p = \infty$ leads to $\tau_p = 0$, i.e., the delay will vanish. Impact of τ_p and ν_p is shown in Fig. 6.12.

6.3 DZT Definition

Zak transformation is an operation that takes a one-variable function as input and produces a two-variable function. Zak transform is a key tool to analyze the delay-Doppler signal processing. Here, a one-dimensional time domain will be the input, whereas the output is a function of delay and Doppler. Depending on the nature of delay and Doppler variables, i.e., continuous or discrete, different variants of Zak transform exist as follows:

Continuous Zak Transform: Let $x(t)$ be a complex-time continuous signal and with τ_p being the delay period, then ZT defined on continuous time and frequency is given by

$$Z_x(\tau,\nu) = \sqrt{\tau_p}\sum_{n=-\infty}^{\infty} x(\tau + nT)\, e^{-j2\pi n\nu\tau_p}, \text{ where } -\infty < \tau < \infty \text{ and } -\infty < \nu < \infty. \tag{6.5}$$

Discrete-time Zak Transform: To obtain the discrete-time Zak transform representation of a signal $x(q)$, the process begins by sampling a continuous signal $x(t)$ at intervals of T/M, resulting in the discretized signal $x(q)$, where $t = qT/M,\quad q = 0, 1, \ldots$. The Zak transform then maps this discrete signal onto a 2D representation, where the delay (time) axis is discrete while the Doppler(frequency) axis remains continuous. The discrete-time Zak transform representation [10] of signal $x(q)$ is given as

$$\tilde{Z}_x(l,\nu) = \sum_{n=-\infty}^{\infty} x(l + nM)\, e^{-j2\pi n\nu}, \text{ where } 0 \le l \le M-1 \text{ and } -\infty < \nu < \infty. \tag{6.6}$$

Discrete Zak Transform: From an implementation perspective, the Zak transform is intended to map a signal onto a 2D grid in discrete delay and discrete Doppler. This mapping is achieved using the Discrete Zak Transform (DZT). Mathematically, it can be represented as follows:

$$Z_x(l,k) = \frac{1}{\sqrt{N}}\sum_{n=0}^{N-1} x(l + nM)\, e^{-j2\pi \frac{n}{N}k} \text{ where } 0 \le l \le M-1 \text{ and } 0 \le k \le N-1. \tag{6.7}$$

The above equation reflects that the processing of the DZT is similar to the Fourier transform of the sub-sampled sequence. If we sample the discrete-time Zak transform (DTZT) at $\nu = \frac{k}{N}$, then we can directly get the DZT, i.e.:

$$Z_x(l,k) = \frac{1}{\sqrt{N}}\tilde{Z}_x\left(l, \frac{k}{N}\right) \tag{6.8}$$

Let us consider discrete-time samples $x(q)$ of length MN. Now, this will be mapped to an $M \times N$ 2D delay-Doppler grid of M bins along the delay and N bins along the Doppler. For a given length of $x(q)$, there will be many possible values of MN for which the product of M and N will be equal to the length of the sequence. Hence, for a given time signal, there will be infinitely many possible Zak representations. But to make this mapping unique, a property called quasi-periodicity was invoked, which will be discussed in the next section. Before going to that, let us consider some test cases of DZT.

Consider the equation

$$Z_x\left(l,k\right)=\frac{1}{\sqrt{N}}\sum_{n=0}^{N-1}x\left(l+nM\right)e^{-j2\pi\frac{n}{N}k}$$

for some specific values of N and M.

Case-1: For $N=1$

$$Z_x\left(l,k\right)=x\left(l\right)$$

It implies that the Zak transform is nothing but the same sequence itself.

Case-2: For $M=1$

$$Z_x\left(l,k\right)=\frac{1}{\sqrt{N}}\sum_{n=0}^{N-1}x\left(l+n\right)e^{-j2\pi\frac{n}{N}k}$$

Further, if $l=0$ then

$$Z_x\left(0,k\right)=\frac{1}{\sqrt{N}}\sum_{n=0}^{N-1}x\left(n\right)e^{-j2\pi\frac{n}{N}k},$$

i.e., the Zak transform is simply the Fourier transform (FT) of the input signal.

Inverse Discrete Zak Transform: IDZT is given by

$$x(q)=x(l+nM)=\frac{1}{\sqrt{N}}\sum_{u=0}^{N-1}Z_x(l,u)e^{j2\pi un/N} \tag{6.9}$$

The above equation can be interpreted as, taking the IDFT along the columns (Doppler axis) and then taking the samples row-wise which is equivalent to performing upsampling by a factor of M and then staggering with $\frac{T}{M}$.

6.3.1 Properties of DZT

In this section, we discuss about some important properties of the DZT. Let us

$$x(t) \xrightarrow{\text{Zak}} Z_x(\tau, \nu)$$

$$y(t) \xrightarrow{\text{Zak}} Z_y(\tau, \nu)$$

$$z(t) \xrightarrow{\text{Zak}} Z_z(\tau, \nu)$$

1. **Linearity:** Consider linear combination of $x(t)$ and $y(t)$ with a and b which are some constants, then the DZT is given by

$$ax(t) + by(t) \leftrightarrow aZ_x(\tau, \nu) + bZ_y(\tau, \nu).$$

2. **Doppler (frequency) shift:** A shift in the spectrum of $x(t)$ will correspond to an equal shift in DZT along the Doppler axis. DZT is periodic along the Doppler axis with a period of ν_p. This is because of the fact that the DZT is the Fourier transform of the sub-sampled sequence. As the Fourier transform is periodic in frequency, so is the DZT, i.e.:

$$e^{j2\pi\nu_0 t}x(t) = e^{j2\pi\nu_0\tau}Z_x(\tau, \nu - \nu_0).$$

   If this shift is an integer multiple of the Doppler period then

$$Z_x\left(\tau, \nu + m\nu_p\right) = Z_x(\tau, \nu) \quad \text{where } m \text{ is some constant.}$$

3. **Delay (time) shift:** Time shift of $x(t)$ will simply reflect the same shift in its DZT along the delay axis. Moreover, DZT is quasi-periodic in delay with a period of τ_p, i.e., in the interval of $\{n\tau_p, (n+1)\tau_p\}$. It is periodic till a phase shift of $e^{j2\pi n\tau_p\nu}$. This property plays a crucial role in defining the uniqueness of the DZT and hence the inverse DZT exists.

$$x(t + \tau_0) \leftrightarrow Z_x(\tau + \tau_0, \nu)$$

   If τ_0 is integer multiple of delay period then

$$Z_x\left(\tau + n\tau_p, \nu\right) = e^{j2\pi\nu n\tau_p}Z_x(\tau, \nu)$$

4. **Multiplication in time:** From Fourier transform perspective, multiplication in time leads to the convolution in frequency. Hence in the DZT, this manifests as convolution in the Doppler variable.

$$x(t).y(t) \leftrightarrow \sqrt{\tau_p} \int_0^{\nu_p} Z_x(\tau, s)\, Z_y(\tau, s - \nu)\, ds$$

5. **Convolution in time:** Convolution in time corresponds to convolution in the delay domain of the DZT, a time-type variable.

$$x(t) \star y(t) \leftrightarrow \frac{1}{\sqrt{\tau_p}} \int_0^{\tau_p} Z_x(\alpha, \nu)\, Z_y(\tau - \alpha, \nu)\, d\alpha$$

6. **Conjugation property:** Conjugation of $x(t)$ corresponds to the Doppler reversal in its conjugate of DZT.

$$x^\star(t) \leftrightarrow Z_x^\star(\tau, -\nu)$$

7. **Symmetric property:** Symmetry of DZT depends on the nature of $x(t)$, i.e., if $x(t)$ is even, then its DZT is also even, and if it is odd, its DZT is also odd.

$$Z_x(\tau, \nu) = \begin{cases} Z_x^\star(-\tau, -\nu) & \text{if } x(t) \text{ is even} \\ -Z_x^\star(-\tau, -\nu) & \text{if } x(t) \text{ is odd} \end{cases}$$

8. **Connection with Ambiguity function:** A cross-ambiguity function of $x(t)$ and $y(t)$ with respect to τ and ν is

$$A_{x,y}(\tau, \nu) = \int_{-\infty}^{\infty} x(t) y^*(t - \tau) e^{-j2\pi\nu(t-\tau)} dt.$$

Now the relation between Zak transform and the ambiguity function is given by

$$Z_x(\tau, \nu) Z_y^*(\tau, \nu) = \sum_{n=-\infty}^{\infty} \sum_{m=-\infty}^{\infty} A_{x,y}(nT, \frac{m}{T}) e^{-j2\pi n\nu T} e^{j2\pi \frac{m}{T}\tau}.$$

Proof of all these properties can be derived from the fundamental equation of the DZT mentioned above [5, 10].

6.3.2 Zak Basis Functions

Till now we have analyzed the DZT-based OTFS, i.e., in discrete domain. In order to see the features of basis functions, let us consider the continuous Zak transform by considering Eq. (6.5) with delay period, i.e., $\tau_p = T$.

$$Z_x(\tau, \nu) = \sqrt{T} \sum_{n=-\infty}^{\infty} x(\tau + nT) e^{-j2\pi n\nu T}, \text{ where } -\infty < \tau < \infty \text{ and } -\infty < \nu < \infty. \tag{6.10}$$

Let us recall that $Z_x(\tau + nT, \nu) = e^{j2\pi\nu nT} Z_x(\tau, \nu)$ and $Z_x(\tau, \nu + m\nu_p) = Z_x(\tau, \nu)$ are the quasi-periodic and periodicity. In addition, the shifts will be directly reflected in the Zak transform, i.e., if $y(t) = x(t - \tau_0)e^{j2\pi\nu_0(t-\tau_0)}$ then

$$Z_y(\tau, \nu) = e^{j2\pi\nu_0(\tau-\tau_0)} Z_x(\tau - \tau_0, \nu - \nu_0). \tag{6.11}$$

Consider $x_{\text{DD}}(\tau, \nu)$ as the transmit signal in the delay-Doppler domain, and let $X_{\text{DD}}(k, l)$ be the information point transmitted on the kth Doppler and lth delay. $x_{\text{DD}}(\tau, \nu)$ is a linearly modulated signal, i.e., $x_{\text{DD}}(\tau, \nu)$ is constructed by linear combinations of a set of MN basis functions weighted by their corresponding information point given as

$$x_{DD}(\tau, \nu) = \sum_{k=0}^{N-1} \sum_{l=0}^{M-1} X_{DD}(k, l)\beta_{k,l}(\tau, \nu), \tag{6.12}$$

where $\beta_{k,l}(\tau, \nu)$ denotes the basis function in the DD domain located at delay l and Doppler k. Let $\tau_l = \frac{lT}{M}$ and $\nu_k = \frac{k}{NT}$. We will be having MN such basis functions defined by set $\mathcal{B}_{\text{DD}}$

$$\mathcal{B}_{\text{DD}} = \left\{\beta_{k,l}(\tau, \nu)\right\} \text{ for } l = 0, 1..M - 1 \text{ and } k = 0, 1..N - 1. \tag{6.13}$$

Each element in this set will be an orthogonal basis function centered at (k, l) with energy spread in the area defined by the inverse of the total bandwidth ($\frac{1}{M\Delta f}$) and the total time duration ($\frac{1}{NT}$) along the delay and Doppler, respectively. Hence, any two adjacent points should be separated by at least $\frac{1}{M\Delta f}$ along the delay and $\frac{1}{NT}$ along the Doppler. Only then will interference be minimized. Ideally, these basis functions will be 2D impulses located at (k, l) for the fundamental rectangle as the area of interest as

$$\beta_{k,l}(\tau, \nu) = \delta(\tau - \tau_l)\delta(\nu - \nu_k) \quad \text{for } 0 < \tau < T \text{ and } 0 < \nu < \Delta f. \tag{6.14}$$

The duration (area in the DD) of an ideal impulse will be zero and hence forces $\frac{T}{M}$ and $\frac{1}{N\Delta T}$ to be zero. For this, M and N should be ∞, implying that we need infinite time and bandwidth. However, practically such a time-frequency (TF) signal cannot be realized. Therefore, we will go for time and band-limited signals by applying time and frequency windowing. Let $w_t(t)$ and $w_f(f)$ be the time and frequency windows defined by $w_t(t) = 0$ for $t \notin [0, NT)$ and $w_f(f) = 0$ for $f \notin [0, M\Delta f)$, where $W_f(f)$ is the frequency spectrum of $w_f(t)$. Thus, time and band-limited basis functions will be

$$\beta_{k,l}^{w_t,w_f}(\tau, \nu) = \left(\beta_{k,l}(\tau, \nu)w_t(t)\right) \star w_f(t) \tag{6.15}$$

with $\beta_{k,l}(\tau, \nu)$ defined in Eq. (6.14). Here, the duration of these window functions in time and frequency will define the energy spread of the basis functions in the delay-Doppler domain.

Going back to the basis functions analysis (need not be impulses), $\mathcal{B}_{DD}$ can be completely constructed by an atom basic function located at (0, 0) by invoking the properties of the Zak transform. Mathematically,

$$\beta_{k,l}(\tau, \nu) = \beta_{0,0}\left(\tau - \tau_l, \nu - \nu_k\right) e^{j2\pi\nu_k(\tau-\tau_l)} \tag{6.16}$$

for $-\infty < \tau < \infty$ and $-\infty < \nu < \infty$. The above relation can be justified from the periodic and the quasi-periodic properties. To get the intuition of Eq. (6.16), take the inverse Zak to see their time-domain relation. Then we get

$$\beta_{k,l}(t) = \beta_{0,0}(t - \tau_l)e^{j2\pi\nu_k(t-\tau_l)}, \tag{6.17}$$

which indicates that the shift of the atom pulse along the delay and Doppler impacts its time-domain signal with a shift equal to the delay and a phase rotation corresponding to Doppler. The set of basis functions that satisfy (6.16) are called TF consistent basis, which has some interesting properties that can be exploited. One such property is the relation between the DD domain matched filtering and the self-ambiguity function of the atom basis, by which we can conclude that the atom pulse is a prototype pulse as

$$\int_0^T \int_0^{\Delta f} \beta_{k_1,l_1}(\tau, \nu)\beta^*_{k_2,l_2}(\tau, \nu)d\tau d\nu = A_\beta(\tau_1 - \tau_2, \nu_1 - \nu_2)e^{j2\pi\nu_2(\tau_1-\tau_2)}, \tag{6.18}$$

where $A_\beta(\tau_1 - \tau_2, \nu_1 - \nu_2)$ denotes the self-ambiguity function of $\beta_{0,0}(t)$ with delay and Doppler offset of $\tau_1 - \tau_2$ and $\nu_1 - \nu_2$.

As far as the DD signal is obeying the quasi-periodicity property, it can be completely characterized by a pulse in the fundamental region of the DD domain. Let $a(\tau, \nu)$ be that pulse. Using this, basis function located at (0, 0) can be written as quasi-periodic extension given by

$$\beta_{0,0}(\tau, \nu) = \sum_{n=-\infty}^{\infty} \sum_{m=-\infty}^{\infty} a(\tau - nT, \nu - m\Delta f)e^{j2\pi n\nu T} \tag{6.19}$$

and any arbitrary basis function with τ_l and ν_k as the delay and Doppler offsets then

$$\beta_{k,l}(\tau, \nu) = \sum_{n=-\infty}^{\infty} \sum_{m=-\infty}^{\infty} \underbrace{a(\tau - nT - \tau_l, \nu - m\Delta f - \nu_k)e^{j2\pi\nu_k(\tau-\tau_l)}}_{\text{twist and shift}} \underbrace{e^{j2\pi n(\nu-\nu_k)T}}_{\text{quasi-periodicity}}. \tag{6.20}$$

The above equation clarifies Zak transform is apt for a TF consistent DD basis functions. If we observe Eq. (6.20), quasi-periodicity and twisted shift are analogous

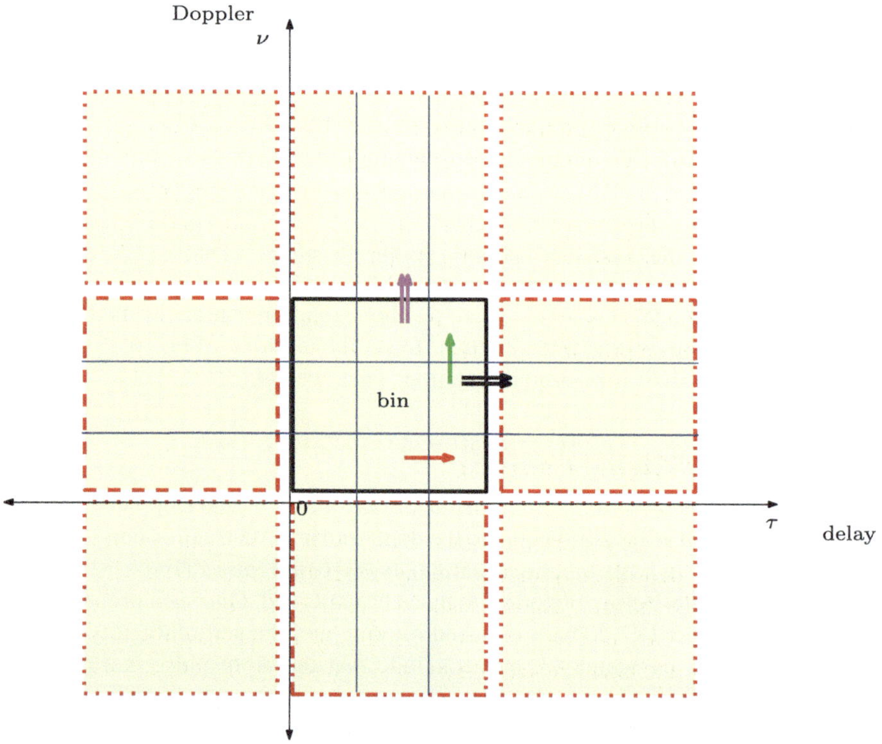

Fig. 6.13 Quasi-periodicity and twisted shift of DD pulses $\beta_{k,l}(\tau, \nu)$

to the property of Zak transform and twisted convolution, respectively, as the interaction of the DD basis with the channel is twisted convolution. Quasi-periodicity indicates the global features of DD domain which is built by infinite replicas of size of the fundamental rectangle. Twist and shift correspond to the local characters of the DD domain, i.e., within the fundamental rectangle.

Figure 6.13 diagrammatically represents the above explanation. We are defining each blue box as a bin and each red box as a grid. The black grid is the fundamental rectangle. Green color arrow ($\uparrow$) indicates the shift along Doppler bins within the fundamental grid (locally) which has the linear shift. Red arrow ($\rightarrow$) indicates the shift along delay bins within the fundamental grid (locally), which is twisted shifted. Violet arrow ($\uparrow\uparrow$) indicates the shift into the Doppler grid (globally) which is periodic. Black arrow ($\rightrightarrows$) indicates shifts along the delay grids (globally), which are quasi-periodic. The time-domain waveform of (6.20) will be

$$\beta_{k,l}(t) = \sqrt{T} e^{j2\pi \nu_k (t-\tau_l)} \sum_{n=-\infty}^{\infty} \int_{-\infty}^{\infty} a(t - \tau_l - nT, \nu) e^{j2\pi n\nu T} d\nu, \quad (6.21)$$

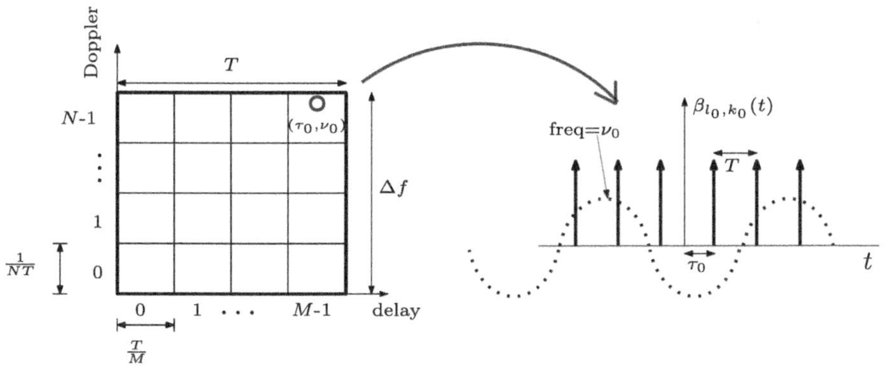

Fig. 6.14 DD domain signal to time-domain signal *pulseones*

and the frequency domain waveform will be

$$\beta_{k,l}(f) = \frac{1}{\sqrt{T}} e^{-j2\pi f \tau_l} \sum_{m=-\infty}^{\infty} \int_{-\infty}^{\infty} a(\tau, f - \nu_k - m\Delta f) e^{-j2\pi f \tau} d\tau. \tag{6.22}$$

Implications in TF domain: From Eqs. (6.21) and (6.22), we can see that the time and frequency signals are a train of pulses separated by T along the time and Δf along the frequency domain. Features like twisted shifts are converted into tones. The structures of Eqs. (6.21) and (6.22) are called *pulseones*, as shown in Fig. 6.14. If we let $T \to \infty$, then the *pulseones* are just separated by their delay, leading to a TDM signal. If we consider $T \to 0$, then the pulses will be separated by Doppler shifts only, resulting in an FDM signal.

6.4 OTFS with DZT

In this section, we will present DZT-based OTFS implementation. Unlike the direct step OTFS where the DD symbols will be converted into the TF and then to the time domain by using multi-carrier modulation, DZT converts the DD domain information into the time domain directly.

DD Grid Formulation: Consider a communication system with total available time of NT and a bandwidth of $M\Delta f$ which is critically sampled, i.e., $T = \frac{1}{\Delta f}$. Let $x_{\mathrm{DD}} \in \mathbb{C}^{MN \times 1}$ be the information symbol vector that needs to be transmitted, where each element will be drawn from a constellation set. X_{DD} be the delay-Doppler grid defined as $x_{\mathrm{DD}} = \mathrm{vec}(\mathrm{X}_{\mathrm{DD}})$ with M bins in the delay axis and N bins in the Doppler axis as shown in Fig. 6.15. $X_{\mathrm{DD}}\,(k, l)$ denotes a sample located on the kth Doppler

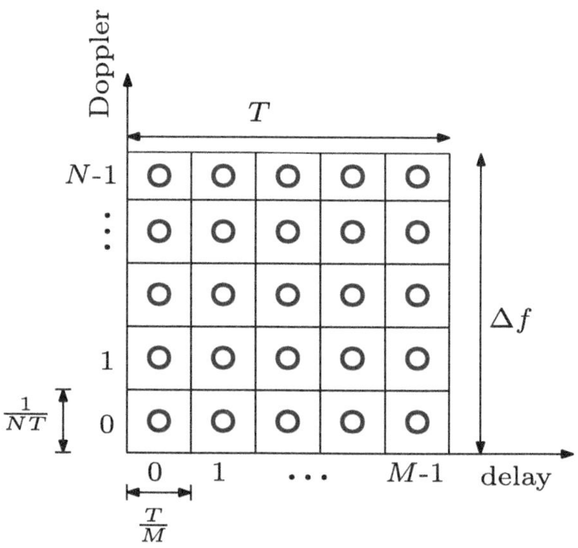

Fig. 6.15 Delay and Doppler grid for OTFS system with $M = 5$ and $N = 5$

and lth delay bin. The delay axis is sampled at a uniform interval of $\frac{1}{M\Delta f}$ which is referred as the delay resolution and the Doppler axis with $\frac{1}{NT}$ referred as Doppler resolution.

System Model: OTFS multiplexes MN information symbols on the fundamental rectangle in the Zak domain, as shown in Fig. 6.16. IDZT maps these information symbols into the time-domain sequence [11, 12]. To preserve the properties of the DZT, it is ensured that the transmitted sequence should be MN periodic in the time domain. To achieve this, a cyclic prefix (CP) of length L_{cp} is added by copying the last L_{cp} samples and then appending them at the beginning of the sequence. Such a CP arrangement guarantees that the sequence $x(q)$ will be MN periodic in the range of $-L_{\text{cp}} \leq q \leq MN - 1$ after CP removal at the receiver. This is sufficient to maintain the properties of DZT over the channel with a maximum delay spread less than L_{cp}.

$$x(q) = \frac{1}{\sqrt{N}} \sum_{n=0}^{N-1} Z_x(k, q \bmod M) \, e^{j2\pi \frac{\lfloor q/M \rfloor}{N} n}, \tag{6.23}$$

where $\lfloor(\cdot)\rfloor$ is the floor function. In addition to that, CP allows the linear convolution of the channel into the circular convolution. Then pulse shaping is performed by considering a transmit filter $g_{\text{tx}}(t)$ as shown in Fig. 6.16. So the transmit filter will be

$$x(t) = \sum_{q=0}^{MN+L_{\text{cp}}-1} x\left(q - L_{\text{cp}}\right) g_{\text{tx}}\left(t - q\frac{T}{M}\right). \tag{6.24}$$

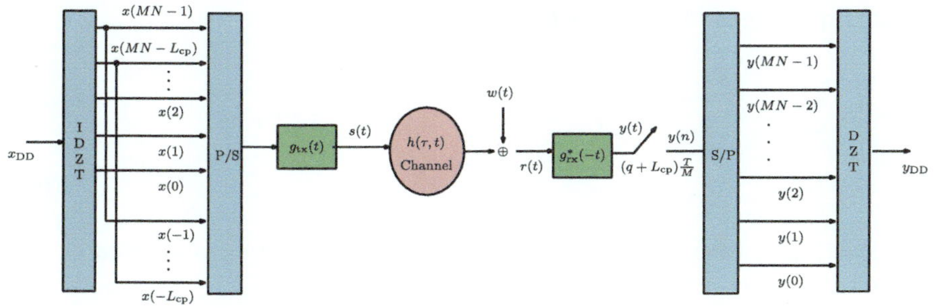

Fig. 6.16 DZT-OTFS system model

Now this transmit signal is passed through a doubly dispersive channel modeled by P discrete paths, where each path is characterized by path delay τ_p, path Doppler as ν_p, and the path gain as h_p. The delay-Doppler spreading function is given as

$$h(\tau, \nu) = \sum_{p=0}^{P-1} h_p \delta\left(\tau - \tau_p\right) \delta\left(\nu - \nu_p\right). \tag{6.25}$$

The received signal is given as

$$r(t) = \int_{-\infty}^{\infty} \int_{-\infty}^{\infty} h(\tau, \nu)\, x(t - \tau)\, e^{j2\pi \nu t} d\tau d\nu + w(t), \tag{6.26}$$

where the noise $w(t)$ is white Gaussian noise with power spectral density (PSD) equal to N_0. Now by substituting Eq. (6.24) in Eq. (6.25), we get

$$r(t) = \sum_{p=0}^{P-1} h_p x\left(t - \tau_p\right) e^{j2\pi \nu_p t} + w(t). \tag{6.27}$$

Here the Doppler shift ν_p is given as $v_p \frac{f_c}{c}$ where v_p is the velocity corresponding to the pth path, f_c is the carrier frequency, and c is the velocity of light. Consider Eq. (6.27), the received signal is the scaled, translated and modulated replicas of the transmitted signal. By substituting Eq. (6.24) in Eq. (6.27), we get

$$r(t) = \sum_{p=0}^{P-1} h_p \sum_{q=0}^{MN+L_{\text{cp}}-1} x\left(q - L_{\text{cp}}\right) g_{\text{tx}}\left(t - q\frac{T}{M} - \tau_p\right) e^{j2\pi \nu_p t} + w(t). \tag{6.28}$$

At the receiver, matched filtering operation is done by a receive filter g_{rx}, i.e.:

$$y(t) = \int r(\tau)\, g_{\text{rx}}^{*}(t - \tau)\, d\tau. \tag{6.29}$$

Using Eq. (6.28) in (6.29), we get

$$y(t) = \sum_{p=0}^{P-1} h_p \sum_{q=0}^{MN+L_{cp}-1} x\left(q - L_{cp}\right) \int_{-\infty}^{\infty} g_{tx}\left(\tau - q\frac{T}{M} - \tau_p\right) e^{j2\pi \nu_p \tau} g_{rx}^{*}(\tau - t)\, d\tau$$
$$+ w(t). \tag{6.30}$$

Considering that the pulse bandwidth is much higher than the maximum Doppler spread and $g_{tx}(t) = g_{rx}(t) = p(t)$ then

$$\int_{-\infty}^{\infty} p\left(\tau - q\frac{T}{M} - \tau_p\right) e^{j2\pi \nu_p \tau} p^{*}(\tau - t)\, d\tau \;\approx\; e^{j2\pi \nu_p \left(q\frac{T}{M} + \tau_p\right)} g\left(t - \tau_p - q\frac{T}{M}\right).$$

With this approximation, Eq. (6.30) can be written as

$$y(t) \;\approx\; \sum_{p=0}^{P-1} h_p \sum_{q=0}^{MN-L_{cp}-1} x\left(q - L_{cp}\right) e^{j2\pi \nu_p \left(q\frac{T}{M} + \tau_p\right)} g\left(t - q\frac{T}{M} - \tau_p\right) + w(t). \tag{6.31}$$

Now this matched filter output is sampled at every multiples of $\frac{T}{M}$ and by discarding the first L_{cp} samples results in

$$y(m) = \sum_{p=0}^{P-1} h_p \sum_{q=0}^{MN-1} x(q)\, g[(m - q)\frac{T}{M} - \tau_p] e^{j2\pi \nu_p \left(\tau_p + q\frac{T}{M}\right)} + w(t). \tag{6.32}$$

After simplification using $\tau_p = \frac{l_p T}{M}$ and $\nu_p = \frac{k_p}{NT}$, we get

$$y(m) = \sum_{p=0}^{P-1} h_p e^{j2\pi \tau_p \nu_p} \sum_{q=0}^{MN-1} x(q)\, e_p(q)\, g_p(m - q) + w(m), \tag{6.33}$$

where $e_p = e^{j2\pi \frac{k_p}{MN} p}$ with $q = 0, 1, 2, \ldots, MN - 1$ is the sequence that captures the effect of the Doppler of the pth path and $g_p(q) = g\left(q - l_p\right)$ is the sequence that captures the delay due to the pth path. Let us consider the effective channel coefficient $h_p^{'} = h_p e^{j2\pi \tau_p \nu_p}$. The received sample at mth instant due to the pth channel will be

$$y_p(m) = \sum_{q=0}^{MN-1} x(q)\, e_p(q)\, g_p(m - q). \tag{6.34}$$

Hence the received sample can be written in the linear combinations as

$$y(m) = \sum_{p=0}^{P-1} h'_p y_p(m) + w(m). \tag{6.35}$$

At the end, the receiver performs the DZT on $y(m)$ to get the received delay-Doppler information.

Delay-Doppler I/O relation: Till now we have studied the DZT-based OTFS based on fundamental communication point of view. Now let us try to have delay-Doppler relation. Consider Eq. (6.35) and from the linearity property of DZT, we can write

$$Z_y[l,k] = \sum_{p=0}^{P-1} h_p Z_{y_p}[l,k] + Z_w[l,k], \tag{6.36}$$

where $Z_w[l,k]$ represent the delay-Doppler noise with its entries from zero-mean Gaussian random variables and $Z_{y_p}[l,k]$ be the DZT of the sequence y_p. Let us write the effective input–output relation by considering the effect of each multipath with their DZT, i.e.:

$$Z_{y_p}[l,k] = \sum_{m=0}^{M-1} \left(\sum_{n=0}^{N-1} Z_x[m,n] Z_{\nu_p}[m,k-n] \right) Z_{\tau_p}[l-m,k], \tag{6.37}$$

where Z_{τ_p} and Z_{ν_p} are the delay and Doppler spreading function where

$$Z_{\nu_p}[l,k] = \frac{1}{\sqrt{N}} e^{j2\pi \frac{k_p}{MN} l} e^{-j\pi \frac{N-1}{N}(k-k_p)} \frac{\sin(\pi(k-k_p))}{\sin(\frac{\pi}{N})(k-k_p)} \tag{6.38}$$

and hence the received samples due to the pth path will be

$$Z_{y_p}[l,k] = e^{j2\pi \frac{k_p}{MN}(l-l_p)} Z_x[l-l_p, k-k_p]. \tag{6.39}$$

Final delay-Doppler input–output relation will be given by

$$Z_y[l,k] = \sum_{p=0}^{P-1} h_p e^{j2\pi \frac{k_p}{MN}(l-l_p)} Z_x[l-l_p, k-k_p] + Z_w[l,k]. \tag{6.40}$$

6.4.1 Matrix Representation of DZT-OTFS

In this section, we will present the matrix form of the above-discussed DZT-OTFS. As mentioned earlier, $X_{\mathrm{DD}} \in \mathbb{C}^{MN}$ is the delay-Doppler matrix that holds the information symbols. Each symbol is drawn from a constellation $\mathbb{A} = a_1, a_2, \ldots, a_Z$ of size Z. Let X_{DT} be the delay-time matrix obtained by taking an N-point IFFT along the Doppler as

$$X_{\text{DT}} = X_{\text{DD}} F_N^H, \tag{6.41}$$

where F_N^H represents N-point inverse FFT matrix given by

$$F_N = \frac{1}{\sqrt{N}} \{\omega_N^{kl}\}_{k,l=0}^{N-1}$$

with $\omega_N = e^{-j2\pi/N}$. Each column of this matrix is DFT orthonormal basis.

A transmit pulse shaping is applied on the delay-time matrix by considering $g_{\text{tx}}(t)$ as the pulse shaping waveform. $G_{\text{tx}} \in \mathbb{C}^{MM}$ is constructed by taking the samples of $g_{\text{tx}}(t)$ spaced equally with an interval of $\frac{T}{M}$, i.e., $G_{\text{tx}} = \text{diag}\,[g_{\text{tx}}(0), g_{\text{tx}}(T/M)\ldots.. g_{\text{tx}}((M-1)T/M)]$. For rectangle pulse shaping, G_{tx} will be identity matrix. By vectorizing (6.41) including the pulse shaping, we get an MN length discrete time-domain sequence with each entry of $\mathbf{s}$ which is equivalent to Eq. (6.4).

$$\mathbf{s} = \text{vec}(G_{\text{tx}} X_{\text{DT}}). \tag{6.42}$$

At the receiver we will receive MN length discrete-time sequence $r(q), q = 0, 1, \ldots MN-1 \in \mathbb{C}^{MN\times 1}$ after interacting with the double dispersive channel. This vector will be converted into the delay-time matrix by considering the samples transmitted at a duration of T sec as one sub-vector of length M along the delay dimension of Y_{DT}. This is done by devectorizing. The received delay-time matrix including receiver pulse shaping G_{rx} will be

$$Y_{\text{DT}} = G_{\text{rx}}(\text{vec}^{-1}(\mathbf{r})_{M,N}), \tag{6.43}$$

with $G_{\text{rx}} = \text{diag}\,[g_{\text{rx}}(0), g_{\text{rx}}(T/M)\ldots..g_{\text{rx}}((M-1)T/M)]$. Received delay-Doppler matrix Y_{DD} is obtained by N point FFT as

$$Y_{\text{DD}} = Y_{\text{DT}} F_N. \tag{6.44}$$

6.4.2 Comparison of Precoded and DZT-Based OTFS

Two popular implementations of OTFS are SFFT based and DZT based. Let us see a comparison between these two. Both are designed to handle high-mobility scenarios by modulating the information in the delay-Doppler domain. For rectangular pulse shaping with integer delay and Doppler case, both of these implementations have the same input–output relation.

OFDM-based OTFS is designed to be implemented on the already existing OFDM modules. Due to this, the entire signal processing, i.e., transmit and receive pulse shaping, multiple access, is done in the TF domain. However, for DZT-based OTFS, all the mentioned processing will be done in the DD domain, which will be more effective compared to OFDM-based OTFS systems. OFDM-based OTFS includes ISFFT-based processing, which involves high complexity. In contrast, DZT uses only M number of N-point FFT/IFFT modules.

Another reason for the robustness of DZT-based OTFS is that, in OFDM-based, we try to find pulses that are perfectly localized in the TF plane. However, this is not possible due to Heisenberg's uncertainty principle. Besides, such a perfectly localized pulse is possible in the DD plane because delay and Doppler are not Fourier transform pairs.

6.5 Conclusion

In this chapter, a brief overview of DZT-based OTFS is presented. Initially, the concepts of radar and communication are reviewed and then their connection with the OTFS was discussed. The governing principles of Zak transform such as quasi-periodicity and crystallization are also highlighted. Later a complete end-to-end input–output relation on OTFS in both TF and DD domains was derived along with their matrix representations. At last, a comparison of OFDM-based OTFS with DZT-based OTFS was highlighted.

References

1. Xu C, Zhang X, Petropoulos P, Sugiura S, Maunder RG, Yang L-L, Wang Z, Yuan J, Haas H, Hanzo L (2023) Optical OTFS is capable of improving the bandwidth-, power-and energy-efficiency of optical OFDM. IEEE Trans Commun
2. Matz G, Bolcskei H, Hlawatsch F (2013) Time-frequency foundations of communications: concepts and tools. IEEE Signal Proc Maga 30(6):87–96
3. Hadani R, Rakib S, Tsatsanis M, Monk A, Goldsmith AJ, Molisch AF, Calderbank R (2017) Orthogonal time frequency space modulation. In: 2017 IEEE wireless communications and networking conference (WCNC). IEEE, pp 1–6
4. Mohammed SK, Hadani R, Chockalingam A, Calderbank R (2022) OTFS-a mathematical foundation for communication and radar sensing in the delay-Doppler domain. IEEE BITS the Inf Theory Mag 2(2):36–55
5. Mohammed SK (2021) Derivation of OTFS modulation from first principles. IEEE Trans Veh Technol 70(8):7619–7636
6. Bondre AS, Richmond CD (2022) Dual-use of OTFS architecture for pulse Doppler radar processing. In: 2022 IEEE radar conference (RadarConf22). IEEE, pp 1–6
7. Lampel F, Alvarado A, Willems FM (2021) Orthogonal time frequency space modulation: a discrete Zak transform approach. arXiv:2106.12828
8. Wei Z, Li S, Yuan W, Schober R, Caire G (2022) Orthogonal time frequency space modulation-part i: fundamentals and challenges ahead. IEEE Commun Lett 27(1):4–8
9. Wang B, Luo W, Zhu J, She X, Chen P (2023) On OTFS and OFDM Radar signal design based on the ambiguity function analysis. In: 2023 IEEE 98th vehicular technology conference (VTC2023-Fall). IEEE, pp 1–5

10. Bolcskei H, Hlawatsch F (1997) Discrete Zak transforms, polyphase transforms, and applications. IEEE Trans Signal Proc 45(4):851–866
11. Yogesh V, Mattu SR, Chockalingam A (2024) Low-complexity delay-Doppler channel estimation in discrete Zak transform based OTFS. IEEE Commun Lett
12. Bhat VS, Harshavardhan G, Chockalingam A (2022) Input-output relation and performance of RIS-aided OTFS with fractional delay-Doppler. IEEE Commun Lett 27(1):337–341

Chapter 7
Visible Light Communication: An Emerging Technology

Anupma Sharma, Prakhar Keshari, Vimal Bhatia, and Ondrej Krejcar

Abstract This chapter provides an overview of the fundamentals of visible light communication (VLC) by thoroughly exploring its technologies, challenges and future prospects. Due to its enormous bandwidth and vast license-free spectrum, the visible light band can be harnessed for data transmission. Owing to this, VLC has emerged as a promising paradigm for next-generation wireless communication that utilizes visible light as a medium for high-speed communication. However, the performance of VLC systems is degraded due to various factors, such as the non-linear effects of light-emitting diodes (LEDs), noise, user mobility, blind spots and atmospheric turbulence, among many others. To mitigate such distortions in realistic VLC systems, different algorithms and architectures for practical deployment are proposed in the literature, which are discussed in detail in the upcoming sections. Furthermore, this chapter also discusses the emerging areas, new technologies, challenges and standards for VLC.

Keywords Optical wireless communication · Free space optics · Visible light communication · Light emitting diode · Line-of-sight · Non-line-of-sight

---

A. Sharma (✉) · P. Keshari · V. Bhatia
SaSg, Department of Electrical Engineering, Indian Institute of Technology Indore, Indore, India
e-mail: phd1901102029@iiti.ac.in

P. Keshari
e-mail: mt2202102010@iiti.ac.in

V. Bhatia
e-mail: vbhatia@iiti.ac.in

V. Bhatia · O. Krejcar
Faculty of Informatics and Management, University of Hradec Kralove, Hradec Kralove, Czech Republic
e-mail: ondrej.krejcar@uhk.cz

S. Sharma et al. (eds.), *Next-Generation Wireless Systems*,
Transactions on Computer Systems and Networks,
https://doi.org/10.1007/978-981-96-3185-8_7

7.1 Introduction

In recent years, the exponential growth in communication technology has created a drastic surge in bandwidth and capacity requirements. The growing need for expanded data and multimedia services has led to congestion within the conventionally utilized radio frequency (RF) spectrum. In response to this challenge, optical wireless communication (OWC) systems have emerged as a pivotal solution, capitalizing on distinctive features such as large bandwidth, license-free spectrum, high data rate, easy and quick deployment, and lower power requirements that set them apart.

OWC systems offer an expansive bandwidth, tapping into a license-free spectrum, thereby mitigating the limitations associated with RF congestion. With the ability to deliver high data rates, OWC stands out as a compelling option due to its attributes, such as rapid and straightforward deployment and lower power requirements. These advantages position OWC as a promising alternative to conventional communication methods.

The mechanism of OWC involves the utilization of an optical carrier to convey information from the transmitter to the receiver through an unguided channel, typically free space or the Earth's atmosphere. As illustrated in Fig. 7.1, OWC can be broadly categorized into three main types: visible light communication (VLC), ultraviolet communication (UVC) and free space optics (FSO). UVC harnesses UV radiation for signal transmission, capable of being dispersed and bounced off particles and aerosols suspended in the air. Owing to the atmospheric scattering of UV radiation, UVC has garnered significant interest for its non-line-of-sight capability, omnidirectional communication links and minimal background noise. These attributes position UVC as a prime choice for discreet, secure communication, particularly in military applications. The outdoor systems, commonly referred to as FSO communication, play a crucial role in addressing the challenges posed by RF congestion. FSO communication systems can be more specifically categorized into terrestrial and space optical links. This classification spans a diverse array of applications, including connections between buildings, communication from the ground to satellites, communication from satellites to the ground and even inter-satellite

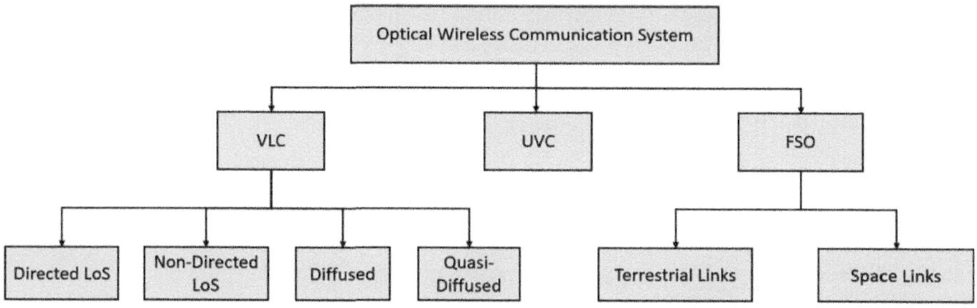

Fig. 7.1 Classification of the optical wireless communication system

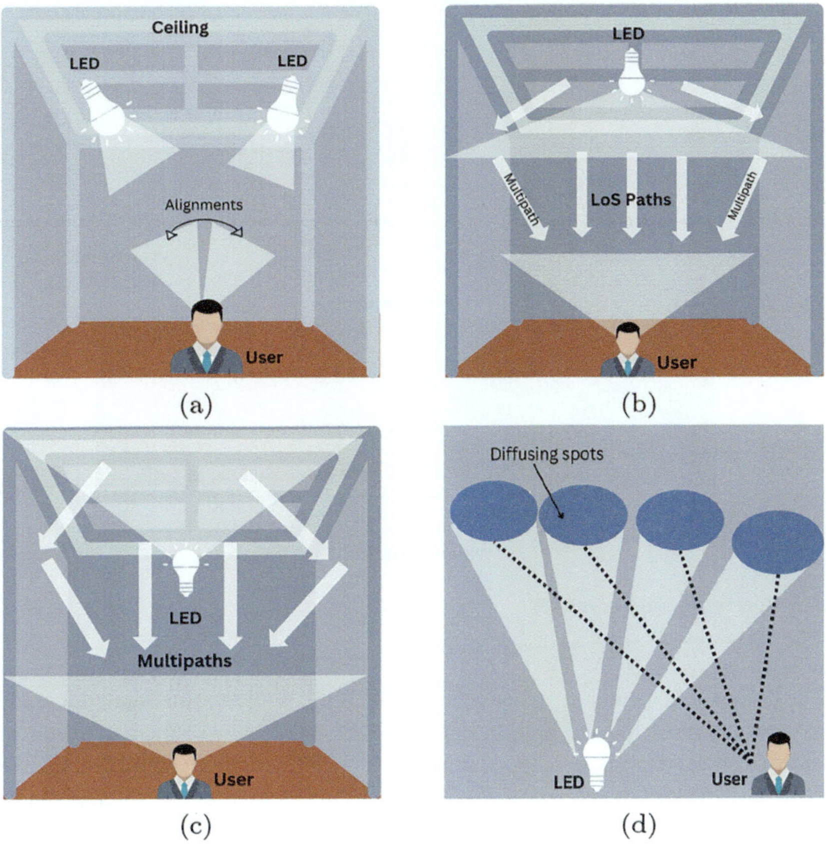

Fig. 7.2 **a** Directed-LoS link; **b** non-directed-LoS link; **c** diffuse link; **d** quasi-diffuse link

communication. This versatility positions FSO as a key player in establishing robust and efficient communication links across various environments and scenarios. The indoor optical systems which include visible light communication systems, are the main focus of this chapter. VLC is categorized into four generic system configurations: directed line-of-sight, non-directed line-of-sight, diffused and quasi-diffused, as illustrated in Fig. 7.2. Figure 7.2a–c illustrates the most common link configurations in VLC. Figure 7.2d displays a quasi-diffuse link utilizing multispot diffusing, initially proposed by Yun and Kavehrad in 1993. This link employs multiple narrow-beam transmitters and an angle diversity receiver consisting of several narrow field-of-view (FoV) detectors directed in different directions. Compared to conventional wide-beam diffuse links (Fig. 7.2c), this system encounters fewer multipaths, resulting in lower path losses and reduced transmission power requirements. However, these advantages come at the cost of increased complexity. While it allows for a higher degree of user mobility compared to line-of-sight (LoS) links (Fig. 7.2a and b), it demands more power under such conditions.

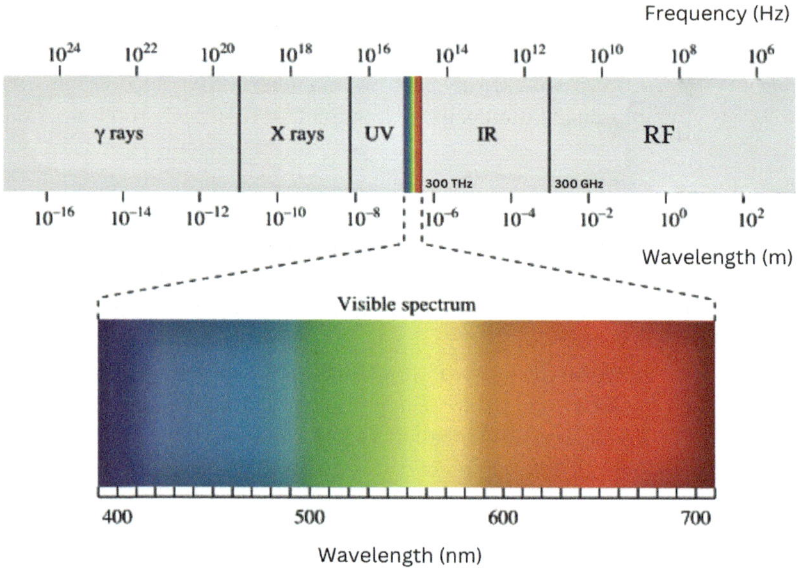

Fig. 7.3 Electromagnetic spectrum

The human eye is most responsive to visible light, with the sun being the predominant natural source of visible light in nature. From the view of the electromagnetic spectrum as shown in Fig. 7.3, the visible spectrum extends from about 390 to 700 nm in terms of wavelength and 430 to 770 THz in terms of frequency, which is ten thousand times higher than radio waves, i.e. 770 THz as compared to a maximum of 300 GHz for radio waves. This suggests that light is capable of transmitting more pulses of data in much less time than radio waves.

Using light rays for transmitting information as a way of communication has been known to mankind for many years. The ancient Romans employed polished metallic plates to reflect sunlight for long-distance signalling. In the 1790s, optical communication systems using semaphore lines were developed. The initial visual telegraph system, created by Claude Chappe in 1792 in France, played a significant role in this historical evolution. Additionally, the US military introduced the Heliograph, a wireless solar telegraph in the early 1800s, utilizing Morse-coded flashes of reflected sunlight through a mirror. Incandescent light bulbs were invented by Thomas Edison in the late 19th century. The bulbs operated by passing an electric current through a tungsten filament, causing it to heat up and emit visible light. The invention of the electric light bulb laid the groundwork for further exploration, ultimately leading to the creation of the signal lamp-a form of an optical communication system devised by Arthur C. W. Aldis for visual signalling. The initial demonstration of information transmission through Visible Light Communication (VLC) occurred in 1880 when Alexander Graham Bell introduced the photophone. This pioneering device facilitated the transmission of voice signals using a focused beam of light. Operating on the principle of modulating sunlight directed onto a mirror through vibrations, the vibrating beam was captured by a detector at the receiving end. The voice signal was then obtained through decoding, akin to the mechanism employed by the telephone with electrical signals. Incandescent bulbs had limitations, such as relatively short lifespans, high energy consumption and the emission of a significant amount of heat. In the mid-20th century, advancements in semiconductor materials began to pave the way for new lighting technologies. The concept of electroluminescence, the emission of light from a material in response to an electric current, was discovered in the 1920s. However, practical light-emitting diode (LED) development took time due to material and technological constraints. The invention and development of LEDs marked a transformative shift from traditional incandescent light bulbs. The development of the first practical LED is attributed to Nick Holonyak Jr., a scientist at General Electric, in 1962. Nick Holonyak Jr. used gallium arsenide phosphide (GaAsP) to create a red LED. LEDs offer several advantages over incandescent bulbs: they use up to 90% less energy, last up to 25 times longer, and have a faster switching rate. During that era, the GaAs semiconductor laser emerged, finding application in optical fibre cables for long-distance communication. The revolutionary advancements in the late 20th century, including the invention of the in-fibre Bragg grating (1990) and optical fibre amplifier (1980), laid the groundwork for a transformative period in the field of telecommunications. In 1970, Corning incorporated successfully developed optical fibres with low attenuation for commercial purposes. Intelligent lighting systems with LEDs have replaced traditional lighting systems. With the introduction of LEDs, indoor communication based on VLC is gaining attention. The invention of LED has again paved the way for new opportunities in the area of VLC.

VLC has become a good complementary technology to RF systems for facilitating high-speed data communication in next-generation wireless systems. It operates by modulating the intensity of LED lamps based on the input signal at an undetectable speed to the human eye. This enables the simultaneous accomplishment of both illumination and communication objectives. Photodiodes or photodetectors (PDs) are employed at the receiver end to convert the optical signal into electrical current. VLC has the following desired features over the conventional RF-based systems:

- **Wide spectrum**: Raising the carrier frequency enhances the information-carrying capacity of a communication system. In RF, the allowable bandwidth can be up to 20% of the carrier frequency. VLC has a license-free, huge bandwidth of 400 THz, which alleviates the spectrum crunch faced by RF communication systems. In VLC, even if the bandwidth is taken to be 1% of carrier frequency ($\sim 10^{16}$ Hz), the allowable bandwidth would be 100 THz, which is 10^5 times that of RF carrier.
- **High directivity**: The gain of an antenna is dependent on its directivity. The superiority of an optical carrier compared to an RF carrier can be observed as

$$\frac{G_{optical}}{G_{RF}} \approx \frac{4\pi/\theta_{optical}^2}{4\pi/\theta_{RF}^2} \tag{7.1}$$

where θ_{RF} and $\theta_{optical}$ are RF and optical beam divergence, respectively. The beam divergence is proportional to λ/D where D is aperture diameter, and λ is carrier wavelength. As the optical wavelength is very small, high directivity and gain are obtained.

- **Cost-effective**: VLC-based systems offer a cost-effective and less complex implementation, leveraging the widespread presence of LEDs in existing infrastructures. The dual-purpose utilization of LEDs for both illumination and data transmission not only streamlines the deployment process but also enhances the overall cost-effectiveness of these systems. This inherent compatibility with prevalent infrastructure positions VLC as an accessible and economical solution for achieving efficient communication alongside illumination in various settings. The seamless integration of data transmission capabilities into the existing lighting framework further underscores the practicality and ease of adopting VLC technology.
- **Energy efficient**: VLC-based systems utilize LEDs as transmitters, which are recognized for their energy efficiency compared to incandescent light bulbs, placing them within the realm of environmentally friendly communication technology.
- **Security**: Unlike traditional wireless communication technologies such as RF that may extend beyond physical barriers, VLC restricts the transmission of information to the illuminated area. VLC has higher security compared to RF-based system as visible light cannot pass through the opaque boundaries and hence prevents eavesdropping. Moreover, VLC systems can be implemented with additional security features, such as encryption and authentication protocols, further fortifying the communication link.
- **Electromagnetic interference**: VLC-based communication systems exhibit a notable advantage in terms of resilience against electromagnetic interference, making them particularly suitable for deployment in environments sensitive to RF disturbances, such as hospitals and aircraft. As a result, VLC systems are less prone to the disruptive effects of competing signals and electromagnetic noise, ensuring a more reliable and undistorted communication environment.
- **Reusability**: VLC-based systems boast superior data density owing to the confined and highly directional nature of visible light beams. This characteristic allows for efficient data transmission along with spatial reusability within a closed environment. The focused nature of VLC beams enables multiple simultaneous communication links without interference, making it an ideal choice for applications requiring high data throughput in crowded spaces. This unique capability enhances the scalability and performance of VLC-based systems, making them well-suited for scenarios where maximizing data density is paramount.
- **High signal-to-noise ratio (SNR)**: As LEDs provide high illumination levels (typically of the order of 400-600 lux), high SNRs are practically achievable for VLC as compared to its RF counterparts. This advantage positions VLC as a favourable choice for scenarios where maintaining high SNRs is crucial for effective communication. The elevated illumination levels from LEDs contribute to the superior signal quality achievable in VLC setups.

Because of the aforementioned benefits, VLC finds applications in diverse fields, including light-fidelity (LiFi) systems, intelligent transportation systems (such as vehicle-to-infrastructure (V2I) communication and vehicle-to-vehicle (V2V) communication), smart lighting, underwater communication, internet of things (IoT) ecosystems and wearable devices. Detailed explanations of these applications will be provided in the following sections.

7.2 System Components and Architecture

A VLC system consists of several key components and follows a specific architecture to enable communication through visible light, as shown in Fig. 7.4. In the VLC system, the optical signals emitted from the LEDs travel through complex communication channels consisting of LoS and non-line-of-sight paths (NLoS). The transmitter is composed of a modulator/encoder, digital-to-analog converter (DAC),

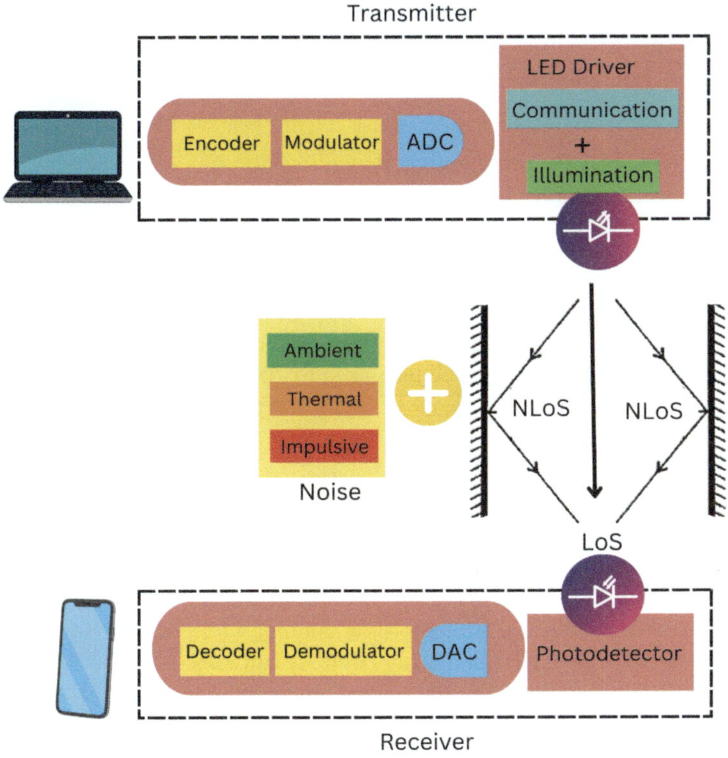

Fig. 7.4 Architecture of VLC systems

and LED luminaires. First, the data bits are modulated using an appropriate modulation technique outlined in IEEE 802.15.7 standard for VLC systems. Further, reflection, diffusion, interference and noise in the channel cause the loss and distortion of the transmitted signal. The PD at the receiver receives the modulated symbols transmitted through nonlinear LEDs. Similar to the transmitter block, the receiver block consists of a demodulator/decoder, analog-to-digital converter (ADC) and PD.

7.2.1 *Light Sources*

The advent of the LED has opened up new possibilities in the domain of VLC. The LED is a semiconductor that has the property that its frequency can be modulated rapidly at such a high speed that turning it on and off is undetectable by human eyes. An LED luminaire consists of an LED driver and an LED lamp. The function of LED driver is to control the amount of current flowing through the LED depending on the type of modulation technique considered. For example, in the conventional on-off keying (OOK) scheme, the light is turned ON and OFF by using an LED driver circuit, where the 'ON' state is represented by the high-intensity level of light, and the 'OFF' state is represented by the low intensity of light. In most deployments, a white LED is generally used for transmission in VLC systems, which produce white light in either of the two ways:

- by using a blue LED coated with yellow phosphorous,
- by using three separate RGB LEDs, which generate white light by mixing red, green and blue light.

Next, the signal is transmitted via a multipath VLC channel, which consists of both LoS and NLoS channel links creating multipath channel. However, the major disadvantage of LED is its limited modulation bandwidth, which is of the order of 3 MHz. This modulation bandwidth can be increased to 20 MHz by a blue filter.

7.2.2 *Modulation Techniques in VLC*

Modulation techniques in VLC involve the implementation of a dimming or biasing circuit with control functionality at the transmitter end. Employing microcontrollers proves to be a cost-effective strategy for encoding, while in more intricate applications, a field programmable gate array (FPGA) can substitute the microcontroller. The FPGA, leveraging digital signal processing techniques, enhances overall performance. At the transmitter side, the encoder transforms data into a modulated message, controlling LED switching to represent the binary data and the specified data rate. Consequently, binary data is translated into a light beam modulated in intensity. VLC systems adopt an intensity modulation and direct detection (IM/DD) approach. In

IM/DD systems, optical intensity must be real-valued and non-negative. As a result of these constraints of IM/DD, same modulation schemes applied in the RF system cannot be employed in the VLC system.

IM/DD is a modulation scheme wherein the modulation occurs by varying the intensity of an optical source. Demodulation is accomplished through the direct detection of the optical carrier, and the signal is restored by employing a PD at the receiver. At the transmitter side, the modulation signal modulates the current of the LED, which varies the instantaneous optical power radiated from the LED. For the receiver, the output signal of the PD is photocurrent. This photocurrent is proportional to the instantaneously received optical power incident on the PD.

OOK and variable pulse position modulation (VPPM) are the conventional modulation techniques outlined in IEEE 802.15.7 standard for VLC systems shown in Figs. 7.5 and 7.6, respectively. However, these modulation techniques are suited for low-data-rate applications, and addressing inter-symbol interference (ISI) caused by multipath effects proves challenging in these modulation schemes. Additionally, these schemes lack support for high modulation formats, hindering the optimal utilization of the available spectrum.

For high-data-rate applications, other modulation techniques are optical-orthogonal frequency division multiplexing (O-OFDM), colour shift keying (CSK), which exploits the diversity of red, green, and blue (RGB) LEDs, and carrier-less amplitude and phase (CAP)-quadrature amplitude modulation (QAM).

The constellation diagram shown in Fig. 7.8 displays real- and complex-valued signals in the IQ plane which displays the in-phase and quadrature components of modulated signals on the real (Re) and imaginary (Im) axes. The constellation diagram of OOK, CSK and 8-CAP-QAM modulation scheme is shown in Fig. 7.8.

- In OOK, the switching of the LED is controlled by turning it 'ON' and 'OFF' according to the bit streams. In OOK, The symbols (s_0 and s_1) are differentiated by the amplitude of the pulse, with high amplitude pulse representing the logical high '1', while low amplitude pulses represent the logical low '0' as shown with the help of an example in Fig. 7.5. OOK constellation in Fig. 7.8a contains two symbols corresponding to two intensity levels of LED.

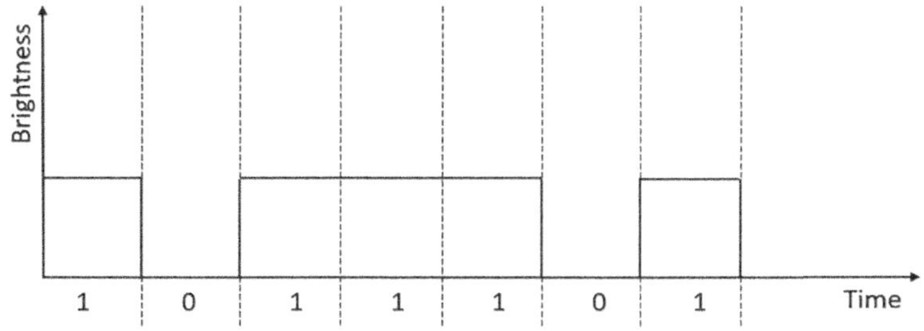

Fig. 7.5 OOK modulation scheme to transmit bit stream '1011101'

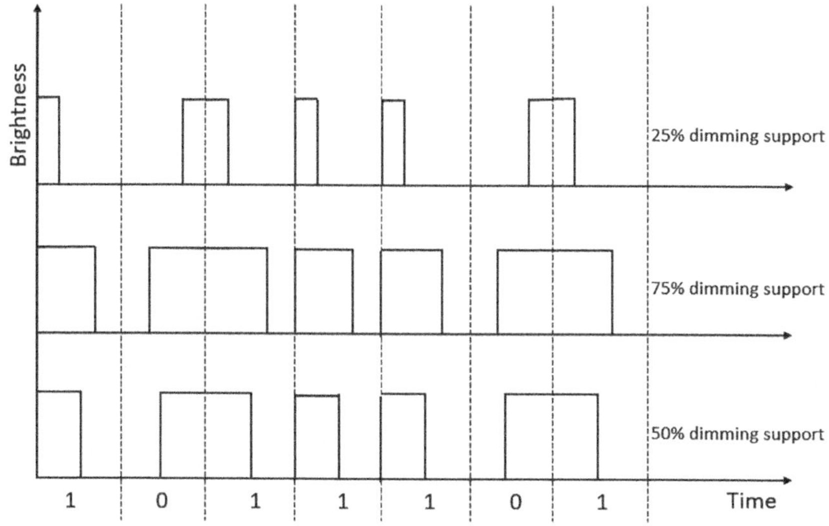

Fig. 7.6 VPP modulation scheme with different dimming rates to transmit bit stream '1011101'

- In VPPM, the information is encoded by varying the position of transmitted pulses within a time slot. The width of the pulse depends on the dimming rate. Figure 7.6 shows the VPPM scheme with different rates to transmit bit stream '1011101'.
- The IEEE 802.15.7 VLC standard outlines the colour bands employed in CSK modulation. In this modulation scheme, the constellation associates each symbol (s_1, s_2 and s_3) with a specific combination of colours encircling a central reference known as the target colour (s_0) as shown in Fig. 7.8b.
- In VLC, it is important to ensure the time domain transmitted signals are real-valued, which implies that the traditional QAM modulation scheme cannot be used for modulation. CAP-QAM is a version of the QAM scheme that suppresses the carrier and does not use the amplitude of two carrier frequency waves as used by QAM. It generates a QAM-modulated signal by combining two filtered PAM signals such that their impulse response forms a 'Hilbert pair' as shown in Fig. 7.7.

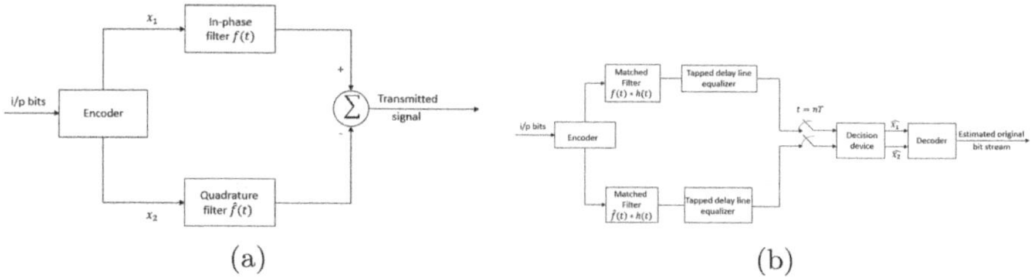

Fig. 7.7 Carrier-less amplitude and phase **a** Transmitter, and **b** Receiver

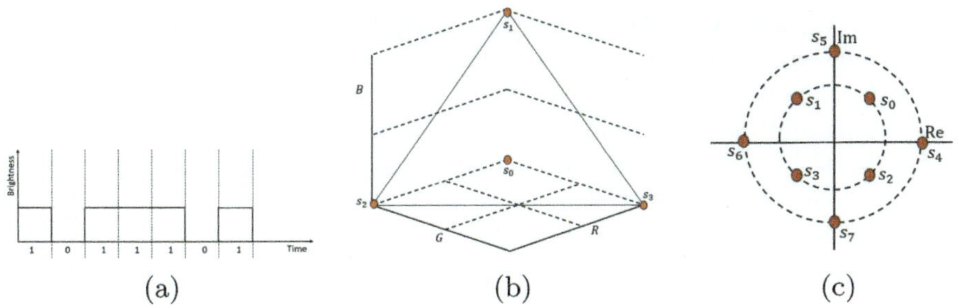

Fig. 7.8 Constellation diagram of **a** OOK, **b** CSK and **c** 8-CAP-QAM modulation scheme

Figure 7.7a shows the CAP transmitter where input bits x_1 and x_2 are encoded and passed through in-phase and quadrature-phase filters respectively and then passed through summer to constitute the transmitted signal. Similarly, Fig. 7.7b shows the CAP receiver which gives estimated input bit stream as output. Figure 7.8c shows the mapping of 8-CAP-QAM symbols $(s_0, s_1, s_2, s_3, s_4, s_5, s_6, s_7)$.

Two prominent O-OFDM schemes are Asymmetrically Clipped Optical Orthogonal Frequency Division Multiplexing (ACO-OFDM) and DC-biased Optical Orthogonal Frequency Division Multiplexing (DCO-OFDM). It is worth noting that, ACO-OFDM exhibits lower spectral efficiency compared to DCO-OFDM.

7.2.3 Photodetectors and Receivers

The signal transmitted by LED can be received by either of the two receivers:

- PD/non-imaging sensor,
- imaging/camera sensor.

PDs are used at the receiver for conversion of optical signal to electrical signal. Visible light signals can also be received by camera sensors, which is a collection of several photodetectors arranged in an integrated circuit and are generally available in the smart mobile devices. However, data rates achieved by imaging sensors are very low due to their low sampling rate. PD is a crucial component in a VLC system, serving as the receiver to capture the modulated light signal. The key functions of the photodetector include:

- **Light Absorption**: When exposed to modulated light, the photodetector absorbs photons, generating electron-hole pairs within its semiconductor material.
- **Conversion to Electrical Current**: The absorbed photons release energy, causing the generation of free electrons and positively charged holes. This process results in the conversion of optical energy into an electrical current.

- **Signal Amplification**: The generated electrical current, representing the modulated information, is often weak. Therefore, amplifiers are employed to strengthen the signal for further processing.
- **Filtering and Demodulation**: The amplified signal may undergo filtering to isolate the desired frequency or information. Demodulation is then applied to extract the original data signal encoded in the light modulation.
- **Output to Receiver**: The final electrical signal, now containing the transmitted data, is sent to the receiver for further processing, decoding and eventual utilization.

7.2.4 Channel Modelling

Channel modelling constitutes the initial and critical phase in the design of a VLC system, aiming for efficiency, reliability and robustness. Numerous endeavours have been made to address VLC channel modelling, as indicated by various studies [1, 2]. The VLC channel models can be divided into two parts: deterministic and stochastic channel models, which we will explain in greater detail in the upcoming subsections.

Deterministic Channel Models Deterministic channel models are usually based on the detailed description of a specific propagation environment, channel scenario, and the position and orientation of the LEDs and photodetectors or users. The ceiling bounce model is used to investigate the effect of multipaths due to reflection of light by various objects in an indoor environment. The channel impulse response (CIR) of the CB model is given as

$$\mathbf{h}(\tau) = \frac{6\alpha^6}{(\tau + \alpha)^7} u(\tau)$$

where $\alpha = 12\sqrt{\frac{11}{13}}D_{RMS}$, D_{RMS} is the root-mean-square (RMS) delay spread of the multipath channel, and $u(\tau)$ is the unit step function. In the study by [3], Monte Carlo ray tracing is employed to assess the channel impulse response (CIR) within an empty room at visible light wavelengths. However, this approach overlooks wavelength dependency, assuming fixed reflectance values for surface materials. Similarly, in [1], the recursive method proposed in [4] is utilized to derive CIR in the visible light band, yet fixed reflectance is presumed. Addressing the impact of wavelength dependency in channel modelling, [5] computes reflectance values as the average of wavelength-dependent coefficients over the visible light band. Notably, [2] stands out as the sole work explicitly accounting for wavelength dependency. Here, a recursive method is employed to determine the CIR of an empty room. Nevertheless, akin to preceding studies, [2] is constrained by assumptions of solely diffuse reflections and ideal Lambertian sources, conditions that may not be universally applicable in practical scenarios (Fig. 7.9).

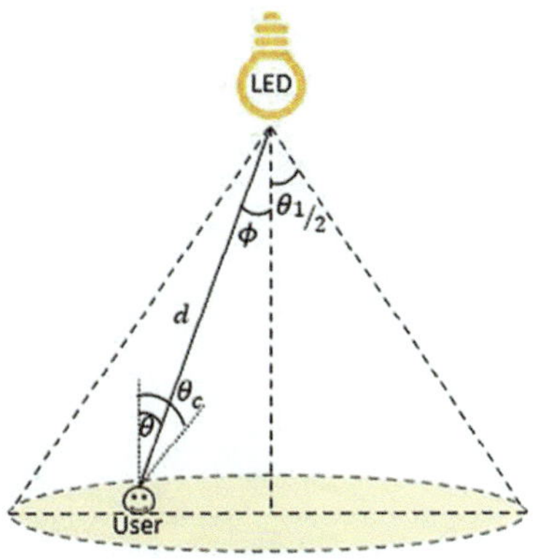

Fig. 7.9 LoS Channel gain model

In [6], the authors introduced an innovative approach to VLC channel modelling, addressing limitations present in previous models. They provide multiple CIRs tailored for diverse indoor environments, which are also incorporated into IEEE 802.15r1 PAN VLC channels. This methodology utilizes the ray tracing functionalities of the commercial optical and illumination design software Zemax®, allowing for a precise depiction of ray interactions emitted from lighting sources within specified confined spaces. Within the simulation environment created using Zemax®, users can define the geometry of the environment, objects present, and specifications of the light sources (LEDs) and receivers (PD). Leveraging the non-sequential ray tracing tool, the software computes detected power and path lengths from the light source to the detector for each ray, considering a specified number of rays and reflections. Figure 7.10 shows the indoor home scenario considered for channel modelling using Zemax® by authors in [6]. The nine yellow circles (L1,. . .,L9) are the nine luminaries, and there are eight test points (T1,. . .,T8) represented by black circles. The Zemax® non-sequential ray tracing program generates an output file containing the detected power and path lengths for each ray from source to detector. This file is then imported into Matlab®, where the information is utilized to represent the CIR as (Fig. 7.11):

$$h(t) = \sum_{i=1}^{N_r} P_i \, \delta(t - \tau_i) \tag{7.2}$$

where P_i is the power of the ith signal, τ_i is the propagation time of the ith ray, $\delta(t)$ is the dirac-delta function, and N_r is the number of receivers at the detector.

Several statistical channel models have been proposed for stationary and uniformly distributed users in prior works, such as [7]. The LoS channel gain, considering the direct path between the transmitter (LED) and the receiver photodetector as shown in Fig. 7.8, is given as follows:

$$h_{LoS} = \frac{m+1}{2\pi d^2} A \cos^m(\phi) \cos(\theta) g(\theta) \tag{7.3}$$

Fig. 7.10 Home scenario for channel modelling using Zemax®

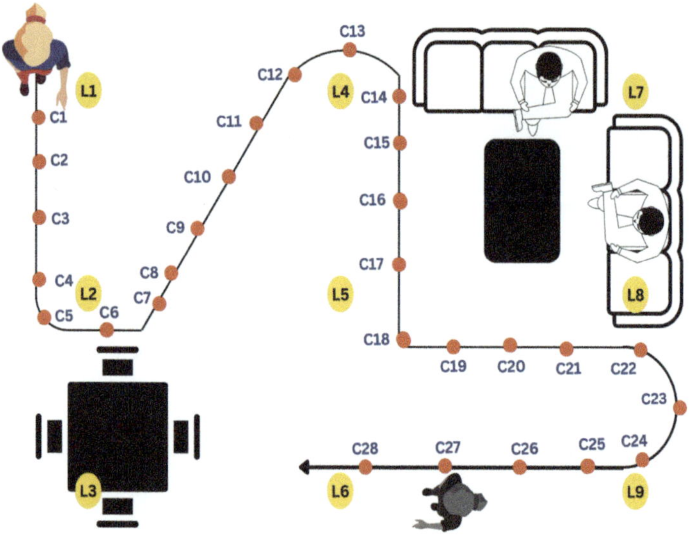

Fig. 7.11 Caption

where d is used to represent the distance between the LED transmitter and the PD, A is the area of the PD, ϕ is the light angle of incidence, θ is the angle of reception of the light at the photodiode, and $g(\theta)$ is the optical concentrator gain expressed as:

$$g(\theta) = \frac{n^2}{\sin(\theta_c)} \tag{7.4}$$

where n is the refractive index of the optical concentrator, θ_c is the field-of-view (FOV), and m is the order of Lambertian emission estimated as

$$m = -\frac{1}{\log_2(\cos(\theta_{1/2}))} \tag{7.5}$$

where $\theta_{1/2}$ is the semi-angle of the LED.

Stochastic Channel Models In stochastic models, the impulse responses of OWC channels are characterized by the law of wave propagation applied to specific LEDs, photodetector and scatterer geometries, which are predefined in a stochastic fashion according to certain probability distributions. The Spherical Model was proposed to model channels utilizing diffused link configurations. The model provides a quick rule of thumb in order to approximate the prediction channel pathloss and $H(f)$ for the higher-order reflections in indoor environments. A simple analytic formula for CIR $h(t)$ is derived from the integrating sphere and scaled to a specific room. While, according to spherical model, the diffused CIR $h_{diff}(t)$ is given as

$$H_{diff}(t) = \frac{H_{diff}}{\tau} e^{(-t/\tau_d)}. \tag{7.6}$$

where τ_d is the decay term and H_{diff} denotes diffused channel gain, which is expressed as

$$H_{diff} = \frac{A_r \rho}{A_{room} \langle \rho \rangle} \tag{7.7}$$

where A_{room} and $\langle \rho \rangle$ denote the room area and the average reflectivity, respectively. In most of the studies, fixed or random incidence angle is considered. However, considering the inherent mobility in wireless networks, a more realistic and non-uniform model for the spatial distribution of users is required.

In recent studies, like those in [8, 9], researchers used the random-way-point (RWP) mobility model to study the SNR in indoor LiFi systems. For the RWP model, the channel's probability distribution function (pdf) is specified as follows:

$$p(h) = \begin{cases} \sum_{l=1}^{4} K_l h^{-\beta_l}, & h_{min} \leq h \leq h_{max}; \\ 0, & \text{otherwise} \end{cases} \tag{7.8}$$

where $K_1 = K[27 + \frac{35D^2}{r_{max}^2} + \frac{8D^4}{r_{max}^4}]$, $K_2 = -K\frac{35}{r_{max}^2}\mathfrak{S}^{\frac{2}{a+3}}$, $K_3 = -K\frac{8}{r_{max}^4}\mathfrak{S}^{\frac{4}{a+3}}$, and $K_4 = -K\frac{16D^2}{r_{max}^4}\mathfrak{S}^{\frac{2}{a+3}}$, $K = \frac{12\mathfrak{S}^{\frac{2}{a+3}}}{73(a+3)r_{max}^2}$, $\beta_1 = \frac{2}{a+3} + 1$, $\beta_2 = \beta_4 = \frac{4}{a+3} + 1$, and $\beta_3 = \frac{6}{a+3} + 1$, where $\mathfrak{S} = b(a+1)D^{a+1}$, $b = \frac{R}{2\pi}$. The line of sight distance of the LED from the user is denoted as D, the effective geometric area of the detector is denoted by R, and r_{max} is the radius of the maximum coverage area. $h_{min} = \frac{\mathfrak{S}}{(r_{max}^2+D^2)^{\frac{a+3}{2}}}$, $h_{max} = \frac{\mathfrak{S}}{D^{(a+3)}}$ and, $a = \frac{-1}{\log(\cos(\phi_{\frac{1}{2}}))}$ where $\phi_{\frac{1}{2}}$ is the half-angle of the fixature of the LED transmitting. However, in [8], they assumed that the device always faces the same way, which might not be realistic. In [9], they assumed that the incidence angle of the light signal is distributed uniformly, which doesn't precisely represent how devices are oriented in realistic scenarios.

The majority of studies on OWC commonly assume that devices always face vertically upward. This assumption might be due to the lack of a proper orientation model or to simplify the analysis. However, this assumption is accurate only for specific devices like laptops with LiFi dongles. In reality, most users use devices like smartphones and tend to hold them comfortably, impacting throughput significantly. Despite some studies considering the impact of random orientation, such as [10], these studies assume predefined models for the random orientation of the receiver. There is a need for further research that carefully analyses the impact of user-specific orientations on throughput in realistic scenarios. However, there is a lack of supporting evidence to validate the presumed models. Notably, none of these studies have taken into account the real-world statistics of device orientation; instead, they predominantly rely on hypothetical moments and assume either a uniform or Gaussian distribution for device orientation.

Recently, experimental measurements were conducted for the first time to characterize the polar and azimuth angles of the user's device in studies such as [11]. The results suggest that the modelling of the polar angle can be efficiently achieved using either a truncated Laplace distribution for stationary users or a truncated Gaussian distribution for mobile LiFi users. Meanwhile, the azimuth angle is observed to adhere to a uniform distribution in both scenarios. In [12], the authors introduced an omnidirectional receiver, which does not take in account the random orientation of devices. Their findings demonstrate that the proposed receiver minimizes SNR fluctuations, leading to a substantial enhancement in user throughput. However, these studies do not the importance of integrating the random spatial distribution of users and the unpredictable orientation of LiFi devices into the analytical framework, as successfully accomplished by the authors in [13].

7.3 Challenges and Open Issues in VLC

The practical deployment of VLC faces several challenges, which can impact its widespread adoption. Various such challenges are discussed in detail in subsequent subsections.

7.3.1 Noise Sources

The dominant noise sources in a typical VLC system, as shown in Fig. 7.4, are given in detail as follows:

Ambient Noise Ambient light noises may arise from sunlight, skylights, incandescent and fluorescent lamps, and other light sources present in the indoor environment. The presence of ambient light gives rise to DC photocurrent, inducing shot noise. The variance of this shot noise originating from ambient light (represented as $\sigma^2_{ambient}$) can be expressed as:

$$\sigma^2_{ambient} = 2qM^2F(I_b + I_x)B \tag{7.9}$$

where q symbolizes the charge of the electron, M corresponds to the average gain of the PD, B represents the bandwidth of the PD, I_b refers to the average photocurrent generated at the PD due to the average optical power received from sunlight, I_x signifies the average photocurrent generated at the PD due to the average optical power received from the LED, and F is characterized as the excess noise calculated as

$$F = rM + \left(2 - \frac{1}{M}\right)(1 - r) \tag{7.10}$$

where r symbolizes the hole-to-electron ionization rate. The ambient light noise can be approximated by Gaussian distribution with mean 0 and variance $\sigma^2_{ambient}$. Ambient light noise has been considered in the literature.

Thermal Noise The variance of the thermal noise ($\sigma^2_{thermal}$) is given by

$$\sigma^2_{thermal} = 4\left(\frac{K_BT}{R_L}\right)F_nB \tag{7.11}$$

where K_B is the Boltzmann constant, T is the temperature (in Kelvin), F_n is the PD noise and R_L is the load resistance (typically of the order of 50 Ω). Thermal noise can be modelled by Gaussian distribution with mean 0 and variance $\sigma^2_{thermal}$.

Impulsive Noise The Middleton Class-A noise model is the most considered statistical noise model-based on Poisson-Gaussian to model impulsive noise induced due to imperfect optical components and other factors such as lightning and sunlight. The Poisson process is employed to depict the likelihood of impulsive noise events occurring, while the Gaussian process is utilized to describe the amplitude distribution of said impulsive noise. The effective variance of the Class-A noise model is determined by

$$\sigma^2_{impulse} = K\frac{\sigma_b^2}{A} \tag{7.12}$$

where $\sigma^2_{impulse}$ is the variance of the impulsive noise and σ_b^2 is the variance of the background noise, K represents the average power of the impulsive noise and the parameter A represents the probability of impulsive noise on the time axis.

Hence, the overall noise in a VLC system is an independent and identically distributed (i.i.d) complex additive white Gaussian noise (AWGN) with zero mean and variance σ^2, where $\sigma^2 = \sigma^2_{ambient} + \sigma^2_{thermal} + \sigma^2_{impulse}$.

7.3.2 Nonlinear Characteristics of LEDs

LEDs are widely deployed in the existing infrastructure and have completely replaced conventional fluorescent and incandescent lamps due to their several desirable features, like lifetime, light density, reliability and energy efficiency. The response of LEDs displays non-linearity due to the nonlinear conversion of current to voltage and current to optical power. Specifically, the characteristics of LEDs become nonlinear in the presence of signals with a substantial dynamic range and at high switching frequencies. LED's nonlinear models are mainly divided into memoryless and memory nonlinear models. The memoryless nonlinear block is modelled by Rapp's model, which can be mathematically written as amplitude-to-amplitude (AM/AM) modelling as follows:

$$f(s) = \frac{s - v_{th}}{(1 + (\frac{s-v_{th}}{i_{sat}})^{2k_f})^{\frac{1}{2k_f}}} \tag{7.13}$$

where i_{sat} is the saturation current, $f(s)$ the output LED intensity, v_{th} is the cut-in voltage of the LED and k_f controls the level of non-linearity, also termed as knee factor. k_f controls the smoothness of transition from linear to the saturation region, as can be seen in Fig. 7.12. Furthermore, the capacitance and conductance of an LED are frequency-dependent, which results in LED's non-linearity with memory effects.

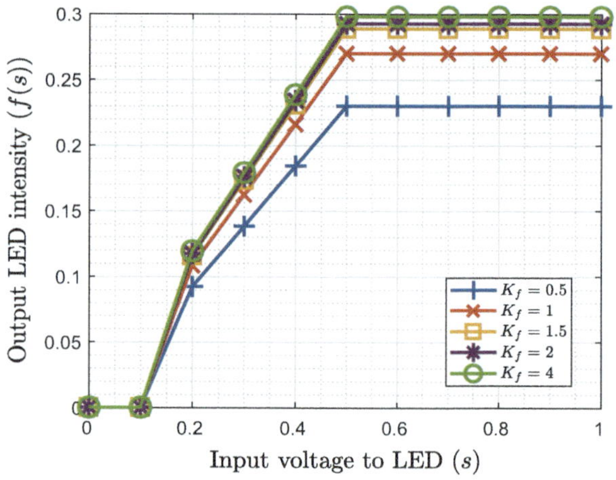

Fig. 7.12 Transfer characteristics of the Rapp's nonlinear model for LED

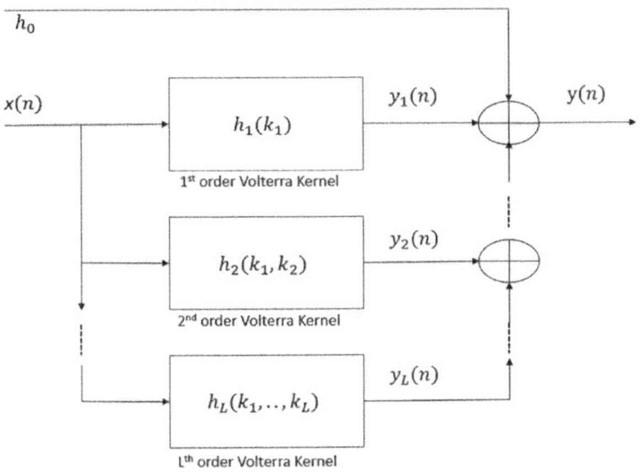

Fig. 7.13 Volterra nonlinear model for LED

There are various models in the literature to describe LED non-linearity with memory effects, which are as follows:

- Volterra model: The Volterra series provides a comprehensive method for modelling a nonlinear system with memory as shown in Fig. 7.13. Given the frequency-dependent nature of LEDs, accounting for memory effects becomes crucial, particularly in scenarios with a broad signal bandwidth. The contribution of the lth order non-linearity to the output signal can be expressed as

$$p_l(n) = \sum_{k_1=0}^{D_l} \cdots \sum_{k_l=0}^{D_l} h_l(k_1, \ldots, k_l) \times x(n-k_1) \cdots x(n-k_l), \tag{7.14}$$

  where D_l is the memory length for the lth order nonlinear kernel, $h_l(\cdot)$ is the lth-order Volterra kernel, h_0 is a constant and M is the nonlinear order.
- Memory polynomial model: The memory polynomial model, is a special case of the Volterra model, incorporates solely diagonal kernels as shown in Fig. 7.14. In contrast to the memory-less polynomial model, the output of this memory model is influenced not only by the current input but also by past input values.
- Wiener model: The Wiener model represents another category within the Volterra model, as shown in Fig. 7.15. It is characterized by a linear time-invariant (LTI) system followed by a memoryless nonlinearity.
- Hammerstein model: The Hammerstein model consists of a nonlinearity without memory followed by a linear time-invariant (LTI) system, which is the opposite of the Wiener model Fig. 7.16. It provides a complexity level similar to that of the Wiener model.

The main limitation of the Volterra model is that its complexity increases exponentially with the increase in nonlinear order and memory length. The special case

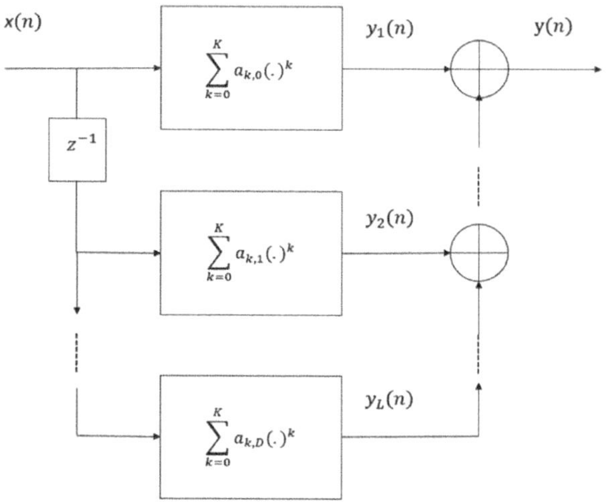

Fig. 7.14 Memory nonlinear model for LED

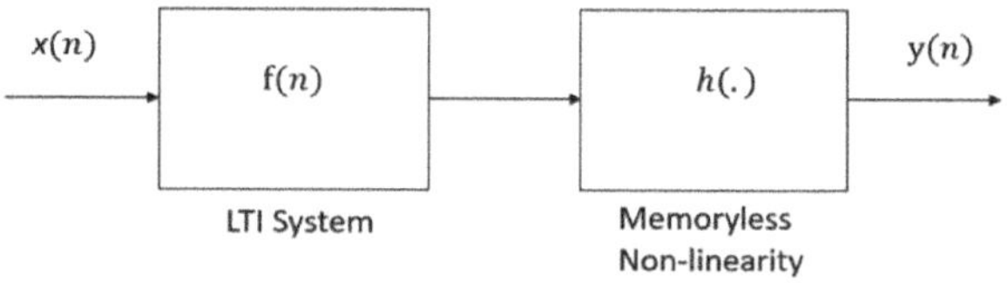

Fig. 7.15 Wiener nonlinear model for LED

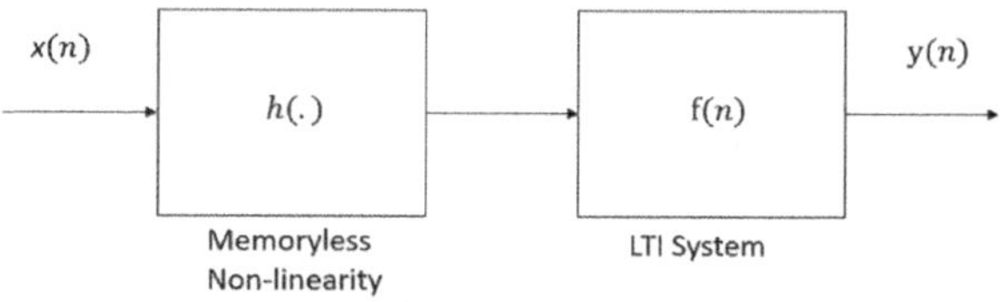

Fig. 7.16 Hammerstein nonlinear model for LED

of the Volterra model is the memory polynomial model, which considers only diagonal kernels. From Bussgang's theorem of decomposition of nonlinear functions, it can be inferred that LED non-linearity results in an uncorrelated additive distortion component at the receiver, which significantly degrades the overall BER performance. Hence, it necessitates appropriate nonlinear signal processing techniques at the receiver to mitigate LED non-linearity.

Prior studies have introduced pre-distortion and post-distortion [14] algorithms for mitigating LED non-linearity. Pre-distortion techniques are known to outperform static inversion, however, they require perfect feedback of the channel state information (CSI) from the receiver to the transmitter in a closed loop. To alleviate the need for precise feedback, authors in [15–17] substituted pre-distorters with open-loop post-distorters based on Volterra and Hammerstein polynomials. However, the Volterra (ex. Volterra least mean square (VLMS)) and Hammerstein based approaches have high computational complexity and suffer from modelling impairments due to truncation of polynomial series. Thus, due to their universal representation, open loop reproducing kernel Hilbert space (RKHS) and classical polynomial series [15] based post-distortion methods have been proposed recently. Among these, RKHS-based post-distorters have demonstrated superior performance compared to polynomial series post-distorters. This improvement is attributed to their advantageous features, such as convexity, universal approximation and computational efficiency achieved through sparsification. RKHS based approaches are computationally simple and provide superior bit error rate (BER) performance. Among the existing RKHS based post-distortion techniques, the kernel least mean square (KLMS) and kernel minimum symbol error rate (KMSER) are the most popular. RKHS based post-distorter based on KLMS/KMSER algorithm delivers better performance over polynomial series based post-distorters. However, KLMS/KMSER based approaches rely on growing dictionary of data sets, which is difficult to implement practically. Furthermore, the feature function of Gaussian kernel allows for approximation as random Fourier feature (RFF). However, conventional feed-forward RKHS-based post-distorters have shown suboptimal performance in scenarios involving multipath VLC channels with high inter-symbol-interference (ISI) and non-Gaussian distortions [18]. To address this limitation, decision feedback equalizer (DFE)-based post-distorters within the RKHS framework are proposed. These DFE-based methods aim to mitigate non-Gaussian processes encountered in VLC systems, particularly those induced by severe ISI and other non-Gaussian additive noises. Additionally, non-minimum mean square error (MMSE) criteria, such as minimum symbol error rate (MSER) and maximum Versoria criterion (MVC), are explored to enhance the performance of RKHS-based post-distorters over VLC channels affected by non-Gaussian distortions [19]. Furthermore, random Fourier features (RFF)-based explicit RKHS methods that eliminate the need for sparsification are proposed. This approach allows post-distortion within a finite memory budget, providing a practical solution for VLC systems. In [20], authors have presented superior performance of RFF based least minimum symbol error-rate post-distorter in RKHS over IEEE 802.15r1 PAN VLC channels. Also, the authors in [21] have proposed an RFF based KMSER decision feedback equalizer (DFE) for VLC. However, the performance of RKHS based techniques whether dictionary-based or RFF is highly sensitive to the choice of kernel-width parameter. Estimating the appropriate kernel-width for convergence adds to the computational complexity. From [22], it is noted that kernel-width sampled from inverse-gamma distribution has same error as compared to the kernel-width obtained by brute-force trial.

7.3.3 *Impact of User Mobility and Multipath Effects*

The performance of a VLC link degrades as a result of mobile receivers or transmitters causing relative motion between the transmitter and the receiver, and the presence of multipath between the transmitter and receiver. In addition to the LoS component,

Table 7.1 Maximum Doppler spread calculated in Hz estimated at different terminal speeds and frequency bands

Frequency band							
Speed v (km/hr)	Cen. freq (Hz)						
	Below 6 GHz	mmWave		sub-THz		THz	
	3 GHz	28 GHz	60 GHz	150 GHz	300 GHz	0.8 THz	1.2 THz
5	14	130	278	695	1390	3706	5559
30	83	778	1668	4170	8339	22238	33356
120	334	3113	6671	16678	33356	88950	133426
300	834	7783	16678	41696	83391	222376	333564
500	1390	12972	27797	69493	138985	370627	555940

multiple reflections among the ceiling, floor and walls of the room for an indoor VLC scenario result in NloS components for a VLC link. Hence, the VLC channel is frequency selective in nature and results in severe ISI, particularly under the high data rate regime, i.e. when the delay spread is larger than the symbol period. In real-world scenarios, users may engage in activities such as reading, talking on a cell phone or shopping in a mall while in motion. This dynamic behaviour introduces a probabilistic nature to the overall VLC system. The effect of user mobility leads to an effective multiplicative distortion that degrades the VLC link due to dispersion in the time domain, causing inter-carrier interference (ICI). From Table 7.1, the high Doppler spread, even with low mobility in the THz range, can be observed. The Doppler spreads observed in Terahertz (THz) channels are significantly larger, by orders of magnitude, compared to those encountered in traditional microwave and millimetre-wave channels. For example, from Table 7.1, at a relative speed of 5 km/hr between the transmitter and the receiver, the Doppler spread is nearly 400 times higher in the THz range as compared to the below-6 GHz range. The same can be extended to a VLC-based system, as it employs visible light with a frequency of 400–800 THz.

Authors in [23, 24] have employed a newly proposed orthogonal time frequency space modulation scheme to mitigate the effects of user mobility and multipath with lower complexity. The literature shows that OTFS is capable of improving the bandwidth, power and energy efficiencies of traditional O-OFDM. Authors in [24] have proposed an orthogonal time frequency space (OTFS) based orthogonal time frequency-multiplexing (OTFM) scheme for a direct current-biased optical wireless communication (DC-OWC) system. The authors have demonstrated that direct current-biased optical orthogonal time frequency space (DCO-OTFS) performs better than the conventional DCO-OFDM in OWC systems. Similar results are obtained for static VLC IEEE 802.15r1 PAN VLC channels in [25]. To model the user mobility for a VLC system, a random way-point channel is considered in [26], whereas a time-varying channel is considered in [27]. These systems eliminate the need for Hermitian symmetry or direct current (DC) bias operations and exhibit improved bit error rate (BER) performance compared to Orthogonal Frequency Division Multiplexing (OFDM). Thus, from the literature, it can be concluded that the proposed algorithm is valid for both indoor and outdoor scenarios.

7.3.4 *Blind Spots: Absence of Direct Link*

Despite VLC's promise as a viable supplement to RF-based communication systems, the performance of a VLC system is significantly limited by the absence of direct links or blockage of the LoS channel, which creates blind spots as shown in Fig. 7.17a. As visible light cannot penetrate obstacles, it is reflected back due to the high penetration loss. In Fig. 7.17a, User 2 and User 3 have direct LoS and NLoS links available, while User 1 is in a blind spot with User 1 and plants blocking the LoS paths. In this context, to facilitate the data requirements of users like User 1, optical reflecting intelligent surface (RIS) has recently been introduced in the literature as a promising solution to overcome the drawbacks of LoS blockages and broaden the coverage area by reconfiguring the propagation environment [28].

RIS is a recent promising technology that uses reflecting surfaces that facilitate the NLoS paths to improve the performance of the wireless communication systems [29]. The mirror array (MA)-based RIS and the metasurface array (MSA)-based RIS are the two most popular reflecting surface designs employed for optical RIS in VLC systems [28]. The cosine of the angle of irradiance for RIS is represented in the form

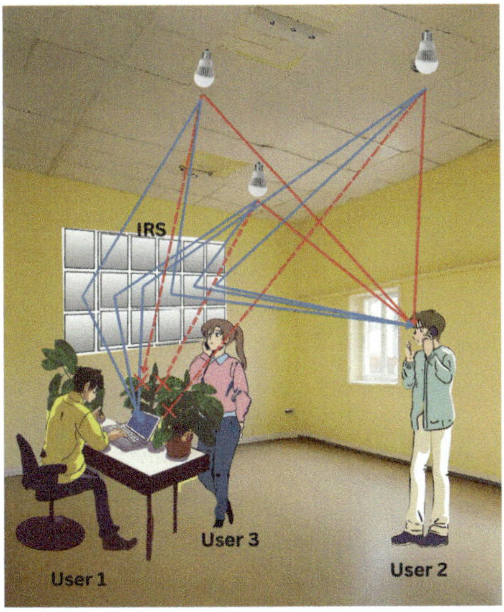

(a)

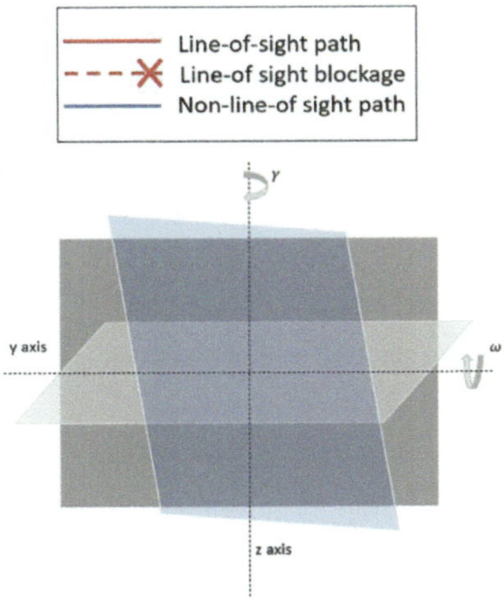

(b)

Fig. 7.17 IRS-aided indoor VLC system model

of the yaw (γ) and roll (ω) angles of the intelligent surface shown in Fig. 7.17b, which are to be optimized for the best performance by maximizing the sum rate. From the rigorous analysis in [28], authors have shown that in the case of MA, the received power gain is always positive and thus better than MSA for the VLC system.

For parallel-moving vehicles in an outdoor environment, the headlamp LED's light generally cannot reach the photodiode installed at the receiver of another parallel vehicle on the road. As a solution to this problem, the MA-based optical RIS-aided VLC system can be employed for parallel vehicles.

7.4 Applications and Use Cases

Based on the transmission range, the application of visible light for communication can be broadly classified, as shown in Fig. 7.18. Some applications of VLC are already exploited in the mainstream commercial market, while others are envisioned for future use. Many companies such as PureLiFi, Oledcomm, Panasonic, Infinity Technology, VLNCommSignify, LumEfficient, Lucibel, LIFX, General Electric, Lightbee, Global LiFi Tech, FireFly Wireless Networks, LVX System, Luciom, Wipro and ZERO.1 are actively working to create and make LiFi products accessible to the general public. Detailed explanations of these applications and use cases can be found in the subsequent subsections.

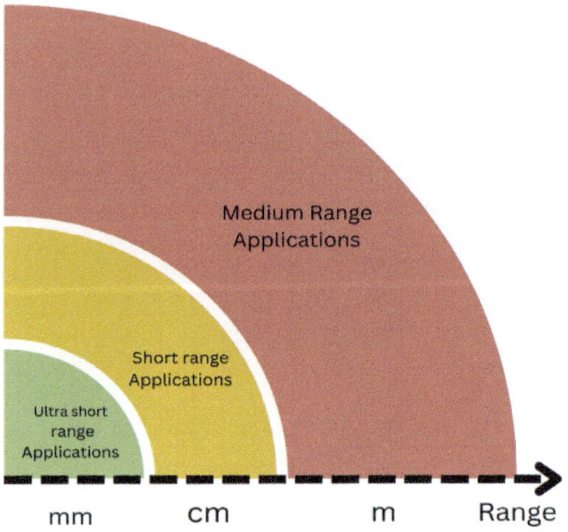

Fig. 7.18 Applications of VLC with transmission range

7.4.1 Ultra Short Range Applications

Due to their advantageous characteristics, including low latency, high bandwidth and low power consumption, optical interconnects have been suggested in the literature as a viable alternative to traditional copper-based electrical interconnects for both inter-chip and intra-chip communication in data centres and systems-on-chip. These chips have found applications in space communication where high data rates, along with immunity to EMI, are required. The optical interconnect market is projected to reach USD 17.1 billion by 2025. The challenge of the optical interconnect market is how efficiently misalignment tolerance can be minimized.

7.4.2 Short Range Applications

Short range VLC applications can be further classified into:

1. Wireless body area network (WBAN)
2. Wireless personal area network (WPAN)

WBAN It involves the use of devices that are wearable and can be used for sensing biochemical information about health signs such as heart rate, blood pressure and body temperature. Traditionally, RF-based WBANs are widely used commercially. However, they can't be employed in hospitals and medical facilities where RF deployment is practical due to the presence of EMI. Also, there are controversial ill-effects of prolonged exposure to RF signals on health. In this context, VLC is a promising alternative to the current RF-based solution. VLC-based solutions are available for the transmission of electrocardiography signals and patient information to health experts simultaneously. With the advancement of organic LEDs, it is possible to integrate VLC-based components into wearable devices.

WPAN includes indoor applications of VLC as shown in Fig. 7.19. These applications include mobile-to-mobile and mobile-to-device communication. M2D communication may include the transfer of data between the user's mobile and devices such as printers, computers and cameras. Recently, research in this area has also included smartphone camera communication, access control systems, IoT applications, etc.

7.4.3 Medium Range Applications

Medium-range applications of VLC are given in Fig. 7.20. The VLC-based systems can be employed in indoor scenarios such as offices, homes, malls or stadiums. As VLC is less prone to EMI, it can be safely used in environments which are prone to EMI, such as hospitals and aircraft. To facilitate these applications, new LiFi products are hitting the market at an accelerated pace, with numerous companies actively

Fig. 7.19 VLC application in WPAN

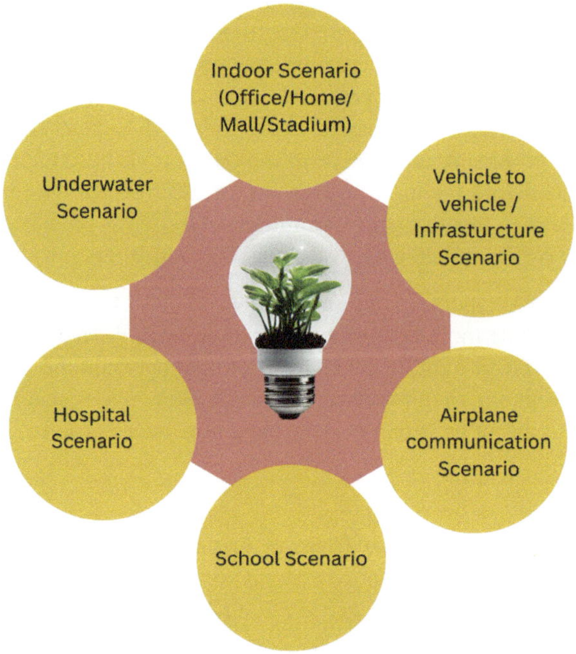

Fig. 7.20 VLC application in medium transmission range

engaged in creating LiFi solutions and enhancing LiFi technology specifically for indoor applications. For ex. MyLiFi® Desk Lamp by LiFi can accommodate up to 7 connections and provide transmission speeds of up to 23 Mbps for downlink and uplink. For installation in hospitals, LiFiCare® LiFi Lamp is available through which medical professionals can gain access to real-time information about patients' medical conditions for accurate and timely monitoring. For offices and indoor home environments, LumiNex LiFi-enabled LED panels are available, which are capable of providing data transmission speeds of up to 60 Mbps for uplink and 70 Mbps for downlink and can cover an area of up to 400 ft^2 (37 m^2) for a 7-ft fixture.

7.5 Case Studies and Implementations

In the past few years, VLC has been employed in the diverse fields of wireless communications owing to its high data rate. The increased demand for green, eco-friendly communication along with high-speed networks for the growing population has created awareness among countries to re-innovate their present and future city planning, supported by sustainable development. This allows the creation of smart cities with smart infrastructure where people, the economy and the environment are seamlessly connected. Some notable marvels of such innovation are mentioned below in detail.

1. LiFi Trials in the Orkney Islands: Ranked at the lowest position among 389 rural towns for having the slowest broadband internet connection speeds, with an average test speed of 6.3 Mb/s, Orkney was selected by 5G RuralFirst for its LiFi trial. The trials yielded promising results, showcasing the successful demonstration of the potential of visible light through LiFi. The experiment resulted from a collaboration between the University of Edinburgh and pureLiFi Ltd., with support from various entities including Orkney Council, CloudNet IT Solutions, and national organizations such as the Northern Lighthouse Board. Additionally, the Department for Digital, Culture, Media, and Sport (DCMS) played a pivotal role in supporting the experiment under the 5G RuralFirst project.
2. 6G Flagship event: Let There Be Light: According to the University of Oulu's Dr. Marcos Katz, there is a 'huge amount of spectrum which is unlicensed, it is free to use, [and that] we have available for light. Also, we can support very high data rates.' During the demonstration, Katz and the team conducted tests to explore the coexistence of radio and light within the same space. They successfully showcased a system capable of seamlessly switching between the two networks, adapting dynamically based on the prevailing circumstances. For example, if one is connected to the VLC network and the light source is interrupted, the system automatically redirects the user to the radio network, and vice-versa.
3. Philips Lighting's LiFi-enabled LEDs: Philips Lighting's LEDs with LiFi capabilities are currently undergoing a trial of LiFi technology at the La Defense

office of the real estate company Icade in Paris. They have demonstrated an uninterrupted LiFi connection in the presence of ambient light, which adds noise and user mobility.
4. The Hegel-Gymnasium : The Hegel-Gymnasium in Stuttgart is presently experimenting with LiFi in classrooms to instruct a diverse array of subjects, including information technology. In Fraunhofer's prototype LiFi setup at the Stuttgart school, data is wirelessly transmitted from LED luminaries through lightwaves to boxes, which are then connected to laptops via Ethernet.
5. The Kyle Academy: The Kyle Academy, situated in Scotland, has initiated the use of LiFi within its classrooms. This LiFi project is a collaborative effort involving PureLiFi and The University of Edinburgh. The deployment of PureLiFi's LiFi-XC system includes setting up eight LED light bulbs with LiFi capabilities on the ceiling. Students are equipped with LiFi-XC Stations, which can be connected to their laptops, enabling fast connectivity through the light-emitting sources.
6. LiFi in Retail: LiFi can guide customers from the moment they step into the store, assisting them in finding particular products, acquiring digital coupons, checking store promotions, verifying stock availability on shelves, engaging with in-store online services using their smart devices, and enhancing their overall shopping experience. LiFi will bring an opportunity for shopping malls and centres to be front runners in gaining an understanding of their customers and ultimately provide a better experience.

7.6 Areas for Further Research and Development

Over the past decade, VLC has evolved from a narrow and limited research domain to a source of academic and commercial promise for the future of wireless communication. This transformation has been fueled by the increasing popularity of LEDs and the growing demand for wireless bandwidth. The surge in VLC research is evident, marked by numerous efforts to standardize the technology during this period.

In the coming years, the trajectory of VLC technology will depend on clearly exploiting its tangible benefits, including its utilization of free, unlicensed spectrum, cost-effectiveness, security features and eco-friendly attributes, especially when compared to RF-based communication. Achieving widespread adoption of VLC by both academia and the market will be pivotal for its future success. Currently, the primary document encapsulating VLC aspects is IEEE 802.15.7.

As VLC gains popularity, the development of new modulation schemes and multiple access mechanisms has become necessary, prompting the need for a revised and updated standard that can cater to higher-speed systems. In summary, the future directions for VLC research are summarized below to explore these areas and address the evolving needs and challenges of VLC technology:

- **Standardization of research**: Over the past decade, significant research has been conducted in the field of VLC. However, a notable challenge for researchers has been the absence of a standardized experimental testbed. Various research groups independently develop their VLC hardware and testbeds, leading to potential interoperability issues among different solutions. Recognizing this challenge, future trends in VLC research are expected to focus on standardization efforts. Such initiatives aim to organize and streamline the research field, fostering greater collaboration and compatibility among diverse VLC implementations.
- **Popularization through the commercial initiative**: The collaboration of academic and commercial endeavours in the realm of VLC marked a noteworthy milestone in 2018. This pivotal moment was symbolized by the official endorsement of LiFi technology by Signify (formerly Philips Lighting), a globally recognized leader in the lighting industry. Looking ahead, the trajectory of VLC will be shaped by these significant commercial developments. Simultaneously, academic initiatives stand to gain from the standardization of light infrastructure, creating a symbiotic relationship between commercial advancements and academic progress in the VLC domain.
- **Optimal transceiver designs**: Exploring optimal transceiver designs and cellular architectures, taking into account the derived channel models, becomes a prospective research avenue. This research aims to address the substantial requirements of 5G and future communication technologies within real-world environments. Investigating and developing efficient transceivers and network structures for VLC systems can pave the way for their deployment in practical scenarios.
- **Uplink communication**: The channel models derived in existing literature primarily focus on downlink communication within indoor LiFi environments. Therefore, future research endeavours should encompass the development of analogous models tailored for uplink transmission scenarios and outdoor environments. Addressing these aspects would contribute to a more comprehensive understanding and optimization of LiFi systems across various communication scenarios.
- **Developing realistic scenarios**: The effects of random receiver orientation, user mobility, multipath between transmitter and receiver, and LED non-linearity should be analysed for a VLC link in detail for both downlink and uplink scenarios.
- **Hybrid systems**: An underlying principle of VLC is its role as a supplementary technology alongside RF-based mechanisms. Looking forward, an emerging trend is the creation of new hybrid VLC systems. This involves collaborative efforts between VLC and wired base network devices, simultaneous transmission within both VLC and RF or Ethernet networks, the exploration of multiple paths and the integration of network coding. These approaches have the potential to enhance the overall quality of communication.

7.7 Conclusion

In conclusion, this chapter has provided a comprehensive exploration of VLC, shedding light on its technologies, challenges and future prospects. VLC has emerged as a promising paradigm for next-generation wireless communication, leveraging the dual functionality of LEDs for both illumination and data transmission. We delved into the foundational components and architecture of VLC systems, emphasizing the significance of channel modelling in their efficient design.

The challenges associated with VLC deployment, including issues with interference, limited range, noise, LED non-linearity and mobility constraints, were discussed. Despite these challenges, researchers and practitioners are actively working on innovative solutions to enhance the reliability and practicality of VLC systems.

Noteworthy advancements in VLC channel modelling were highlighted, with a focus on recent approaches that overcome limitations present in earlier models. The incorporation of realistic simulations using commercial software like Zemax® has enabled a more accurate description of channel characteristics in diverse indoor environments.

As VLC continues to evolve, it holds immense potential in various applications, ranging from LiFi systems to smart lighting, intelligent transportation, underwater communication, IoT ecosystems and wearable devices. The integration of spatial modulation and ongoing research in VLC technologies, such as the suppression of LED non-linearities, reflects a commitment to addressing existing challenges and pushing the boundaries of VLC capabilities.

In essence, VLC stands at the forefront of innovative communication solutions, and its continued development promises a future where visible light becomes a versatile and efficient medium for data transmission. The insights gained from this chapter provide a foundation for understanding the current state and potential advancements in VLC technology.

References

1. Komine T, Nakagawa M (2004) Performance evaluation of visible-light wireless communication system using white LED lightings. In: Proceedings. ISCC, (2004) ninth international symposium on computers and communications (IEEE Cat. No. 04TH8769), vol 1. IEEE, pp 258–263
2. Lee K, Park H, Barry JR (2011) Indoor channel characteristics for visible light communications. IEEE Commun Lett 15(2):217–219
3. O'Brien DC, Zeng L, Le-Minh H, Faulkner G, Walewski JW, Randel S (2008) Visible light communications: challenges and possibilities. In: IEEE 19th international symposium on personal, indoor and mobile radio communications, pp 1–5
4. Barry J, Kahn J, Krause W, Lee E, Messerschmitt D (1993) Simulation of multipath impulse response for indoor wireless optical channels. IEEE J Sel Areas Commun 11(3):367–379
5. Long S, Khalighi M-A, Wolf M, Bourennane S, Ghassemlooy Z (2014) Channel characterization for indoor visible light communications. In: (2014) 3rd international workshop in optical wireless communications (IWOW). IEEE, pp 75–79

6. Miramirkhani F, Uysal M (2015) Channel modeling and characterization for visible light communications. IEEE Photonics J 7(6):1–16
7. Yin L, Popoola WO, Wu X, Haas H (2016) Performance evaluation of non-orthogonal multiple access in visible light communication. IEEE Trans Commun 64(12):5162–5175
8. Gupta A, Garg P (2018) Statistics of SNR for an indoor VLC system and its applications in system performance. IEEE Commun Lett 22(9):1898–1901
9. Arfaoui MA, Soltani MD, Tavakkolnia I, Ghrayeb A, Assi C, Haas H, Safari M (2019) SNR statistics of indoor mobile VLC users with random device orientation. In: Proceedings of IEEE international conference on communications (ICC), pp 1–6
10. Wang Z, Yu C, Zhong W-D, Chen J (2011) Performance improvement by tilting receiver plane in m-qam ofdm visible light communications. Opt Express 19(14):13 418–13 427
11. Soltani MD, Purwita AA, Zeng Z, Haas H, Safari M (2019) Modeling the random orientation of mobile devices: Measurement, analysis and lifi use case. IEEE Trans Commun 67(3):2157–2172
12. Chen C, Soltani MD, Safari M, Purwita AA, Wu X, Haas H (2019) An omnidirectional user equipment configuration to support mobility in LiFi networks. In: Proceedings of IEEE international conference on communications (ICC), pp 1–6
13. Arfaoui MA, Soltani MD, Tavakkolnia I, Ghrayeb A, Assi CM, Safari M, Haas H (2021) Measurements-based channel models for indoor LiFi systems. IEEE Trans Wireless Commun 20(2):827–842
14. Mitra R, Bhatia V (2016) Adaptive sparse dictionary-based kernel minimum symbol error rate post-distortion for nonlinear LEDs in visible light communications. IEEE Photonics J 8(4):1–13
15. Stepniak G, Siuzdak J, Zwierko P (2013) Compensation of a VLC phosphorescent white LED nonlinearity by means of Volterra DFE. IEEE Photon Technol Lett 25(16):1597–1600
16. Qian H, Yao SJ, Cai SZ, Zhou T (2014) Adaptive postdistortion for nonlinear LEDs in visible light communications. IEEE Photonics J 6(4):1–8
17. Wang Y, Tao L, Huang X, Shi J, Chi N (2015) Enhanced performance of a high-speed WDM CAP 64 VLC system employing Volterra series-based nonlinear equalizer. IEEE Photonics J 7(3):1–7
18. Mitra R, Bhatia V, Jain S, Choi K (2021) Performance analysis of random Fourier features-based unsupervised multistage-clustering for VLC. IEEE Commun Lett 25(8):2659–2663
19. Mitra R, Bhatia V (2017) Finite dictionary techniques for MSER equalization in RKHS. Signal Image Video Process 11(5):1863–1811
20. Mitra R, Jain S, Bhatia V (2020) Least minimum symbol error rate based post-distortion for VLC using random Fourier features. IEEE Commun Lett 24(4):830–834
21. Mitra R, Kaddoum G, Bhatia V (2021) Hyperparameter-free transmit-nonlinearity mitigation using a kernel-width sampling technique. IEEE Trans Commun 69(4):2613–2627
22. Oliva JB, Dubey KA, Wilson AG, Póczos B, Schneider JG, Xing EP (2015) Bayesian nonparametric kernel-learning. In: International conference on artificial intelligence and statistics
23. Sharma A, Shukla VB, Bhatia V (2022) ZA-LMS based sparse channel estimator for VLC-OTFS. In: IEEE international conference on advanced networks and telecommunications systems (ANTS), pp 1–6
24. Zhong J, Zhou J, Liu W, Qin J (2020) Orthogonal time-frequency multiplexing with 2D Hermitian symmetry for optical-wireless communications. IEEE Photonics J 12(2):1–10
25. Sharma A, Jain S, Mitra R, Bhatia V (2021) Performance analysis of OTFS over multi-path channels for visible light communication. In: Proceedings of international conference on communications, signal processing, and their applications (ICCSPA), pp 1–6
26. Aalo VA, Mukasa C, Efthymoglou GP (2016) Effect of mobility on the outage and BER performances of digital transmissions over Nakagami- *m* fading channels. IEEE Trans Veh Technol 65(4):2715–2721

27. Miramirkhani F, Narmanlioglu O, Uysal M, Panayirci E (2017) A mobile channel model for VLC and application to adaptive system design. IEEE Commun Lett 21(5):1035–1038
28. Abdelhady AM, Salem AKS, Amin O, Shihada B, Alouini M-S (2021) Visible light communications via intelligent reflecting surfaces: metasurfaces vs mirror arrays. IEEE Open J Commun Soc 2:1–20
29. Kumar D, Singya PK, Krejcar O, Choi K, Bhatia V (2023) Performance of IRS-Aided FD two-way communication network with imperfect SIC. IEEE Trans Veh Technol 72(4):5491–5496

Chapter 8
Advancements in Wireless Communication with Zero Energy Devices

Mohit Kumar Sharma and Anuj Abraham

Abstract This chapter explores the recent approaches toward achieving zero-energy consumption in wireless communication systems, focusing on the integration of zero energy devices (ZED) in the wireless communication systems. First, we delve into the design of wireless communications with ZEDs powered by natural energy sources and highlight the challenges associated with it. We elaborate on the design of efficient power control policies which not only optimize the energy consumption but also facilitate the uncoordinated operation of these wireless links. Next, we introduce the captivating domain of radio frequency (RF) energy harvesting (EH) which holds significant promise as a key enabler for zero-energy devices, harnessing ambient electromagnetic signals to power communication devices and sensor networks. We provide insights into the principles, challenges, and potential applications of RF energy harvesting, by offering a comprehensive overview of wireless power transfer (WPT) and its generalization, termed as simultaneous wireless information and power transfer (SWIPT). By drawing the insights into the interplay of wireless communications and natural energy harvesting, and unraveling the possibilities of RF energy harvesting (controlled source), the conceptual frameworks presented herein provide a foundation for the design and implementation of energy-efficient, sustainable, and eco-friendly wireless communication systems.

Keywords Energy harvesting communications · Wireless communications · Zero energy devices · Wireless power transfer

M. K. Sharma (✉)
Digital Science Research Center, Technology Innovation Institute, Abu Dhabi, United Arab Emirates
e-mail: mohit.sharma@tii.ae

M. K. Sharma · A. Abraham
Technology Innovation Institute, Abu Dhabi, United Arab Emirates
e-mail: anuj.abraham@tii.ae

S. Sharma et al. (eds.), *Next-Generation Wireless Systems*,
Transactions on Computer Systems and Networks,
https://doi.org/10.1007/978-981-96-3185-8_8

8.1 Introduction

The sixth generation (6G) wireless networks envision to enable several enticing applications such as digital twin and tactile internet. These applications will rely on large number of interconnected sensors, which imposes challenges not only for maintenance but also in terms of energy consumption. For instance, as per the study undertaken in [1], the standby energy consumption of various smart applications will reach 46 TWh in year 2025, with an annual increase rate of 20%. In addition, the maintenance cost for replacing and recharging the batteries for a large number of nodes will be prohibitive. The development of ZED is a promising technology to alleviate these obstacles.

Depending on the fact whether a battery is required for their operation, the ZEDs can be classified as *active* and *passive*. In particular, the passive ZEDs operates without energy storage and uses only ambient signals to transmit their data, e.g., in backscatter communications [2]. On the other hand, the active ZEDs, which are the focus of this chapter, use a battery to store energy harvested from the environmental sources such as solar, thermal, vibrations, RF. Table 8.1 illustrates various EH sources with respect to their EH density, and characteristics. In comparison to conventional communication systems, which operate with a consistently available but limited energy source, an EH system which harvests energy from the uncontrolled energy sources has unlimited, yet time-varying energy availability. Consequently, the performance assurances and design approaches established for traditional communication systems may not be directly adaptable to networks featuring ZEDs, presenting a distinct design challenge.

In particular, the functioning of a ZED, relying solely on harvested energy, adheres to the energy conservation principle. This principle dictates that, at any given moment, the cumulative energy consumption of a node must not exceed the total energy harvested by the node up to that specific point in time. This essential limitation is known

Table 8.1 Energy density and characteristics of various EH sources

Energy source	Power density	Characteristics
Solar	$100\,\mathrm{mW/cm^2}$	Uncontrolled, partially predictable
Indoor lights	$10\text{–}100\,\mu\mathrm{W/cm^2}$	Partially controllable, predictable
Mechanical vibrations	$30\,\mathrm{mW}$	Controllable, predictable
Ambient RF	$1\text{–}10\,\mu\mathrm{W/cm^2}$	Controllable, partially predictable
Thermal	$1\text{–}10\,\mathrm{mW/cm^2}$	Uncontrollable, unpredictable

as the energy neutrality constraint (ENC) and can be expressed mathematically, in the context of a time-slotted system, as follows:

$$\sum_t E_h(t) \leq \sum_t P_c(t) \tag{8.1}$$

where $E_h(t)$ and $P_c(t)$ denote the EH rate and energy consumption rate, including both the transmit and the circuit power, at a time instant t.

As the energy harvested from the sources which cannot be *controlled* is sporadic and random in the nature, having an energy storage aids in matching the energy harvesting rate with the rate of energy consumption. The matching between the harvesting and consumption rate is attained through a power management policy $\mathcal{P} \triangleq \{P_c(t)\}$. In practice, ZEDs are equipped with a battery which can only store a finite amount of energy. Therefore, for a ZED with finite size battery, the constraint in (8.1) can be expressed, in terms of energy level in the battery, as

$$B_t = \min\{(B_{t-1} + E_h(t) - P_c(t))^+, B_{\max}\}, \tag{8.2}$$

where $(x)^+ = \max\{0, x\}$, and $B_{\max}$ and B_t denote the battery size and energy level in the battery at time instant t, respectively. A carefully designed power management policy (PMP) balances out the tradeoff between the utility and the energy outages, due to battery under/over-flow, by taking into account the statistics of the EH source and the utility function.

On the other hand, in the RF-based EH mechanism the wireless signal has a dual functionality, serving both as a carrier of information and as a potential power source. This imposes a tradeoff between the amount of harvested energy and the information transfer. In particular, the receiving node harvests energy from the information signal, either through a power-splitting mechanism or rate-splitting mechanism. The challenge lies in optimizing the system parameters to strike an equilibrium between the rate of data transmission, crucial for communication efficiency, and the harvested energy, vital for sustaining the operation of the receiver ZED. Achieving an optimal rate-energy tradeoff in this paradigm, termed as SWIPT, involves careful management of resources, such as power allocation, signal processing techniques, and modulation schemes, to effectively balance the conflicting requirements of data communication and energy harvesting. This delicate equilibrium is pivotal for enhancing the overall performance and sustainability of wireless systems capable of simultaneous information and power transfer.

In the rest of the chapter, we describe these design trade-offs in detail for wireless communications with both uncontrolled and controlled EH mechanisms. In Sects. 8.2 and 8.3, we elaborate on the modeling and design challenges pertaining to communication systems design for uncontrolled and controlled EH sources. In Sect. 8.4, we present a discussion on the future research directions.

8.2 Wireless Communications with Uncontrolled EH Sources

In this section, we discuss the design of a PMP for a point-to-point EH link between two ZEDs, to illustrate the design challenges involved in wireless communications with uncontrolled EH sources. We examine a scenario in which a ZED transmits its delay-insensitive saturated traffic data to another ZED over an additive white Gaussian noise (AWGN) channel. The goal is to develop a PMP to maximize the average sum-throughput of this point-to-point link between ZEDs, accounting for the energy cost of decoding at the receiving ZED, while the transmitter and receiver do not have information about the energy level and the actions of each other, i.e., operating in an *uncoordinated fashion*.

As shown in Fig. 8.1, the transmitting ZED sends its data to the receiver, in a time-slotted fashion, over an AWGN communication channel. In a slot, the ZEDs harvest energy according to an stationary and ergodic discrete-valued harvesting process with joint probability mass function $P_{\mathcal{E}^t,\mathcal{E}^r}(E^t, E^r)$, where random variables $\mathcal{E}^t$ and $\mathcal{E}^r$ denote the amount of energy harvested by the transmitter and receiver, respectively, and E_n^t and E_n^r denote a realization of these random variables in the nth slot, respectively. The harvesting rates at the transmitter and receiver are $\mu_t \triangleq \mathbb{E}\left[\mathcal{E}^t\right]$ and $\mu_r \triangleq \mathbb{E}\left[\mathcal{E}^r\right]$, respectively. At each ZED, the harvested energy is stored in a leakage-free, *finite* capacity battery. Both the transmitting and receiving ZED knows

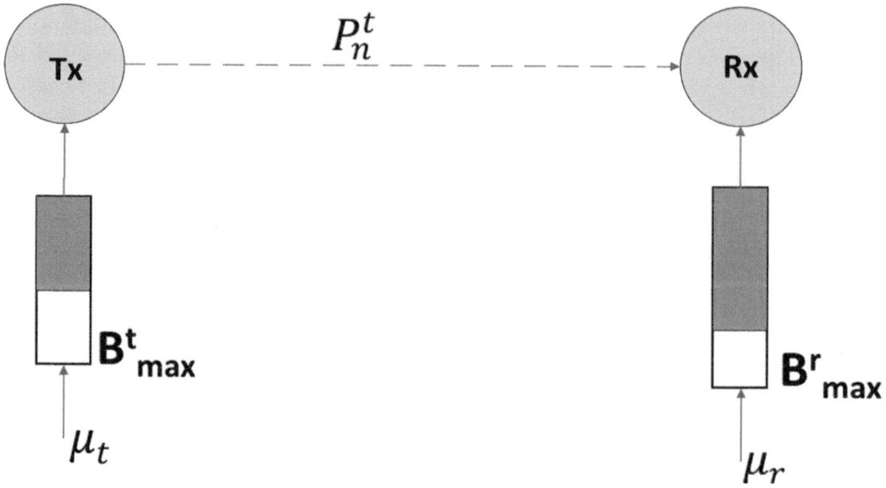

Fig. 8.1 System model for an uncoordinated point-to-point EH link with a ZEDs transmitter and receiver. The average harvesting rates for the transmitter and the receiver are represented by μ_t and μ_r, respectively. Further, the battery capacities for the transmitter and the receiver are denoted as $B_{\max}^t$ and $B_{\max}^r$, respectively

the harvesting *rates* of both the ZEDs; any other information such as the instantaneous battery state and energy usages is not known to the other ZED.

The energy used by the receiver in the nth slot is denoted by $P_n^r \in \{0, R\}$, i.e., if it decides to turn *on* in a slot it consumes R units of energy; and it does not incur any energy cost in the sleep mode. On the other hand, the power control policy at the transmitter is continuous valued. Let P_n^t denote the amount of energy used by the transmitter in the nth slot. The evolution of the battery at each ZED can be expressed as follows $B_{n+1}^k = \min\{[B_n^k + E_n^k - P_n^k]^+, B_{\max}^k\}$, where $B_{\max}^k$, B_n^k and E_n^k denote the battery capacity, battery level, and the energy harvested at the start of the nth slot, for kth ZED, respectively, for $k \in \{t, r\}$.

The data rate achieved with transmit power P_n^t is well approximated by the capacity expression $\mathcal{R}(P_n^t) \triangleq \frac{1}{2}\log(1 + P_n^t)$[3, 4]. Without loss of generality, we set the power spectral density of the AWGN at the receiver as unity. Due to the uncoordinated operation, P_n^k depends only on the ZED's own battery state, B_n^k, the energy harvested by it in the current slot, E_n^k, and the slot index n. Under this model, the time-averaged sum-throughput can be written as

$$\mathcal{T} = \frac{1}{N}\sum_{n=1}^{N} \mathbb{1}_{\{P_n^r \neq 0\}} \cdot \log(1 + P_n^t), \tag{8.3}$$

where $\mathbb{1}_{\{P_n^r \neq 0\}}$ is an indicator variable which equals one if the receiver is *on* in the nth slot, and takes the value zero otherwise. This captures the fact that data is received successfully only if the receiver is *on* in that slot. Our goal is to find a decentralized PMP that maximizes the time-averaged sum-throughput. The optimization problem can then be formulated as

$$\max_{\{P_n^t, P_n^r,\ \text{for } n=1,2,\ldots\}} \liminf_{N\to\infty} \mathcal{T}, \tag{8.4}$$

$$\text{s.t.}\quad 0 \leq P_n^t \leq B_n^t, \tag{8.5}$$

$$\text{and}\quad 0 \leq P_n^r \leq B_n^r,\ P_n^r \in \{0, R\},\ \text{for all } n. \tag{8.6}$$

In the above optimization problem, the constraints (8.5) and (8.6) ensure that, in an any given slot, the maximum energy used by a ZED does not exceed the energy available in its battery.

In order to maximize the sum-throughput in (8.3), an optimal policy needs to ensure that both the nodes are *on* simultaneously to ensure a successful data transmission. In addition, the transmit power must be chosen optimally at the transmitter, using the local information only. The uncoordinated nature of the problem makes it challenging to achieve the optimal throughput. In the next section, we derive an upper bound on achievable throughput, which serves as a benchmark for the performance of any policy.

8.2.1 Maximum Achievable Throughput

The upper bound on the sum-throughput, presented in Lemma 8.1 below, is derived by considering a genie-aided system where the ZEDs are equipped with infinite-sized batteries and have non-causal information about the energy arrivals.

Lemma 8.1 *The long-term time-averaged sum-throughput of an uncoordinated point-to-point link with ZEDs satisfies*

1. $\liminf_{T\to\infty} \mathcal{T} \le \frac{1}{2}\log(1+\mu_t)$, *if* $\frac{R}{\mu_r} \le 1$,
2. $\liminf_{T\to\infty} \mathcal{T} \le \frac{\mu_r}{2R}\log\left(1+\frac{R\mu_t}{\mu_r}\right)$, *if* $\frac{R}{\mu_r} > 1$,

where μ_t *and* μ_r *denote the harvesting rate of the transmitter and receiver, respectively, and* R *denotes the energy required per slot for decoding data at the receiver.*

The detailed proof of the above result can be found in [5]. It is interesting to note that the expression for the upper bound in Case 1 is the same as the expression for the capacity of a point-to-point communication system where the receiver is connected to the mains [6] and only transmitter is a ZED. In particular, the Case 1 corresponds to the setting where the average harvesting rate at the receiver exceeds R, the energy consumed by it per slot when it is *on*. Thus, the battery state of the receiver has a positive drift[1] even if it remains *on* in all slots, i.e., the receiver is energy *unconstrained* and can remain *on* in every slot, provided the receiver is equipped with an infinitely large battery. Therefore, the PMP design exercise needs to focus only on the PMP design for transmitting ZED. On the other hand, Case 2 corresponds to a scenario where the receiver is energy constrained, i.e., the average energy harvested in a slot is less than the energy consumed in one slot. Consequently, the receiver can only turn *on* intermittently. To avoid wastage of energy, the transmitters must not transmit when the receiver is *off*. However, this requires the transmitters to know the battery state at the receiver, which is not available in an uncoordinated operation. In the following, we present asymptotically optimal policies for both the scenarios in Lemma 8.1.

8.2.2 PMP with Energy Unconstrained Receiver

In this section, we present an asymptotically optimal PMP for the scenario represented by Case 1 above. Here, asymptotic optimality implies the optimality as the battery size at the transmitter and the receiver goes to infinity. As in this case the EH rate at the receiver, μ_r, is more than the energy required for decoding in a slot, R, the receiver can always remain *on*, provided it has an infinitely large energy storage.

[1] Note that, the difference between the average energy harvesting and consumption rates is termed as the battery *drift*.

This holds true even in the scenario when the battery size at the receiver is finite but sufficiently large. More precisely, it is known that, when $\mu_r > R$, the probability that the receiver does not have sufficient energy to turn *on*, $\Pr[B_n^r < R]$, decays exponentially with $B_{\max}^r$ [7].

Consequently, the PMP in this case only need to be concerned with the energy management at the transmitter ZED. Hence, when $\mu_r > R$, the optimal policy for the transmitter, denoted by $\mathcal{P}_u^t \triangleq \{P_n^t, n = 1, 2, \ldots\}$, is the same as the one proposed in [8], which is as follows:

$$P_n^t = \begin{cases} \mu_t + \delta^t, & B_n^t \geq \frac{B_{\max}^t}{2}, \\ \min\{B_n^t, \mu_t - \delta^t\}, & B_n^t < \frac{B_{\max}^t}{2}, \end{cases} \tag{8.7}$$

where $\delta^t = \beta^t \sigma_t^2 \frac{\log B_{\max}^t}{B_{\max}^t}$. Here, σ_t^2 denotes the asymptotic variance of the harvesting process at the transmitter, and $\beta^t \geq 2$ is a constant. It follows from [8, Theorem 1] that the throughput achieved by this policy converges to the upper bound in Case *1* of Lemma 8.1 at the rate $\Theta\left(\left(\frac{\log B_{\max}^t}{B_{\max}^t}\right)^2\right)$.[2] Thus, for a network with an EH transmitter and an energy unconstrained receiver, the policy $\mathcal{P}_u^t$, along with the policy under which the receiver is always *on*, is *asymptotically optimal.* We note that the policy is *fully uncoordinated*, i.e., it operates using the knowledge about the local battery state and does not require the information about the battery state or action taken by the other ZED.

Figure 8.2 shows the average per slot throughput when the receiver is energy unconstrained. The performance of the proposed policy is evaluated by computing the time-averaged throughput using Monte Carlo simulations of the system. The harvesting processes at the transmitter and receiver are assumed to be distributed according to spatially and temporally independent and identically distributed Bernoulli distribution with harvesting probabilities ρ_t and ρ_r, respectively. We note that the policy given in (8.7) achieves the upper bound derived in Lemma 8.1. In this case, the harvesting rate at the transmitter completely determines the average throughput performance.

In the next section, we consider the Case 2 of Lemma 8.1. Intuitively, in this case, emulating a policy similar to $\mathcal{P}_u^t$, given in (8.7), at both the ZEDs can help in achieving a throughput close to upper bound. In particular, the policy $\mathcal{P}_u^t$ can be directly adapted to be used at the transmitter. On the other hand, the receiver can implement a policy similar to $\mathcal{P}_u^t$ by *deterministically* turning *on* in one out of every $\lfloor N_r \rfloor$[3] or $\lceil N_r \rceil$ slots, where $N_r \triangleq \frac{R}{\mu_r}$, depending on whether the battery at the receiver is above or below the half-full mark, $\frac{B_{\max}^r}{2}$, respectively. To facilitate coordination, the receiver sends one-bit of feedback to the transmitter whenever its battery level crosses the halfway mark. This allows the transmitter to determine the index of the next slot in which the receiver will be *on*. However, at the receiver, the drifts in lower and upper halves,

---

[2] $f_n = \Theta(g_n)$ implies that both f_n and g_n goes to zero at the same rate.

[3] This results in a negative drift in the energy level of the battery, since $\lfloor N_r \rfloor \mu_r - R \leq 0$.

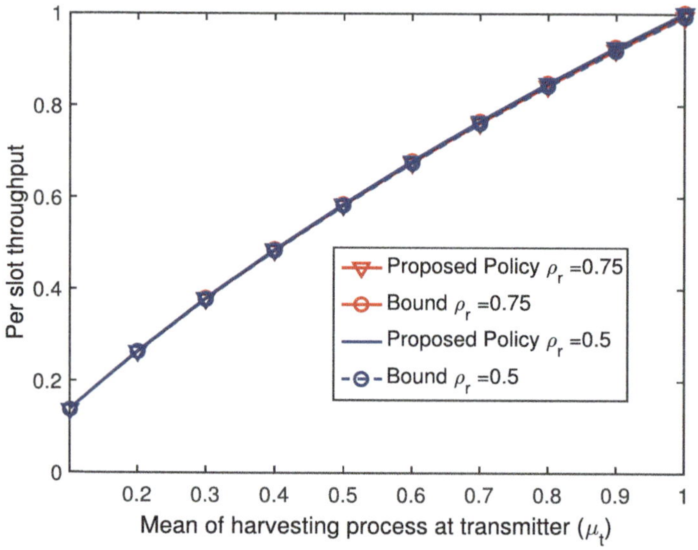

Fig. 8.2 Performance of the policy $\mathcal{P}_u^t$ for a point-to-point link with an energy unconstrained receiver: The policy presented in (8.7) achieves the upper bound presented in Lemma 8.1. The experiment was conducted with $R = 0.5$ and $B_{\max}^t = B_{\max}^r = 50$

$\lceil N_r \rceil - N_r$ and $N_r - \lfloor N_r \rfloor$, respectively, do not decay with the size of its battery. As a consequence, the throughput achieved by this policy can have a one-bit gap from the upper bound in Lemma 8.1. Therefore, we present a time-dilation-based policy that leads to asymptotically optimal performance by enabling finer control over the battery drifts.

8.2.3 PMP with Energy Constrained Receiver

In this section, we present a time-dilation-based policy that requires *occasional* one-bit feedback and asymptotically achieves the upper bound presented in Lemma 8.1. Before presenting the time-dilation-based policy, we first briefly describe a policy which asymptotically achieves a throughput within one-bit of the upper bound. The operation of the policy can be described as follows. The receiver sends a one-bit feedback whenever the battery level crosses the half-full mark, which is received without any error and delay. The one-bit feedback enables the transmitter to track whether the receiver's battery is more than or less than half full. Further, the receiver executes a deterministic policy in either half of the battery state; and the transmitter follows the receiver's policy and transmits only in slots where the receiver is also *on*. In the slots when the receiver is *off*, the transmitter accumulates the energy prescribed by the policy $\mathcal{P}_{\mathrm{u}}^{t}$ in a supercapacitor and uses the accumulated energy for transmission in the next slot when the receiver turns on.[4]

Mathematically, the policy is described as follows. The energy accumulated in the supercapacitor by the end of nth slot, given that the transmitter does *not* transmit in that slot, is given by

$$\mathcal{C}_n = \begin{cases} \mathcal{C}_{n-1} + \mu_t + \delta^t, & \text{if } B_n^t \geq \frac{B_{\max}^t}{2}, \\ \mathcal{C}_{n-1} + \min\{\mu_t - \delta^t, B_n^t\}, & \text{if } B_n^t < \frac{B_{\max}^t}{2}, \end{cases} \tag{8.8}$$

and $\mathcal{C}_n = 0$, if data is transmitted in the nth slot, and $\delta^t = \beta^t \sigma_t^2 \frac{\log B_{\max}^t}{B_{\max}^t}$. The size of the supercapacitor is chosen such that it is sufficient to store the energy accumulated between two consecutive data transmissions. Let $\mathbb{1}_{\mathcal{R}^+}$ denote an indicator function which takes the value one if $B_n^r \geq \frac{B_{\max}^r}{2}$ and zero otherwise. Also, let $N_r^+ \triangleq \lfloor \frac{R}{\mu_r} \rfloor$, and $N_r^- \triangleq \lceil \frac{R}{\mu_r} \rceil$. In the nth slot, the transmitter follows the policy $\mathcal{P}_n^t$ given by

$$P_n^t = \begin{cases} \mathcal{C}_{n-1} + \mu_t + \delta^t, & \text{if } B_n^t \geq \frac{B_{\max}^t}{2}, n = N_{\text{on}} + N_r^+ \mathbb{1}_{\mathcal{R}^+} + N_r^- \mathbb{1}_{\mathcal{R}^-}, \\ \mathcal{C}_{n-1} + \min\{\mu_t - \delta^t, B_n^t\}, & B_n^t < \frac{B_{\max}^t}{2}, n = N_{\text{on}} + N_r^+ \mathbb{1}_{\mathcal{R}^+} + N_r^- \mathbb{1}_{\mathcal{R}^-}, \\ 0 & \text{otherwise.} \end{cases} \tag{8.9}$$

In the above, $\mathbb{1}_{\mathcal{R}^-} \triangleq 1 - \mathbb{1}_{\mathcal{R}^+}$, and N_{on} denotes the previous slot when the transmitter and receiver were scheduled to turn *on*. It is initialized to zero at the first slot ($N_{\text{on}} = 0$ when $n = 0$), and at any slot index n satisfying $n = N_{\text{on}} + N_r^+ \mathbb{1}_{\mathcal{R}^+} + N_r^- (1 - \mathbb{1}_{\mathcal{R}^+})$, the transmitter and receiver make an attempt if they have energy, and N_{on} is set to $N_{\text{on}} = n$, i.e., it is updated to the current slot index. The policy (8.9) is derived using the policy given in (8.7), i.e., in each slot, the transmitter computes the energy prescribed by $\mathcal{P}_u^t$ for that slot and transfers the energy from the battery to the supercapacitor. In a slot when the receiver is *on*, the transmitter uses all the energy accumulated in the supercapacitor till that slot to transmit its data. On the other hand, the policy at the receiver is given as

$$P_n^r = \begin{cases} R, & B_n^r \geq \frac{B_{\max}^r}{2}, n = N_{\text{on}} + N_r^+ \\ R, & R \leq B_n^r < \frac{B_{\max}^r}{2}, n = N_{\text{on}} + N_r^-, \\ 0 & \text{otherwise.} \end{cases} \tag{8.10}$$

The receiver's policy $\mathcal{P}_n^r \triangleq \{P_n^r\}_{n=1}^N$ also emulates the policy $\mathcal{P}_u^t$ given in (8.7). Specifically, the receiver executes a policy similar to $\mathcal{P}_u^t$ by turning on after

[4] The consequence of using a supercapacitor to temporarily store energy is that the battery energy consumption at the transmitter depends only on its own battery state; specifically, it is independent of the policy at the receiver.

$N_r^+ \triangleq \lfloor \frac{R}{\mu^r} \rfloor$ slots (resulting in a negative drift in the battery state) if the battery is more than half full, otherwise it turns on after $N_r^- \triangleq \lceil \frac{R}{\mu_r} \rceil$ slots (resulting in a positive drift in the battery state).

The throughput achieved by the above-described policy is within *1 bit* of the upper bound when the battery capacity is large. In addition, the probability of battery discharge decays polynomially with the battery size. It is interesting to note that the one-bit gap in the throughput arises because of the receiver's policy. As, the drift in the receiver's battery remains fixed at $\delta_r^- = N_r^- - N_r$ when $B_n^r < B_{\max}^r/2$ and $\delta_r^+ \triangleq N_r - N_r^+$ when $B_n^r \geq B_{\max}^r/2$, irrespective of the value of $B_{\max}^r$. In order to close the gap, we need finer control over the battery drift at the receiver. We need it to be of the order $O(1/B_{\max}^r)$, similar to that at the transmitter. This can be achieved using time-dilation, as described next.

Optimal Throughput via Time-dilation: The key idea behind time-dilation is to spread the drift δ_r^+ and δ_r^- at the receiver over a larger number of slots, resulting in a smaller per-slot drift. That is, instead of (8.10), which operates in batches of $\lfloor \frac{R}{\mu_r} \rfloor$ or $\lceil \frac{R}{\mu_r} \rceil$ slots, we consider a policy that operates in batches of $\lfloor \frac{Rf(B_{\max}^r)}{\mu_r} \rfloor$ and $\lceil \frac{Rf(B_{\max}^r)}{\mu_r} \rceil$ slots, where $f(\cdot) > 1$ is a time-dilation function. For example, if $f(B_{\max}^r)$ is an integer, the time-dilated policy turns the receiver *on* for $f(B_{\max}^r)$ slots out of $\lfloor \frac{Rf(B_{\max}^r)}{\mu_r} \rfloor$ slots if the battery at the receiver is more than half full, and it turns the receiver on for $f(B_{\max}^r)$ slots out of $\lceil \frac{Rf(B_{\max}^r)}{\mu_r} \rceil$ if the battery is less than half full. This results in a drift of

$$\delta_{r,f}^+(B_{\max}^r) = f(B_{\max}^r)N_r - \left\lfloor \frac{Rf(B_{\max}^r)}{\mu_r} \right\rfloor \tag{8.11}$$

$$\delta_{r,f}^-(B_{\max}^r) = \left\lceil \frac{Rf(B_{\max}^r)}{\mu_r} \right\rceil - f(B_{\max}^r)N_r \tag{8.12}$$

over $\lfloor \frac{Rf(B_{\max}^r)}{\mu_r} \rfloor$ and $\lceil \frac{Rf(B_{\max}^r)}{\mu_r} \rceil$ slots, respectively. Hence, the per-slot drift is given by

$$\delta_{\text{eff}}^+ = \frac{\delta_{r,f}^+(B_{\max}^r)}{\lfloor \frac{Rf(B_{\max}^r)}{\mu_r} \rfloor} \quad \text{and} \quad \delta_{\text{eff}}^- = \frac{\delta_{r,f}^+(B_{\max}^r)}{\lceil \frac{Rf(B_{\max}^r)}{\mu_r} \rceil}. \tag{8.13}$$

The transmit policy is still determined according to (8.9). Furthermore, with the help of the one-bit feedback, the transmitter can ensure that it transmits only in the $f(B_{\max}^r)$ slots when the receiver is *on*. The dynamics of the policy under time dilation is similar to the dynamics of the policy described in the previous section, as long as the dilation function $f(B_{\max}^r)$ is sub-linear in $B_{\max}^r$. It can be shown that the gap between the upper bound and the throughput achieved by the time-dilation policy can be made to approach zero as the battery size at the receiver gets large. Also, the policy operates over a longer time window and has a smaller per-slot drift and lower rate of crossing the half-full mark, resulting in a smaller feedback overhead.

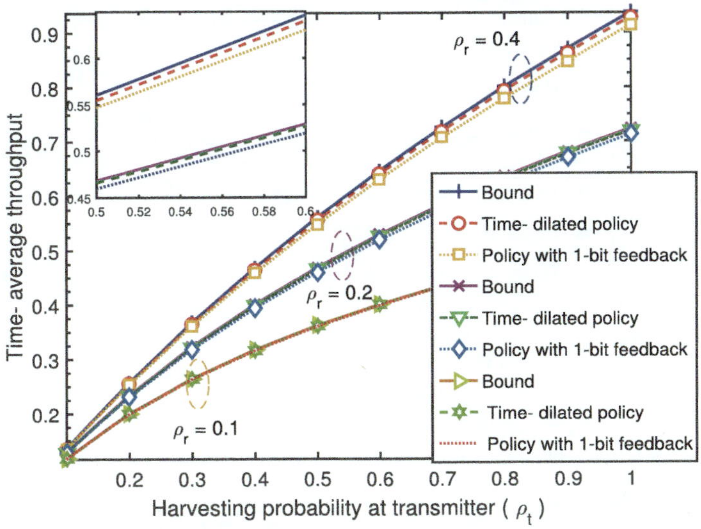

Fig. 8.3 Energy constrained receiver: The policy $\mathcal{P}^c$ with occasional one-bit feedback, achieves a throughput close to upper bound. The time-dilation further improves its performance. The result corresponds to time-dilation $f(\cdot) = 100$. Other parameters are $R = 0.5$ and $B^t_{\max} = B^r_{\max} = 1000$

The performance of the above-described policy with and without time-dilation is shown in Fig. 8.3 by plotting the average per slot throughput when the receiver is energy constrained. The performance of policy, which requires an occasional one-bit feedback, is benchmarked against the upper bound. We see that the throughput of the policy is very close to the upper bound. The figure also shows that the time-dilated policy further closes the gap to the upper bound. In Fig. 8.4, we study the impact of the battery size at the two nodes on the performance of the policy for a system with an energy constrained receiver. The per slot throughput achieved by the policy is near-optimal even with small capacity batteries.

Further Extensions: To achieve a truly uncoordinated operation, it is possible to dispense even the one-bit feedback. In order to do this, we can design a policy which uses a similar policy architecture as the above-described time-dilation-based policy, the only difference being the policy adopted at the receiver. In particular, receiver turn *on* deterministically in every $N_r = \lceil \frac{Rf(B^r_{\max})}{\mu_r} \rceil$. The performance of such a policy is shown in Fig. 8.5. It can be observed that the fully uncoordinated policy achieves a throughput performance close to time-dilation-based policy. The performance loss of such policy, compared to the upper bound, is characterized in [9].

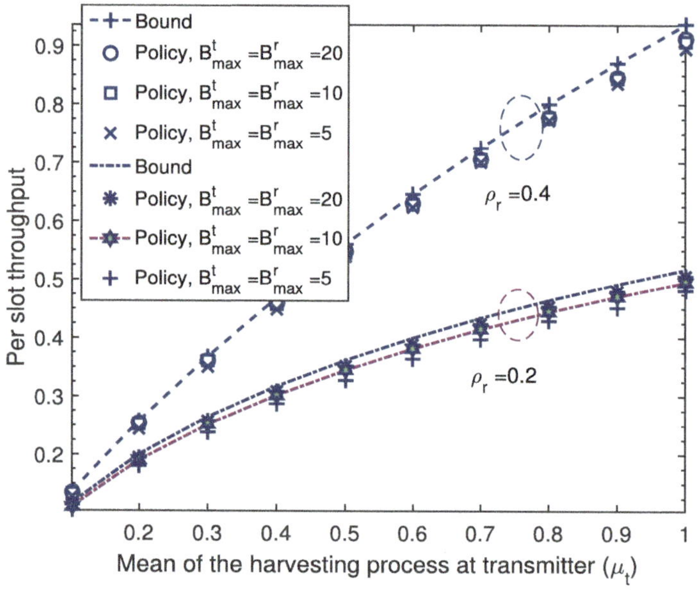

Fig. 8.4 Impact of battery size on the throughput of policy, for $R = 0.5$

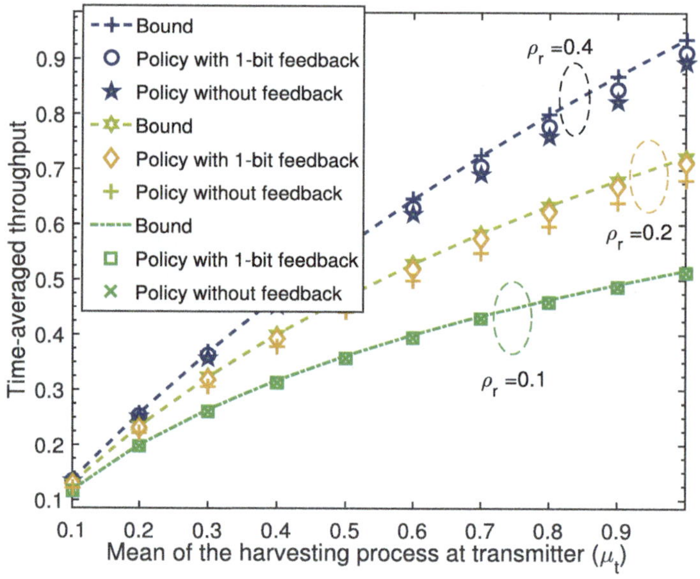

Fig. 8.5 The fully uncoordinated policy achieves a throughput close to that of the policy for $\rho_r = 0.1, 0.2$, and 0.4, respectively. Other parameters: $B^t_{\max} = B^r_{\max} = 50$, $R = 0.5$

8.3 Wireless Communications with Controlled EH Sources

In contrast to the design of wireless communications for the ZEDs which harvest energy from natural, but uncontrolled sources, the design with controlled EH sources, does not need to deal with the randomness of energy arrivals. Specifically, the energy harvesting by leveraging the RF signals present in the environment makes it possible to control the amount and timing of the harvested energy. The energy harvesting from ambient RF signals can be characterized by two distinct operating paradigms, termed as wireless information transfer (WIT) and WPT, with WIT focusing solely on communication and WPT concentrating solely on power transfer through RF signals. A SWIPT based design offers a flexible solution that can seamlessly transition between these extremes. For instance, in a SWIPT based system an RF signal can carry information for a high-power device, while simultaneously charging a low-power device. Nonetheless, to develop a fully functional SWIPT paradigm would require to address several challenges across diverse disciplines, ranging from the design of efficient RF solutions, circuits, antennas, rectifiers, and power management units, to develop communications and signal processing techniques to enhance the harvesting efficiency. In the following, we first review the basics of RF based energy harvesting systems and discuss the fundamental challenges involved in the design of communications and signaling scheme for efficient SWIPT based systems.

8.3.1 Transmit and Receive Signal Model for WPT

As shown in Fig. 8.6, a generic point-to-point WPT system is composed of an energy transmitter (ET) and energy receiver (ER). At the ET a direct current (DC) signal is up-converted to RF domain, which is then transmitted over a wireless channel to the ER. At the ER, the received RF domain signal is converted back to the DC using a rectifier coupled with the receiver antenna. The converted DC power is either used for directly powering a low-power circuit or is stored in a battery, with the help of a power management unit. The end-to-end power conversion efficiency of the WPT process can be expressed as

$$\eta = \frac{P^r_{\mathrm{dc}}}{P^t_{\mathrm{dc}}} = \underbrace{\frac{P^t_{\mathrm{rf}}}{P^t_{\mathrm{dc}}}}_{\eta_t} \cdot \underbrace{\frac{P^r_{\mathrm{rf}}}{P^t_{\mathrm{rf}}}}_{\eta_c} \cdot \underbrace{\frac{P^r_{\mathrm{dc}}}{P^r_{\mathrm{rf}}}}_{\eta_r}, \tag{8.14}$$

where η_t, η_c, and η_r denote the energy conversion efficiencies at the transmitter, over wireless transmission, and at the receiver, respectively.

To better understand the impact of various system parameters on these conversion efficiencies, we consider the WPT over a point-to-point multi-input multi-output (MIMO) link. In this setup, the ET and ER are equipped with M and K transmit antennas, respectively, and the signal is transmitted over N orthogonal frequency

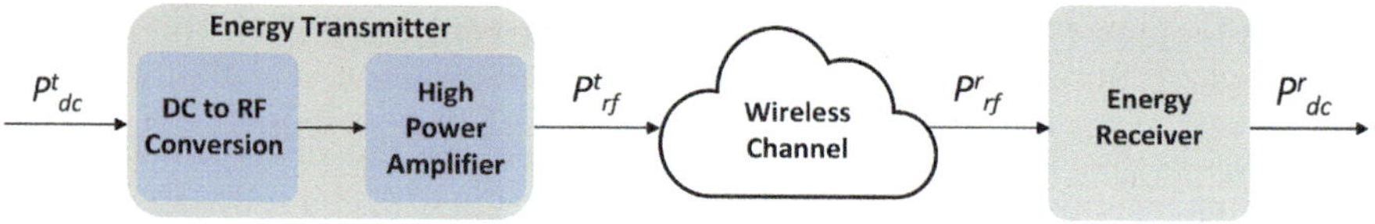

Fig. 8.6 A generic WPT system

sub-bands. The nth frequency sub-band has carrier frequency f_n with bandwidth f_w, $n = 0, \ldots, N-1$. All the subcarriers are evenly spaced with an inter-carrier spacing $f_w \le \Delta f$, such that $f_n = f_0 + n\Delta f$. At time t, the WPT signal transmitted on antenna m, $x_{\text{rf},m}(t)$, is a multi-carrier modulated waveform with frequencies f_n, with $n = 0, 1, \ldots, N-1$, carrying independent symbols. Therefore, the input signal to the high power amplifier (HPA) of the mth antenna is given as

$$x_{\text{in,m}} = \sqrt{2}\mathcal{R}\left[\sum_{n=0}^{N-1} x_{m,n}(t)e^{j2\pi f_n t}\right], \tag{8.15}$$

where $\mathcal{R}[\cdot]$ denote the real part of a complex number, and the baseband equivalent signal $x_{m,n}(t)$, with bandwidth $\left[-\frac{f_w}{2}, \frac{f_w}{2}\right]$, can be expressed as

$$x_{m,n}(t) = \sum_{j=-\infty}^{\infty} x_{m,n,j} \cdot \text{sinc}[f_w t - j], \tag{8.16}$$

where $\text{sinc}(t) = \frac{\sin(\pi t)}{\pi t}$, and $x_{m,n,j}$ denotes a power carrying complex symbol at time j and is modeled as an independently and identically distributed (i.i.d.) random variable.

Next, the input signal on each antenna m, $x_{\text{in,m}}(t)$, is amplified by a HPA and filtered by a band-pass filter to produce the transmit WPT signal, expressed as

$$x_{\text{rf},m}(t) = \sqrt{2}\mathcal{R}\left[\textstyle\sum_{n=0}^{N-1} x_{\text{rf},m,n}(t)e^{j2\pi f_n t}\right],$$
$$\text{where, } x_{\text{rf},m,n}(t) = \textstyle\sum_{j=-\infty}^{\infty} x_{\text{rf},m,n,j} \cdot \text{sinc}[f_w t - j]$$

In practice, the relationship between the input and output signal of the HPA, $x_{\text{in},m}(t)$ and $x_{\text{rf},m}(t)$, is nonlinear and is given as

$$f_{HPA}(x_{in}, m)(t) = \frac{G}{\left(1 + \left(\frac{G|x_{\text{in},m}(t)|}{A_s}\right)^{2\beta}\right)^{\frac{1}{2\beta}}} x_{\text{in},m}(t),$$

where G denotes the amplification gain, and A_s and β denote the output saturation voltage and smoothness of the transition from the linear region to saturation, respectively. In particular, the HPA operates in the linear regime if the input voltage is significantly smaller than A_s, otherwise, it operates in the nonlinear regime. The above relationship accounts for coupling across frequencies, amplitude, and phase distortion. This nonlinear relationship leads to $P_{\text{rf}}^t = \eta_t(x_{\text{in},m}(t))P_{\text{DC}}^t$. Note that, the energy conversion efficiency η_t is a nonlinear function of input signal $x_{\text{in},m}(t)$.

The generated WPT signal, $x_{\text{rf},m}(t)$ is transmitted over a multipath wireless channel, with L multipaths. The signal received at the kth receive antenna, $1 \leq k \leq K$, from the mth antenna, can be expressed as

$$y_{\text{rf},k,m}(t) = \sqrt{2}\mathcal{R}\left[\sum_{\ell=0}^{L}\sum_{n=0}^{N-1}\alpha_\ell x_{\text{rf},m,n}(t-\tau_\ell)e^{j2\pi f_n(t-\tau_\ell+\theta_{k,m,n,\ell})}\right], \quad (8.17)$$

$$\approx \sqrt{2}\mathcal{R}\left[\sum_{n=0}^{N-1}h_{k,m,n}x_{\text{rf},m,n}(t)e^{j2\pi f_n t}\right]. \quad (8.18)$$

In the above, $h_{k,m,n} = \sum_{\ell=0}^{L-1}\alpha_\ell e^{-j2\pi f_n \tau_\ell + \theta_{k,m,n,\ell}}$ denotes the baseband channel frequency response between transmit and receive antenna m and n, respectively, at frequency f_n. Here, α_ℓ and τ_ℓ denotes the pathloss and delay of the ℓth path, $1 \leq \ell \leq L$, respectively. Further, $\theta_{k,m,n,\ell}$ denote the phase shift induced by the ℓth path between the transmit antenna m and receive antenna k, over the nth sub-band. The composite signal received at the receive antenna k is formed of superposition of the signals received from all M transmit antennas, and can be written as

$$y_{\text{rf},k}(t) = \sqrt{2}\mathcal{R}\left[\sum_{n=0}^{N-1}\mathbf{h}_{k,n}\mathbf{x}_{\text{rf},n}(t)e^{j2\pi f_n t}\right], \quad (8.19)$$

with $\mathbf{h}_{k,n} \triangleq \left[h_{k,1,n}, \ldots, h_{k,M,n}\right]$ denotes the channel between the M transmit antennas and the kth receive antenna, and $\mathbf{x}_{\text{rf},n} \triangleq \left[x_{\text{rf},1,n}(t), \ldots, x_{\text{rf},M,n}\right]^T$. Here, $[\cdot]^T$ denotes the transpose of a vector.

Using the above signal model, in the following, we derive the expression for P_{dc}^t, P_{rf}^t, and P_{rf}^r.

8.3.2 Expressions for Transmit and Receive Power

Using (8.15), the power at the input of the HPA at the transmitter can be written as

$$P_{\text{dc}}^t = \sum_{n=0}^{N-1}\text{Tr}(\mathbf{P}_n) = \text{Tr}(\mathbf{P}), \quad (8.20)$$

where $\text{Tr}(\cdot)$ denotes the trace of a matrix, and $\mathbf{P} = \text{diag}\{\mathbf{P}_0, \ldots, \mathbf{P}_{N-1}\}$. Here, the positive semi-definite input covariance matrix $\mathbf{P}_n$ is defined as $\mathbf{P}_n \triangleq \mathbb{E}[\mathbf{x}_n(t)\mathbf{x}_n^H(t)] \in \mathbb{C}^{M\times M}$ with $\mathbf{x}_n^H(t)$ denoting the Hermitian transpose of

$\mathbf{x}_n^H(t) \triangleq [x_{1,n}(t), \ldots, x_{M,n}(t)]^T$ denoting the signal vector across the M antennas. Further, (8.20) can be rewritten as

$$P_{dc}^t = \sum_{n=0}^{N-1} P_n = \text{Tr}(\mathbf{P}), \tag{8.21}$$

where $P_n \triangleq \text{Tr}(\mathbf{P}_n)$ denotes the transmit power in the sub-band n. Similarly, the total average transmit power can by written as sum of transmit powers across all the transmit antennas,

$$P_{\text{rf}}^t = \sum_{m=1}^{M} \mathbb{E}\{[x_{\text{rf},m}(t)]^2\}, \text{ such that } P_{\text{rf}}^t \leq P. \tag{8.22}$$

On the other hand, the total RF power received can be written as sum of power received across the antennas,

$$P_{\text{rf}}^r = \sum_{k=1}^{K} \mathbb{E}\{[y_{\text{rf},k}(t)]^2\} = \sum_{k=1}^{K}\sum_{n=1}^{N-1} \mathbb{E}\{|\mathbf{h}_{k,n}\mathbf{x}_{\text{rf},n}(t)|^2\}. \tag{8.23}$$

As noted above, the received RF signal is leveraged to extract the DC power. The amount of extracted DC power, P_{dc}^r depends on the architecture deployed at the receiver. The commonly used architectures are described in the following.

8.3.3 Energy Receiver Architecture

At the receiver, a rectenna is used to harvest DC energy from the received RF signal. A rectenna consists of a rectifier which can be tuned for a specific operating frequency, input power level, and a particular input waveform. A single series rectifier, shown in Fig. 8.7, represent the most basic form of a rectifier. The output DC voltage of the rectifier v_{out} can be expressed as a nonlinear function of the received RF signal $y_{\text{rf}}(t)$, as follows,[5]

$$v_{\text{out}} = f_{\text{rect.}}(y_{\text{rf}}(t)) = \sum_{\{p\geq 2, \text{s.t.} p \text{ is even}\}}^{n_0} \psi_p \mathbb{E}\left[y_{\text{rf}}, (t)^p\right] \tag{8.24}$$

[5] At a multi-antenna receiver, either DC combining or RF combining is used for combining the rectifier output or RF received signal, respectively. In RF combining, the combined received signal is fed to the rectifier.

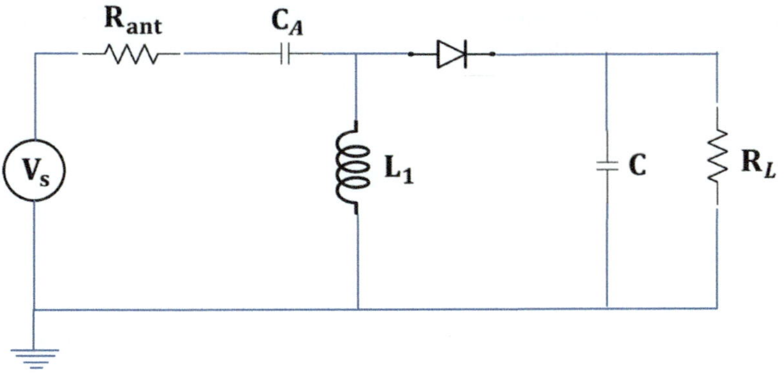

Fig. 8.7 Single series rectifier

where $\psi_p = \frac{R_{\text{ant}}^{p/2}}{p!(\eta v_t)^{(p-1)}}$ and R_{ant} denotes the antenna resistance. Further, v_t and η denote the thermal voltage and the ideality factor, and $\mathbb{E}\{\cdot\}$ denotes the averaging over the randomness of the received signal, $y_{\text{rf}}(t)$. From (8.24), the harvested DC power can be written as

$$P_{\text{dc}}^{r} = \frac{v_{\text{out}}^{2}}{R_L}. \tag{8.25}$$

From above, we observe that the output voltage is not only a function of the input signal average power $P_{\text{rf}}^{r} = \mathbb{E}\{y_{\text{rf}}(t)^2\}$, but also of higher order moments $\mathbb{E}\{y_{\text{rf}}(t)^p\}$. Moreover, using Jensen's inequality, for $p \geq 2$ and even,

$$\mathbb{E}\{y_{\text{rf}}(t)^p\} \leq \left(\mathbb{E}\{y_{\text{rf}}(t)^2\}\right)^{\frac{p}{2}} = (P_{\text{rf}}^{r})^{\frac{p}{2}},$$

which results in

$$v_{\text{out}} \geq \sum_{\{p \geq 2, \text{s.t.} p \text{ is even}\}}^{n_0} \psi_p (P_{\text{rf}}^{r})^{\frac{p}{2}} \tag{8.26}$$

This illustrates a key role of choosing input signals with large $\mathbb{E}\left[y_{\text{rf}}(t)^p\right]$. Indeed, for a multi-tone receive signal, $y_{\text{rf}}(t)$, $\mathbb{E}\left[y_{\text{rf}}(t)^4\right]$ scales proportional to $N(P_{\text{rf}}^{r})^2$, therefore for a sufficiently large N, it can be easily established that $\mathbb{E}\left[y_{\text{rf}}(t)^4\right] > (P_{\text{rf}})^2$. Due to this, the power conversion efficiency at the receiver, η_r, depends on the shape of the input signal. In addition, the lower bound in (8.26) shows that the efficiency, η_r, increases with the input power level, for the practical regime of operation. This highlights the importance of the modulation and signal design, along with waveform design, in improving the overall efficiency of the WPT.

It is important to note that the optimization of overall efficiency $\eta = \eta_t \eta_c \eta_r$ is not possible by optimizing individual efficiency factors, η_t, η_c, η_r, independently of each other, as they are coupled with each other due to non-linearities discussed above.

For instance, as observed above, η_r is a function of the received signal shape and power and thus, in turn, depends on the transmit signal as well as wireless channel. Likewise, η_c also depends on the wireless channel and the transmit signal. Therefore, the optimization problem can be written as

$$\max_{p(\mathbf{x}_0,\ldots,\mathbf{x}_{N-1})} \quad \eta\left(\mathbf{x}_0, \ldots, \mathbf{x}_{N-1}\right)$$
$$\text{subject to } P_{\text{rf}}^t \leq P.$$

In the above, the feasible set consists of distributions $p(\mathbf{x}_0, \ldots, \mathbf{x}_{N-1})$ satisfying the power constraint $P_{\text{rf}}^t \leq P$. Solving the above optimization problem has been investigated with approaches, including beamforming, waveform design, etc. [10], including the SWIPT design. In contrast, to wireless power transfer, the SWIPT design approach aims to maximize the tradeoff between the amount of information and the power transfer. In the following, we briefly discuss these design aspects of SWIPT systems.

8.3.4 Simultaneous Wireless Information and Power Transfer

As shown in Figs. 8.8 and 8.9, in addition to the information decoder, the receiver in a SWIPT system consists of an EH block. A SWIPT receiver can be implemented in two possible ways, namely, time-splitting (TS) and energy splitting (ES) architecture. As shown in Fig. 8.8, for the TS receiver architecture, the received signal partitioned across information decoder (ID) and EH block, in orthogonal time-slots. In contrast, the ES receiver architecture splits the incoming signal in the power-domain to provide a fraction of the power for energy harvesting, and using the remaining signal for

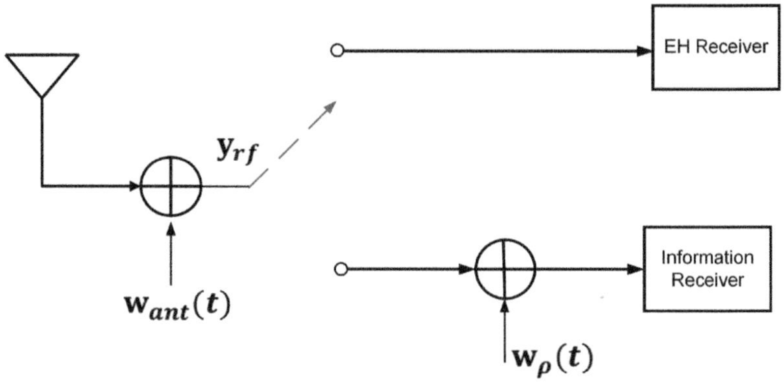

Fig. 8.8 Time-splitting receiver architecture

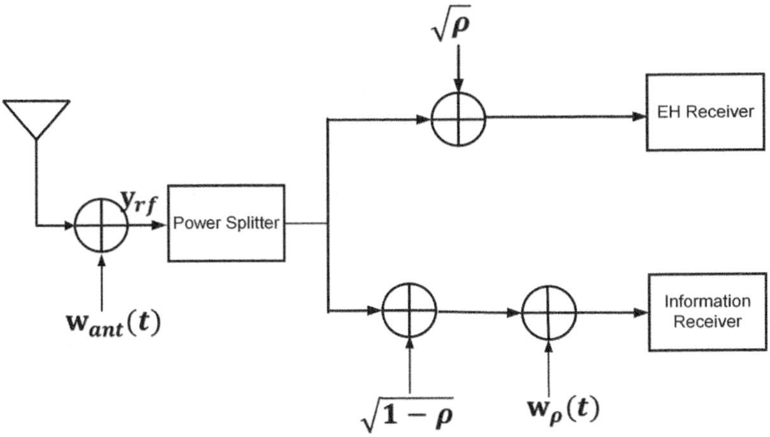

Fig. 8.9 Energy-splitting receiver architecture

information decoding. Splitting the received signal leads to a natural information-energy tradeoff. Note that, for both the TS and ES architectures, different rate-energy pairs can be realized by adjusting the fraction of splitting. Specifically, the rate-energy region is a collection of all the rate-energy tuples, (R, E), which can be achieved simultaneously. Mathematically, the rate-energy region can be expressed as:

$$\mathcal{R}_{R-E}(P) \triangleq \bigcup_{p(\mathbf{x}_0,\ldots,\mathbf{x}_{N-1}),\, P^t_{\mathrm{rf}} \leq P} \left\{ (R, E) | R \leq \sum_{n=0}^{N-1} I(X_n, Y_n),\, E \leq P^r_{\mathrm{DC}}(\mathbf{x}_0, \ldots, \mathbf{x}_{N-1}) \right\},$$

where $I(X_n, Y_n)$ denote the mutual information between the channel input and the channel output. One possible way to identify the rate-energy region is through finding the capacity of a complex AWGN channel with an average transmit and receiver power constraints $P^t_{\mathrm{rf}} \leq P$ and $P^r_{\mathrm{DC}}(\mathbf{x}_0, \ldots, \mathbf{x}_{N-1}) \geq E'$, respectively, where $E' > 0$. The optimization problem can be written as

$$\sup_{p(\mathbf{x}_0,\ldots,\mathbf{x}_{N-1})} \sum_{n=0}^{N-1} I(X_n, Y_n), \tag{8.27}$$

$$\text{subject to } P^t_{\mathrm{rf}} \leq P, \tag{8.28}$$

$$P^r_{\mathrm{dc}}(\mathbf{x}_0, \ldots, \mathbf{x}_{N-1}) \geq E'. \tag{8.29}$$

Solving the above optimization problem, essentially requires designing an information carrying signal which can balance the task of harvesting a minimum amount of energy from the RF signal while simultaneously maximizing the information rate.

Achieving this requires to address several interesting questions, such as type of modulation, waveform, coding, and input distribution to be used. For further details on this, we refer the readers to [10].

8.4 Conclusion and Future Research Directions

In conclusion, this chapter has elucidated the symbiotic relationship between wireless communication and zero-energy devices, showcasing the potential for sustainable and energy-efficient solutions by introducing power control policies for uncoordinated operations of a point-to-point wireless link with natural energy harvesting transmitter and receiver nodes, and for the optimization of energy consumption in wireless communication systems. Moreover, the RF EH has opened new frontiers for wireless power transfer and simultaneous wireless information and power transfer (SWIPT). The conceptual framework presented here underscores the fundamental trade-offs involved in harvesting ambient electromagnetic signals to power communication devices and sensor networks, paving the way for the development of zero-energy devices.

Looking ahead, several exciting avenues require further exploration and research in this domain. First and foremost, the refinement and real-world implementation of the proposed power control policies merit thorough investigation, necessitating comprehensive testing, and validation in diverse environments. Additionally, the integration of machine learning algorithms for adaptive power control, capable of dynamically responding to changing environmental conditions, presents an intriguing area for future research.

The future trajectory of research in RF energy harvesting calls for a deeper understanding of the challenges associated with SWIPT, exploring advanced signal processing techniques, such as beamforming and cooperative communication, could further enhance the efficiency and reliability of energy harvesting systems. Investigating the impact of these technologies on the performance of wireless networks and the development of practical energy-efficient protocols will be crucial for their successful deployment in real-world applications.

In summary, this chapter not only sheds light on the current state of wireless communication with zero-energy devices, but also charts a course for future endeavors in this promising field. The convergence of communication technology and sustainable energy solutions holds the key to shaping a future where wireless devices seamlessly operate with minimal impact on the environment, marking a significant step toward achieving the vision of zero-energy communication systems.

References

1. Kyburz R (2017) Energy efficiency of the internet of things, technology and energy assessment report. IEA 4E EDNA, Technical Report
2. Niu J-P, Li GY (2019) An overview on backscatter communications. J Commun Inf Netw 4(2):1–14
3. Dong Y, Farnia F, Özgür A (2015) Near optimal energy control and approximate capacity of energy harvesting communication. IEEE J Sel Areas Commun 33(3):540–557
4. Xu J, Zhang R (2014) Throughput optimal policies for energy harvesting wireless transmitters with non-ideal circuit power. IEEE J Sel Areas Commun 32(2):322–332
5. Sharma MK, Murthy CR, Vaze R (2019) Asymptotically optimal uncoordinated power control policies for energy harvesting multiple access channels with decoding costs. IEEE Trans Commun 67(3):2420–2435
6. Inan HA, Ozgur A (2016) Online power control for the energy harvesting multiple access channel. In: 2016 14th international symposium on modeling and optimization in mobile, ad hoc, and wireless networks (WiOpt), pp 1–6
7. Sharma MK, Murthy CR (2017) On the design of dual energy harvesting communication links with retransmission. IEEE Trans Wireless Commun 16(6):4079–4093
8. Srivastava R, Koksal CE (2013) Basic performance limits and tradeoffs in energy-harvesting sensor nodes with finite data and energy storage. IEEE/ACM Trans Netw 21(4):1049–1062
9. Sharma MK, Murthy CR, Vaze R (2017) On distributed power control for uncoordinated dual energy harvesting links: Performance bounds and near-optimal policies. In: 2017 15th international symposium on modeling and optimization in mobile, ad hoc, and wireless networks (WiOpt), pp 1–8
10. Clerckx B, Huang K, Varshney LR, Ulukus S, Alouini M-S (2021) Wireless power transfer for future networks: signal processing, machine learning, computing, and sensing. IEEE J Sel Top Signal Proc 15(5):1060–1094

Chapter 9
Integrated Sensing and Communications: Fundamentals, Recent Developments, and Prospects in 6G

Jyotsna Rani, Kuntal Deka, and Sanjeev Sharma

Abstract In the advent of the 6G era, the convergence of physical and digital domains mark an evolution towards "perceptual networking." This innovative approach marks a transformative stride towards network awareness, where wireless systems not only transmit data but also perceive and interpret the surrounding physical environment. This convergence extends across diverse platforms, from vehicles and drones to urban surveillance and agricultural equipment. To enable this transition, the wireless industry advocates for networks endowed with sensing capabilities akin to radar systems, termed Integrated Sensing and Communications (ISAC). ISAC embodies a paradigm shift from conventional communication-centric networks to systems capable of perceiving and interpreting their environment. This chapter explores the coexistence, collaboration, and integrated design of wireless communication and radar sensing, showcasing ISAC as a pivotal technology in future wireless networks. Through synergistic co-design strategies and unified waveform approaches, ISAC promises optimized performance and enhanced system efficiency, positioning it as a revolutionary catalyst in the domain of wireless communications.

Keywords Integrated sensing and communication · Orthogonal frequency-division multiplexing · Orthogonal time frequency space modulation · Intelligent reflecting surfaces · Near-field communications

9.1 Introduction

In the upcoming 6G era, we foresee a progressive merging of the physical and digital domains, marking a significant step towards realization. This convergence will extend from various platforms like vehicles and drones to surveillance systems in

J. Rani · K. Deka (✉)
IIT Guwahati, Guwahati, India
e-mail: kuntaldeka@iitg.ac.in

S. Sharma
IIT (BHU) Varanasi, Varanasi, India

S. Sharma et al. (eds.), *Next-Generation Wireless Systems*,
Transactions on Computer Systems and Networks,
https://doi.org/10.1007/978-981-96-3185-8_9

urban areas and agricultural equipment in rural settings. To facilitate this transition, the wireless industry advocates for the incorporation of a capacity for "perceiving" the physical environment in future networks, akin to the sensing capabilities of radar systems. This ability stands out as a defining characteristic of next-generation networks, departing from the traditional emphasis solely on communication functionalities. This forward-looking perspective has prompted the scientific community to reconsider the architectures of existing communication infrastructures and devices, giving rise to the emerging discipline termed "Integrated Sensing and Communications (ISAC)."

Over the years, the development of wireless communication and radar sensing (C&S) has progressed along separate but parallel paths, with limited points of convergence. However, there are shared elements such as signal processing algorithms, devices, and, to some extent, system architecture. As a result, there has been a recent upsurge in research interest focused on exploring the coexistence, collaboration, and integrated design of these two systems. ISAC embodies a design paradigm and associated enabling technologies that intricately blend sensing and communication systems. This fusion is strategically designed to maximize the utilization of wireless resources, promoting efficiency while unlocking the potential for mutual benefits between sensing and communication functionalities. ISAC represents a synergy where the seamless interplay of sensing and communication enables networks to not only transmit data but also perceive and interpret the environment, comparable to the multifunctionality of radar systems. It not only enhances wireless resource utilization but also opens avenues for leveraging the strengths of each system to create more sophisticated and responsive technological solutions. The underlying principle of ISAC rests on the notion that radio signals can serve a dual purpose: not only transmitting communication data from the sender to the recipient but also concurrently functioning as a means to extract valuable information from the surrounding physical environment. This dual functionality represents a harmonious integration, where the same wireless signals used for smooth communication can also be utilized to extract insights and data from the surrounding environment. This capability enhances the overall efficiency and versatility of the integrated system. Hence, ISAC offers dual advantages compared to dedicated sensing and communication functionalities. Firstly, it leverages Integration Gain, optimizing the utilization of congested resources by effortlessly merging sensing and communication capabilities. Furthermore, and notably intriguing, ISAC achieves Coordination Gain, enabling a balanced dual-functional performance and facilitating mutual assistance between sensing and communication components. This collaborative approach underscores the multifaceted advantages inherent in the integration of these two fundamental functionalities.

In essence, ISAC materializes through the collaborative utilization of spectral and hardware resources. This synergistic design, exemplified by cohabitation strategies involving interference management, aims to achieve integration gain. At the same time, ISAC can be approached from a co-design standpoint, where communication and sensing functionalities perfectly merge within unified waveforms. This approach optimizes performance and promotes coordination gain, allowing for enhanced system efficiency. These attributes position ISAC as a pivotal technology, garnering recognition in the realm of next-generation wireless networks, specifically in the context of advanced standards such as 5G-Advanced and the envisioned 6G communications.

9.1.1 Classification of ISAC Technologies

Based on design priority and fundamental signal formats, existing ISAC systems can be categorized into three distinct groups:

- *Radar-Centric Design*: Radar-centric design entails integrating communication capabilities into primary radar systems or easily embedding communication functions within radar operations.
- *Communication-Centric Design*: Communication-centric design involves integrating radio/radar sensing capabilities into primary communication systems or seamlessly incorporating radar features into communication setups.
- *Joint Design and Optimization*: In case of joint design and optimization of ISAC, it encompasses technologies that operate without constraints related to underlying systems and signals, allowing for integrated and optimized solutions without limitations.

A block diagram representation of the classification of three types of ISAC systems, in terms of their technical scope, has been presented in Fig. 9.1.

In the first two categories, the primary emphasis in design and research lies in determining methods to integrate the additional function based on the signal formats of the primary system. The primary principle is to achieve this integration without causing significant impact on the primary system. Minor adjustments and optimizations can be applied to both the system and its signals to facilitate this integration. The third category entails designing and optimizing the signal waveform, system,

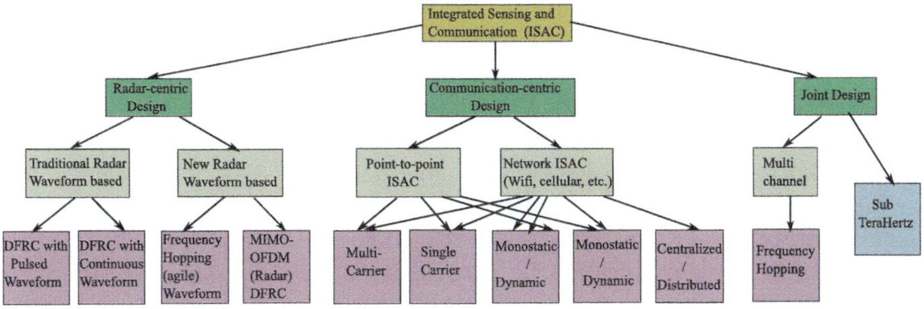

Fig. 9.1 A classification of the three types of ISAC systems

and network architecture without biasing towards either communication or sensing. In the subsequent section, we explore each of these technologies in detail.

Radar-Centric Design The radar-centric design methodology involves introducing alterations in radar waveforms, exploiting these distinctions to modulate communication information [1]. Radar systems, particularly in military contexts, demonstrate impressive long-range capabilities, spanning distances of up to several hundred kilometers. Consequently, a notable advantage of integrating communication within radar systems is the potential to establish extended-distance communication with significantly reduced latency compared to satellite communication. It's crucial to recognize that the attainable data rates in such systems are frequently restricted by inherent limitations in the radar waveform.

In [2], the authors have successfully implemented an innovative system that amalgamates radar and communication. This integration is established on a software-defined radar platform, cleverly utilizing radar pulses for communication purposes. Research findings from [3] further emphasize the potential for creating communication networks in both fixed and mobile radar applications within military and aviation settings. Unmanned aerial vehicles, equipped with airborne radar, can employ adaptive transmit signals to concurrently sense a scene and transmit the acquired data to a ground base station. The core aim of such systems remains steadfast: to forge perfectly integrated, secure, and long-range communications with minimal latency within the preexisting radar infrastructure. These all-encompassing systems are widely acknowledged as dual-function radar communications (DFRC), particularly within the context of Joint communication and radar/radio sensing (JCAS).

The radar-centric design of ISAC can be categorized into two main groups: (a) traditional radar waveform-based, and (b) new radar waveform-based. Traditional radar waveform-based ISAC encompasses DFRC with pulsed waveform and DFRC with continuous waveform. New radar waveform-based ISAC includes frequency hopping/agile waveform and MIMO-OFDM (Radar) DFRC.

Communication-Centric Design In contrast to traditional radar waveforms utilizing pulse or chirp signals, modern approaches favor a continuous waveform for data transmission to achieve higher throughput. Notably, communication waveforms like orthogonal frequency-division multiplexing (OFDM) signals, widely adopted in 4G/5G cellular and WiFi standards, are preferred for enhancing data transmission capabilities. However, the application of communication signals to sensing is constrained, often demonstrating a high peak-to-average power ratio (PAPR).

The use of a cyclic prefix (CP) to alleviate intersymbol interference (ISI) in communication systems may introduce Doppler ambiguity in radar sensing [4]. Furthermore, the information-bearing waveform can lead to increased sidelobe gain, reducing the detection dynamic range. Consequently, it is crucial to implement additional signal preprocessing techniques to effectively address these challenges.

In [5], OFDM-based radars in LTE and 5G NR mobile networks' base stations are investigated for radar and sensing applications. The research emphasizes the excellent radar/sensing performance of 5G NR waveforms and proposes targeted solutions for the challenging transmitter-receiver isolation in shared-antenna OFDM radar deployments. In [6], optimization of OFDM waveforms within a full-duplex radar transceiver and communication transmitter combination is carried out, highlighting its significance in efficient radio frequency spectrum management for beyond 5G networks.

Exploration of a joint vehicular communication-radar system is presented in [7], where an IEEE 802.11ad-based radar system operating at 60 GHz is introduced for long-range radar applications in vehicles. This system facilitates a unified waveform for automotive radar alongside potential millimeter-wave vehicular communication, promoting hardware reuse. Another area of research has focused on leveraging mmWave communication in Vehicle-to-Everything (V2X) scenarios to achieve high data rates for applications like infotainment and augmented reality-assisted driving. To address the challenge of configuring large antenna arrays with high training overhead in mmWave communication, the utilization of infrastructure-mounted sensors in future smart cities is proposed. Additionally, in [8], the time–frequency waveform design problem for an OFDM dual-functional radar-communications (DFRC) system is addressed, aiming to optimize subcarrier powers for simultaneous communication and target parameter estimation.

The communication-centric design of ISAC can be categorized into two main groups: (a) point-to-point ISAC and (b) network ISAC, including technologies like WiFi and cellular networks. These categories can then be further refined based on several factors including the use of multi-carrier or single-carrier signals, the system's architecture (such as monostatic or dynamic), and its operational structure (whether it's centralized or distributed).

Joint Design and Optimization In the dynamic realm of wireless communication systems, there is a growing recognition of the need for innovative strategies to tackle the challenges posed by spectrum congestion. The exploration of spectral bands traditionally reserved for radar systems emerges as a promising avenue for mitigating these issues. This transformative journey is characterized by pioneering efforts, such as the development of dual-function radar-communications systems, which harmoniously blend radar and communication functionalities. Complementing this, novel multibeam frameworks are introduced, facilitating a smooth integration of communication and radar sensing. Furthermore, cutting-edge millimeter-wave joint radar-communications systems underscore the potential of advanced super-resolution algorithms in optimizing performance. These initiatives collectively reflect a paradigm shift. They offer a glimpse into the future where joint design and optimization redefine the landscape of wireless communication, promising enhanced efficiency and expanded possibilities.

In [9], a beamforming strategy is proposed for a unified multi-input–multi-output (MIMO) radar-communication (RadCom) system, where a single device serves as both a radar and a communication base station (BS), allowing simultaneous communication with downlink users while detecting radar targets. An optimal waveform design for a MIMO base station is presented in [10], accommodating both line-of-sight and non-line-of-sight channels. Weighted optimizations are conducted in this work, achieving a versatile balance between radar and communication performance and introducing streamlined low-complexity algorithms for practical implementation. In [11], a millimeter-wave joint radar-communications (JRC) system involving a bi-static automotive radar and vehicle-to-vehicle communications is considered, demonstrating the effectiveness of low-complexity JRC super-resolution algorithms. The potential of beamforming in JCAS is highlighted in [12] by introducing a novel multibeam framework utilizing steerable analog antenna arrays. The framework employs key innovations in multibeam technology, supporting fixed subbeams for communication and packet-varying scanning subbeams for sensing simultaneously from a single transmitting array. Additionally, joint transmit beamforming for a multiple-input-multiple-output (MIMO) radar and multiuser MIMO communication transmitter is explored in [13].

In the joint design framework for ISAC, it is classified into (a) Multi-channel systems, which includes techniques like frequency hopping, and (b) sub terahertz systems.

9.2 System and Channel Models

Heading into the realm of dual-functional radar-communication systems involves exploring optimal waveform design. In this context, the emphasis lies on developing waveforms that serve the dual purpose of both radar and communication functions. This task involves finding the most effective and efficient waveforms capable of fulfilling the requirements of both radar operations, such as target detection and tracking, and communication tasks, ensuring perfect integration of these dual functionalities within the system [10].

Here, we examine a dual-functional Multiple Input Multiple Output (MIMO) Radar-Communication (RadCom) system. This system is capable of simultaneously emitting radar probing waveforms towards specified targets while transmitting communication symbols to users in the downlink. As illustrated in Fig. 9.2, the integrated system is equipped with a Uniform Linear Array (ULA) consisting of N antennas. This array serves K single-antenna users while concurrently detecting radar targets.

9.2.1 Communication Model

Let $\mathbf{X}$ be the transmitted signal matrix, where $\mathbf{X} = [\mathbf{x}_1, \mathbf{x}_2, \ldots, \mathbf{x}_L] \in \mathbb{C}^{N \times L}$, with L denoting the length of the communication frame. The channel matrix $\mathbf{H}$ is defined as $\mathbf{H} = [\mathbf{h}_1, \mathbf{h}_2, \ldots, \mathbf{h}_K]^T \in \mathbb{C}^{K \times N}$, and the noise matrix $\mathbf{Z} = [\mathbf{z}_1, \mathbf{z}_2, \ldots, \mathbf{z}_L] \in \mathbb{C}^{K \times L}$ follows a distribution $\mathbf{z}_j \sim \mathcal{CN}(0, N_0 I_N)$ for all j.

The received symbol matrix at the downlink user can thus be expressed as:

$$\mathbf{Y} = \mathbf{HX} + \mathbf{Z}. \tag{9.1}$$

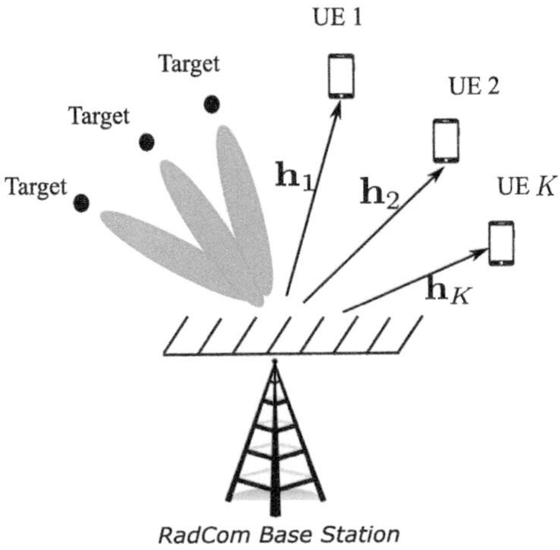

Fig. 9.2 ISAC system model

Building on insights from [14], we adopt three assumptions for our analysis. Firstly, the transmitted signal matrix **X** serves as a dual-functional waveform, fulfilling both radar and communication operations. This implies that each communication symbol also serves as a snapshot of a radar pulse. Secondly, we assume that the downlink channel **H** experiences flat Rayleigh fading and maintains stability throughout a communication frame or radar pulse. Lastly, we assume that the channel **H** is perfectly estimated using pilot symbols.

Assume **S** to be the specified constellation symbol matrix $\mathbf{S} \in \mathbb{C}^{K \times L}$ for the downlink users. Also, each user's entry in matrix **S** is assumed to be randomly selected from the same constellation. The above expression in (9.1) can be reformulated as follows:

$$\mathbf{Y} = \mathbf{S} + (\mathbf{HX} - \mathbf{S}) + \mathbf{Z}. \tag{9.2}$$

In (9.2), $(\mathbf{HX} - \mathbf{S})$ refers to the interference that occurs in the downlink transmission of a communication system when multiple users are simultaneously receiving signals from a common base station or access point known as MUI. The collective Multi-User Interference (MUI) energy can be quantified as:

$$P_{MUI} = \left\| \mathbf{HX} - \mathbf{S}^2 \right\|_F, \tag{9.3}$$

where F is the Frobenius norm and is used to calculate the square root of the sum of the squares of all the elements.

It has been demonstrated in [15] that the Multi-User Interference (MUI) energy in Eq. (9.3) directly influences the achievable sum-rate of the downlink users. The Signal-to-Interference-plus-Noise Ratio (SINR) per frame for the ith user is given by:

$$\gamma_i = \frac{\mathbb{E}\left[|s_{i,j}|^2\right]}{\mathbb{E}\left[|\mathbf{h}_i^T \mathbf{x}_j - s_{i,j}|^2\right] + N_0}. \tag{9.4}$$

In this context, $s_{i,j}$ represents the entry at the (i, j)th entry position of matrix $\mathbf{S}$. Also, $\mathbb{E}$ signifies the ensemble average conducted with respect to the time index.

Hence, the attainable sum-rate of the users can be expressed as:

$$R = \sum_{i=1}^{N} \log_2(1 + \gamma_i). \tag{9.5}$$

In the context of a given constellation with a constant energy level, the power of the useful signal, denoted as $\mathbb{E}|s_{i,j}|^2$ remains fixed. Consequently, the sum-rate can be optimized by minimizing the Multi-User Interference (MUI) energy. The inherent fixed nature of the useful signal's power implies that enhancing the overall communication performance is achieved by mitigating the impact of interference, specifically the MUI energy.

9.2.2 Radar Model

In the realm of MIMO radar, the strategic adoption of uncorrelated waveforms stands out as a well-established technique, presenting a pathway to attain higher Degrees of Freedom (DoFs) in comparison to traditional phased-array radar systems [16, 17]. The design of the beampattern within MIMO radar is intricately linked to the covariance matrix of the probing signals. Authors in [17] have emphasized that the integration of spatial and temporal waveform aspects in the design of transmit beamforming introduces substantial complexity. A more versatile form of transmit beamforming is demonstrated in this work which is achieved through a decoupled approach, utilizing orthogonal (uncorrelated) waveforms and a multi-rank transmit beamformer. In [18], authors have explored the potential of combining sparse code multiple access (SCMA) and faster-than-Nyquist (FTN) signaling to enhance spectral efficiency in next-generation wireless communication systems. Here, the integrated uplink FTN-SCMA system introduces intentional inter-symbol interference (ISI) to achieve higher spectral efficiency.

In this context, our attention is directed towards the design of the dual-functional waveform matrix $\mathbf{X}$. This matrix serves a dual role in the system, acting as a crucial element for both radar and communication operations. The focus is particularly on shaping the spatial characteristics of $\mathbf{X}$, which is fundamentally defined by its spatial covariance matrix. The spatial covariance matrix plays a key role in determining how the waveforms are distributed and correlated across different spatial dimensions. The design intricacies of this matrix have significant implications for the overall performance and capabilities of the system in handling both radar and communication functionalities. The spatial covariance matrix denoted by $\mathbf{R}$ is expressed as:

$$\mathbf{R}_X = \frac{1}{L}\mathbf{X}\mathbf{X}^H. \tag{9.6}$$

In order to guarantee that $\mathbf{R}_X$ is positive-definite, we introduce the assumption that $L \geq N$, and this assumption is made without any loss of generality. In mathematical terms, $\mathbf{R}_X$ being positive-definite means that all its eigenvalues are positive, and by ensuring $L \geq N$, we aim to maintain this positive-definiteness property. This condition helps to ensure the reliability and stability of the spatial covariance matrix $\mathbf{R}_X$ in the system's design.

Furthermore, the transmit beampattern for the RadCom system is a representation of how the transmitted energy is spatially distributed in the desired direction. It characterizes the directional sensitivity of the transmitted signals and is a crucial aspect of system design. The specific form and characteristics of the transmit beampattern play a significant role in determining the system's performance, including its ability to effectively communicate and probe targets in a radar and communication integrated system. The transmit beampattern in a dual-functional radar-communication system can be expressed as:

$$\mathbf{B}_d(\theta) = \mathbf{a}^H(\theta)\mathbf{R}_X\mathbf{a}(\theta). \tag{9.7}$$

In the given expression, θ represents the detection angle at which the system is detecting or probing the target. Also, $\mathbf{a}(\theta) = \left[1,\ e^{j2\pi\Delta\sin(\theta)},\ \ldots,\ e^{j2\pi(N-1)\Delta\sin(\theta)}\right] \in \mathbb{C}^{M\times 1}$ is the steering vector associated with the transmit antenna array in the RadCom system. The steering vector provides a mathematical representation of how the transmitted signal is spatially directed or focused, depending on the detection angle θ. In this context, Δ represents the spacing between adjacent antennas in the array. The normalization by the wavelength is a crucial aspect, as it standardizes the antenna spacing relative to the signal's wavelength. When Δ is normalized by the wavelength, it allows for a consistent measure that takes into account the propagation characteristics of the transmitted signal. This normalization ensures that the steering vector is effectively aligned with the spatial properties of the signal, facilitating proper beamforming and directional control of the transmitted energy.

The optimization problem outlined in this context are strategically designed to address the challenge of minimizing Multi-User Interference (MUI) within the framework of Multiple Input Multiple Output (MIMO) radar systems. MUI arises when multiple users transmit signals simultaneously, leading to interference that can impede the radar system's ability to accurately detect and differentiate between targets. The primary aim of these optimization problems is to develop solutions that effectively mitigate the adverse effects of MUI. By tailoring the optimization objectives and constraints to the unique characteristics of MIMO radar systems, these problems seek to strike a balance between various considerations. This includes optimizing the spatial configuration of antennas, shaping transmitted waveforms, and accounting for the inherent complexity of MIMO radar operations.

In essence, the overarching goal is to enhance the performance and reliability of MIMO radar systems in scenarios with multiple users. This becomes particularly crucial in applications where radar systems need to serve dual functions, such as both probing targets and facilitating communication. The optimization problems provide a systematic approach to achieving this balance, ultimately improving the radar system's ability to operate efficiently and deliver accurate target detection in the presence of interference from multiple users.

9.3 ISAC-Assisted Waveform Design

The initial phase in deploying RadCom (Radar and Communication) systems involves the crucial task of identifying waveforms that exhibit compatibility for both wireless communication and radar applications. This selection process is pivotal and hinges on meeting the specific requirements of the radar system, particularly in relation to tasks such as determining object range, Doppler shifts, and angular offsets.

The selection of these waveforms is contingent upon the unique demands of radar functionalities, with a primary focus on achieving accurate object range measurements, capturing Doppler shifts for velocity assessments, and determining angular offsets for precise directional information [19]. In all three cases, the dynamic range and resolution of the selected waveforms become critical factors. A high dynamic range ensures that the radar system can detect objects over a wide span of distances, while resolution is imperative for distinguishing between closely spaced objects. Furthermore, in scenarios characterized by dense traffic, the chosen waveform must exhibit robustness against interference and noise. This robustness is essential for maintaining the integrity of radar operations amidst challenging and crowded communication environments. Ultimately, the identification and selection of waveforms that meet these specific criteria lay the foundation for the successful implementation of RadCom systems, ensuring their effectiveness in diverse and dynamic operational scenarios.

In the context of RadCom systems, where wireless communication and radar functions coexist, the waveform selection process involves a detailed consideration of the specific requirements for both communication and radar applications. For communication purposes, the key parameters that demand attention are the data rate and the bit error rate. These factors are crucial in determining the efficiency and reliability of the communication link. Achieving a high data rate is often essential for supporting various communication services, and a low bit error rate ensures the accuracy of data transmission.

On the radar side, the chosen waveform must exhibit robustness against interference, noise, and distortion arising from multipath propagation. These challenges are common in real-world scenarios and can significantly impact the radar system's performance. Robust waveforms ensure that the radar can effectively detect and identify targets amidst adverse environmental conditions. To provide high performance in terms of bit rate for communication, it's advantageous to select a continuous signal with a sufficiently high symbol rate. This ensures that a significant amount of information can be transmitted within a given timeframe, supporting the demands of modern communication systems. Importantly, there is no need to encrypt or protect the information content, as the transmitted data is intended to be openly received by all RadCom systems in the neighborhood. Therefore, the waveform selection process involves a delicate balance between meeting communication requirements, such as data rate and bit error rate, and addressing the challenges inherent in radar applications, including interference, noise, and multipath propagation. The ultimate goal is to choose a waveform that caters to the dual objectives of reliable communication and effective radar functionality within the RadCom system.

In the domain of automotive radar applications, specific waveforms, notably FM-CW (Frequency Modulated Continuous Wave), and to a lesser extent pulsed-FM, are predominantly employed [20]. These waveforms have proven to be particularly effective in automotive radar systems due to their specific characteristics [4]. The key features of these waveforms are highlighted below:

- *Frequency Modulated Continuous Wave (FM-CW)*: This waveform involves the continuous transmission of a signal with a frequency that varies linearly over time. FM-CW radar is known for its ability to provide accurate range measurements and has found widespread use in automotive applications, where precise distance estimation is crucial for collision avoidance and adaptive cruise control.
- *Pulsed-FM*: Pulsed-FM, although less commonly used in automotive radar, involves the transmission of short pulses with frequency modulation.
  It is to be noted that while in FM-CW radar, a continuous signal is transmitted with a frequency that linearly varies over time whereas pulsed-FM radar involves the transmission of short pulses with frequency modulation.

For broader radar applications beyond the automotive industry, a diverse set of waveforms are recognized, including:

- *Chirp*: A signal that varies its frequency over time. Chirp waveforms are versatile and find applications in radar for their ability to provide range and Doppler information simultaneously.
- *Pseudorandom*: Waveforms generated with pseudorandom sequences. These waveforms offer benefits in terms of anti-jamming capabilities and increased resolution.
- *Staggered*: Staggered waveforms involves transmitting multiple pulses with varying parameters such as frequency or phase. Staggered waveforms enhance resolution and mitigate range ambiguity.
- *Phase and Frequency Coding*: These waveforms utilize variations in phase or frequency to encode information. They are employed for applications requiring increased robustness against interference and improved target discrimination.

The choice of waveform depends on the specific requirements of the radar application, including factors such as range resolution, Doppler sensitivity, resistance to interference, and the ability to handle different target scenarios. The diverse array of available waveforms allows radar systems to be tailored to the specific challenges and objectives of a given application. In the realm of digital communications, particularly in the context of mobile communications, the design and application of waveforms are primarily characterized by two key concepts: single-carrier waveforms and multi-carrier waveforms, often referred to as Orthogonal Frequency Division Multiplexing (OFDM). The concepts of single-carrier and multi-carrier waveforms are further illustrated below:

- *Single-Carrier Waveforms*: This concept involves the transmission of data using a single-carrier frequency. In single-carrier waveforms, discrete phase modulation techniques are commonly employed, including Binary Phase Shift Keying (BPSK), Quadrature Phase Shift Keying (QPSK), and higher-order schemes such as 8 Phase Shift Keying (8 PSK). These modulation techniques enable the encoding of information in the phase of the carrier signal.
- *OFDM Waveforms*: Multicarrier waveforms, notably Orthogonal Frequency Division Multiplexing (OFDM), utilize multiple closely spaced carrier frequencies to transmit data concurrently. OFDM has become prevalent in modern digital communication standards due to its ability to mitigate the impact of frequency-selective fading and improve spectral efficiency.
- *OTFS Waveforms*: Another multicarrier waveform is Orthogonal Time Frequency Space (OTFS) waveform. OTFS waveform is an emerging paradigm that leverages the principles of time–frequency space for improved performance in challenging communication environments. Unlike traditional waveforms, OTFS exploits the entire space–time channel response, providing advantages in scenarios with high mobility, fading, and multipath propagation. This waveform concept has garnered attention for its ability to mitigate the impact of time-varying channels, making it particularly promising for applications in dynamic and rapidly changing communication environments. The OTFS framework offers a unique perspective by jointly considering time and frequency characteristics, contributing to enhanced robustness and reliability in challenging communication scenarios. Additionally, OTFS demonstrates efficacy in mitigating the effects of Delay-Doppler spread, further enhancing its suitability for real-world communication systems. delay-Doppler analysis in radar systems enables simultaneous estimation of target range and velocity, crucial for tracking and target discrimination. In communication systems, managing delay-Doppler spread is vital for signal integrity and minimizing data loss, employing techniques like equalization and adaptive modulation for reliable wireless communication in dynamic environments.

  Both OFDM and OTFS waveforms offer unique benefits in communication systems. OFDM excels in robustness to frequency-selective fading, high spectral efficiency, and resistance to interference. On the other hand, OTFS is designed to address challenges in time-varying channels, with advantages in scenarios involving high mobility and enhanced spatial-temporal resolution. In the next section, we delve into a more detailed discussion, exploring the specific strengths and applications of each waveform concept.

9.3.1 OFDM-Based Waveform Design

Orthogonal Frequency Division Multiplexing (OFDM) has risen to prominence as a pivotal technology, offering an effective solution to the complex challenges encountered in modern communication systems, particularly those arising from frequency-selective fading and inter-symbol interference. At the core of the OFDM framework lies a sophisticated architecture where a multitude of orthogonal subcarriers, meticulously spaced to prevent interference, collaboratively establish a resilient communication channel. This inherent resilience to frequency-selective fading and the ability to adapt to dynamic channel conditions contribute significantly to OFDM's prominence, particularly in the realm of high-speed data transmission.

The technology's prowess extends beyond its foundational role in communication, finding application in prominent standards such as the 5G New Radio (NR) [21] and the IEEE 802.11 standard [22]. Its widespread adoption is attributed to several noteworthy advantages, including elevated spectral efficiency, robustness against multipath fading, and a straightforward implementation approach.

The versatility of the OFDM waveform proves to be a key asset, especially concerning the allocation of time and frequency resources. This adaptability has found valuable application in radar target detection, as evidenced by studies [23–25]. The technology's seamless transition between communication and radar functionalities underscores its utility across diverse domains, showcasing both versatility and robustness [26].

An OFDM multicarrier signal can be conceptualized as a parallel stream comprising multiple single-carrier signals, each featuring orthogonal carrier waveforms and modulated with distinct transmit data. Mathematically, this signal can be formally expressed as [19]

$$s(t) = \sum_{\mu=0}^{N_{\text{sym}}-1} \left(\sum_{n=0}^{N_{\text{c}}-1} a(\mu N_{\text{c}} + n) \exp(j2\pi f_n t) \text{rect}\left(\frac{t - \mu T_{\text{OFDM}}}{T_{\text{OFDM}}} \right) \right). \tag{9.8}$$

Here, μ represents the individual OFDM symbol index within the total number of N_{sym} OFDM symbols. This index distinguishes different OFDM symbols in the sequence. n denotes the individual subcarrier index from a total of N_{c} subcarriers. Each subcarrier corresponds to a specific frequency within the OFDM system. a is the complex modulation symbols, they carry the information that is modulated onto the individual subcarriers for transmission. f_n denotes the individual subcarrier frequency. Each subcarrier has a unique frequency determined by its index. T_{OFDM} represents the total OFDM symbol duration, composed of an elementary symbol duration T and a guard interval duration T_G. The guard interval is a period of time added between consecutive symbols to mitigate the effects of multipath propagation. And, rect$\left(\frac{t}{T_{\text{OFDM}}}\right)$ is applied to shape the signal in the time domain, defining the duration over which each symbol is transmitted.

The intricate spectral characteristics of OFDM signals unfold over time, shaping a dynamic interplay between individual subcarriers and OFDM symbols. As depicted in Fig. 9.3, the strategic use of a considerable number of subcarriers results in a meticulously divided frequency spectrum. This strategic distribution contributes to a quasi-constant power spectral density, endowing the signal with a remarkable uniformity across both time and frequency domains. This quasi-constant nature is particularly beneficial, as it imparts a noise-like quality to the output signal. This noise-like appearance is a key attribute that enhances the signal's resilience in complex channel environments. The signal's ability to mimic the characteristics of noise bolsters its robustness against interference and other challenges. Thus, the dynamic and quasi-constant spectral behavior, illustrated in the Fig. 9.3, stands as a foundational element in the efficacy of OFDM systems, ensuring reliable and resilient communication over a diverse range of conditions.

For the individual subcarriers to achieve orthogonality and avoid interference, a crucial condition must be satisfied:

$$f_n = n\Delta f = \frac{n}{T}, \quad n = 0, \ldots, N_c - 1. \tag{9.9}$$

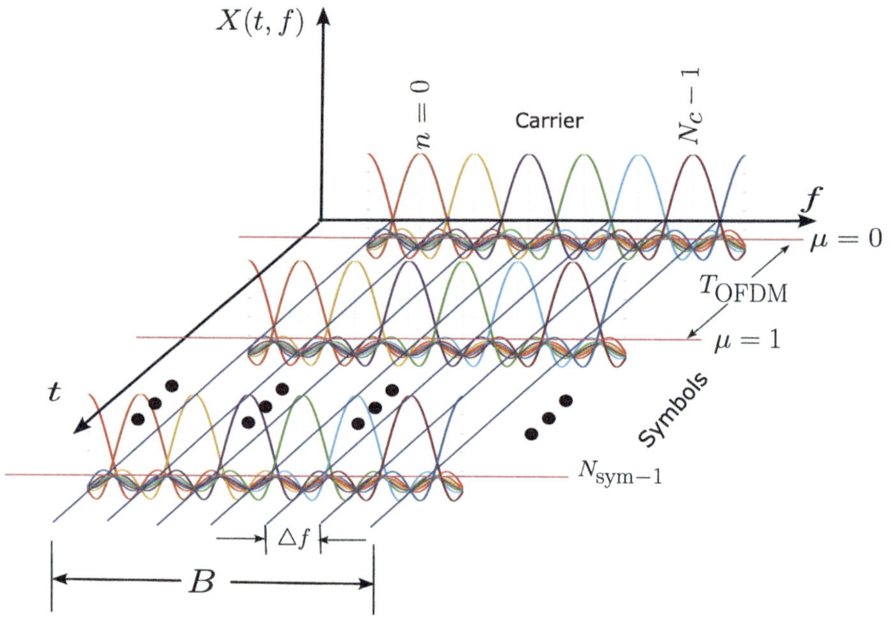

Fig. 9.3 Time and frequency structure of an OFDM signal

The separation Δf represents the difference in frequency between adjacent subcarriers. Therefore, as Δf increases, the subcarriers become more spaced out in the frequency domain. The relationship expressed as $\Delta f = \frac{1}{T}$ signifies that the subcarrier separation is inversely proportional to the elementary symbol duration T. This implies that as the duration of each symbol decreases, the subcarriers are spaced farther apart in the frequency spectrum. Compared to the symbol duration of a single-carrier transmission at the same data rate, the elementary OFDM symbol duration is increased by the factor N_c. This augmentation plays a crucial role in enhancing the signal's resilience against fast fading effects. In a single-carrier transmission system, the symbol duration is typically determined by the reciprocal of the bandwidth, and as a consequence, fast fading phenomena can adversely impact the signal quality. However, in OFDM, the use of multiple subcarriers allows for a more efficient utilization of the available bandwidth. The increase in the elementary symbol duration by a factor of N_c is a distinctive feature of OFDM. This extended symbol duration in OFDM offers inherent advantages, particularly in scenarios characterized by rapid variations in the channel conditions (fast fading). The longer symbol duration provides a form of temporal diversity, mitigating the impact of fast fading and contributing to the robustness of the OFDM signal.

The guard interval, also known as the cyclic prefix, plays a crucial role in mitigating the effects of multipath propagation in OFDM systems. It involves a partial cyclic repetition of the time-domain signal and is characterized by a duration T_G corresponding to the maximum expected multipath delay. By incorporating a cyclic prefix, OFDM systems achieve effective compensation for multipath propagation effects. This is particularly valuable in environments where signals may experience reflections and delays due to obstacles or varying transmission paths. The cyclic prefix is strategically added at the beginning of the OFDM symbol and occupies the time interval $[0, T_G]$ within the total OFDM symbol duration T_{OFDM}. During reception, the receiver focuses its evaluation on the interval $[T_G, T_{OFDM}]$, effectively excluding the portion containing the cyclic prefix. This selective evaluation ensures that the receiver observes a signal interval free from inter-symbol interference induced by multipath propagation [27]. In essence, the cyclic prefix acts as a protective buffer, allowing the OFDM system to maintain robust performance in the presence of multipath effects.

OFDM-ISAC System: Here, we take a look into an advanced algorithm for joint range and Doppler estimation tailored for OFDM signals [4]. This algorithm stands out for its simplicity and intuitive representation as it directly engages with the transmitted and received modulation symbols, diverging from the conventional approach of dealing with baseband signals. For clarity, we term this methodology as processing in the "Modulation Symbol" domain. This methodology adeptly harnesses the two-dimensional signal structure inherent in the OFDM signal, ensuring an efficient implementation. Here, the user data is entirely eliminated during processing, leading to exceptionally high peak-to-side-lobe ratios. This is achieved by mitigating the impact of imperfect autocorrelation properties in the baseband signal on the estimation process. A closely related concept can be applied for Doppler estimation in parallel, specifically in the modulation symbol domain [28].

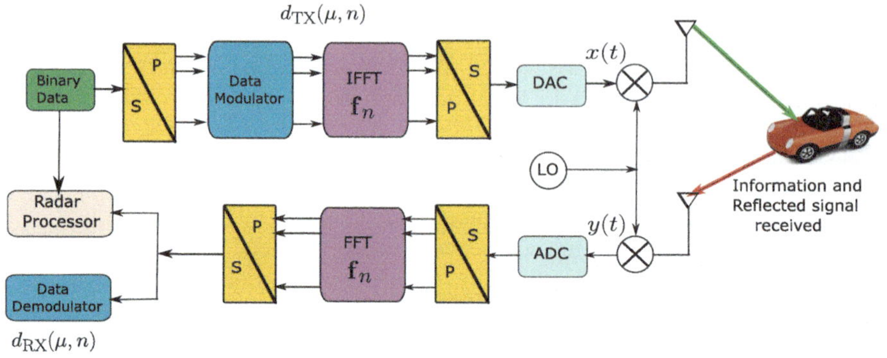

Fig. 9.4 Block scheme of the OFDM system architecture

Next, we explore the combination of both estimators into a two-dimensional estimator in the modulation symbol domain. Figure 9.4, illustrates a block diagram of the OFDM system architecture. The binary user data is segregated into parallel streams and then mapped onto complex-valued PSK symbols, yielding the modulation symbol sequence $d_{\mathrm{Tx}}(\mu, n)$. Through a block-wise inverse fast Fourier transform (IFFT) followed by a subsequent parallel-to-serial conversion, the time-domain signal $x(t)$, consisting of the orthogonal carrier waveforms f_n, is computed. Subsequently, this signal is converted into an analog signal and transmitted at the carrier frequency, following a mixing process with a local oscillator signal. In the receiver, the inverse order of steps is executed. The received modulation symbols $d_{\mathrm{Tx}}(\mu, n)$ are retrieved from the received baseband signal $y(t)$ through a fast Fourier transform (FFT) operation. Unlike traditional OFDM radar algorithms that primarily focus on utilizing the baseband signals $x(t)$ and $y(t)$, the upcoming discussions will introduce an algorithm capable of radar measurements based solely on the knowledge of the modulation symbols $d_{\mathrm{Tx}}(\mu, n)$ and $d_{\mathrm{Rx}}(\mu, n)$.

Now, consider an OFDM signal, as defined in (9.8) which encounters reflection from an object positioned at a range R with a Doppler shift f_D resulting from relative movement. The resulting received signal in this case can be characterized as follows:

$$
\begin{aligned}
y(t) = & \sum_{\mu=0}^{N_{\mathrm{sym}}-1} \sum_{n=0}^{N_{\mathrm{c}}-1} A(\mu, n) d_{\mathrm{Tx}}(\mu N_{\mathrm{c}} + n) \\
& \times \exp\left(j2\pi f_n \left(t - \frac{2R}{c_0} \right) \right) \exp\left(j2\pi f_D t \right) \\
& \times \operatorname{rect}\left(\frac{t - \mu T_{\mathrm{OFDM}} - 2R/c_0}{T_{\mathrm{OFDM}}} \right),
\end{aligned} \tag{9.10}
$$

where c_0 is the speed of light. Note that the symbols $d_{\text{Tx}}(\mu, n)$ represent the user data symbols, and unlike spread spectrum transmission, these symbols do not undergo any specific coding.

To better illustrate the contrasting effects of range and Doppler, (9.10) can be reformulated in a rearranged manner. This restructuring allows for a clearer depiction of how changes in range and Doppler contribute to the received signal. The restructuring of (9.10) gives:

$$\begin{aligned} y(t) = & \sum_{\mu=0}^{N_{\text{sym}}-1} \exp(j2\pi f_D t) \\ & \times \sum_{n=0}^{N_c-1} A(\mu, n) d_{\text{Tx}}(\mu N_c + n) \exp(-j2\pi f_n \frac{2R}{c_0}) \\ & \times \exp(j2\pi f_n) \operatorname{rect}\left(\frac{t - \mu T_{\text{OFDM}} - 2R/c_0}{T_{\text{OFDM}}}\right). \end{aligned} \tag{9.11}$$

The receiver extracts individual modulation symbols corresponding to a single OFDM symbol by exclusively monitoring the received signal during the fundamental OFDM symbol duration T_{OFDM}. This process involves observing the received signal over a time interval equal to the elementary duration of one OFDM symbol, T_{OFDM}, to recover the modulation symbols associated with that specific symbol. With the appropriate selection of the guard interval duration T_G, the receiver consistently captures samples from the same OFDM symbol. Consequently, the time shift introduced by the rectangular function in (9.10) and (9.11) can be disregarded.

Regarding the individual modulation symbols, it is evident that for a constant OFDM symbol index n, the Doppler frequency does not distinctly influence the various modulation symbols within the same OFDM symbol. This implies that within a given OFDM symbol, the Doppler frequency remains uniform across different modulation symbols. However, the Doppler frequency induces a consistent phase shift across all subcarriers, assuming that the bandwidth of the OFDM signal is significantly smaller than the carrier frequency. This implies a uniform impact of Doppler frequency on each subcarrier within the OFDM signal, maintaining a consistent phase shift across the entire signal bandwidth.

Conversely, when considering a fixed subcarrier index n, the Doppler effect introduces a linear phase shift between consecutive modulation symbols on that specific subcarrier. This linear phase shift corresponds to the change in phase caused by the Doppler frequency over the entire duration of the transmitted OFDM symbol, which is $2\pi f_D T_{\text{OFDM}}$.

Furthermore, the distance to the reflecting object does not affect the phase shift between consecutive modulation symbols on a specific subcarrier. A critical observation is that the range and Doppler effects induced by a reflecting object have completely orthogonal influences on the modulation symbols. While the range introduces a linear phase shift solely along the frequency axis, the Doppler similarly introduces a linear phase shift, but exclusively along the time axis. This orthogonality persists, assuming that the observation duration is sufficiently short for the reflecting object to remain within one range resolution cell. Consequently, it becomes feasible to recover range and Doppler independently through an appropriately designed processing algorithm.

Let us now distinguish between the Doppler shift in communication and radar sensing. In scenarios involving vehicles, it's anticipated that the received signal will experience a shift in frequency. This phenomenon occurs due to the relative speed between communication devices or between the radar platform and objects reflecting the signal. Specifically, for a communication link between two platforms in motion with a relative velocity v_{rel}, the Doppler frequency shift f_D at the receiver is determined by:

$$f_{\text{D, comm}} = \frac{v_{\text{rel}}}{\lambda} = \frac{v_{\text{rel}} f_c}{c_0}, \tag{9.12}$$

where λ is the wavelength, c_0 is the speed of light and f_c is the frequency of the carrier wave. Now, in applications related to radar sensing, a two way propagation occurs resulting in Doppler shift occurring twice [29], therefore, we have:

$$f_{\text{D, radar}} = \frac{2v_{\text{rel}}}{\lambda} = \frac{2v_{\text{rel}} f_c}{c_0}. \tag{9.13}$$

To quantify the influence of a moving object on the received modulation symbols $d_{\text{Rx}}(n)$, considering the subcarrier spacing from Eq. (9.9) and the relationship between relative velocity and Doppler frequency from Eq. (9.13), we can express this influence on the received modulation symbol $d_{\text{Rx}}(n)$ as:

$$d_{\text{Rx}}(\mu N_c + n) = A(\mu, n) d_{\text{Tx}}(\mu N_c + n) \exp\left(-j2\pi n \Delta f \frac{2R}{c_0}\right) \exp\left(j2\pi \mu T_{\text{OFDM}} \frac{2v_{\text{rel}} f_c}{c_0}\right). \tag{9.14}$$

To provide a more illustrative representation, consider the modulation symbols arranged in a matrix, where each column corresponds to an OFDM symbol and each row corresponds to a subcarrier. This rearrangement applies to both the transmitted and received modulation symbols. In this matrix view, the structure of the modulation symbols becomes evident, offering a clearer understanding of the signal's organization, as shown below:

$$\begin{bmatrix} d(0) & d(N_c) & \dots & d((N_{\text{sym}} - 1)N_c) \\ d(1) & d(N_c + 1) & \dots & d((N_{\text{sym}} - 1)N_c + 1) \\ \vdots & \vdots & \ddots & \vdots \\ d(N_c - 1) & d(2N_c - 1) & \dots & d(N_{\text{sym}} N_c - 1) \end{bmatrix}. \tag{9.15}$$

The modulation symbols received after reflecting off the moving object can be expressed as:

$$(\mathbf{D}_{\mathrm{Rx}})_{(\mu,n)} = A(\mu,n)(\mathbf{D}_{\mathrm{Tx}})_{(\mu,n)}.(\boldsymbol{k}_{\mathrm{R}} \otimes \boldsymbol{k}_{\mathrm{D}})_{(\mu,n)}. \tag{9.16}$$

Here,

$$\boldsymbol{k}_R = \left[0 \ \exp\left(-j2\pi \Delta f \tfrac{2R}{c_0}\right) \ldots \exp\left(-j2\pi (N_c - 1)\Delta f \tfrac{2R}{c_0}\right)\right] \tag{9.17}$$

and

$$\boldsymbol{k}_D = \left[0 \ \exp\left(j2\pi T_{\mathrm{OFDM}} \tfrac{2v_{\mathrm{rel}} f_c}{c_0}\right) \ldots \exp\left(j2\pi (N_{\mathrm{sym}} - 1)\Delta T_{\mathrm{OFDM}} \tfrac{2v_{\mathrm{rel}}}{f_0}\right)\right]. \tag{9.18}$$

To facilitate the estimation of range and relative velocity, a dedicated processor is employed. In the initial stage of processing, the transmitted information is extracted from the received information symbols. This extraction is accomplished through element-wise complex division, generating the dyadic product of the two vectors responsible for conveying range and Doppler information. This intricate operation effectively isolates the influence of the transmitted information, enabling subsequent focused processing specifically tailored to analyze the distinct range and Doppler characteristics of the received signals.

To create a processor for estimating range and relative velocity, the initial step involves eliminating the transmitted information obtained from the received symbols. The processor carries out element-wise complex division, producing the dyadic product of two vectors that contain both range and Doppler information. This operation effectively isolates the essential details related to the range and relative velocity, forming the basis for further processing steps in the algorithm to be discussed later. The division operation is implemented as shown below[1]:

$$(\mathbf{D}_{\mathrm{div}})_{\mu,n} = \frac{(\mathbf{D}_{\mathrm{Rx}})_{\mu,n}}{(\mathbf{D}_{\mathrm{Tx}})_{\mu,n}} = A(\mu,n)(\boldsymbol{k}_R \otimes \boldsymbol{k}_D). \tag{9.19}$$

The element-wise expression for (9.17) and (9.18) can be re-written as:

$$k_R(n) = \exp(-j2\pi n \Delta f \frac{2R}{c_0}) \tag{9.20}$$

and

$$k_D(\mu) = \exp(j2\pi \mu T_{\mathrm{OFDM}} \frac{2v_{\mathrm{rel}} f_c}{c_0}). \tag{9.21}$$

[1] Dyadic product is a mathematical operation involving two vectors, resulting in a second-order tensor or matrix. Given two vectors **a** and **b**, their dyadic product $\mathbf{a} \otimes \mathbf{b}$ is defined as: $\mathbf{a} \otimes \mathbf{b} = \mathbf{a}\mathbf{b}^T$, where $\mathbf{b}^T$ denotes the transpose of vector **b**.

In (9.20), the range R introduces a linear phase shift among the modulation symbols along the frequency axis. To determine the range to the reflecting object, a practical approach is to perform the inverse discrete Fourier transform (IDFT) of $k_R(n)$. This process involves transforming the frequency-domain information represented by $k_R(n)$ back into the time domain, providing a representation from which the range information can be extracted. Thus, $r(k)$ is given by:

$$\begin{aligned} r(k) = \mathrm{IDFT}[k_R(n)] &= \frac{1}{N_c}\left[\sum_{n=0}^{N_c-1} k_R(n) e^{j\frac{2\pi}{N_c}nk}\right]_{k=0}^{N_c-1} \\ &= \frac{1}{N_c}\left[\sum_{n=0}^{N_c-1} e^{-j2\pi n\Delta f \frac{2R}{c_0}} e^{jnk\frac{2\pi}{N_c}}\right]_{k=0}^{N_c-1}. \end{aligned} \tag{9.22}$$

In (9.22), it is observed that the two exponential terms cancel each other out, yielding unity. This cancelation occurs under certain condition:

$$k = \left[\frac{2R\Delta f N_c}{c_0}\right]_{k=0}^{N_c-1}. \tag{9.23}$$

Specifically, at the corresponding index k in the time response $r(k)$, a peak will manifest, provided that the specified condition in (9.23) is satisfied.

Similar to the process described earlier, the calculation of v_{rel} can be addressed by employing a discrete Fourier transform on (9.21). Thus, we have:

$$\begin{aligned} v(l) &= \mathrm{DFT}\left[\{k_D(\mu)\}\right] \\ &= \left[\sum_{\mu=0}^{N_{\mathrm{sym}}-1} k_D(\mu) e^{-j\mu l\frac{2\pi}{N_{\mathrm{sym}}}}\right]_{l=0}^{N_{\mathrm{sym}}-1} \\ &= \left[\sum_{\mu=0}^{N_{\mathrm{sym}}-1} e^{j2\pi\mu T_{\mathrm{OFDM}}\frac{2v_{\mathrm{rel}} f_c}{c_0}} e^{-j\mu l\frac{2\pi}{N_{\mathrm{sym}}}}\right]_{l=0}^{N_{\mathrm{sym}}-1}. \end{aligned} \tag{9.24}$$

The nullification of the exponential terms and their constructive combination result in a distinct index given by:

$$l = \left[\frac{2v_{\mathrm{rel}} f_c T_{\mathrm{OFDM}} N_{\mathrm{sym}}}{c_0}\right]_{l=0}^{N_{\mathrm{sym}}-1}. \tag{9.25}$$

Algorithm for Range and Doppler Estimation in OFDM Radar The entire range and Doppler estimation in OFDM radar can be summarized in three processing steps as described below:

Input: Received modulation symbol matrix $\mathbf{D}_{Rx}$ (size: $N_{\text{sym}} \times N_{\text{c}}$)
Processing Steps:

1. **Step 1:** Element-wise complex division as per (9.19)
   - For each element $D_{\text{Rx}}(i, j)$ in D_{Rx}, perform complex division according to

$$(D_{\text{div}})_{(i,j)} = \frac{D_{\text{Rx}}(i, j)}{D_{\text{Tx}}(i, j)^*} = A(i, j) \cdot e^{j2\pi k_R(i,j)} \cdot e^{j2\pi k_D(i,j)}$$

where * denotes the complex conjugate.
Optional Step: Apply windowing functions to D_{div} along both dimensions.

2. **Step 2:** Discrete Fourier Transform (DFT) as per (9.24).
   - Compute the DFT of every row in D_{div}, resulting in a matrix F_{row}.
3. **Step 3:** Inverse Discrete Fourier Transform (IDFT) as per (9.22).
   - Compute the IDFT for every column in F_{row}, resulting in the final matrix $R_{\text{DopplerRange}}$.

Output: The matrix $R_{\text{DopplerRange}}$ represents a two-dimensional radar image in range and Doppler.

Note that the entire algorithm is highly practical, making it suitable for real-world applications, as it relies on the computation of discrete Fourier transforms, a standard and efficiently implementable procedure. Notably, the Fourier transforms involve normalizations that can be flexibly chosen. It is important to mention that the peak value in the radar image doesn't serve as a measure for processing gain. Instead, during each Fourier transform, the signals undergo a coherent and deterministic process, leading to a total power gain of N^2, while noise, being a stochastic quantity, only experiences a power gain of N. This characteristic enhances the algorithm's robustness and effectiveness in extracting meaningful information from the received signals.

Processing gain is an important performance indicator. The processing gain can be defined as the increase in signal-to-noise ratio (SNR) achieved by coherent and deterministic signal processing techniques, such as discrete Fourier transforms, which result in a power gain of N^2 for signals and N for noise, enhancing the algorithm's robustness and effectiveness in extracting meaningful information from received signals. Observe from (9.22) that the processing gain G_P in the range domain N_{sym}. Similarly, as indicated by (9.24), the processing gain in the Doppler domain is given by:

$$G_P = N_c N_{\text{sym}}. \tag{9.26}$$

This processing gain is comparable to what would be attained with a correlation processor operating under optimal conditions, aligning directly with the count of correlated samples. The algorithm based on modulation symbols delivers equivalent performance in terms of resolution and processing gain when compared to a correlation processor.

Also, it is observed that in Step 2 of the algorithm, the DFT is applied to every row in the matrix D_{div}, resulting in a matrix F_{row}. This step mirrors the process of transforming from the delay-Doppler domain to the delay profile domain in OTFS. Subsequently, in Step 3, the IDFT is applied to every column in the matrix F_{row}, yielding the final matrix $R_{\mathrm{DopplerRange}}$. This operation corresponds to the process of transforming from the delay profile domain back to the time–frequency-domain in OTFS. Hence, the sequential application of Step 2 and Step 3 closely resembles the process of Inverse symplectic finite Fourier transform (ISFFT) in OTFS systems.

9.3.2 OTFS-Based Waveform Design

Fundamentals of OTFS The Orthogonal Time Frequency Space (OTFS) modulation has garnered significant attention owing to its distinctive capability of offering more reliable communications when compared to the established OFDM modulation. This heightened interest is particularly pronounced in high-mobility environments [30]. In scenarios characterized by high mobility, the traditional OFDM modulation may encounter challenges such as inter-carrier interference and fading effects. OTFS modulation addresses these issues by simultaneously leveraging both the time and frequency dimensions in a manner that goes beyond the capabilities of conventional approaches. OTFS modulation operates in the delay-Doppler (DD) domain, providing a unique and effective way to combat the challenges posed by high-mobility scenarios. By doing so, it introduces a new perspective that allows for improved signal reception and decoding even in dynamic and challenging environments. The enhanced reliability offered by OTFS modulation has sparked considerable interest and research, positioning it as a promising solution for next-generation wireless communication systems.

Benefits of OTFS over OFDM: In contrast to the ISAC system based on OFDM, the OTFS-ISAC system introduces a novel approach by facilitating direct interaction between transmitted signals and channel responses. This interaction occurs in a unified DD domain, encompassing both radar and communication functionalities. Moreover, the OTFS-ISAC system exhibits resilience to delay and Doppler spreads, offering a comprehensive solution to the challenges encountered in dynamic communication and radar scenarios. The OTFS modulation, by operating in the DD domain, enables a more unified and efficient treatment of signals in comparison to traditional OFDM-based systems. This unique characteristic makes OTFS an appealing signal waveform for fully realizing the potential of ISAC systems. The direct interaction in the unified DD domain aligns with the requirements of both radar and communication functionalities, positioning OTFS as a promising candidate to advance the capabilities of ISAC systems. Works carried out in [31] have established that

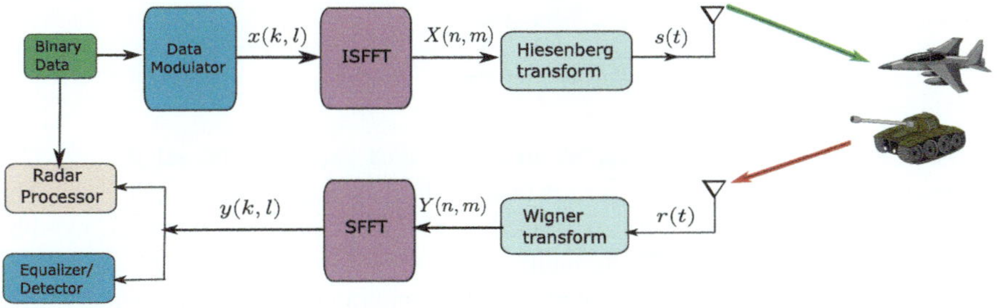

Fig. 9.5 OTFS-based radar system architecture

the OTFS modulation surpasses the conventional OFDM scheme in robustness for velocity estimation, emphasizing its potential for enhanced performance in dynamic communication and radar applications.

Implementation of OTFS Modulation: The two prevalent methodologies for implementing OTFS modulation delve into the symplectic finite Fourier transform (SFFT) [32] and the discrete Zak transform (DZT) [33, 34]. The SFFT and DZT approaches are foundational techniques employed to realize OTFS modulation, each offering distinct advantages and characteristics within the domain of signal processing and communication systems. Wei et. al. have detailed both the modulation techniques in [35]. In this chapter, we direct our focus towards an in-depth exploration of the Symplectic Finite Fourier Transform (SFFT)-based modulation technique.

At the Transmitter: The transceiver architecture for OTFS based on the SFFT is illustrated in Fig. 9.5. In our assumptions, we have a certain bandwidth denoted as B_{OTFS} and a designated time duration T_{OTFS} both defined in the *Time–Frequency* (TF) domain, at our disposal to accommodate a singular OTFS frame [35]. The allocated bandwidth B_{OTFS} is partitioned into M subcarriers with a subcarrier spacing $\Delta f = \frac{B_{\text{OTFS}}}{M}$. Simultaneously, the specified time duration T_{OTFS} is subdivided into N time slots, each having a slot duration $T = \frac{T_{\text{OTFS}}}{N}$. This division and allocation provide the framework for constructing and organizing the constituents of an OTFS frame, facilitating the transmission and reception processes within the system. At the transmitter (Tx) side, the information symbol for the (l, k)th Delay-Doppler Resource Element (DDRE) is denoted as $X_{DD}[l, k]$, where $l \in \{0, \ldots, M-1\}$ is the delay index and $k \in \{0, \ldots, N-1\}$ is the Doppler index. Each element in this grid, $X_{DD}[l, k]$, carries specific information tailored to the delay and Doppler characteristics, contributing to the overall transmitted signal in the OTFS modulation scheme.

The symbols represented in the DD domain $X_{DD}[l, k]$ undergo a conversion process, transforming them into the TF domain signal $X_{TF}[m, n]$. This transformative operation is executed using the *inverse symplectic finite Fourier transform* (ISFFT) [36] as shown below:

$$X_{TF}[m,n] = \frac{1}{\sqrt{NM}} \sum_{k=0}^{N-1} \sum_{l=0}^{M-1} X_{DD}[l,k] e^{j2\pi\left(\frac{nk}{N} - \frac{ml}{M}\right)}. \tag{9.27}$$

Here, m represents the subcarrier index, it takes values in the set $\{0, \ldots, M-1\}$ and n represents the slot index, it takes values in the set $\{0, \ldots, N-1\}$. This transformation mathematically expressed in (9.27) effectively maps the symbols from the DD domain to the TF domain.

The OTFS transmit signal, denoted as $s(t)$, is generated by employing multicarrier modulation on the TF domain signal $X_{TF}[m,n]$, which is alternatively referred to as the *Heisenberg transform* [32], as shown below:

$$s(t) = \sum_{n=0}^{N-1} \sum_{m=0}^{M-1} X_{TF}[m,n] g_{TX}(t - nT) e^{j2\pi m \Delta f (t-nT)}. \tag{9.28}$$

In the above expression, g_{TX} represents the pulse shaping filter applied at the Tx. This filter is essential for shaping the transmitted signal influencing its spectral characteristics. The convolution of $X_{TF}[m,n]$ and $g_{TX}(t - nT)$ in (9.28), accounts for the pulse shaping process. The subsequent term $e^{j2\pi m \Delta f (t-nT)}$ introduces the carrier frequency modulation associated with the specific carrier index m and its corresponding frequency spacing Δf. Heisenberg transform captures both the time and frequency information simultaneously and can effectively address challenges related to high-mobility environments. It can therefore provide superior performance compared to the traditional OFDM modulation. This simultaneous representation in both time and frequency domains enhances the effective reliability of OTFS-based communication systems.

Channel Modeling: Following the transmission of the signal $s(t)$, it undergoes parallel-to-serial conversion and digital-to-analog conversion processes. Subsequently, the received signal $r(t)$ can be expressed as [37]:

$$r(t) = \int\int h(\tau, \nu) s(t - \tau) e^{j2\pi\nu(t-\tau)} d\tau d\nu + z(t). \tag{9.29}$$

In communication channels, especially those encountered in radar and wireless communication systems, the environment often features only a limited number of reflectors or scattering objects. These objects contribute to the propagation of signals, introducing delays and Doppler shifts in the received signals. Due to the sparse nature of these reflectors, the channel can be efficiently characterized in the DD domain using only a small number of parameters. This sparse representation simplifies the modeling and analysis of the channel's behavior, allowing for effective communication or radar signal processing.

Given the sparse nature of the channel representation, it is advantageous to express the channel response $h(\tau, \nu)$, in a specific form. In this context, sparsity implies that only a limited number of reflectors contribute significantly to the channel's behavior in terms of delays τ and Doppler shifts ν. This sparse representation aids in efficiently capturing the essential characteristics of the channel, facilitating simplified and computationally more manageable signal processing. The delay-Doppler domain can be represented as shown below:

$$h(\tau, \nu) = \sum_{i=1}^{P} h_i \delta(\tau - \tau_i)\delta(\nu - \nu_i). \tag{9.30}$$

In this expression, $\delta(.)$ represents the Dirac delta function, and the parameters P, h_i, τ_i, and ν_i correspond to the number of propagation paths, the complex gain, delay, and Doppler shift associated with the ith path, respectively.

The delay and Doppler-shift parameters for the ith path can be expressed as follows:

$$\tau_i = l_i \frac{1}{M\Delta f} \tag{9.31}$$

and

$$\nu_i = k_i \frac{1}{NT}. \tag{9.32}$$

Here, NT represents the total time duration, and $M\Delta f$ signifies the bandwidth attributed to the transmitted signal frame. For simplicity in mathematical calculations, we consider l_i and k_i to be integers. In other words, τ_i and ν_i are assumed to be the integer multiples of $\frac{1}{M\Delta f}$ and $\frac{1}{NT}$, respectively. The accommodation of fractional delay and Doppler shifts has been established in methodologies discussed in [38, 39].

As per [40], it is customary to take $T\Delta f = 1$, $\tau_{\max} = \frac{(L-1)T}{M}$, where $\tau_{\max}$ is the maximum delay. The channel is considered to be underspread, i.e., $l_i < M$ and $k_i < N$. Therefore, after sampling the received signal $y(t)$ at a rate $f_s = M\Delta f = \frac{M}{T}$, and discarding the cyclic prefix, we have:

$$r(n) = \sum_{i=1}^{P} h_i e^{j2\pi \frac{k_i(n-l_i)}{MN}} s\left([n - l_i]_{MN}\right) + w(n). \tag{9.33}$$

Here, it is observed that a vector $\{r(n)\}_{n=0}^{MN-1}$ is formed in (9.33) for which the sample entries are from (9.29), (9.30).

Rewriting (9.33) in vector form, we get:

$$\mathbf{r} = \mathbf{H}\mathbf{s} + \mathbf{w}. \tag{9.34}$$

Here, $\mathbf{H}$ is a square matrix of size $MN \times MN$ and is expressed as:

$$\mathbf{H} = \sum_{i=1}^{P} h_i \prod^{l_i} \Delta^{k_i}. \tag{9.35}$$

$\prod$ is a permutation matrix given by:

$$\prod = \begin{bmatrix} 0 \dots 0 \; 1 \\ 1 \dots 0 \; 0 \\ \vdots \; \ddots \; \vdots \; \vdots \\ 0 \dots 1 \; 0 \end{bmatrix}_{M \times N} . \tag{9.36}$$

Also, is $MN \times MN$ diagonal matrix with entries $[z^0, z^1, \dots, z^{MN-1}]$. Here, each of the diagonal entries are obtained from the expression, $z = e^{\frac{j2\pi}{MN}}$.

It is to be observed that in the context of (9.29), the matrices $\prod$ and play crucial roles in representing the delays and Doppler shifts, respectively. Specifically, $\prod$ is responsible for introducing cyclic shifts to the transmitted signal vector s, where each path contributes a cyclic shift of l_i steps, denoted by $\prod^{l_i}$. On the other hand, Δ^{k_i} models the modulation of the signal vector s by a carrier at the frequency k_i. Together, these matrices collectively capture the temporal and frequency characteristics introduced by different propagation paths in the channel.

At the Receiver: At the receiver, the inversion of transmitter operations involves transforming the received signal samples denoted as $\mathbf{r}$ into time–frequency-domain symbols $\mathbf{R}$, where vector elements are folded back into a matrix and is given by: $\mathbf{R} = \text{vec}^{-1}(r)$. Subsequently, these symbols undergo a transformation into delay-Doppler domain symbols $\mathbf{Y}$ through the following operations, thereby completing the intricate signal processing chain at the receiver.

$$\mathbf{Y} = \mathbf{F}_M^H (\mathbf{F}_M \mathbf{G}_{rx} \mathbf{R}) \mathbf{F}_N. \tag{9.37}$$

In the reception process, we first execute an M-point Fast Fourier Transform (FFT), succeeded by a Symplectic Finite Fourier Transform (SFFT). Here, $\mathbf{G}_{\text{rx}} \in \mathbb{C}^{M \times M}$ denotes the filter implemented at the receiver, utilizing the pulse-shaping waveform $g_{\text{rx}}(t)$. Subsequently, the received signal can be compactly represented in vectorized form as:

$$\begin{aligned} \mathbf{y} &= (\mathbf{F}_N \otimes \mathbf{G}_{\text{rx}})\mathbf{r} \\ &= (\mathbf{F}_N \otimes \mathbf{G}_{\text{rx}})\mathbf{H}(\mathbf{F}_N^H \otimes \mathbf{G}_{\text{tx}})\mathbf{x} + (\mathbf{F}_N \otimes \mathbf{G}_{\text{rx}})\mathbf{w} \\ &= \mathbf{H}_{\text{eff}}\mathbf{x} + \tilde{\mathbf{w}}. \end{aligned} \tag{9.38}$$

It is to be noted that the received signal expressed in (9.38) is the input–output relation in DD domain. $\mathbf{H}_{\text{eff}}$ is the effective channel matrix and $\tilde{\mathbf{w}}$ is the noise vector.

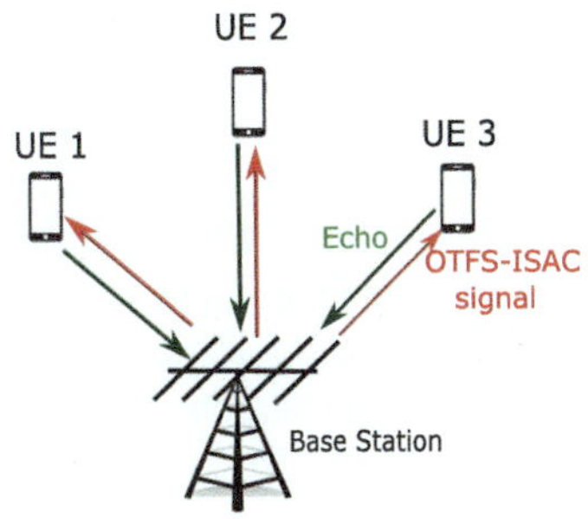

Fig. 9.6 OTFS ISAC system model

In case of rectangular waveforms, $\mathbf{G}_{\text{tx}}$ and $\mathbf{G}_{\text{rx}}$ will be identity matrices because the transmitted signal experiences minimal distortion, the noise samples are often independent and identically distributed (iid), resulting in a scalar covariance matrix for $\tilde{\mathbf{w}}$. Note that the variance of each noise component is equal, indicating a uniform distribution of noise across the signal dimensions.

OTFS-ISAC System Consider a comprehensive system configuration consisting of a monostatic base station (BS) and P designated targets, as depicted in Fig. 9.6. To simplify matters, we assume that all these targets also serve as communication user equipments (UEs), a common scenario in vehicular networks. The BS comprises a transmit uniform linear array (ULA) with N_t antennas and a separate receive ULA with N_r antennas. We presume satisfactory isolation between the transmit and receive arrays to prevent any interference from sensing echoes with the transmitted signals. Furthermore, each communication target is assumed to be equipped with N_u antennas to ensure effective reception of communication signals.

Communication Model In our system, we adopt a DD domain grid of dimensions $M \times N$, where $l \in \{0, \ldots, M-1\}$ and $k \in \{0, \ldots, N-1\}$ represent the delay and Doppler indices, respectively.

For the ith target, the information symbols are represented as $X^i_{\text{DD}}[l, k]$. As explained earlier, these symbols undergo a transformation to the time–frequency (TF)-domain through the inverse symplectic finite Fourier transform (ISFFT), resulting in $X_{i,\text{TF}}[m, n]$, where m and n denote the subcarrier and time slot indices.

Upon completing the ISFFT, we obtain a representation of the signal in the TF domain. Subsequently, by employing multicarrier modulation techniques, the transmitted signal to the ith target can be expressed as:

$$X^i_{\text{TF}}[m, n] = \text{ISFFT}\left(X^i_{\text{DD}}[l, k]\right). \tag{9.39}$$

Therefore, through the implementation of multicarrier modulation, the signal directed towards the ith target is expressed as:

$$s_i(t) = \sum_{n=0}^{N-1} \sum_{m=0}^{M-1} X^i_{\text{TF}}[m, n] g_{\text{tx}}(t - nT) e^{j2\pi m \Delta f (t-nT)}. \tag{9.40}$$

In (9.40), the variable g_{tx} corresponds to the transmit shaping pulse, Δf represents the frequency spacing, and T signifies the symbol duration. It is important to emphasize that to preserve the orthogonality condition, there must be adherence to the relationship $T\Delta f = 1$. This condition ensures that the frequency and time components maintain orthogonality, a crucial aspect in multicarrier modulation systems.

Now, in the initial phase of target detection, the BS can leverage the antenna array to generate a broad beam or an omni-directional signal, facilitating the detection of all relevant targets. Subsequently, during the tracking mode, the BS configures multiple beams for both information transmission and continuous monitoring of the target positions.

Through a beamforming matrix $\mathbf{F} \in \mathbb{C}^{N_t \times P}$, the multibeam signal is expressed as:

$$\widetilde{\mathbf{s}}(t) = \mathbf{F}\mathbf{s}(t). \tag{9.41}$$

In the expression $\mathbf{s}(t) = [s_1(t), \ldots, s_P(t)]^T$, representing the transmitted signal to all P targets, the ith column of matrix $\mathbf{F}$, denoted by $\mathbf{f}_i = \sqrt{\frac{p_i}{N_t}}\mathbf{a}_{N_t}(\widetilde{\theta}_i)$, is utilized to allocate power p_i and direct the transmitted signal towards the intended direction $\widetilde{\theta}_i$. Here, the steering vector is given by:

$$\mathbf{a}_{N_t}(\theta_i) = [1, e^{j\pi \sin\theta_i}, \ldots, e^{j(N_t-1)\pi \sin\theta_i}]^T. \tag{9.42}$$

This steering vector helps in directing the beamforming towards the specific direction associated with the target. Due to the asymptotic orthogonality of the massive antenna array, post-transmit beamforming, the communication channel to the ith target becomes asymptotically line-of-sight (LoS) dominated and can be represented as:

$$\mathbf{C}_{i_{\mathrm{DD}}}(\tau, \nu) = h_i \mathbf{a}_{N_u}(\theta_i)\mathbf{a}_{N_t}^{\mathrm{H}}(\theta_i)\delta(\tau - \tau_i)\delta(\nu - \nu_i). \tag{9.43}$$

Here, h_i represents the channel gain in (9.43), given by $h_i = \sqrt{\frac{c_0}{4\pi f_c d_i^2}}$, where c_0 denotes the signal propagation speed, f_c is the carrier frequency, d_i is the range distance, τ_i is the delay, and ν_i is the Doppler shift.

Now, after traversing the communication channel [41], the signal received from the ith target can be described as follows:

$$y_i(t) = h_i \mathbf{u}_i^H \mathbf{a}_{N_u}(\theta_i)\mathbf{a}_{N_t}^{\mathrm{H}}(\theta_i)\mathbf{f}_i s_i(t - \tau_i)e^{j2\pi\nu_i(t-\tau_i)} + z_i(t), \tag{9.44}$$

where $\mathbf{u}_i \in \mathbb{C}^{N_u \times 1}$ denotes the receive beamformer, and $z_i(t)$ represents the time-domain noise signal.

We finally employ a receive filter $g_{tx}(t)$, and following OTFS demodulation, the received OTFS signal is given by:

$$Y_{\mathrm{DD}}^i[l,k] = h_i \mathbf{u}_i^{\mathrm{H}} \mathbf{a}_{N_u}(\theta_i)\mathbf{a}_{N_t}(\theta_i) f_i X_{\mathrm{DD}}^i[(l - l_i)_M, (k - k_i)_N] + Z_{\mathrm{DD}}^i[l,k], \tag{9.45}$$

where $k_i = \nu_i N_T, l_i = \tau_i M \Delta f$ are integers. The term $Z_i^{\mathrm{DD}}[l,k]$ represents additional white Gaussian noise with a power spectral density of N_0.

Communication System Algorithm: The OTFS-based ISAC communication model can be summarized in an algorithm as mentioned below:

Step 1: System Initialization

1. Set system parameters: P (Number of targets), N_t (number of transmit antennas), N_r (number of receive antennas), N_u (number of antennas at each target).
2. Choose transmit shaping pulse g_{tx}, symbol duration T, and frequency spacing Δf such that $T\Delta f = 1$ to maintain orthogonality.
3. Initialize the DD domain grid with $M \times N$, where $l \in \{0, \ldots, M-1\}$ and $k \in \{0, \ldots, N-1\}$ represent delay and Doppler indices.

Step 2: Transmission to Targets

1. Transform information symbols $X_i^{\mathrm{DD}}[l,k]$ to TF domain via ISFFT, yielding $X_i^{\mathrm{TF}}[m,n]$.
2. Perform multicarrier modulation to obtain the transmitted signal $s_i(t)$ for the ith target using the formula:

$$s_i(t) = \sum_{n=0}^{N-1} \sum_{m=0}^{M-1} X_i^{\mathrm{TF}}[m,n] g_{\mathrm{tx}}(t-nT) e^{j2\pi m\Delta f(t-nT)}.$$

Step 3: Beamforming and Channel Model

1. Apply the beamforming matrix $\mathbf{F}$ to express the multibeam signal as $\widetilde{s}(t) = \mathbf{F}s(t)$, where $s(t) = [s_1(t), \ldots, s_P(t)]^T$.
2. Model the communication channel to the ith target as $C_i^{\mathrm{DD}}(\tau,\nu) = h_i \mathbf{a}_N^u(\theta_i)\mathbf{a}_{N_t}^T(\theta_i)\delta(\tau-\tau_i)\delta(\nu-\nu_i)$.

Step 4: Received Signal Processing

1. After passing through the channel, the received signal $y_i(t)$ for the ith target is given by:

$$y_i(t) = h_i \mathbf{u}_i^H \mathbf{a}_N^u(\theta_i)\mathbf{a}_{N_t}^T(\theta_i)\mathbf{f}_i s_i(t-\tau_i)e^{j2\pi\nu_i(t-\tau_i)} + z_i(t).$$

   where $\mathbf{u}_i$ is the receive beamformer and $z_i(t)$ is the time-domain noise signal.
2. Apply a receive filter $g_{\mathrm{rx}}(t)$ and perform OTFS demodulation to obtain the received OTFS signal $Y_i^{\mathrm{DD}}[l,k]$.

Sensing Model The sensing channel, denoted as $H(t,\tau)$, is represented by the following equation:

$$H(t,\tau) = \sum_{i=1}^{K} \gamma_i \mathbf{a}_{Nr}(\theta_i)\mathbf{a}_{Nt}^H(\theta_i)\delta(\tau-\eta_i)e^{j2\pi\nu_i t}. \tag{9.46}$$

Here, γ_i, η_i, and ν_i represent the reflected coefficient, delay, Doppler associated with the ith target, respectively, and K is the total number of transmit antennas.

Now, the received sensing echoes at the base station can be expressed as follows:

$$\mathbf{r}(t) = \sum_{i=1}^{K} \gamma_i \mathbf{a}_{N_r}(\theta_i)\mathbf{a}_{N_t}^H(\theta_i)\mathbf{f}_i s_i(t - \eta_i)e^{j2\pi \nu_i t} + w(t). \tag{9.47}$$

Considering the influence of measurement noise represented by $w(t)$ and the asymptotic orthogonality properties of massive MIMO receive antenna arrays, specifically $\mathbf{a}_{Nr}^H(\theta_i)\mathbf{a}_{Nr}(\theta_i') \approx 0$ for $\theta_i \neq \theta_i'$, the interference between different targets in the sensing echoes can be effectively disregarded. This implies that the base station (BS) can differentiate between various targets based on their distinct angles-of-arrival (AoAs). As a result, the sensing echo corresponding to the ith target can be separated from the overall received signal $r(t)$ using a dedicated receive beamformer $b_i = \mathbf{a}_{Nr}(\tilde{\theta}_i)$. This separation can be expressed as:

$$r_i(t) = \gamma_i \mathbf{b}_i^H \mathbf{a}_{N_r}(\theta_i)\mathbf{a}_{N_r}^H(\theta_i)\mathbf{f}_i s_i(t - \eta_i)e^{j2\pi \nu_i t} + w(t). \tag{9.48}$$

Here, $w(t)$ signifies the measurement noise, and $\mathbf{b}_i = \mathbf{a}_{N_r}(\tilde{\theta}_i)$ is the receive beamformer for the ith target.

It is important to observe that the angular parameter (θ_i) can be deduced from the receive beamforming process through a comparison of gains achieved for different beam directions. Consequently, (9.48) can be further simplified as:

$$r_i(t) = G_a s_i(t - \eta_i)e^{j2\pi \nu_i t} + w_i(t). \tag{9.49}$$

Here, G_a represents the composite array gain associated with the beamforming for the ith target, and $s_i(t - \eta_i)$ is the signal delayed by η_i. Incorporating an ideal receive filter and employing OTFS demodulation, the input–output relationship in the DD domain is established as:

$$R_{\mathrm{DD}}^i[l,k] = G_a \sum_{k'=0}^{N-1}\sum_{l'=0}^{M-1} H_{\mathrm{DD}}^i[l',k'] X_{\mathrm{DD}}^i[(l-l')_M, (k-k')_N] + \mathbf{W}_{\mathrm{DD}}^i[l,k]. \tag{9.50}$$

Here, $H_{\mathrm{DD}}^i[l',k']$ represents the gain of the ith target at the DD grid (bin) with indices l', k' corresponding to the delay of $\frac{l'}{M\Delta f}$ and Doppler of $\frac{k'}{NT}$. The term $X_{\mathrm{DD}}^i[(l-l')_M, (k-k')_N]$ denotes the input signal for the ith target, and $\mathbf{W}_{\mathrm{DD}}^i[l,k]$ is the associated noise component.

Finally, in matrix form, the (9.51) is expressed as:

$$\mathbf{r}_{\mathrm{DD}}^i = \mathbf{G}_a \mathbf{X}_{\mathrm{DD}}^i \mathbf{h}_{\mathrm{DD}}^i + \mathbf{w}_{\mathrm{DD}}^i. \tag{9.51}$$

Here, $\mathbf{r}_i^{\mathrm{DD}}$ represents the stacked vector of all received samples, $\mathbf{X}_i^{\mathrm{DD}}$ is the DD domain representation of the transmitted signal, $\mathbf{h}_i^{\mathrm{DD}}$ is the estimate of the sensing channel, and $\mathbf{w}_i^{\mathrm{DD}}$ is the vector of noise samples. $H_{\mathrm{DD}}^i[l', k']$ is the $(Nk + l')$th element in $\mathbf{h}_i^{\mathrm{DD}}$, representing the gain associated with the delay and Doppler indices of the ith target.

Sensing Model Algorithm The OTFS-based ISAC sensing model can be summarized in an algorithm as mentioned below:

Input: Transmit signal $\mathbf{s}(t)$, channel matrix $\mathbf{H}(t, \tau)$, noise $w(t)$.

1. **Step 1:** Formulate transmit signal:

$$\mathbf{s}(t) = \begin{bmatrix} s_1(t) \\ \vdots \\ s_P(t) \end{bmatrix}.$$

2. **Step 2:** Compute received signal:

$$\mathbf{r}(t) = \sum_{i=1}^{K} \gamma_i \mathbf{a}_{Nr}(\theta_i)^{\mathrm{H}} \mathbf{a}_{Nt}(\theta_i) \tilde{s}(t - \eta_i) e^{j2\pi \nu_i (t-\eta_i)} + \mathbf{w}(t).$$

3. **Step 3:** Apply receive beamformer:

$$\mathbf{b}_i = \mathbf{a}_{Nr}(\widetilde{\theta}_i), \quad \text{where } \widetilde{\theta}_i \text{ is the estimated angle of arrival.}$$

4. **Step 4:** Extract target signal:

$$\mathbf{r}_{\mathrm{DD}}^i = \mathbf{G}_a \mathbf{X}_{\mathrm{DD}}^i \mathbf{h}_{\mathrm{DD}}^i + \mathbf{w}_{\mathrm{DD}}^i.$$

5. **Step 5:** Stack received samples:

$$\mathbf{r}_i^{\mathrm{DD}} = \left[r_i[0, 0] \cdots r_i[M-1, N-1] \right]^{\mathrm{T}}.$$

Output: Received signal $\mathbf{r}(t)$.

After receiving the signal, parameters such as reflected coefficient, delay, and Doppler shift for each target are estimated. Sensing echoes at the BS, accounting for measurement noise, are expressed with these parameters. Massive MIMO receive antenna arrays aid in distinguishing targets based on their angles-of-arrival (AoAs). Subsequently, receive beamforming techniques extract the sensing echo of each target, streamlining processing by focusing on relevant target information.

9.4 IRS-Assisted ISAC System Design

Intelligent Reflecting Surfaces (IRS) refer to a cutting-edge technology that has gained prominence in wireless communication systems. Also known as reconfigurable intelligent surfaces (RIS) or smart mirrors, IRS are essentially surfaces embedded with tunable elements, such as sensors or controllers, that can dynamically manipulate the reflection and absorption of electromagnetic waves, including radio frequencies and microwave signals. The primary goal of IRS is to enhance wireless communication by intelligently redirecting signals to improve coverage, reduce interference, and improve the overall performance of wireless networks. By fine-tuning the properties of the reflective surfaces in real-time, IRS can effectively optimize the signal strength and quality in different environments. IRS has recently captured significant attention from both academic and industrial circles. It is acknowledged as a promising approach for improving communication performance in various systems. This interest has extended to ISAC systems, where the incorporation of IRS technology is seen as a means to further optimize communication effectiveness.

Researchers have investigated the integration of IRS into ISAC systems to address communication challenges. The application of IRS in mitigating multi-user interference was investigated to enhance overall communication performance while aligning with radar waveform requirements [42]. This approach demonstrates the adaptability of IRS in managing interference scenarios. In the domain of millimeter-wave band communication systems, researchers in [43] concentrated on maximizing the sum rate of communication users while respecting radar waveform constraints. This was accomplished through comprehensive optimization of IRS phase shifts, communication beamformers, and radar signal covariance matrix. The results suggest that IRS can play a pivotal role in improving communication efficiency at higher frequency bands. Additionally, in [44], researchers explored the joint design of constant-modulus waveform and discrete IRS phase shifts. The aim was to minimize the mean squared error (MSE) of communication users (CUs) under constraints dictated by the Cramer–Rao bound (CRB) for direction of arrival (DoA) estimation. This approach showcases the potential of IRS in contributing to improved estimation accuracy in communication systems. Additionally, researchers explored IRS-enabled sensing designs [45] aimed at minimizing the DoA estimation error in terms of the CRB. This underscores the versatility of IRS technology, not only in communication optimization but also in enhancing sensing capabilities for more accurate parameter estimation.

The effective performance of ISAC systems relies heavily on the quality of the direct link between the base station and the target of interest, particularly in radar detection and sensing applications. However, in non-line-of-sight (NLoS) areas where direct links are unavailable, realizing optimal ISAC functionality becomes challenging, as highlighted in several recent works. Fortunately, IRS technology presents a solution by reconfiguring the wireless propagation environment. In situations where direct links are not available, IRS can enhance wireless communication rates and establish sensing channel links between the BS and the target [46]. IRS can have a beneficial impact on reflected sensing signals, mitigate interference, and provide additional degrees of freedom to mitigate severe channel fading effects. In [47], researchers investigated an IRS-assisted ISAC system, leveraging IRS to establish new channel links for enhanced communication and sensing performance. Moreover, a hybrid IRS model, integrating both active and passive elements, was examined in [48] to support radar and communication, with the goal of maximizing the worst-case target illumination power. Additionally, [49] investigated the use of IRS to enhance target detection capabilities. Minimizing cross-correlation coefficients between radar signals was emphasized as a means to achieve better radar detection performance [50]. In the context of IRS-assisted ISAC system design, [51] considered constraints on radar signal cross-correlation patterns, acknowledging that IRS could introduce additional interference for sensing purposes. Furthermore, [52] investigated IRS-assisted ISAC systems aimed at preventing information leakage. They utilized IRS to establish a virtual link for target sensing and to counteract eavesdropping.

It is to be noted that all the mentioned studies are concentrated on environments devoid of impediments, commonly described as clutter-free settings. However, a notable research void emerges, as there has been minimal exploration of IRS-assisted ISAC designs in environments characterized by blockages or obstacles. In these obstructed settings, where physical barriers may disrupt signal propagation, there is a need to investigate the impact of IRS technology on ISAC system performance. This research gap underscores the importance of understanding how IRS can effectively operate and enhance communication and sensing capabilities in the presence of blockages, introducing an additional layer of complexity to the design considerations for these intelligent systems.

In an environment with obstacles, the radar sensing performance is often influenced by diverse elements such as trees, tall buildings, and cars [49]. The reflected signal power from these obstacles may surpass the power of signals reflected from the intended targets, leading to a substantial degradation in target detection reliability. Effectively detecting and sensing multiple targets in such obstructed environments pose a significant challenge for ISAC system designs [53, 54]. In addressing this issue, a study [55] utilized multiple IRSs to alleviate interference between radar and communication systems within an obstructed environment. The obstacles, often referred to as signal-dependent interference, can be mitigated by leveraging prior knowledge of the obstacles, such as the covariance matrix associated with them [56]. This approach involves adopting the clutter's characteristics to suppress its impact [57]. In the context of maintaining communication quality of service in cluttered environments, two studies [58, 59] introduced transmit beamforming strategies to maximize the Signal-to-Interference-plus-Noise Ratio (SINR) for radar signals, particularly with the aid of IRS. This approach seeks to improve overall radar performance by optimizing signal quality in the presence of obstacles or interference.

9.4.1 System Model

As depicted in Fig. 9.7, we investigate an ISAC system aided by IRS in an environment with obstacles. The system comprises a multi-antenna base station (BS) assisted by an IRS, facilitating the transmission of unicast communication signals for a group of K single-antenna communication users (CUs) and dedicated radar sensing signals for a total of L targets. The IRS is strategically deployed to enhance the propagation of both communication and radar sensing signals.

The base station (BS) is equipped with a Uniform Linear Array (ULA) comprising M antennas, while the IRS incorporates a Uniform Planar Array (UPA) with N passive reflecting elements. In this scenario, Q obstacles are located between the BS and the L sensing targets, introducing interference that affects the BS's target sensing performance. It's worth noting that the Q obstacles are positioned farther away from the K communication users (CUs) compared to the L targets, resulting in a minimal interference effect on the CUs compared to its impact on the L targets [60].

To define the sets involved, the CU set is denoted as $\mathcal{K} \in \{1, \ldots, K\}$, the obstacle set as $\mathcal{Q} \in \{1, \ldots, Q\}$, and the set of sensing targets as $\mathcal{L} \in \{1, \ldots, L\}$. In this scenario, we assume that the L targets are located in the Non-Line-of-Sight (NLoS) region of the base station (BS). This implies that the direct channels between the BS and the L sensing targets are effectively negligible due to blockage caused by the Q obstacles.

In this scenario, each communication user (CU), denoted by k in the set $\mathcal{K}$, transmits information symbols s_k using a transmit beamforming vector $\mathbf{w}_k$. The infor-

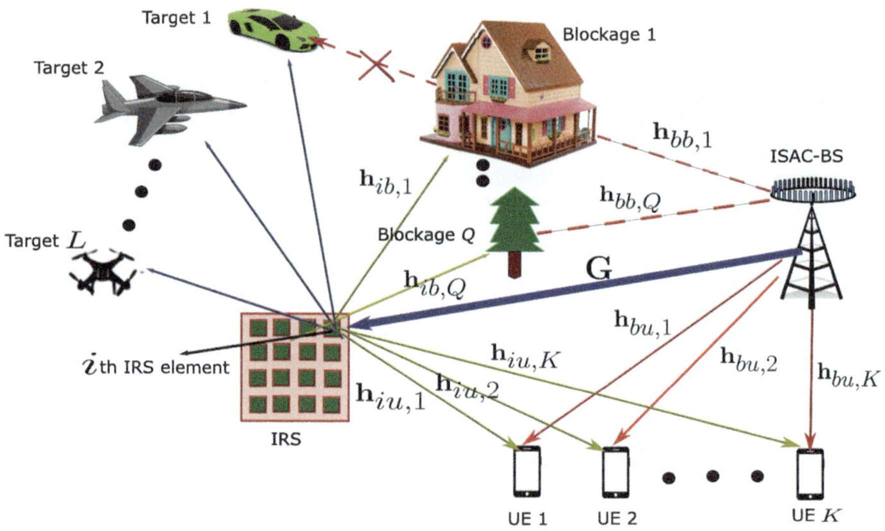

Fig. 9.7 Illustration of the IRS-assisted ISAC system in an environment with blockages

mation symbols are assumed to be independent and identically distributed random variables with zero mean and unit variance, expressed as $s_k \sim \mathcal{CN}(0, 1)$ for each k in $\mathcal{K}$. The base station (BS) transmits a dedicated sensing signal $\mathbf{x}_0$ with zero mean, indicated as $E[\mathbf{x}_0] = 0$. The sensing covariance matrix $\mathbf{R}_0 = E[\mathbf{x}_0\mathbf{x}_0^H]$ has a general rank, meaning its rank can range from 0 to M, where M is the number of antennas at the BS.

In this context, the ISAC transmit signal $\mathbf{x}$ from the BS is modeled as the sum of the dedicated sensing signal $\mathbf{x}_0$ for L targets and the summation over all CUs k of their respective communication signals $\mathbf{w}_k s_k$. This representation captures the combined transmission of sensing signals for target detection and communication signals from the CUs.

Therefore, the integrated transmission signal from the BS in the ISAC system, combining sensing and communication components can be expressed mathematically as:

$$\mathbf{x} = \mathbf{x}_0 + \sum_{k=1}^{K} \mathbf{w}_k s_k. \tag{9.52}$$

In accordance with (9.52), the limitation on the transmit power of the BS is articulated as follows:

$$\mathbb{E}[\|\mathbf{x}\|^2] = \text{tr}(\mathbf{R}_0) + \sum_{k=1}^{K} \|\mathbf{w}_k\|^2 \leq P_0. \tag{9.53}$$

Here, the expected transmit power $\mathbb{E}[\|\mathbf{x}\|^2]$, where the expectation is taken with respect to the randomness of $\mathbf{x}_0$ and s_k, is constrained by the sum of the trace of the sensing covariance matrix $\mathbf{R}_0$ and the summation over all CUs k of the squared norm of their respective transmit beamforming vectors $\|\mathbf{w}_k\|^2$. This overall transmit power is required to be less than or equal to P_0, where P_0 represents the maximum budget for the BS's transmit power. This condition ensures that the total transmit power, considering both sensing and communication components, complies with the specified power constraint.

9.4.2 Communication Models for Type-I and Type-II Communication Users in the Presence of an IRS

Based on the capability to cancel the sensing signal $\mathbf{x}_0$, two types of communication users (CUs) are considered: Type-I and Type-II CUs.

Type-I CUs:

- In Type-I CUs, the sensing signal $\mathbf{x}_0$ is known globally.
- Each Type-I CU has the capability to fully cancel the interference caused by $\mathbf{x}_0$.

Type-II CUs:

- Type-II CUs are likely characterized by a different capability regarding the sensing signal $\mathbf{x}_0$.
- Unlike Type-I CUs, Type-II CUs may not have the ability to completely cancel the interference generated by $\mathbf{x}_0$.

Let $\mathbf{G} \in \mathbb{C}^{N\times M}$ denote the complex-valued channel matrix from the base station (BS) to the IRS. Additionally, let $h_{bu,k} \in \mathbb{C}^{M\times 1}$ and $h_{iu,k} \in \mathbb{C}^{N\times 1}$ denote the complex-valued channel vectors from the BS and the IRS to communication user (CU) k, respectively.

Denoted by $\phi_n \in (0, 2\pi]$ is the phase shift value of each passive reflecting element $n \in \{1, \ldots, N\}$ at the IRS.

Let $\Phi = \text{diag}(e^{j\phi_1}, \ldots, e^{j\phi_N}) \in \mathbb{C}^{N\times N}$ denote the IRS's phase shifting matrix. In the IRS-assisted ISAC system under consideration, each CU $k \in \mathcal{K}$ receives the BS's signal via both the direct BS-CU link and the cascaded BS-IRS-CU link. The received signal $y_k \in \mathbb{C}$ of CU k is modeled as:

$$y_k = h_{CU,k}^{\mathrm{H}} x + n_k \quad \forall k \in \mathcal{K}, \tag{9.54}$$

where $h_{\mathrm{CU},k} = \mathbf{G}^{\mathrm{H}}\Phi^{\mathrm{H}} h_{iu,k} + h_{bu,k}$ is the effective channel vector from the BS to CU k, and $n_k \sim \mathcal{CN}(0, \sigma_k^2)$ denotes the additive white Gaussian noise (AWGN) at the kth CU's receiver.

For Type-I CUs, the sensing signal $\mathbf{x}_0$ is globally known, and each Type-I CU can fully cancel the interference generated by $\mathbf{x}_0$. Denote γ^I as the Signal-to-Interference-plus-Noise Ratio (SINR) of Type-I CU $k \in \mathcal{K}$:

$$\gamma_k^I = \frac{\|h_{CU,k}^{\mathrm{H}} \mathbf{w}_k\|^2}{\sum_{j=1, j\neq k}^{K} \|h_{CU,k}^{\mathrm{H}} \mathbf{w}_j\|^2 + \sigma_k^2}. \tag{9.55}$$

For Type-II CUs, the sensing signal $\mathbf{x}_0$ is not globally known, and each Type-II CU cannot eliminate the interference power caused by $\mathbf{x}_0$. Denote γ^{II} as the SINR of Type-II CU $k \in \mathcal{K}$:

$$\gamma_k^{II} = \frac{\|h_{CU,k}^{\mathrm{H}} \mathbf{w}_k\|^2}{\sum_{j=1, j\neq k}^{K} \|h_{CU,k}^{\mathrm{H}} \mathbf{w}_j\|^2 + \|h_{CU,k}^{\mathrm{H}} \mathbf{x}_0\|^2 + \sigma_k^2}. \tag{9.56}$$

9.4.3 Sensing Model Enhancement for L Targets with IRS Assistance

In the IRS-assisted ISAC framework, the IRS can adjust the phases of its reflecting elements to reconfigure a sensing link (i.e., the BS-IRS-target cascaded channel link) for direct target sensing service. We consider a three-dimensional (3D) scenario where a Uniform Planar Array (UPA) is employed on the IRS, enabling it to support both sensing and communication tasks alongside the BS. Each target $l \in L$ is characterized by its direction, denoted as $\theta_l = \{\phi_l, \theta_l\}$, where $\phi_l \in (-\frac{\pi}{2}, \frac{\pi}{2})$ and $\theta_l \in (-\frac{\pi}{2}, \frac{\pi}{2})$ represent the azimuth and elevation angles relative to the IRS, respectively. Moreover, d_x and d_y denote the inter-element distances along the x-axis and y-axis of the IRS, respectively.

For each target $l \in L$, the IRS's array response with azimuth angle θ_l and elevation angle ϕ_l is given by[2]:

$$\alpha(\theta_l) = \alpha_{l,x} \otimes \alpha_{l,y}, \tag{9.57}$$

where

- $\alpha_{l,x} = [1, e^{-j\tilde{d}_x}, \ldots, e^{-j\tilde{d}_x}(N_x - 1)]^T$, $\alpha_{l,y} = [1, e^{-j\tilde{d}_y}, \ldots, e^{-j\tilde{d}_y}(N_y - 1)]^T$
- $\tilde{d}_x = \frac{2\pi d_x}{\lambda \sin\theta_l \cos\phi_l}$, $\tilde{d}_y = \frac{2\pi d_y}{\lambda \sin\theta_l \sin\phi_l}$
- λ denotes the carrier wavelength
- $N = N_x N_y$ is the number of IRS reflecting elements with N_x and N_y being the numbers of the IRS reflecting elements along the x-axis and y-axis, respectively.

In ISAC systems, both the communication signals $\{s_k\}_{k\in\mathcal{K}}$ and the dedicated sensing signal $\mathbf{x}_0$ can be employed to illuminate the $\mathcal{L}$ targets to fulfill the quality of service requirement in sensing. The beampattern gain $\rho(\theta_l)$ for target $l \in L$ located at the Angle of Departure (AoD) angle θ_l with respect to the IRS is given by:

$$\rho(\theta_l) = |h^{\mathrm{H}}_{\mathrm{target},l}\mathbf{R}_0 + \sum_{k=1}^{K} \mathbf{w}_k \mathbf{w}_k^{\mathrm{H}} h_{\mathrm{target},l}|^2. \tag{9.58}$$

Furthermore, we introduce the cross-correlation pattern constraint for the ISAC signal to sense the L targets with different angles. Let P_{cross} represent the mean-squared cross-correlation pattern for the L targets. Following the approach in [61], we have:

$$P_{\mathrm{cross}} = \frac{2}{L(L-1)} \sum_{l=1}^{L-1} \sum_{i=l+1}^{L} P(\theta_l, \theta_i), \tag{9.59}$$

where $P(\theta_l, \theta_i) = |h^{\mathrm{H}}_{\mathrm{target},l}\mathbf{R}_0 + \sum_{k=1}^{K} \mathbf{w}_k \mathbf{w}_k^{\mathrm{H}} h_{\mathrm{target},i}|$, $l \neq i \in L$, and $\mathbf{R}_0 = E[\mathbf{x}_0 \mathbf{x}_0^{\mathrm{H}}]$ which is the sensing covariance matrix.

---

[2] The Kronecker product, denoted by the symbol $\otimes$, is a mathematical operation that combines two matrices to form a larger matrix. Given two matrices A and B, if A is of size $m \times n$ and B is of size $p \times q$, then the Kronecker product $A \otimes B$ results in a new matrix of size $(m \times p) \times (n \times q)$. The elements of the resulting matrix are formed by multiplying each element of A by the entire matrix B.

9.4.4 Blockage Modeling for Q Blockages

In this scenario, blockages do not emit signals spontaneously but serve as passive reflectors, receiving signals from the BS and reflecting echo signals back, leading to the degradation of the BS's target sensing performance [62]. The interference effect of blockages on the CUs is considered negligible [63]. Following established conventions in ISAC literature [59, 62–64], only the interference effect of each blockage $q \in \mathcal{Q}$ on the L targets is considered, while disregarding its impact on the K CUs.

Let $\mathbf{g}_{bc,q} \in \mathbb{C}^{M\times 1}$ represent the complex-valued channel vector from the base station (BS) to blockage $q \in \mathcal{Q}$, and $\mathbf{g}_{ic,q} \in \mathbb{C}^{N\times 1}$ denote the complex-valued cascaded channel vector from the BS to blockage q via the IRS. The interference power caused by each blockage $q \in \mathcal{Q}$ is expressed as:

$$P_{\text{blockage},q} = |\mathbf{g}_{\text{blockage},q}^{\text{H}} \mathbf{R}_0 + \sum_{k=1}^{K} \mathbf{w}_k \mathbf{w}_k^{\text{H}} \mathbf{g}_{\text{blockage},q}|^2, \tag{9.60}$$

where $\mathbf{g}_{\text{blockage},q} = G^{\text{H}} \Phi^{\text{H}} \mathbf{g}_{ic,q} + \mathbf{g}_{bc,q}$ defines the effective channel vector from the BS to blockage $q \in \mathcal{Q}$.

9.4.5 Future Directions for IRS-Assisted ISAC Systems

As a future direction in the domain of IRS-assisted ISAC systems, investigation into advanced optimization algorithms is crucial for enhancing the efficiency and accuracy of solving non-convex optimization problems. Techniques such as convex relaxation, machine learning approaches, and metaheuristic algorithms hold promise for improving solution quality and convergence speed. Additionally, conducting robustness analysis is essential to evaluate the performance of proposed IRS-assisted ISAC designs under various practical scenarios, including imperfect channel state information, hardware impairments, and environmental changes. Assessing system robustness against uncertainties and disturbances is critical for real-world deployment reliability. Moreover, extending the optimization framework to address multi-objective optimization objectives, considering trade-offs between conflicting performance metrics such as energy efficiency, spectral efficiency, and fairness among communication users, presents an important research direction. Developing efficient algorithms capable of handling multi-objective optimization problems while maintaining a balance between different optimization goals is imperative. Comprehensive experimental studies, including simulations and testbed implementations, are necessary to validate proposed IRS-assisted ISAC designs in practical wireless communication environments. Real-world experimentation will provide valuable insights into performance characteristics, practical challenges, and potential deployment issues of the proposed solutions. Furthermore, exploring novel resource allocation strategies, including power allocation, subcarrier allocation, and time–frequency resource scheduling, can optimize the utilization of available resources and improve overall system performance. Investigating dynamic resource allocation schemes that adapt to changing network conditions and user requirements can lead to more efficient and flexible communication systems. Finally, investigating the integration of IRS-assisted ISAC designs with emerging technologies such as artificial intelligence, edge computing, and the Internet of Things holds significant promise. Exploring synergies between IRS-assisted ISAC and other cutting-edge technologies could unlock new opportunities for enhancing communication efficiency, scalability, and reliability. By addressing these future research directions, we can advance the state-of-the-art in IRS-assisted integrated sensing and communication systems, paving the way for their widespread deployment and adoption in next-generation wireless networks.

9.5 ISAC in Near-Field Environment

ISAC stands out as an emerging technology for the upcoming sixth-generation (6G) wireless networks, where the simultaneous execution of sensing and communication functions is achieved by utilizing the identical spectrum and hardware framework [65]. In response to the escalating demands for communication and sensing performance, the evolution of future ISAC systems is envisioned to incorporate extremely large-scale antenna arrays (ELAAs) operating at high frequencies, such as millimeter wave (mmWave) and terahertz (THz). These frequencies are essential for attaining both high communication capacity and precise sensing resolution [65, 66].

However, this anticipated transition towards ELAAs and higher frequencies comes with a notable consequence: a substantial transformation in the electromagnetic properties of the wireless environment. This shift, from conventional planar-wave propagation to spherical-wave propagation, introduces an inevitable near-field effect [67–69]. This effect challenges the existing ISAC designs, which predominantly rely on the far-field assumption [13, 70, 71]. Consequently, there arises a compelling need to reevaluate and redesign ISAC systems to align with the altered dynamics of real wireless environments.

Despite posing a challenge, the near-field effect brings about a host of new opportunities. The spherical-wave propagation in the near-field introduces an additional dimension related to distance. This novel dimension has the potential to revolutionize sensing applications by enabling joint estimation of both distance and angle [72]. Moreover, in the realm of communications, the near-field effect can be leveraged to mitigate interference, offering a promising avenue for enhancing the reliability and efficiency of wireless communication systems [73].

In essence, while the shift towards ELAAs and higher frequencies necessitates a rethinking of ISAC designs, the near-field effect introduces exciting possibilities for advancing the capabilities of wireless networks in terms of both sensing and communication. This dynamic interplay between challenges and opportunities highlights the evolving landscape of wireless technology, particularly as it pertains to the envisioned 6G networks.

9.5.1 System Model

As depicted in Fig. 9.8, we consider a narrowband near-field ISAC system consisting of an N-antenna dual-functional base station, where $N = 2\tilde{N} + 1$, K single-antenna communication users, and one sensing target. We assume a monostatic sensing configuration at the base station. Moreover, the base station utilizes a uniform linear array (ULA) with an antenna spacing of d, resulting in an aperture of $D = (N - 1)d$. The boundary between the near-field and the far-field regions can be determined by the Rayleigh distance $2D^2/\lambda$ [67], where λ represents the wavelength of the signal.

We consider that both the communication users and the sensing target are positioned within the near-field proximity of the base station, indicating that their distance from the base station is shorter than $2D^2/\lambda$. To capitalize on the benefits offered by the near-field, employing electronically steerable antennas (ELAAs) and operating at high frequencies are crucial to enlarge the near-field area. We delve into the discussion of optimal fully digital antennas to establish an upper performance limit

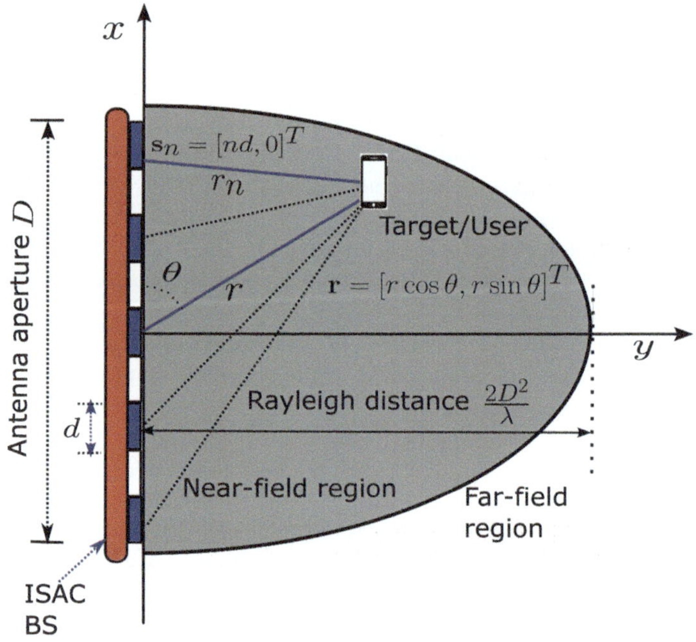

Fig. 9.8 Illustration of the near-field ISAC system [74]

for the near-field ISAC system in this chapter. Nevertheless, this approach can be expanded to encompass hybrid digital and analog antennas, which are deemed more energy-efficient for ELAAs operating at high frequencies, as elaborated in [74].

9.5.2 Near-Field Channel Model for ISAC

We initiate by presenting the channel model for the near-field ISAC system under consideration. To maintain a broad applicability, we position the origin of the coordinate system at the ULA center of the base station. Consequently, the coordinate of the nth element of the ULA is denoted as $s_n = [nd, 0]^T$, where $n \in \{-\tilde{N}, \dots, \tilde{N}\}$.

Let's contemplate a communication user or a sensing target situated at a distance r and an angle θ from the ULA center. Its coordinate is expressed as $\mathbf{r} = [r\cos\theta, r\sin\theta]^T$. The distance from the nth antenna element to this user or target ($r_n(r,\theta)$) can be computed as follows:

$$r_n(r,\theta) = \sqrt{r^2 + n^2d^2 - 2rnd\cos\theta}. \tag{9.61}$$

Within the Fresnel region of the near-field, defined as $1.2D \le r \le 2D^2/\lambda$, the channel gain of each link connecting the antenna elements to the user or target is nearly uniform. Consequently, the channel gain for all links can be computed using the free-space path loss of the central link, represented by $\tilde{\beta} = \sqrt{\frac{\rho_0}{r}}$, where $\rho_0 = \frac{\lambda}{4\pi}$ denotes the path loss at the reference distance of 1 meter. Thus, the channel between the nth antenna element and the user or target ($h_n(r,\theta)$) can be expressed as:

$$h_n(r,\theta) = \tilde{\beta} e^{-j\frac{2\pi}{\lambda} r_n(r,\theta)}, \tag{9.62}$$

where $\beta = \tilde{\beta} e^{-j\frac{2\pi}{\lambda} r}$ denotes the complex channel gain.

The near-field channel vector $\mathbf{h} \in \mathbb{C}^{N\times 1}$ connecting the base station to the user or target can be represented as:

$$\mathbf{h} = [h_{-\tilde{N}}(r_k,\theta_k), \dots, h_{\tilde{N}}(r_k,\theta_k)]^T = \beta\mathbf{a}(r,\theta), \tag{9.63}$$

where $\mathbf{a}(r,\theta)$ denotes the near-field array response vector. Now, the nth element, $\forall n \in \{-\tilde{N}, \dots, \tilde{N}\}$, of $\mathbf{a}(r,\theta)$ is given by:

$$[\mathbf{a}(r,\theta)]_n = e^{-j\frac{2\pi}{\lambda}(r_n(r,\theta) - r)}. \tag{9.64}$$

The near-field array response vector transitions to the far-field counterpart under the assumption that $D \ll r$, and employing the first-order Taylor approximation $\sqrt{1+x} \approx 1 + \frac{1}{2}x$ for $x = \frac{1}{r^2}(n^2d^2 - 2rnd\cos\theta)$ in the distance expression (9.61).

In this scenario, the nth element of the far-field array response vector is expressed as:

$$[\mathbf{a}_{\text{far}}(\theta)]_n = e^{-j\frac{2\pi}{\lambda}(-nd\cos\theta)}. \tag{9.65}$$

Let r_k, θ_k, and β_k denote the distance, angle, and complex channel gain of user k, respectively. The near-field communication channel vector $\mathbf{h}_k \in \mathbb{C}^{N\times 1}$ between the BS and user k can thus be modeled as:

$$\mathbf{h}_k = \beta_k \mathbf{a}(r_k, \theta_k). \tag{9.66}$$

Unlike communication, target sensing using the monostatic sensing setup depends on the echo signal received at the base station. Consequently, the round-trip channel must be taken into account. Let r_s, θ_s, and β_s represent the distance, angle, and complex channel gain of the sensing target, respectively. The near-field round-trip channel matrix $\mathbf{G} \in \mathbb{C}^{N\times N}$ for target sensing is expressed as:

$$\mathbf{G} = \beta_s \mathbf{a}(r_s, \theta_s)\mathbf{a}^T(r_s, \theta_s). \tag{9.67}$$

9.5.3 Advantages of Near-Field Communication and Sensing

Near-field communication and sensing offer distinct advantages over traditional far-field approaches, with implications for both channel reliability and spatial awareness as discussed below:

I. Near-Field Communication Advantages

- As indicated by Eq. (9.66), the near-field communication channel is affected by both the distance and angle of communication users.
- This observation stands in stark contrast to the far-field communication channel, which relies solely on the angle.
- Consequently, even when users align in the same direction, their differentiation in the distance domain minimizes inter-user interference.

II. Near-Field Sensing Advantages

- It is observed in (9.67), the near-field sensing channel integrates both distance and angle information.
- In opposition to far-field sensing systems limited to angle estimation, near-field sensing systems empower the joint estimation of both distance and angle.
- This signifies a significant leap forward, enabling a more comprehensive understanding of the spatial dynamics within the sensing environment.

9.5.4 ISAC Model for Fully Digital Antennas

This subsection focuses on the ISAC model tailored for fully digital antennas, wherein each antenna at the base station (BS) is linked to a dedicated radio-frequency (RF) chain. Throughout a coherent time block of duration T, the communication channels and sensing target parameters remain relatively stable. At the commencement of each coherent time block, conventional channel estimation techniques are utilized to ascertain the communication channels. Following this, for the duration of the remaining time within the coherent block, the BS transmits a combined communication and sensing signal at time t:

$$x[t] = \sum_{k \in K} \mathbf{f}_k c_k[t] + s[t]. \tag{9.68}$$

Here, $\mathbf{f}_k \in \mathbb{C}^{N \times 1}$ represents the fully digital beamformer transmitting the information symbol $c_k[t] \in \mathbb{C}$ to user k, $s[t]$ denotes the dedicated sensing signal aimed at achieving complete sensing degrees of freedom, and $\mathcal{K} = \{1, \ldots, K\}$.

The information symbols are assumed to be independently distributed with unit power, i.e., $\mathbb{E}[c_k[t]c_i^*[t]] = 1$ if $k = i$ and $\mathbb{E}[c_k[t]c_i^*[t]] = 0$ otherwise. Let $\mathbf{R}_s = \mathbb{E}[\mathbf{s}[t]\mathbf{s}^H[t]]$ represent the covariance matrix of the dedicated sensing signal. The covariance matrix of the sensing signal $\mathbf{R}_x = \mathbb{E}[\mathbf{x}[t]\mathbf{x}^H[t]]$ can be expressed as:

$$\mathbf{R}_x = \sum_{k \in \mathcal{K}} \mathbf{f}_k \mathbf{f}_k^H + \mathbf{R}_s. \tag{9.69}$$

1. Near-Field Communication Model The received communication signal at user k is expressed as:

$$y_k[t] = \mathbf{h}_k^T \mathbf{f}_k c_k[t] + \sum_{i \neq k} \mathbf{h}_k^T \mathbf{f}_i c_i[t] + \mathbf{h}_k^T \mathbf{s}[t] + z_k[t], \tag{9.70}$$

Here, $z_k[t] \sim \mathcal{CN}(0, \sigma_k^2)$ represents the additive white Gaussian noise (AWGN). The achievable communication rate for user k is defined as:

$$R(\mathbf{f}_k, \mathbf{R}_x) = \log_2 \left(1 + \frac{|\mathbf{h}_k^T \mathbf{f}_k|^2}{\mathbf{h}_k^T \mathbf{R}_x \mathbf{h}_k^* - |\mathbf{h}_k^T \mathbf{f}_k|^2 + \sigma_k^2}\right). \tag{9.71}$$

2. Near-field Sensing Model The received echo signal at the BS for target sensing is given by:

$$\mathbf{y}_s[t] = \mathbf{G}\mathbf{x}[t] + \mathbf{z}_s[t]. \tag{9.72}$$

Here, $\mathbf{z}_s[t] \sim \mathcal{CN}(\mathbf{0}_N, \sigma_s^2 \mathbf{I}_N)$ denotes the additive white Gaussian noise (AWGN). The objective of sensing is to estimate the target parameters from the received echo signal samples over the entire coherent time block ($\mathbf{Y}_s = [\mathbf{y}_s[1], \ldots, \mathbf{y}_s[T]]$). As discussed in Remark 2, joint distance and angle estimation can be achieved using the near-field sensing channel **G**. For this purpose, the classic multiple signal classification (MUSIC) algorithm is employed, leveraging the orthogonality between signal subspaces. The CramÃƒÂ©r–Rao bound (CRB) is utilized as the performance metric for target sensing, offering a lower bound on the Mean Square Errors (MSEs) for distance ($\epsilon_{r_s}^2$) and angle ($\epsilon_{\theta_s}^2$). The CRB matrix is expressed as:

$$\text{CRB}(\mathbf{R}_x, \mathbf{G}, \sigma_s^2) = (\mathbf{J}_{11} - \mathbf{J}_{12}\mathbf{J}_{22}^{-1}\mathbf{J}_{12}^T)^{-1}, \tag{9.73}$$

where detailed expressions for $\mathbf{J}_{11}$, $\mathbf{J}_{12}$, and $\mathbf{J}_{22}$ and the CRB matrix derivation are provided in Appendix A [74]. Specifically, $\epsilon_{r_s}^2 \geq [\text{CRB}]_{1,1}$ and $\epsilon_{\theta_s}^2 \geq [\text{CRB}]_{2,2}$.

9.5.5 Future Directions for Near-Field ISAC Systems

A novel approach for fully digital antennas can be explored, starting with the resolution of the optimization problem, which currently poses challenges due to the intricate nature of the CRB matrix. To overcome this complexity, efforts will be made to reformulate the problem into a more manageable form by introducing an auxiliary matrix. Leveraging semidefinite relaxation (SDR) techniques and introducing auxiliary variables, the goal will be to transform the non-convex constraints into convex forms, thereby enabling the formulation of a convex optimization problem. This innovative approach is expected to provide valuable insights into the theoretical performance bounds of fully digital antennas, laying the groundwork for further exploration into the potential of hybrid digital and analog antenna systems.

Moreover, future investigations can delve into optimizing hybrid digital and analog antennas using a heuristic two-stage optimization framework. In the initial stage, emphasis will be placed on designing the analog beamformer to maximize array gain, followed by optimizing the digital beamformers and dedicated sensing signal. By refining the analog beamformer design, subsequent optimization of the digital components aims to emulate the problem formulation for fully digital antennas. Through extensive numerical simulations, the effectiveness of the proposed near-field integrated sensing and communication (ISAC) framework will be examined, with a specific emphasis on elucidating the trade-offs between communication rate and sensing performance. These insights are anticipated to illuminate the promising potential of near-field ISAC systems and pave the way for future research endeavors aimed at further enhancing their capabilities.

9.6 Potential Applications of ISAC

The capacity for sensing functionality and the concurrent capability of the network to collect sensory data from the environment are considered crucial facilitators for learning and fostering intelligence in the envisioned smart world. These capabilities are poised to have widespread utility across a myriad of scenarios that demand location and environment awareness, showcasing the vast potential applications of ISAC [65]. Few of the applications are listed down below:

- **Smart Cities**: ISAC plays a crucial role in smart city initiatives by integrating sensors with communication networks. This integration enables real-time monitoring and management of urban infrastructure, including efficient traffic flow, environmental conditions, and public safety.
- **Healthcare**: In healthcare, ISAC facilitates remote patient monitoring by integrating sensors with communication devices. This combination ensures timely and accurate health data collection, leading to improved patient care.
- **Industrial IoT (IIoT)**: ISAC contributes to the Industrial Internet of Things (IIoT) by enabling real-time monitoring of industrial equipment and processes. This integration enhances efficiency, reduces downtime, and improves overall operational effectiveness.
- **Environmental Monitoring**: ISAC finds applications in environmental monitoring, collecting data on air quality, water conditions, and ecological factors. The integration of sensing and communication enhances our ability to monitor and address environmental challenges.
- **Transportation**: ISAC enhances intelligent transportation systems by enabling vehicle-to-vehicle (V2V) and vehicle-to-infrastructure (V2I) communication. This integration enhances safety, traffic management, and communication reliability.
- **Agriculture**: Precision agriculture benefits from ISAC by integrating sensors for monitoring soil conditions, crop health, and irrigation needs. The combination of sensing and communication technologies contributes to more efficient agricultural practices.
- **Disaster Response and Management**: ISAC is instrumental in disaster response, providing real-time communication and sensing capabilities. The integration of these functionalities improves situational awareness and aids in effective disaster management.
- **Wireless Networking**: ISAC optimizes wireless networks by adapting communication links based on environmental conditions and user requirements. This adaptability improves overall network performance, ensuring reliable and efficient communication.
- **Security and Surveillance**: In security and surveillance applications, ISAC integrates communication and sensing devices for effective monitoring. This integration enhances security measures and provides a comprehensive surveillance solution.

- **Consumer Electronics**: ISAC enhances consumer electronics by integrating sensors for context-aware applications. This integration improves the capabilities of smart devices, creating a more personalized and efficient user experience.

Some more future uses of ISAC include vehicle-to-everything communication, smart home systems, advanced manufacturing with intelligent capabilities, remote sensing technologies, environmental monitoring solutions, and interactive interfaces for human–computer interaction.

9.7 Conclusions

In this book chapter, a comprehensive exploration of ISAC systems has been carried out with a focus on the waveform design across various paradigms. The integration of sensing and communication functionalities has presented novel challenges, met with innovative solutions in waveform optimization. From the robust capabilities of ISAC-OFDM to the unique advantages brought by ISAC-OTFS, and the collaborative potential of IRS-assisted systems, each one can showcase a dynamic landscape of integrated systems. The exploration of near-field ISAC communications provided insights into the spatial dynamics, emphasizing the significance of proximity in these interactions. As we proceed to navigate these investigations, the synergy between sensing and communication opens avenues for enhanced performance and efficiency.

As we look ahead to the prospective era of 6G networks, several challenges and exciting possibilities emerge. The challenges encompass critical issues such as spectrum utilization, where efficient allocation becomes pivotal amid the increasing demand for high data rates. Energy efficiency is another concern, given the surge in connected devices, necessitating the development of sustainable technologies. Security and privacy considerations gain prominence with the proliferation of connected devices, demanding robust measures to safeguard user data. On the flip side, 6G holds promising prospects. Terahertz communication is poised to unlock unprecedented data rates and ultra-low latency, revolutionizing applications like augmented reality. Artificial Intelligence (AI) will be a linchpin, driving dynamic adaptability and optimizing resource allocation in 6G networks. Integration of satellite communication promises ubiquitous connectivity, particularly in remote areas, while holographic technologies may introduce immersive communication experiences. Lastly, quantum communication security could become a standard feature, addressing cybersecurity concerns with unbreakable encryption through quantum key distribution. The future of 6G appears to be marked by both challenges and groundbreaking innovations.

References

1. Ma D, Shlezinger N, Huang T, Liu Y, Eldar YC (2020) Joint radar-communication strategies for autonomous vehicles: combining two key automotive technologies. IEEE Signal Process Mag 37(4):85–97
2. Rossler CW, Ertin E, Moses RL (2011) A software defined radar system for joint communication and sensing. In: 2011 IEEE RadarCon (RADAR), pp 1050–1055
3. Chiriyath AR, Paul B, Bliss DW (2017) Radar-communications convergence: coexistence, cooperation, and co-design. IEEE Trans Cognit Commun Netw 3(1):1–12
4. Sturm C, Wiesbeck W (2011) Waveform design and signal processing aspects for fusion of wireless communications and radar sensing. Proc IEEE 99(7):1236–1259
5. Baquero Barneto C, Riihonen T, Turunen M, Anttila L, Fleischer M, Stadius K, Ryynänen J, Valkama M (2019) Full-duplex OFDM radar with LTE and 5G NR waveforms: challenges, solutions, and measurements. IEEE Trans Microwave Theory Tech 67(10):4042–4054
6. Liyanaarachchi SD, Riihonen T, Barneto CB, Valkama M (2021) Optimized waveforms for 5G–6G communication with sensing: theory, simulations and experiments. IEEE Trans Wireless Commun 20(12):8301–8315
7. Kumari P, Choi J, González-Prelcic N, Heath RW (2018) IEEE 802.11ad-Based radar: an approach to joint vehicular communication-radar system. IEEE Trans Veh Technol 67(4):3012–3027
8. Keskin MF, Koivunen V, Wymeersch H (2021) Limited feedforward waveform design for OFDM dual-functional radar-communications. IEEE Trans Signal Process 69:2955–2970
9. Liu F, Masouros C, Li A, Sun H, Hanzo L (2018) MU-MIMO communications with MIMO radar: from co-existence to joint transmission. IEEE Trans Wireless Commun 17(4):2755–2770
10. Liu F, Zhou L, Masouros C, Li A, Luo W, Petropulu A (2018) Toward dual-functional radar-communication systems: optimal waveform design. IEEE Trans Signal Process 66(16):4264–4279
11. Dokhanchi SH, Mysore BS, Mishra KV, Ottersten B (2019) A mmWave automotive joint radar-communications system. IEEE Trans Aerosp Electron Syst 55(3):1241–1260
12. Zhang JA, Huang X, Guo YJ, Yuan J, Heath RW (2019) Multibeam for joint communication and radar sensing using steerable analog antenna arrays. IEEE Trans Veh Technol 68(1):671–685
13. Liu X, Huang T, Shlezinger N, Liu Y, Zhou J, Eldar YC (2020) Joint transmit beamforming for multiuser MIMO communications and MIMO radar. IEEE Trans Signal Process 68:3929–3944
14. Liu F, Masouros C, Li A, Sun H, Hanzo L (2018) MU-MIMO communications With MIMO radar: from co-existence to joint transmission. IEEE Trans Wireless Commun 17(4):2755–2770
15. Mohammed SK, Larsson EG (2013) Per-antenna constant envelope precoding for large multi-user MIMO systems. IEEE Trans Commun 61(3):1059–1071
16. Bechter J, Rameez M, Waldschmidt C (2017) Analytical and experimental investigations on mitigation of interference in a DBF MIMO radar. IEEE Trans Microw Theory Tech 65(5):1727–1734
17. Friedlander B (2012) On transmit beamforming for MIMO radar. IEEE Trans Aerosp Electron Syst 48(4):3376–3388
18. Yuan W, Wu N, Zhang A, Huang X, Li Y, Hanzo L (2020) Iterative receiver design for FTN signaling aided sparse code multiple access. IEEE Trans Wireless Commun 19(2):915–928
19. Sturm C, Wiesbeck W (2011) Waveform design and signal processing aspects for fusion of wireless communications and radar sensing. Proc IEEE 99(7):1236–1259
20. Skolnik MI (2008) Radar handbook, vol 3rd. McGraw-Hill, New York
21. Yli-Kaakinen J, Loulou A, Levanen T, Pajukoski K, Palin A, Renfors M, Valkama M (2021) Frequency-domain signal processing for spectrally-enhanced CP-OFDM waveforms in 5G new radio. IEEE Trans Wireless Commun 20(10):6867–6883
22. Rahbari H, Krunz M (2017) Exploiting frame preamble waveforms to support new physical-layer functions in OFDM-based 802.11 systems. IEEE Trans Wireless Commun 16(6):3775–3786

23. Tigrek RF, De Heij WJA, Van Genderen P (2012) OFDM signals as the radar waveform to solve doppler ambiguity. IEEE Trans Aerosp Electron Syst 48(1):130–143
24. Wang W-Q (2013) Mitigating range ambiguities in high-PRF SAR with OFDM waveform diversity. IEEE Geosci Remote Sens Lett 10(1):101–105
25. Cui Y, Jing X, Mu J (2022) Integrated sensing and communications via 5G NR waveform: performance analysis. In: ICASSP 2022 - 2022 IEEE international conference on acoustics, speech and signal processing (ICASSP), pp 8747–8751
26. Sturm C, Pancera E, Zwick T, Wiesbeck W (2009) A novel approach to OFDM radar processing. In: 2009 IEEE Radar Conference, pp 1–4
27. Engles M (2002) Wireless OFDM systems: how to make them work? Kluwer Academic, Boston, MA
28. Sturm C, Braun M, Zwick T, Wiesbeck W (2010) A multiple target doppler estimation algorithm for OFDM based intelligent radar systems. In: The 7th European radar conference, pp 73–76
29. Mahafza BR, Radar RB (2009) Signal analysis and processing using MatLab. Chapman and Hall/CRC, FL
30. Wei Z, Yuan W, Li S, Yuan J, Bharatula G, Hadani R, Hanzo L (2021) Orthogonal time-frequency space modulation: a promising next-generation waveform. IEEE Wirel Commun 28(4):136–144
31. Raviteja P, Phan KT, Hong Y, Viterbo E (2019) Orthogonal time frequency space (OTFS) modulation based radar system. In: 2019 IEEE radar conference (RadarConf), pp 1–6
32. Raviteja P, Phan KT, Hong Y, Viterbo E (2018) Interference cancellation and iterative detection for orthogonal time frequency space modulation. IEEE Trans Wireless Commun 17(10):6501–6515
33. Mohammed SK (2021) Derivation of OTFS modulation from first principles. IEEE Trans Veh Technol 70(8):7619–7636
34. Lampel F, Alvarado A, Willems F (2016) Orthogonal time frequency space modulation: a discrete Zak transform approach. arXiv:2106.12828
35. Wei Z, Li S, Yuan W, Schober R, Caire G (2023) Orthogonal time frequency space modulation-part I: fundamentals and challenges ahead. IEEE Commun Lett 27(1):4–8
36. Wei Z, Yuan W, Li S, Yuan J, Ng DWK (2021) Transmitter and receiver window designs for orthogonal time-frequency space modulation. IEEE Trans Commun 69(4):2207–2223
37. Hadani R, Rakib S, Tsatsanis M, Monk A, Goldsmith AJ, Molisch AF, Calderbank R (2017) Orthogonal time frequency space modulation. In: 2017 IEEE wireless communications and networking conference (WCNC), pp 1–6
38. Fish A, Gurevich S, Hadani R, Sayeed AM, Schwartz O (2013) Delay-doppler channel estimation in almost linear complexity. IEEE Trans Inf Theory 59(11):7632–7644
39. Raviteja P, Phan KT, Jin Q, Hong Y, Viterbo E (2018) Low-complexity iterative detection for orthogonal time frequency space modulation. In: 2018 IEEE wireless communications and networking conference (WCNC), pp 1–6
40. Raviteja P, Hong Y, Viterbo E, Biglieri E (2019) Practical pulse-shaping waveforms for reduced-cyclic-prefix OTFS. IEEE Trans Veh Technol 68(1):957–961
41. Yuan W, Wei Z, Yuan J, Ng DWK (2020) A simple variational bayes detector for orthogonal time frequency space (OTFS) modulation. IEEE Trans Veh Technol 69(7):7976–7980
42. Wang X, Fei Z, Zheng Z, Guo J (2021) Joint waveform design and passive beamforming for RIS-assisted dual-functional radar-communication system. IEEE Trans Veh Technol 70(5):5131–5136
43. Zhu Z, Li Z, Chu Z, Sun G, Hao W, Xiao P, Lee I (2024) Intelligent reflecting surface assisted integrated sensing and communications for mmwave channels
44. Wang X, Fei Z, Huang J, Yu H (2022) Joint waveform and discrete phase shift design for RIS-assisted integrated sensing and communication system under Cramer-Rao bound constraint. IEEE Trans Veh Technol 71(1):1004–1009
45. Song X, Xu J, Liu F, Han TX, Eldar YC (2022) Intelligent reflecting surface enabled sensing: Cramér-Rao lower bound optimization. In: 2022 IEEE Globecom workshops (GC Wkshps), pp 413–418

46. Buzzi S, Grossi E, Lops M, Venturino L (2022) Foundations of MIMO radar detection aided by reconfigurable intelligent surfaces. IEEE Trans Signal Process 70:1749–1763
47. Song X, Zhao D, Hua H, Han TX, Yang X, Xu J (2022) Joint transmit and reflective beamforming for IRS-assisted integrated sensing and communication. In: 2022 IEEE wireless communications and networking conference (WCNC), pp 189–194
48. Sankar RP, Chepuri SP (2022) Beamforming in hybrid RIS assisted integrated sensing and communication systems. In: 2022 30th European signal processing conference (EUSIPCO), pp 1082–1086
49. Xing Z, Wang R, Yuan X (2023) Joint active and passive beamforming design for reconfigurable intelligent surface enabled integrated sensing and communication. IEEE Trans Commun 71(4):2457–2474
50. Bose A, Khobahi S, Soltanalian M (2020) Efficient waveform covariance matrix design and antenna selection for MIMO radar. arXiv:2002.06025
51. Hua M, Wu Q, He C, Ma S, Chen W (2023) Joint active and passive beamforming design for IRS-aided radar-communication. IEEE Trans Wireless Commun 22(4):2278–2294
52. Hua M, Wu Q, Chen W, Dobre OA, Swindlehurst AL (2024) Secure intelligent reflecting surface-aided integrated sensing and communication. IEEE Trans Wireless Commun 23(1):575–591
53. Liu R, Li M, Liu Q, Swindlehurst AL (2022) Joint waveform and filter designs for STAP-SLP-based MIMO-DFRC systems. IEEE J Sel Areas Commun 40(6):1918–1931
54. Su N, Liu F, Wei Z, Liu Y-F, Masouros C (2022) Secure dual-functional radar-communication transmission: exploiting interference for resilience against target eavesdropping. IEEE Trans Wireless Commun 21(9):7238–7252
55. He Y, Cai Y, Mao H, Yu G (2022) RIS-assisted communication radar coexistence: joint beamforming design and analysis. IEEE J Sel Areas Commun 40(7):2131–2145
56. Chen C-Y, Vaidyanathan PP (2009) MIMO radar waveform optimization with prior information of the extended target and clutter. IEEE Trans Signal Process 57(9):3533–3544
57. Wang F, Li H (2021) Power allocation for coexisting multicarrier radar and communication systems in cluttered environments. IEEE Trans Signal Process 69:1603–1613
58. Liu R, Li M, Liu Y, Wu Q, Liu Q (2022) Joint transmit waveform and passive beamforming design for RIS-aided DFRC systems. IEEE J Sel Top Signal Process 16(5):995–1010
59. Wei T, Wu L, Mishra KV, Shankar MRB (2023) Multi-IRS-aided doppler-tolerant wideband DFRC system. IEEE Trans Commun 71(11):6561–6577
60. Liao C, Wang F, Lau VKN (2023) Optimized design for IRS-assisted integrated sensing and communication systems in clutter environments. IEEE Trans Commun 71(8):4721–4734
61. Fuhrmann DR, San Antonio G (2008) Transmit beamforming for MIMO radar systems using signal cross-correlation. IEEE Trans Aerospace Electr Syst 44(1):171–186
62. Wang F, Li H, Fang J (2022) Joint active and passive beamforming for IRS-assisted radar. IEEE Signal Process Lett 29:349–353
63. Liu R, Li M, Liu Y, Wu Q, Liu Q (2022) Joint transmit waveform and passive beamforming design for RIS-aided DFRC systems. IEEE J Sel Top Signal Process 16(5):995–1010
64. He Y, Cai Y, Mao H, Yu G (2022) RIS-assisted communication radar coexistence: joint beamforming design and analysis. IEEE J Sel Areas Commun 40(7):2131–2145
65. Liu F, Cui Y, Masouros C, Xu J, Han TX, Eldar YC, Buzzi S (2022) Integrated sensing and communications: toward dual-functional wireless networks for 6g and beyond. IEEE J Sel Areas Commun 40(6):1728–1767
66. Mu X, Wang Z, Liu Y (2023) NOMA for integrating sensing and communications towards 6G: a multiple access perspective. IEEE Wireless Commun 1–8
67. Selvan KT, Janaswamy R (2017) Fraunhofer and Fresnel Distances: Unified derivation for aperture antennas. IEEE Antennas Propag Mag 59(4):12–15
68. Björnson E, Sanguinetti L, Wymeersch H, Hoydis J, Marzetta T (2019) Massive MIMO is a reality – what is next? Five promising research directions for Antenna arrays. arXiv:1902.07678
69. Liu Y, Xu J, Wang Z, Mu X, Hanzo L (2023) Near-field communications: what will be different? arXiv:2303.04003

70. Liu F, Liu Y-F, Li A, Masouros C, Eldar YC (2022) Cramér-Rao bound optimization for joint radar-communication beamforming. IEEE Trans Signal Process 70:240–253
71. Wang Z, Mu X, Liu Y (2023) STARS enabled integrated sensing and communications. IEEE Trans Wireless Commun 22(10):6750–6765
72. Huang Y-D, Barkat M (1991) Near-field multiple source localization by passive sensor array. IEEE Trans Antennas Propag 39(7):968–975
73. Zhang H, Shlezinger N, Guidi F, Dardari D, Imani MF, Eldar YC (2022) Beam focusing for near-field multiuser MIMO communications. IEEE Trans Wireless Commun 21(9):7476–7490
74. Wang Z, Mu X, Liu Y (2023) Near-field integrated sensing and communications. arXiv:2302.01153

Chapter 10
Unmanned Aerial Vehicle Assisted Wireless Communications: Enabling Connectivity and Applications

Soumendu Das, Nagendra Kumar, and Dharmendra Dixit

Abstract The chapter provides an extensive exploration of the role of Unmanned Aerial Vehicles (UAVs) in enhancing wireless communication systems. Beginning with an introduction, the chapter delves into the classification and attributes of UAVs, outlining their varied functions and capabilities. Subsequently, it discusses the integration of UAVs into wireless networks, presenting an overview of different UAV types, communication interfaces, and attributes of the UAVs-Base Station (BS) control link. Addressing the communication challenges in UAV systems, the chapter explores the complexities associated with limited bandwidth, non-line-of-sight operations, latency, and security concerns. Furthermore, it elucidates UAV communication and control architectures, elucidating how data and commands are exchanged between ground stations and UAVs, or among UAVs themselves. Highlighting applications of UAV communication, the chapter underscores their utility in various sectors such as disaster response, surveillance, and agriculture. It further analyzes the performance of communication scenarios involving single UAV, deriving outage probability and average symbol error rate, and presenting numerical and simulation results. Finally, the chapter discusses future directions and challenges in the field, outlining potential avenues for research and development to further leverage the capabilities of UAV-assisted wireless communications.

Keywords Unmanned Aerial Vehicles · Wireless communications · Base stations · Outage probability · Latency · Security concerns

S. Das · N. Kumar
National Institute of Technology Jamshedpur, Adityapur, Jamshedpur 831014, Jharkhand, India
e-mail: 2021rsec009@nitjsr.ac.in

N. Kumar
e-mail: nagendrakumar.ece@nitjsr.ac.in

D. Dixit (✉)
Motilal Nehru National Institute of Technology Allahabad, Teliarganj, Prayagraj 211004, Uttar Pradesh, India
e-mail: dharmendradixit@mnnit.ac.in

S. Sharma et al. (eds.), *Next-Generation Wireless Systems*,
Transactions on Computer Systems and Networks,
https://doi.org/10.1007/978-981-96-3185-8_10

10.1 Introduction

The rapid progress of fifth-generation (5G) and beyond networks, combined with Unmanned Aerial Vehicles (UAVs), has unlocked a realm of thrilling opportunities spanning various applications and leading-edge technologies. These UAV-enabled communications systems have been proposed as a pivotal element in revolutionizing the way we connect and communicate [1–3]. An UAV or drone is an aircraft that operates without a human pilot on board. Instead, it is typically controlled remotely by a human operator or autonomously by onboard computers. The utilization of flying platforms, such as UAVs or drones, is experiencing rapid expansion. These UAVs offer various advantageous features, including mobility, flexibility, and adaptive altitude, which make them highly suitable for numerous applications in wireless systems. On the one hand, by functioning as aerial BSs, UAVs have the potential to improve coverage, capacity, reliability, and energy efficiency of wireless networks. On the other hand, UAVs can also serve as flying mobile terminals within a cellular network, expanding their versatility and capabilities in wireless communications [3]. To provide a comprehensive understanding of UAVs can function effectively as airborne wireless BSs, this section presents an overview of several potential applications for deploying these wireless-focused UAVs. These applications encompass a diverse range of scenarios, including both near-term use cases like public safety scenarios and hotspot coverage, as well as more forward-looking applications such as utilizing UAVs as caching devices or facilitators for the Internet of Things (IoT).

10.2 Classification and Attributes of UAV

An UAV is a remotely piloted or autonomous aircraft that operates without a human pilot onboard. These vehicles are equipped with sensors, cameras, and other payloads for various purposes such as reconnaissance, surveillance, data collection, and aerial photography [2]. It is essential to tailor the choice of UAV to the task at hand and the desired outcomes, taking into account factors such as quality of service (QoS) demands, environmental conditions and regulatory constraints. For effective integration of UAVs into wireless networking applications, it is necessary to carefully consider their capabilities and operating altitude [4].

Broadly, UAVs are classified into two categories based on altitude: high altitude platforms (HAP) and low-altitude platforms (LAP). HAPs typically operate above 17 km while maintaining quasi-stationary positions, while LAPs range in altitude from tens of meters to a few kilometers, providing agility and flexibility in movement. Under U.S. federal aviation regulations, LAP-drones are allowed to fly to a maximum altitude of 400 feet without requiring a permit. Their rapid deployment makes them ideal for urgent scenarios such as emergencies. Unlike HAPs, LAPs are capable of storing ground sensor data and can be recharged or replaced quickly. In contrast, HAPs boast extended endurance, are designed for longer operations lasting several

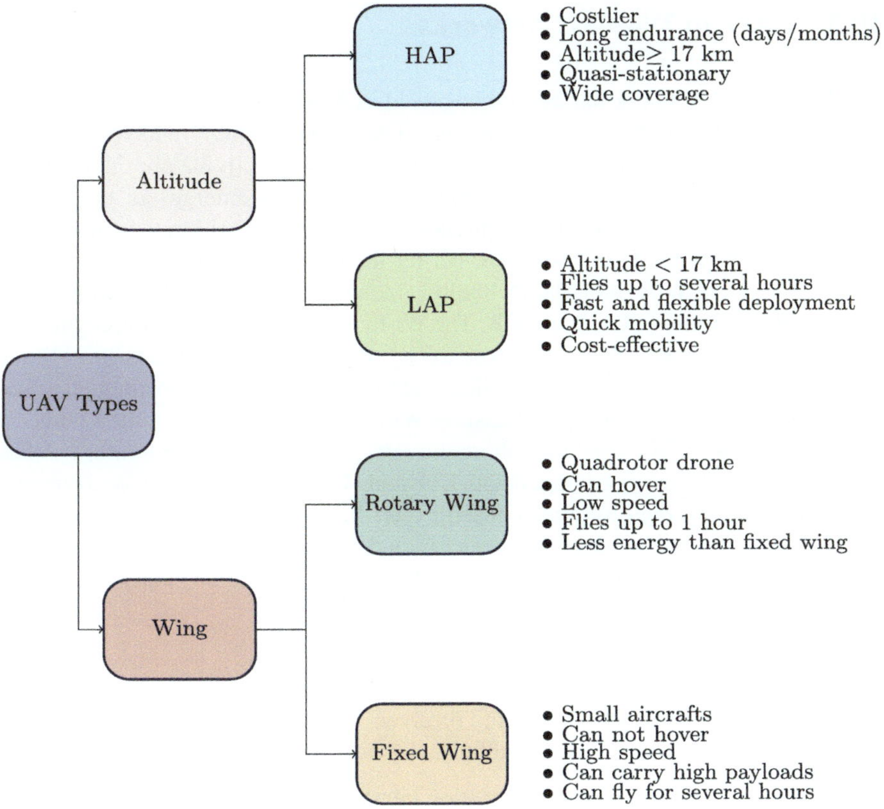

Fig. 10.1 UAV types and its attributes

months, and are preferred for widespread wireless coverage over vast areas. However, their deployment is expensive and more time-consuming than LAP [5, 6].

UAVs are also divided into two main types based on the wing of the UAV: fixed-wing and rotary-wing. Fixed-wing UAVs, resembling small planes, carry more weight, achieve higher speeds, and must maintain forward momentum to remain airborne. In contrast, rotary-wing UAVs, like quadrotor drones, have the ability to hover above a specified area and remain stationary. In Fig. 10.1, an overview of different UAV types, functions and capabilities is presented. It is worth noting that the flight duration of a UAV is affected by factors such as its energy source (battery, fuel, etc.), type, weight, speed, and trajectory.

10.3 UAVs in Wireless Networks

The UAV-based network has been recognized as a dynamic network having nodes with high mobility and a dynamic topology. These nodes exhibit random changes in their positions as a result of their constant movement in both 2D and 3D domains. Hence, the use of appropriate communication technology emerges as a significant concern for this network. Various technologies may be used within the communication network between UAVs and their BS for the purpose of data transmission and control, including Wi-Fi, cellular technology including Long-Term Evolution (LTE), and the 4th generation (4G) standards. The Wi-Fi technology is widely recognized as a prevalent kind of communication technology. It operates on the IEEE 802.11 (a, b, and b/g) specifications and use the unlicensed frequency spectrum spanning from 2.4 to 5 GHz. The primary obstacles faced by Wi-Fi technologies are limited range for communication and the need for a direct line-of-sight (LoS) connection. In contrast, cellular-connected UAVs equipped with LTE and 4G technologies provide a potential solution to address the limitations posed by Wi-Fi technology. These UAVs enable communication across greater distances, beyond the constraints of LoS connections. Within the context of the UAV-based network, the communication interfaces of UAVs include a total of four distinct channel interfaces:

1. Inter UAV Interface,
2. Interface between UAVs and ground BSs,
3. Interface between UAV and ground receiver,
4. Interface between UAVs and satellites.

Figure 10.2 shows possible communication interfaces of UAVs in wireless networks. Furthermore, there exists two primary type of communication links:

1. The data communication link.
2. The control link.

The use of datalink facilitates the transmission and reception of data in both uplink and downlink transmission. The transmission of data may be facilitated by the use

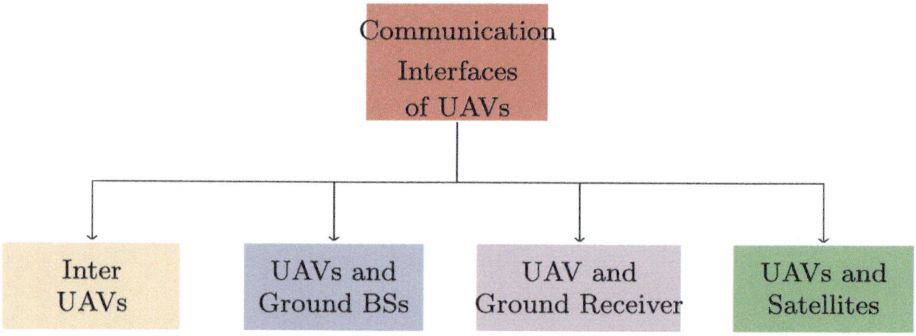

Fig. 10.2 Communication interfaces of UAVs

of microwave or millimetre wave (mmWave) spectrum bands. The International Telecommunication Union (ITU) has established the frequency bands within the microwave spectrum as spanning from 1 to 6 gigahertz (GHz) [7]. In contrast, the mmWave technology has the capability of achieving improved wireless communications and significantly increased data transmission speeds. Furthermore, it is capable of exhibiting frequencies that span a range of up to 300 GHz. The Control link provides a dependable means for ensuring the safety of UAVs throughout their operations. This connection facilitates the transmission of control information at a reduced data rate between the UAVs and a ground-based control station, as well as across different UAVs [2]. The primary attributes of the control link between UAV and BSs are as follows:

1. Full-duplex communication,
2. High-reliability connections,
3. Low latency response,
4. Low data rates,
5. Secure connections.

Figure 10.3 lists attributes of the UAV-BS control link.

The potential disruption of control link connectivity for UAVs might lead to severe and perhaps disastrous consequences. Hence, the International Civil Aviation Organization (ICAO) mandates that the connectivity of UAV control links must operate within a designated and safeguarded frequency range. Consequently, the ITU has granted authorization for certain segments within the L and C bands to be allocated for UAVs. In the L-band, the authorized frequencies range from 960 to 977 megahertz (MHz), while in the C-band, the authorized frequencies range from 5030 to 5091. In addition, it is necessary for control link to function in both LoS and Non-Line-of-Sight (NLoS) scenarios, which entails the allocation of a 34 MHz spectrum for LoS communications and a 56 MHz spectrum for NLoS link. In contrast, the data connections used in UAV-based networks need higher data rates in comparison to the requirements of control links. In search and rescue operations, it is essential for UAVs to transmit high-resolution films and photos to the ground station. The range

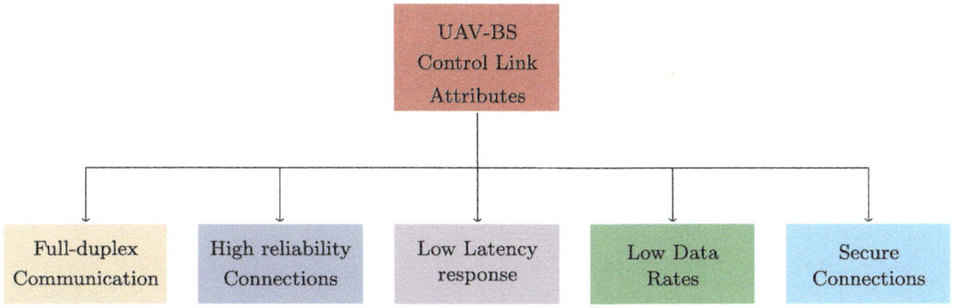

Fig. 10.3 Attributes of the UAV-BS control link

of data rates varies from a few megabits per second (Mbps) to beyond 30 Mbps. Furthermore, the use of UAVs as airborne BSs has the potential to achieve data rates above tens of gigabits per second (Gbps) in both downlink and backhaul networks.

10.3.1 UAV-UAV Interface

The LoS channel component is primarily responsible for governing the connection between UAVs. The use of mmWave and 5G communications may effectively enhance the UAV-UAV link for LoS, enabling the establishment of high-capacity backhaul networks for UAV-UAV communication. Furthermore, it is worth noting that the UAVs inside a UAV-based network often operates under circumstances of continuous movement and mobility. Hence, UAVs have the capability to establish communication links with other UAVs either directly or indirectly via the establishment of multi-hop communication channels with other UAVs. The concept of air-to-air (ATA) communication channel is widely used in UAV networks for the purpose of establishing a backhaul connection between numerous UAVs, enabling communication among them. Hence, the communication channels used for UAV-UAV communications are based on Air Traffic Control (ATC) standards. In this context, ATA has similarities to free-space communication, particularly in relation to the LoS situation. In contrast, the use of UAVs as relay nodes is a common application within the realm of UAV-UAV communication. UAVs have the capability to establish communication among themselves by using microwave and mmWave bands. These bands include a range of frequencies, spanning from several hundred MHz to a few GHz supporting 4G (700 MHz–6 GHz) as well as LTE technologies. In the case of 5G networks, the frequencies used extend to tens of GHz, specifically beyond 20 GHz.

10.3.2 UAV-Ground BS Interface

When establishing air-to-ground (ATG) communication, it is essential to carefully evaluate appropriate technologies that can provide uninterrupted and reliable links for both data and control links across a range of UAV applications. In [8], the authors offer a comprehensive analysis of four distinct communication technologies:

1. Satellite communication,
2. Ad hoc network,
3. Cellular network, and
4. Direct link communication.

Satellite Communication: Seamless communication between UAVs and Ground Base stations is essential for their effective operation in diverse scenarios, encompassing both LoS and NLoS environments. While LoS conditions often rely on ground control stations, NLoS scenarios necessitate satellite connections, each presenting unique challenges influenced by varying parameters and frequencies, thereby impacting latency. Satellite communication emerges as a critical component, particularly in remote or uncovered areas, facilitating reliable control and payload transmission. The preference for low earth orbit (LEO) satellites in UAV operations is driven by their constrained frequency bands and superior latency performance compared to Geostationary Orbit counterparts. Frequencies designated for NLoS satellite communication, including L-band, C-band, Ku-band, and Ka-band, are crucial for optimizing connectivity. Moreover, the dynamic movement of UAVs underscores the need for continuous enhancements in physical layer configurations to maintain robust communication links.

Ad hoc Network: In times of emergency, when conventional communication infrastructure like wired and cellular networks is compromised, Emergency Ad hoc Networks (EANs) become indispensable. These networks establish communication links between the nearest working base station and users, acting as emergency responders and facilitating a high-speed link to transmit vital information. UAVs play an important role in this infrastructure, serving as temporary mobile base stations to establish wireless communication links with the nearest operational cellular network access points [9]. By integrating UAV networks with existing infrastructure, regardless of damage, a robust disaster information system can swiftly disseminate crucial information, ensuring rapid response and coordination in the face of adversity. Ad hoc networks are formed on-the-fly, allowing UAVs to establish direct communication links with each other or with ground stations as needed, without the necessity of fixed infrastructure like cell towers. A decentralized UAV-based Ad Hoc network is explained in Sect. 10.5.2. This flexibility enables UAVs to quickly adapt to changing conditions and environments, making them well-suited for emergency response scenarios where traditional communication networks may be unavailable or damaged.

Communication Through Cellular Network: The use of cellular-enabled UAVs for networking purposes shows great potential, particularly when incorporating the upcoming 5G technology. It is anticipated that this network will provide a data throughput of around 10 Gbits/sec, accompanied by a round-trip latency of below 1 ms. The aforementioned data rate has the capability to meet the demands of real-time UAVs applications, including the transmission of high-resolution movies in real-time. The implementation of 5G and mmWave technologies presents novel radio options and enables the exploration of intelligent spectrum management possibilities. In the context of a network using UAVs, the use of the 5G spectrum may facilitate the allocation and reservation of time and frequency resources inside the cellular band. Hence, the cellular network has the capability to serve as the control and data connection for UAVs. Additionally, it has the capability to provide comprehensive control coverage for UAVs. The UAV-based communication system employs two distinct streams. One must establish control over the UAV in order to ensure its operation at a low rate, while maintaining high dependability and strong links. The alternative method for data streaming runs at a much elevated pace. The responsiveness of the remote control system for UAVs is a critical prerequisite in the context of UAV-centric networks. The transmission delay necessitates the use of short time intervals and frequent changes in direction inside the half-duplex channel. The system provided by the authors in [10] use chirp spread-spectrum modulation in conjunction with a correlator-based de-modulator for the control link, while using high data rate orthogonal frequency division multiplexing (OFDM) modulation for the data transmission. OFDM employs the same bandwidth as spread-spectrum modulation yet efficiently utilizes its whole bandwidth. In a more specific manner, the system comprises of a single uplink connection that connects a ground station with the UAV, as well as two separate downlink connections. One of these downlink connections is dedicated to control, while the other is used for data transmission. In the context of datalink connections with high rate requirements, it is common to use OFDM for data transmission. This modulation technique allows for efficient utilization of the available bandwidth by dividing it into multiple orthogonal subcarriers. On the other hand, for control purposes, spread-spectrum chirp modulation is often employed. This modulation scheme spreads the signal throughout a wide frequency range, providing robustness.

Direct Link Communication: The establishment of a direct communication channel between the UAVs and the ground station necessitates a LoS connection. Typically, this kind of communication takes place inside an unlicensed frequency range, such as the 2.4 GHz band. The primary problem encountered by the operation of this connection in urban and dense-urban areas is the presence of blockage effects caused by buildings, trees, and other impediments. These obstructions have a substantial negative impact on the dependability and data rate of the transmission. Furthermore, the use of this particular connection is not feasible for the extensive deployment of UAVs across expansive regions, particularly in cases where NLoS communications are involved. Hence, the use of cellular network communication may be employed as a means to address these difficulties.

10.3.3 UAV-Ground Receiver Interface

UAVs have shown notable efficacy in facilitating wireless coverage for terrestrial users, particularly in scenarios characterized by disasters or densely populated gatherings. ATG channel models serve the purpose of establishing a communication link between an airborne BS and users on the ground. In recent times, a number of studies have been undertaken with UAVs in order to establish ATG communication link. The classification of these devices may be determined by their communication technology as well as operating frequencies, which include 4G/LTE, 5G, and Wi-Fi. The 4G/LTE technology functions inside the microwave frequency ranges, while the 5G technology works within the mmWave bands. On the other hand, Wi-Fi utilizes the IEEE 802.11 standards. The frequency band is widely recognized as a fundamental resource of any communication system. The impact and alteration of the phenomenon may be influenced by the various frequency bands. In [4], the authors presented a statistical propagation model for ATG communication in low-altitude platforms (LAPs) across several kinds of habitats, including rural, urban, sub-urban, and dense metropolitan scenarios. In a 4G network, three distinct frequency bands, namely 700, 2000, and 5800 MHz are generally used. Numerous studies have used this paradigm to facilitate wireless connectivity for terrestrial users across various contexts. The aforementioned approach is used for the purpose of delivering wireless connectivity via the utilization of UAVs in various scenarios such as emergency circumstances, search and rescue operations, densely populated gatherings, and supporting ground stations in facilitating wireless coverage for diverse applications. Frequency division multiple access (FDMA) is a widely used method for channel access in the communication between UAVs and ground users. The UAV allocates the whole channel capacity among users, ensuring that each user receives an equitable distribution of bandwidth. An alternative statistical propagation channel model has been proposed by Feng et al. [11] in which an ATG channel paradigm was created specifically for urban environments, whereby UAVs function as aerial BSs that offer wireless access for ground nodes. The model works within the frequency range of 200, 1000, 2000, 2500, and 5000 MHz. The use of mmWave bands is crucial for the 5G network due to its ability to provide substantial bandwidths and facilitate high-speed communication channels. The use of mmWave/5G communication network in UAV applications and consumer use cases presents an opportunity to obtain the demanding data rate needs of real-time applications and high-definition video transmission. However, when considering control link communications, it is not advisable to use mmWave as an alternative because to its high attenuation, limited range, and dependence on strong LoS links. In the context of mmWave frequency ranges, there exists a multitude of spectrum resources that may be effectively used to meet the needs for high data rates and facilitate the implementation of 5G communications. Several studies have conducted on the transmission models and channel characteristics of ATG communication in the mmWave frequency range. The researchers in [12] conducted a comprehensive examination and evaluation of the propagation properties of the ATG mmWave channel. This investigation focused on two distinct frequency bands: the licensed 28 GHz band and the unlicensed 60 GHz band. The analysis for several contexts, including rural, urban, sub-urban, and over-seas, was conducted using ray-tracing tools. This allowed for the simulation of the received signal strength (RSS) and root mean square delay spread (RMS-DS) of multipath components in relation to the communication system being proposed. The use of omnidirectional or directional antennas on a UAV-mounted Wi-Fi network shows potential as a viable method for delivering ATG coverage and establishing real-time connections for users in situations when the terrestrial network becomes entirely unavailable during catastrophic events. The integration of commercial off-the-shelf IEEE 802.11 radio equipments with UAVs enables the development of an ATG communication link and facilitates the study of the propagation channel statistics associated with this model. The researchers in [13] conducted an analysis on the attributes of the wireless connection between an UAV equipped with an antenna, a commercially available wireless radio, and the ground terminal. Subsequently, the researchers conducted measurements to determine the route loss exponent associated with the ATG propagation channel, whereas 5.24 GHz frequency band was utilized considering the IEEE 802.11a standard. Additionally, UAV used in the study had a varying altitude between 20 and 120 m. The model works within the frequency range of 200, 1000, 2000, 2500, and 5000 MHz. The use of mmWave bands is crucial for the 5G network due to its ability to provide substantial bandwidths and facilitate high-speed communication channels. The use of mmWave/5G communication network in UAV applications and consumer use cases presents an opportunity to obtain the demanding data rate needs of real-time applications and high-definition video transmission. However, when considering control link communications, it is not advisable to use mmWave as an alternative because to its high attenuation, limited range, and dependence on strong LoS links. In the context of mmWave frequency ranges, there exists a multitude of spectrum resources that may be effectively used to meet the needs for high data rates and facilitate the implementation of 5G communications. Several studies have conducted on the transmission models and channel characteristics of ATG communication in the mmWave frequency range. The researchers in [12] conducted a comprehensive examination and evaluation of the propagation properties of the ATG mmWave channel. This investigation focused on two distinct frequency bands: the licensed 28 GHz band and the unlicensed 60 GHz band. The analysis for several contexts, including rural, urban, sub-urban, and over-seas, was conducted using ray-tracing tools. This allowed for the simulation of the RSS and RMS-DS of multipath components in relation to the communication system being proposed. The use of omnidirectional or directional antennas on a UAV-mounted Wi-Fi network shows potential as a viable method for delivering ATG coverage and establishing real-time connections for users in situations when the terrestrial network becomes entirely unavailable during catastrophic events. The integration of commercial off-the-shelf IEEE 802.11 radio equipments with UAVs enables the development of an ATG communication link and facilitates the study of the propagation channel statistics associated with this model. The researchers in [13] conducted an analysis on the attributes of the wireless connection between an UAV equipped with an antenna, a commercially available wireless radio, and the ground terminal. Subsequently, the researchers conducted measurements to determine the route loss exponent associated with the ATG propagation channel, whereas 5.24 GHz frequency band was utilized considering the IEEE 802.11a standard. Additionally, UAV used in the study had a varying altitude between 20 and 120 m.

10.3.4 UAV-Satellite Interface

The command-and-control link plays an important role in ensuring the safety of UAVs by facilitating reliable communication and the interchange of commands between the ground control station and the UAV. This link is designed to operate effectively in both LoS and NLoS scenarios. The use of a ground control station is often employed for LoS conditions, while satellite communication connections are often employed for NLoS circumstances. The LoS and NLoS links, which consist of two-channel components, exhibit distinct channel parameters and operating frequencies, resulting in differing latency ranges. Consequently, the control link encounters a significant obstacle in meeting the requirements for establishing highly reliable and safe connections. Therefore, communicating via satellite is a viable method for establishing control link along with payload communication in areas that are remote or lack coverage. LEO is considered as the optimal option for utilizing UAV, Navigation, and payload control links. LEO operates within limited frequency bands and exhibits lower latency in comparison to geosynchronous earth orbit (GEO). Consequently, LEO enables the autonomous functions of UAVs to be supported efficiently, facilitating extensive coverage areas spanning hundreds of kilometers through a direct link between the UAV and ground station, the authors provide an overview of the forthcoming frequency bands being used for satellite communications. These bands are intended to be assigned for NLoS communications. The frequency bands discussed include L-band (850–2000 MHz), C-band (5–6 GHz), Ku-band (12–18 GHz), and Ka-band (above 26 GHz). Furthermore, the dynamic motion of UAVs results in the alteration of antenna orientations, potentially leading to the attenuation of a signal strength. Hence, it is essential to enhance the existing physical layer configuration in order to meet the standards. The use of OFDM in LTE is deemed unsuitable for this particular mode of communication. Several other modulation techniques, such as filter bank multicarrier (FBMC) along with orthogonal chirp spread spectrum (OCSS), may be explored as potential solutions. These advanced modulation schemes provide a more compact design and have the potential to be efficiently used for NLoS communications in ATG satellite communication. One of the primary obstacles encountered in UAV-based communication systems is to the scarcity of research and tests undertaken on radio links and the physical layer. Therefore, it is essential to explore radio interfaces, modulation techniques, as well as physical layers in the context of communication systems using UAVs. Table 10.1 gives a comparison of various aspects of each interface.

10.3.5 Communication Standards and Protocols for UAVs

Wireless standards plays an important role since UAV nodes interchange and/or report data regarding factors such as position, velocity, and energy consumption. The following address several technologies for wireless communication that are being developed to assist cooperative UAV applications.

3GPP Cellular Networks: The 5G New Radio standard includes additional frequency bands referred to as mmWave bands, which are specifically intended for adapting to the needs of UAV applications. These frequencies are 26, 28, 39 GHz (licensed), and 60 GHz (unlicensed, used in Wi-Fi), which are situated within the Ka along with V bands of the EM spectrum. According to a measurement study conducted by Qualcomm, it has been observed that an UAV has the capability to be

Table 10.1 A comparison of various aspects of each interface

Aspect	UAV-UAV interface	UAV-ground base station interface	UAV-ground receiver interface	UAV-satellite interface
Communication link type	UAV to UAV	UAV to ground BS	Ground receiver to UAV	Ground station to UAV
Communication scenario	LoS and NLoS	LoS and NLoS	LoS and NLoS	LoS and NLoS
Channel components	LoS channel	LoS channel	LoS and NLoS channels	LoS and NLoS channels
Frequencies	mmWave, 5G	4G/LTE, 5G, Wi-Fi	4G/LTE, 5G, Wi-Fi	L-band, C-band, Ku-band, Ka-band
Coverage area	Local and long-range	Local and long-range	Local and long-range	Global
Latency	Varies	Varies	Varies	Varies
Suitable applications	UAV networking, data sharing	Data and control transmission, networking	ATG communication, networking	Command-and-control, payload communication
Challenges	Blockage effects, mobility	Blockage effects, mobility	Attenuation, mobility	Scarcity of research, signal attenuation

operated beyond visual line of sight (BVLoS) conditions. Specifically, the UAV can be managed at a maximum distance of 6 kilometers from the cellularBS that provides connectivity, while also maintaining an altitude of up to 120 m above ground level. In recent times, the specifications of 5G have shown significant promise in facilitating the utilization of UAV applications. This is mostly due to the inherent benefits of reduced latency and increased data transmission rates.

Wireless Local Area Networks: The IEEE 802.11a standard use OFDM as a modulation technique to achieve high data rates and improve signal robustness against multipath effects. However, it has a lower transmission range when compared to the IEEE 802.11b standard, which employs Direct Sequence Spread Spectrum (DSSS) as its modulation scheme. The IEEE 802.11g type was developed with the intention of incorporating the favorable attributes of its preceding standards in terms of data transfer rate and coverage distance. Conversely, the IEEE 802.11n standard has been enhanced by the use of Multiple Input Multiple Output (MIMO) technology, which capitalizes on the benefits derived from simultaneous transmission and reception utilizing multiple antennas. The IEEE 802.11p standard has been specifically developed to enhance vehicle-to-everything (V2X) communications. Its primary objective is to provide improved support for various intelligent transport services and applications. The IEEE 802.11s standard is suggested as a means to effectively harness the advantages of mesh networking. Moreover, the IEEE 802.11ah standard has been specifically developed to address the demands of IoT applications. This is achieved by enhancing the coverage area via the utilization of sub-GHz frequency bands and MIMO technology.

Wireless Personal Area Networks: Communication modules that adhere to the IEEE 802.15.4 specification are frequently observed within the market for UAV accessories. The primary objective of the IEEE 802.15.4 standard is to provide personal area networks having direct-to-device communication capabilities, characterized by low power consumption and cost-effectiveness. Nevertheless, the data rate of these devices is often constrained when compared to those falling in the IEEE 802.11 family. In contrast, the IEEE 802.15.4a, sometimes referred to as Ultra-Wide Band (UWB), offers a superior data transmission rate compared to the 2006 IEEE 802.15.4 standard. One example is the UWB transceiver developed by DecaWave, which asserts a maximum data throughput of 6.8 Mbps and an enhanced communication range of up to 290 m, operating at a rate of 110 kilobits per second (Kbps) with a 10% Packet Error throughput (PER) [14]. Schmidt et al. in [15] conducted an experimental investigation on the performance of UWB technology in comparison to ZigBee, a wireless communication protocol based on the 802.15.4 standard. The experiments were conducted in an indoor setting with multipath propagation conditions. The results indicate that UWB technology exhibits superior performance compared to ZigBee in regards to range, data throughput, as well as packet loss rate. Khawaja et al. [16, 17] conducted a study on the propagation channels of UAVs for ATG communication. Their research focused on the UWB channel model, specifically examining path loss, multipath effects, and the characterization of A2G channels at different distances and altitudes of UAVs.

Low-Power Wide-Area Networks: Cycleo, a company bought by Semtech [18], released a series of standards known as Long Range (LoRa) along with Long-Range Wireless Area Networks (LoRaWAN), with a focus on applications in long-range communication. The LoRa protocol stack encompasses the physical layer components, whereas the LoRaWAN protocol focuses on the data connection and higher protocol layers. LoRa technology has the capability to operate in both licensed frequency bands (169 MHz, 868 MHz) as well as unlicensed frequency bands (433 MHz, 914 MHz, and 2.4 GHz). This versatility makes it suitable for both planned and ad hoc installations. The LoRa technology utilizes Chirp Spread Spectrum (CSS) modulation, which is a spectrum expansion method derived from DSSS. In this approach, a narrow band signal with a certain data rate is chipped at a higher rate, referred to as the chip rate. Subsequently, the chipped signal is modulated into a variable frequency signal, often known as a chirp. The LoRa technology exhibits a reduction in throughput as the transmission distance increases. Table 10.2 provides a broad overview of the key features and characteristics of each type of wireless network.

Table 10.2 A broad overview of the key features and characteristics of each type of wireless network

Feature	3GPP cellular Networks	Wireless local area networks	Wireless personal area networks	Low-power wide area networks
Standard	3GPP standards (e.g., LTE, 5G NR)	IEEE 802.11 standards (e.g., 802.11a, 802.11ac)	IEEE 802.15.4 standards (e.g., 802.15.4, 802.15.4a)	LoRaWAN
Frequency bands	Various (e.g., mmWave, Sub-6 GHz)	2.4 GHz, 5 GHz	2.4 GHz, 915 MHz, 868 MHz	433 MHz, 868 MHz, 915 MHz, 2.4 GHz
Range	Medium to Long	Short to Medium	Short	Long
Data rates	High	Medium to High	Low to Medium	Low
Power consumption	Moderate to High	Moderate	Low	Low
Coverage area	Wide	Limited	Limited	Wide
Modulation techniques	OFDM, QPSK, QAM, etc.	OFDM, DSSS, MIMO, etc.	DSSS, UWB, etc.	Chirp Spread spectrum
Application examples	Cellular phones, mobile broadband	Wi-Fi routers, hotspots	Bluetooth devices, Zigbee, etc.	Smart meters, IoT devices

10.4 Communication Challenges in UAV Systems

10.4.1 Overview

UAVs encounter a range of communication challenges that can significantly impact their performance and reliability.

Limited Bandwidth and Range: UAVs rely on wireless communication systems to transmit commands and receive real-time data. The limited available bandwidth and communication range can hinder the transmission of large amounts of data, especially in scenarios where UAVs operate at long distances from the control station.

NLoS Operations: UAVs might be required to operate beyond the LoS, for example, in urban environments or remote areas. Establishing and maintaining communication links under such conditions is challenging due to obstacles like buildings, hills, and vegetation that can obstruct the signal.

Latency and Real-Time Requirements: In applications where real-time data is crucial, such as surveillance or remote medical assistance, communication latency becomes a critical challenge. Delays in transmitting commands and receiving data can lead to incorrect decision-making or missed opportunities.

Security and Privacy Concerns: UAV communication systems are vulnerable to hacking, jamming, and interception, posing security risks both in terms of the UAV's control and the sensitive data it might be transmitting. Ensuring the confidentiality and integrity of communication is essential.

10.4.2 Mobility and Dynamic Network Topology

UAVs are frequently deployed in dynamic and unpredictable environments, such as disaster response, search and rescue missions, or agricultural monitoring. Rapid changes in the environment due to UAV's mobility, can potentially disrupt communication links, making it difficult to adapt and ensure continuous control and data transmission. The mobility of UAVs and the dynamic nature of their network topology play vital roles in ensuring efficient communication, coordination, and mission execution. Here are key considerations for both aspects:

Adaptive Flight Paths: UAVs often need to dynamically adjust their flight paths based on mission requirements, environmental conditions, or real-time data. Adaptive flight planning allows UAVs to optimize routes, avoid obstacles, and respond to changing circumstances during the mission.

Collaborative Maneuvers: In scenarios where multiple UAVs are involved, collaborative maneuvers become essential. UAVs should be able to coordinate their movements to achieve mission objectives collectively. This requires communication and synchronization among UAVs to avoid collisions and optimize overall efficiency.

Emergency Response: Mobility is crucial in emergency situations where UAVs may need to quickly change their position or route to respond to a developing event. Rapid and adaptive mobility features are essential for effective emergency response missions.

Decentralized Communication: UAVs often operate in environments where a centralized communication infrastructure may not be available. A dynamic network topology allows UAVs to establish and maintain communication links in a decentralized manner, adapting to changes in the network environment.

Mesh Networking: UAVs can form a mesh network, where each UAV acts as a node that can communicate with other nodes within its range. This allows for reliable and dynamic communication links, even if direct LoS communication is obstructed.

Dynamic Node Discovery: The ability of UAVs to dynamically discover and connect to other nodes in the network is crucial for collaborative missions. Dynamic node discovery enables UAVs to adapt to changes in the composition of the network and establish communication links as needed.

Self-Healing Networks: UAVs may encounter communication disruptions due to various factors. A dynamic network topology with self-healing capabilities allows the network to autonomously reconfigure itself to restore communication links and maintain mission continuity.

10.4.3 Limited Energy and Computing Resources

UAVs often operate on limited onboard power sources. Several factors contribute to the necessity of managing limited energy and computing resources in UAVs:

Size and Weight Restrictions: UAVs are designed to be compact and lightweight for increased maneuverability and versatility. However, these constraints impose limitations on the size and weight of onboard components, including energy storage systems and computing hardware.

Power Consumption Efficiency: Energy efficiency is a critical consideration for UAVs, as the power required to sustain flight significantly impacts the overall mission duration. UAVs need to balance the energy consumed by propulsion systems, avionics, and other onboard equipment to optimize their operational efficiency. Solar power can also be utilized to mitigate the power demands of long-range UAVs.

Flight Duration and Range: The energy density of available batteries limits the amount of power that can be carried onboard, directly affecting the UAV's flight duration and operational range. UAVs must navigate within the constraints of their energy capacity, influencing the scope and duration of missions.

Payload Considerations: UAVs often carry various payloads, such as cameras, sensors, or communication devices. The power requirements of these payloads need to be factored into the overall energy budget. Balancing payload capacity with energy constraints is crucial for mission success.

Computational Processing on Board: Onboard computing resources are constrained by the need for lightweight and compact systems. Processing power is essential for tasks such as navigation, real-time data analysis, and decision-making. However, the challenge is to integrate powerful computing capabilities within the limited space and power available on a UAV.

Real-Time Decision-Making: Certain UAV applications, especially those involving autonomous flight and data processing, require real-time decision-making capabilities. Limited onboard computing resources can pose challenges in achieving the necessary processing speed for timely and accurate decision-making.

10.4.4 Interference and Spectrum Management

There exist certain problems such as Network Congestion, scarcity of bandwidth, interference and signal loss.

Network Congestion: In scenarios where multiple UAVs operate simultaneously within the same airspace, the communication network can become congested. Coordinating the communication channels and managing the data traffic to avoid collisions and delays is a significant challenge.

Interference and Signal Loss: UAVs often operate in environments with other electronic devices and sources of electromagnetic interference. This interference can lead to signal degradation, loss of connection, or even complete communication failure, jeopardizing the control and safety of the UAV.

10.5 UAV Communication and Control Architectures

A communication and control architecture describes how data and control commands are transferred between the ground station and an UAV or between UAVs. In this section, different communication and control architectures for UAVs are discussed.

10.5.1 Centralized Control and Communication Architectures

Figure 10.4 illustrates a centralized design for UAV communication, whereby a central node, referred to as the ground station, serves as the hub connecting all UAVs. This network topology is quite prevalent. Within this network architecture, each UAV establishes a direct connection with the ground station to send and receive the command-and-control data. It is important to note that the UAVs do not establish direct connections with one another. The whole of the network is centralized around the ground station, necessitating the use of the ground station as a relay for communication between two UAVs. Given that all UAVs are directly linked to the ground station, it is reasonable to anticipate a little delay in the transmission of command-and-control data between the ground station and a UAV. Nevertheless, the data transmission between two UAVs is subject to a comparatively extended delay due to the need of routing the data via the ground station. Additionally, the need for efficient long-distance communication between ground stations and UAVs necessitates the use of sophisticated radio transmission devices with sufficient amount of transmission power. However, the implementation of such devices in medium or small UAVs may prove impractical owing to limitations imposed by their size and

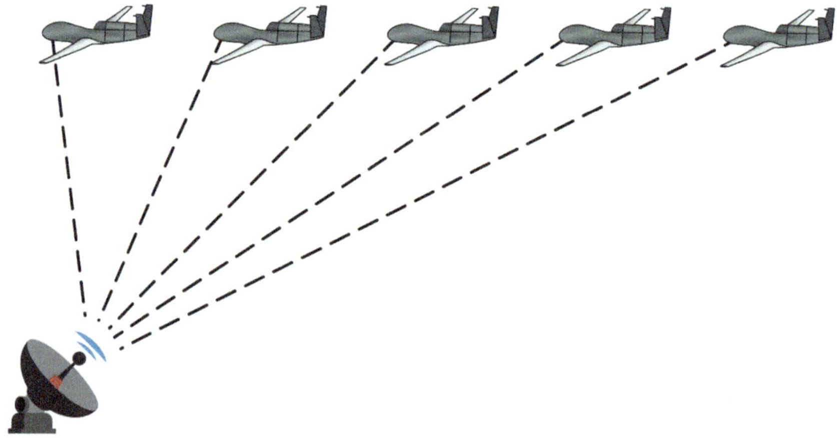

Fig. 10.4 Centralized UAV control and communication

payload capacity. Furthermore, within the context of the centralized UAV communication architecture, it is important to acknowledge that the ground station may introduce a vulnerability to the UAV network due to its potential to become a single point of failure. For instance, in the event that the ground station has any issues, it will result in the disruption of the whole UAV network. Therefore, it may be stated that the aforementioned communication architecture lacks robustness.

10.5.2 Decentralized Control and Communication Architectures

In a decentralized communication architecture, the presence of a central node is not necessary, enabling direct or indirect communication between two UAVs. Therefore, the data which is not intended for the ground station may be efficiently directed via an UAV as an alternative to use the ground station. Three different decentralized communication architectures are discussed below.

UAV Ad Hoc Network: An UAV ad hoc network is illustrated in Fig. 10.5. In general, an UAV ad hoc network doesn't need a pre-existing infrastructure, and each UAV in the network will take part in sending data to other UAVs. Basically, a backbone UAV acts as a gateway for the ad hoc network, passing information between the ground station and the other UAVs. The gateway UAV will then need two radio channels, one to talk to other UAVs and the other to talk to the ground station. Since only one UAV is needed to communicate with the ground station in this UAV ad hoc network, the network can cover a much larger area. Also, because many UAVs move close to each other, the transmission device in a UAV can be inexpensive and light, which

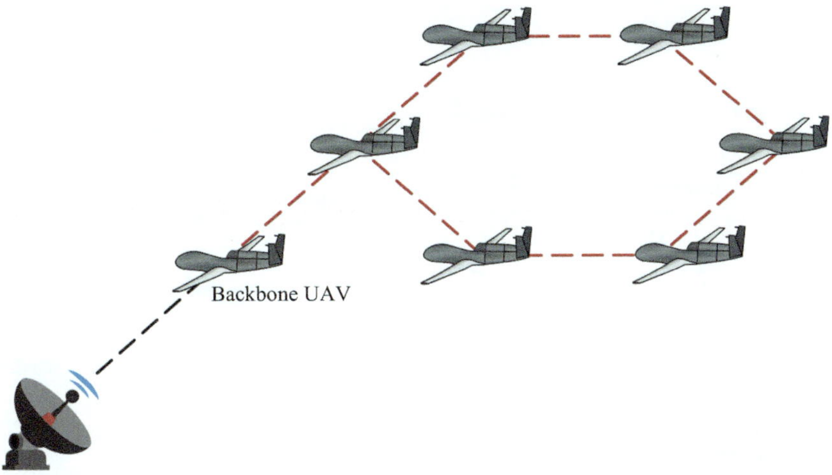

Fig. 10.5 Decentralized UAV based Ad Hoc Network

makes it better for small or medium-sized UAVs. To make sure the network stays connected, however, all UAVs need to move in the same way, with the same speed and heading. So, the ad hoc network design is especially good for linking up a group of similar UAVs for activities like continuous monitoring. For some tasks, there are a lot of UAVs that need to be sent out. These UAVs often come in different shapes and sizes, from big to small, and carry different payloads. In a cluster consisting multi-type of UAVs, the same type UAVs tend to be close to each other and forms the cluster, while UAVs of different types tend to be farther apart. So, these UAVs can be put into more than one group based on their types. It seems that a large number of UAVs forming a single UAV ad hoc network may not be efficient or useful due to the traffic patterns and network control overhead that would be caused. In fact, UAVs of different types can be linked together using either a multi-group UAV network or a multi-layer UAV ad hoc network.

Multi-group UAV Network: In this network, a group of UAVs forms an UAV ad hoc network, and each group's backbone UAV connects to the ground station as shown in Fig. 10.6. Intra-group communications (communications between members of the same group) are done through an UAV ad hoc network, which is explained in UAV Ad Hoc Network. Inter-group communications (communications between members of different groups) are done through the backbone UAVs and the ground station. It is important to note that the multi-group UAV network design can be thought of as a network that combines a centralized UAV network and UAV ad hoc networks. This communication system is good for a task that involves a lot of UAVs with different ways of flying or talking to each other. This architecture is still not very strong, because it is only semi-centralized in nature.

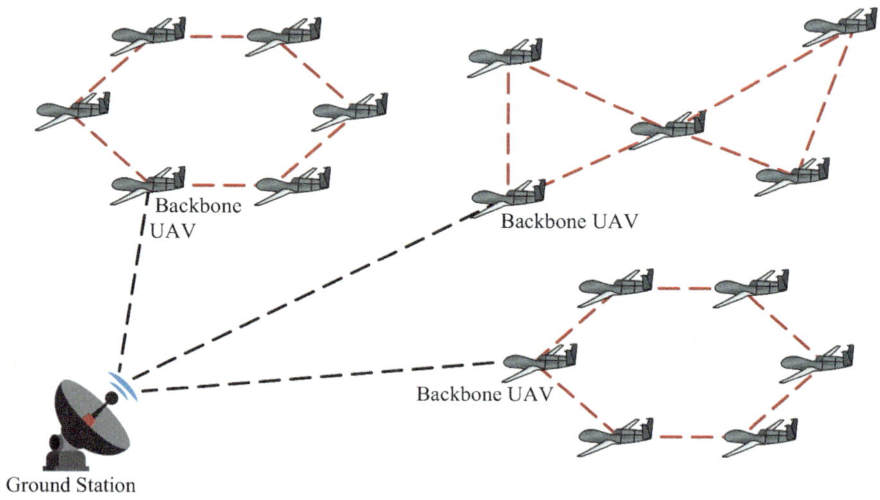

Fig. 10.6 Decentralized Multi-group UAV Network

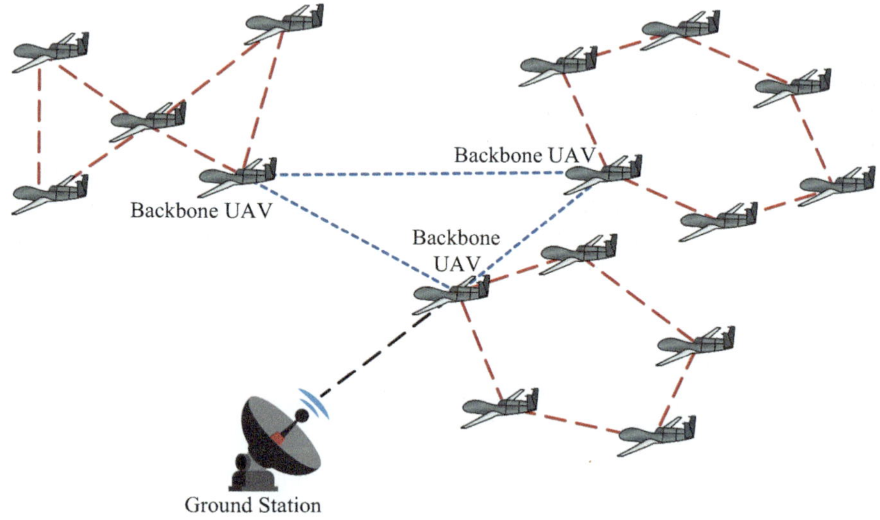

Fig. 10.7 Decentralized Multi-layer UAV Ad Hoc Network

Multi-layer UAV Ad Hoc Network: A multi-layer UAV ad hoc network is another way to connect different groups of different types of UAVs together. Figure 10.7 shows an example of a UAV ad hoc network with more than one layer. A UAV ad hoc network is made up of the UAVs in a single group. This network is the lowest layer of a multi-layer UAV ad hoc network design. The backbone UAVs from all of the groups make up the upper layer of the UAV ad hoc network. In the

Table 10.3 A comparison of centralized and decentralized UAV control and communication architectures

Aspect	Centralized architectures	Decentralized architectures
Presence of central node	Central node (ground station) serves as hub for all UAVs	No central node required, allowing direct or indirect UAV communication
Communication pathways	UAVs establish direct connections with ground station	Direct or indirect communication between UAVs, bypassing ground station when unnecessary
Flexibility of communication routes	Communication routed through ground station	More flexible communication routes between UAVs
Reliance on ground station	High reliance on ground station for data transmission	Reduced reliance on ground station, increasing network robustness
Potential vulnerabilities	Ground station serves as single point of failure	Network less vulnerable to single point failures
Delay in data transmission	Potential delays due to routing through ground station	Reduced delays as communication can be direct between UAVs
Implementation challenges	Requires sophisticated radio transmission devices	Simplified implementation, with fewer hardware requirements

multi-layer UAV ad hoc network, only one backbone UAV is directly linked to the ground station. This is the key difference from the multi-group UAV network. In this network, two groups of UAVs don't have to go through the ground station to talk to each other. The ground station only processes the information data that is meant for it. This makes the ground station's computing and transmission loads much lighter. The multi-layer UAV ad hoc network design is a key part of how the one-to-many UAV operation mode is put into action. A multi-layer UAV ad hoc network is also reliable because there is no single point of failure. Table 10.3 presents a comparison of various aspects between centralized and decentralized UAV control and communication architectures.

10.6 Applications of UAV Communication

Numerous applications leverage UAV communication as their foundation. These encompass a wide array of use cases where UAVs facilitate effective communication solutions. The versatility of UAV communication contributes to its applicability in diverse contexts, spanning emergency response, agriculture, environmental monitoring, and telecommunication expansion, etc.

10.6.1 Surveillance and Reconnaissance

There exist several desirable attributes pertaining to UAV surveillance applications, particularly in context of military or homeland security scenarios. The compact dimensions of the UAV not only reduces its physical vulnerability to attacks but also enhance its ability to conduct covert patrolling operations. The inclusion of high quality pictures in the dataset allows more precise data post-processing techniques, leading to enhanced accuracy in drawing inferences and obtaining insights. One notable attribute of the UAV is its low energy consumption, which enables it to sustain extended operational duration, hence facilitating the execution of more intricate tasks. Hence, in order to get more precise and expedient data while maintaining confidentiality, it is important to contemplate the utilization of sophisticated sensor technologies. Examples of advanced sensor technologies include nano-sensors, which are characterized by their tiny size, ultra-high-resolution image sensors that provide correct data, and energy efficient sensors that use minimal amounts of energy. In order to enhance the effectiveness and precision of multi-UAV cooperation algorithms and data post-processing algorithms, it is possible to use sophisticated machine learning techniques, such as deep learning, to get superior performance and expedited reaction times. In particular, the use of multi-UAV collaboration yields greater advantages compared to single UAV surveillance. These benefits include an expanded surveillance scope, enhanced error tolerance, and expedited job completion time. Nevertheless, the implementation of multi-UAV surveillance necessitates the use of increasingly sophisticated algorithms for data collecting, exchange, and processing. The processing of a large volume of data is a time-consuming and highly intricate task. The use of sophisticated machine learning techniques, such as deep learning algorithms, has the potential to enhance the UAV system's ability to derive more accurate and timely findings. Deep learning algorithms have the potential to provide novel discoveries and enhance the accuracy and brevity of analytical outputs by using enhanced data processing models. Free space optical (FSO) technology can be effectively used in UAVs within the armed services, particularly in situations when military wireless communications need safe transfer of information in battlefield environments. The use of remote sensing UAVs enables the efficient dissemination of substantial quantities of photos and videos to military forces, mostly in real-time. UAVs designed for near Earth observation have the potential to be used for the purpose of capturing high-resolution photographs of surface contours. This may be achieved via the utilization of synthetic aperture radar (SAR) and light detection and ranging (LiDAR) technologies. Aerial sensors have the capability to transfer the data they gather to the command center via the use of onboard satellite communication sub-systems, using FSO technology.

10.6.2 Real-Time Monitoring of Road Traffic

The comprehensive automation of the transportation system cannot be achieved just via the automation of vehicles. Indeed, it is necessary to automate several aspects of the comprehensive transportation system, including the functions performed by field support teams, traffic police, road surveyors, and rescue teams. Intelligent and dependable UAVs have the potential to facilitate the automation. UAVs have emerged as a promising technological solution for traffic monitoring, enabling the collection of valuable data pertaining to road conditions. In contrast to conventional monitoring systems like loop detectors, surveillance video cameras, and microwave sensors, UAVs provide a cost-effective solution for monitoring extensive and uninterrupted road segments, as well as the ability to concentrate on particular road sections. The data produced by sensor technology has a certain level of aggregation, which limits its ability to accurately capture the specific trajectories of individual vehicles within the traffic flow. The use of this data in examining individual driving behavior, as well as in the calibration and validation of simulation models, is impeded by their inherent constraints. In addition, it is important to note that catastrophes have the potential to do significant harm to communications infrastructure, and power systems. Failures of this kind might lead to a total absence of the capacity to manage and gather data pertaining to the transportation network.

10.6.3 Environmental Monitoring and Disaster Management

UAVs play an important role in disaster management across various scenarios, offering versatile capabilities for assessment, response, and recovery efforts.

Pre-disaster Assessment: UAVs equipped with Interferometric Synthetic Aperture (InSAR) and LiDAR technologies is used to detect subtle changes in the Earth's surface, such as fault lines and crustal deformation, which may precede earthquakes and landslides. The United States Geological Survey (USGS) utilizes LiDAR to create three-dimensional bare-earth maps, identifying fault lines and aiding in ongoing monitoring efforts. Moreover, UAVs can monitor areas prone to land slippage, supplementing traditional methods of geological assessment.

Infrastructure Inspection and Maintenance: UAVs are programmed to fly on predetermined routes over lifeline systems, collecting baseline data and assisting in infrastructure maintenance. For instance, during wildfire events, UAVs conduct aerial surveys to assess the condition of power lines, telecommunication towers, and water supply networks, identifying potential vulnerabilities and facilitating timely repairs.

Earthquake Management: UAVs are deployed to conduct rapid damage assessments, particularly in inaccessible or hazardous areas after earthquakes. Moreover, micro UAVs are utilized for low-level assessments of structural integrity, while satellite and airborne imagery is used to supplement ground-based assessments, providing comprehensive coverage of affected areas. Additionally, UAVs were employed during the 2011 Tohoku earthquake in Japan to assess the damage to the Fukushima nuclear reactors, highlighting their critical role in disaster response.

Wildfire Management: UAVs play a vital role in wildfire management by providing real-time aerial reconnaissance and monitoring capabilities. Projects such as Project First Response Experiment (FiRE) and the Wildfire Research and Applications Partnership (WRAP) leverage UAVs to collect tactical wildfire imagery, supporting firefighting efforts and enhancing situational awareness. UAV systems equipped with radiometric sensors are capable to autonomously detect wildfires and conduct closer inspections of incident locations.

Flood Management: UAVs complement existing data sources for flood hazard mapping and mitigation efforts. Automated systems for landslide detection using fixed-wing UAVs are being developed in Taiwan, while small multi-rotor UAVs are utilized for mapping landslides on road embankments. Moreover, UAV are utilized for rapid damage assessments following floods, aiding in the prioritization of recovery activities.

Man-Made Disasters: UAVs are invaluable tools for assessing and mitigating the impact of man-made disasters, including chemical spills and structural failures. Recent studies in the United States and Germany have demonstrated the effectiveness of UAVs in detecting hazardous substances and developing three-dimensional models of chemical plumes. Additionally, UAVs are deployed for structural inspections and environmental monitoring in hazardous environments.

Search and Rescue Operations: UAVs are utilized in remote and challenging environments for search and rescue operations. Projects such as the European CLOSE-SEARCH Project and the Australian UAV Outback Challenge showcase the potential of UAVs for locating missing persons and coordinating rescue efforts. Moreover, UAVs equipped with advanced sensors and imaging technology facilitate rapid search operations and enhance situational awareness for rescue teams.

Weather Events Management: UAVs contribute to the management of weather-related disasters by providing real-time data for monitoring and response efforts. During cyclones and hurricanes, UAVs gather atmospheric data and track storm progression, aiding in the prediction of storm paths and landfall locations. Additionally, UAVs support drought management by monitoring water resources and vegetation health, enabling authorities to implement timely drought mitigation measures.

10.6.4 Precision Agriculture

UAVs have shown their efficacy in the context of precision agriculture by facilitating crop management and monitoring, detecting weeds, scheduling irrigation, detecting diseases, spraying pesticides, collecting data from ground sensors such as moisture levels and soil qualities and reduction in use of herbicides. The use of UAVs in precision agriculture is an economically efficient over conventional manned aircraft and UAVs are time-efficient technological solution that has the potential to enhance crop yields, increase production in farms, and improve overall profitability. Numerous applications of UAVs in precision agriculture are shown in Fig. 10.8 and discussed in the following:

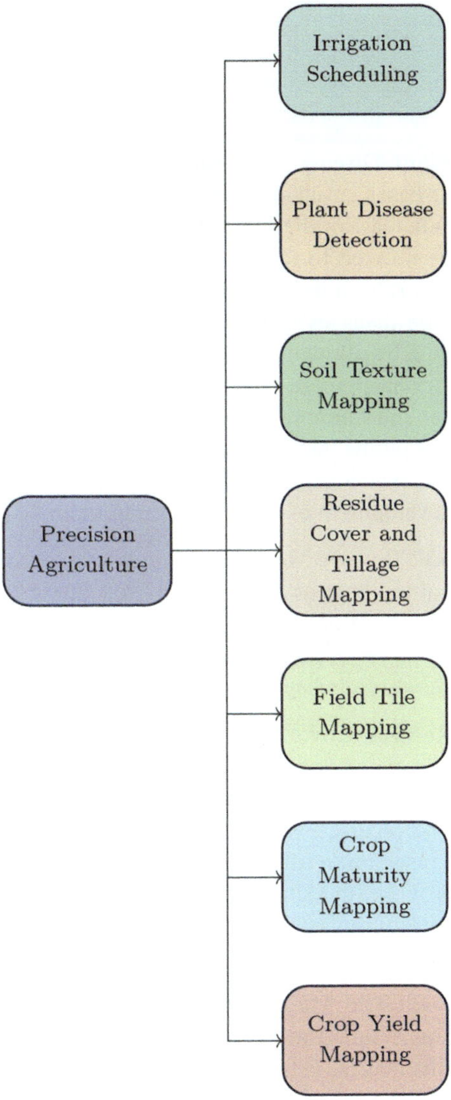

Fig. 10.8 Numerous applications of UAVs in precision agriculture

Irrigation Scheduling: There are four important factors that must be monitored to determine if irrigation is required:

1. Water content present in the soil.
2. Crop water requirement, which represents the quantity of water required for optimal crop growth.
3. Total quantity of rainfall.
4. Effectiveness of the irrigation system.

Using UAVs to measure soil moisture, plant-based temperature, and evapotranspiration, it is possible to quantify these variables. For example, the spatial distribution of surface soil moisture can be estimated using high-resolution multi-spectral imagery acquired by UAVs when combined with ground sampling. Further, by analyzing the thermal UAV images, the crop water stress index can also be approximated to identify water-stressed areas.

Plant Disease Detection: It is observed that crop damage associated with plant diseases result in approximately $33 billions annually in the United States only. Whereas, in the era of technological advancements, artificial intelligence (AI) can be utilized to identify the plant diseases from images. UAVs can play an important role in thermal remote sensing to monitor the spatial and temporal patterns of crop diseases in presymptomatically during various disease development phases, thereby allowing producers to reduce crop losses. UAVs have the capability to use thermal remote sensing techniques in order to observe and analyze the geographical and temporal distribution of agricultural illnesses prior to the manifestation of symptoms throughout different stages of disease development. Consequently, farmers may potentially mitigate crop losses. For example, detection of early stage of development of soil-borne fungus can be identified through aerial thermal images.

Soil Texture Mapping: Certain soil features, including soil texture, have the potential to serve as indicators of soil quality, hence exerting an impact on crop yield. Therefore, the use of thermal images acquired by UAVs could facilitate the quantification of soil texture at a regional scale. This can be achieved by detecting the variations in land surface temperature within a reasonably uniform climate condition.

Residue Cover and Tillage Mapping: The use of crop leftovers plays a crucial role in soil conservation as it serves to establish a layer of protection on agricultural fields, effectively safeguarding the soil from the erosive forces of wind and water. The precise evaluation of crop residue is essential for the effective execution of conservation tillage methodologies. The authors in conducted a study wherein they provided evidence that airborne thermal pictures had the potential to identify over 95% of the variation in crop residue cover quantity, in contrast to the use of visible and near infrared images which accounted for 77% of the variation.

Field Tile Mapping: Tile drainage systems are used to effectively eliminate surplus water from agricultural fields, hence offering both economic and ecological advantages. Efficient monitoring of tile drains may enhance the ability of farmers and natural resource managers to effectively avoid any negative environmental and economic consequences. Thermal imagery taken from UAVs provide further possibilities in the mapping of field tiles by means of sensing temperature variations within a certain area.

Crop Maturity Mapping: UAVs have shown to be a viable technological solution for monitoring crop maturity in order to identify the optimal time for harvesting, especially in cases when the whole cultivation area cannot be harvested within the available timeframe. For example, visual and infrared photos obtained by UAVs from barley trial sites located in Lundavra, Australia, were used to delineate two key development phases of barley. The analysis of these images yielded a classification accuracy of 83.5%.

Crop Yield Mapping: Accurate and timely assessment of crop production is essential for farmers due to several factors, such as crop insurance, strategic planning of harvest and storage needs, and effective cash flow management. In a study conducted by researchers, photographs taken from UAVs were used to assess the yield and overall biomass of rice cultivation in Thailand. Images taken by UAVs were used to forecast maize grain yields throughout the initial to intermediate phases of crop development in Germany. A range of sensors used in the context of UAV-based precision agriculture. These sensors include digital camera, multi-spectral camera, hyperspectral camera, thermal camera, spectrometer, 3D camera, LiDAR, SONAR.

Challenges with UAVs in Precision Agriculture: In reality, several challenges are incorporated with UAVs including but not limited to sensors, impact of adverse weather, payload capacity, and flight time.

Suboptimal while Expensive Thermal Sensors: Thermal cameras have suboptimal resolution capabilities and are characterized by a relatively high cost. The cost of thermal cameras varies between $2000 and $50,000, contingent upon their quality and usefulness. Furthermore, most of the thermal cameras possess a resolution of 640 pixels by 480 pixels. Also, thermal images can be affected by atmospheric moisture, shooting distance, and different sources of emitted and reflected thermal radiation. Hence, the calibration of aerial sensors plays a crucial role in obtaining surface temperatures of objects that are scientifically valid. The temperature measurements obtained by aerial sensors may be influenced during the observation at various phases of crop development. During the first stages of the plant growth cycle, measurements of temperature may be subject to the impact of reflectance radiating from the surface of the soil.

Bad Weather Conditions: The deployment of UAVs in precision agriculture applications poses a significant difficulty in the bad weather conditions, including but not limited to strong winds, heavy rainfall, severe storms, and dense fog. Under such circumstances, UAVs may encounter operational challenges that affects the successful completion of their assigned tasks of collection of data under such circumstances.

Limited Payload and Short Flight-Time: Another challenging issue is that lightweight UAVs can't carry heavy payloads. This makes it hard for UAVs to carry an integrated system with many sensors such as high-resolution cameras and thermal cameras at once. UAVs often suffer from limited battery capacity, typically spanning less than 60 minutes. Hence, the power constraints associated with UAVs provide a significant obstacle in their use and to cover large regions, frequent return trips to the charging station are required in order to replenish their power reserves.

10.6.5 Delivery and Logistics

In the context of a UAV-based goods delivery system, an UAV has the ability to traverse the distance between a designated pick-up point and a designated delivery location. The UAV is outfitted with a control processor and a Global Positioning System (GPS) module. The system gets a transaction packet corresponding to the delivery operation, which includes the GPS coordinates and the identification of a package docking device linked to the order. Upon the arrival of UAV at the designated delivery location, the control processor verifies the correspondence between the identifier of the package docking device and the device identifier specified in the transaction packet. Subsequently, it proceeds to execute the package transfer operation and transmits a confirmation message to the individual or entity who initiated the order, signifying the successful completion of the operation. In the event that the identification of a package docking device located at the delivery site does not correspond with the device identifier specified in the transaction packet, the UAV communication components initiate the transmission of a request using a short-range network, such as Bluetooth or Wi-Fi. The request may include the device identification or network address of the package docking device. Assuming that the package docking device remains within the range of UAV communication, it is seen that the device, which has a network address, transmits a signal providing the address of a different place. Subsequently, the package docking device has the capability to communicate revised GPS coordinates to the UAV. The UAV is redirected to the revised destination by using the modified GPS coordinates.

10.6.6 Communication Infrastructure Support

UAVs has the capability to carry different types of communication payloads to

UV Communications for UAV Networks: Ultraviolet (UV) communication is resistant to obstruction or shadowing thanks to strong molecule and aerosol scattering at these wavelengths. The influence of noise from the environment in the UV-C sub band (200–280 nm) is minimal due to ozone absorption in higher atmospheric layers. Therefore, a UAV outfitted with a UV transceiver with a wide field of vision may communicate successfully in a NLoS situation. In [19], the authors examine how well a ground network of sensors uplink performs when a UAV regularly patrols the region to collect data from serial sensing nodes. The required signal is sent by the sensor that is nearest to the flying UAV, while received signals from other nodes are considered as interference.

FSO Communication for UAV: The use of LoS FSO communication has gained popularity due to the use of unlicensed point-to-point (PtP) narrow beams. The use of UAVs in conjunction with FSO technology enables the provision of wireless connection to distant areas where access to 3G or 4G networks is limited or nonexistent. The integration of terrestrial and aerial networks may be facilitated by the use of UAVs, which can provide last mile wireless connectivity to critical regions such as battlefields and disaster relief zones. These areas need high bandwidth and enhanced accessibility. As an example, Facebook intends to provide wireless access to distant regions by using solar-powered high altitude UAVs that use FSO communications. In situations when the employment of UAVs is neither feasible nor cost-effective, geostationary earth orbit as well as low earth orbit satellites may be used as an alternative means to provide wireless connection to terrestrial users via the utilization of FSO links. One of the primary obstacles encountered by UAV-FSO technology pertains to the increased likelihood of vertical FSO connection obstruction resulting from various meteorological factors. The authors in [20] present many methodologies to address this issue. One approach involves employing an adaptive algorithm that regulates the transmit power based on prevailing weather conditions. Additionally, this algorithm has the capability to modify other system parameters, such as the incident angle associated with the FSO transceiver, in order to counteract any degradation in the link. For instance, during adverse weather conditions, it is advisable to employ powerful vertical FSO beams, whereas, under favorable weather conditions, low-power beams may be utilized. The second approach involves using a system optimization algorithm to optimize the placement of the UAVs. This may include positioning the UAV below clouds in a geographical region with little turbulence.

In case of an intensity modulation/direct detection (IM/DD) FSO system, the photo detector (PD) is sensitive to variations in the strength of the optical signal it receives. The primary source of noise at the PD is shot noise resulting from background illumination. The channel coefficient between the FSO source and PD, h, is subject to many phenomena and may be represented as (10.1).

$$h = \eta \, h_p \, h_a \, h_g, \tag{10.1}$$

where η represents the responsivity of the PD and h_p, h_a, and h_g correspond to the atmospheric loss, atmospheric turbulence-induced fading, and geometry and misalignment losses (GML), respectively. Fluctuations (jitters) in the position and orientation of a hovering UAV adversely affect the reliability of the FSO link with the stationary node. These fluctuations arise from various sources, including dynamic wind load, random air disturbances in the surrounding atmosphere, and internal vibrations induced by the rotation of the UAV's propellers. Consequently, the performance of the tracking system, which is tasked with aligning the beam with the receiver lens at the stationary node, is directly impacted by these variances. The deterministic atmospheric loss, (h_p), over the FSO transmission channel distance (L) is caused by the absorption and scattering of light by particles that exist in the atmosphere can be written as (10.2)

$$h_p = 10^{-kL/10}, \tag{10.2}$$

where k is the weather dependent attenuation constant. Typically, k can be selected as $\{0.43, 4.2, 20, 42.2, 125\} \times 10^{-3}\ \text{m}^{-1}$ for clear air, haze, light fog, moderate fog, and heavy fog, respectively. Atmospheric turbulence is a stochastic phenomenon caused by variations in temperature and pressure of the atmosphere. Typically, turbulence conditions can be classified into three regimes based on the Rytov variance $\left(\sigma_R^2\right)$ as weak when $\sigma_R^2 < 0.3$, moderate when $0.3 \le \sigma_R^2 < 5$, and strong turbulence when $\sigma_R^2 \ge 5$. The log-normal distribution is used to model FSO systems for weak turbulence conditions and the Gamma-Gamma distribution is considered for moderate to high turbulence conditions. The pdf of h_a for moderate to strong channel conditions [21, Eq. (11)] can be expressed as

$$f_{h_a}(h) = \frac{2(\alpha\beta h)^{\frac{\alpha+\beta}{2}}}{\Gamma(\alpha)\Gamma(\beta)h} K_{\alpha-\beta}\left(2\sqrt{(\alpha\beta h)}\right), \tag{10.3}$$

where α and β are fading parameters of large- and small-scale fluctuations and can be expressed as (10.4) and (10.5), respectively, whereas $K_\nu(\cdot)$ is the modified Bessel's function of the second kind and order ν.

$$\alpha = \left[exp\left(\frac{0.49\sigma_R^2}{\left(1 + 1.11\sigma_R^{12/5}\right)^{7/6}}\right) - 1\right]^{-1}, \tag{10.4}$$

$$\beta = \left[exp\left(\frac{0.51\sigma_R^2}{\left(1 + 0.69\sigma_R^{12/5}\right)^{5/6}}\right) - 1\right]^{-1}. \tag{10.5}$$

Further, for weak turbulence condition, the pdf of h_a [21, Eq. (9)] can be expressed f as

$$f_{h_a}(h) = \frac{1}{h\sigma_a\sqrt{2\pi}} exp\left[-\frac{(Ln(h) - \mu_a)^2}{2\sigma_a^2}\right], \tag{10.6}$$

where μ_a and σ_a are mean and variance of the random atmospheric turbulence-induced fading (h_a). Further, due to the instabilities caused by wind flow and vibration in the UAV due to rotating propellers, misalignment occurs in between FSO transmitter and receiver. The probability density function (PDF) of resulting position and orientation fluctuations of the UAV [22, Eq. (20)] can be expressed as

$$f_{h_g}(h) = \frac{\bar{\omega}}{A_0}\left[\frac{h}{A_0}\right]^{\frac{(1+q^2)\bar{\omega}}{2q}-1} \times I_0\left[-\frac{(1-q^2)\bar{\omega}}{2q} ln\left(\frac{h}{A_0}\right)\right], \quad \forall\,(0 < h \leq A_0), \tag{10.7}$$

where $I_0(\cdot)$ is the zero-order modified Bessel's function of first kind. A_0 represents the highest proportion of collected optical power by the lens of the receiver when FSO transmitter and receiver are exactly aligned.

Intelligent Reflecting Surface Based Aerial V2X Network: UAV and intelligent reflecting surface (IRS) can be combined to create a UAV-enabled aerial IRS (UAIRS), which has enormous promise for facilitating V2X communication. UAIRS may accomplish a full-angle reflection and reduce the performance deterioration of a V2X network in a challenging propagation environment. Therefore, the attainable data rate can be increased by achieving signal phase synchronization at the receiver by real-time IRS reflecting coefficient adjustment. Additionally, it is possible to use an IRS-based spatial equalizer to reduce the inter-symbol interference brought on by multipath fading. In addition, UAV real-time position adjustment offers mobile vehicles continuous LoS connectivity. Additionally, UAIRS may provide intelligent coverage extension at the boundary of the BS or UAV coverage zone in accordance with the needs of the vehicle user [23].

10.7 Performance Analysis

Considering a communication scenario in which the receiver is situated on the ground and the transmitter is contained within a UAV. Additionally, it is assumed that the transmitter, receiver and their significant scatterers in their vicinity are in motion, whereas the two local regions of scatterers surrounding both transmitter and receiver are at considerable distance as demonstrated in Fig. 10.9. Therefore, it is anticipated that a received envelope in such a communication environment would be doubly scattered as a consequence of keyhole propagation. Furthermore, it is logical to hypothesize that shadowing is present as a result of obstacles, such as structures, positioned between the transmitter and the receiver. Double-scattered propagation therefore coexists with large-scale fading (or shadowing) in the channel model under consideration. The random fluctuations that cause shadowing are represented by the Inverse Gamma distribution. Therefore, the overall shadowing can be modeled with double inverse Gamma (DIG) distribution. Hence, the PDF of double shadowing [24, Eq. (6)]can be expressed as

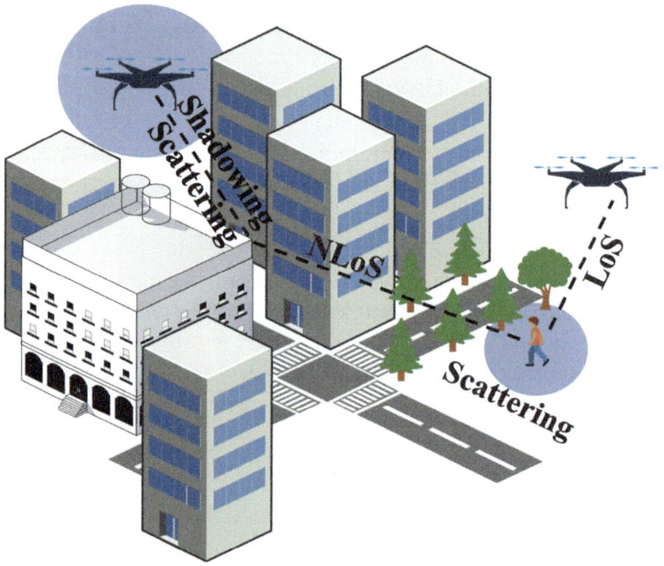

Fig. 10.9 UAV-based double shadowing and double scattering (DS-DS) system model

$$f_{DIG}(h) = \frac{2\bar{\gamma}^{\left(\frac{\alpha_1+\alpha_2}{2}\right)}}{\Gamma(\alpha_1)\Gamma(\alpha_2)} h^{-\left(\frac{\alpha_1+\alpha_2}{2}+1\right)} K_{\alpha_2-\alpha_1}\left(2\sqrt{\frac{\bar{\gamma}}{h}}\right), \tag{10.8}$$

where $\bar{\gamma}$ is the normalized average signal to noise ratio (SNR). Similarly, the random fluctuations that cause random scattering surrounding to the transmitter and receiver are considered to follow nakagami-m distribution, individually. Therefore, the overall scattering can be modeled with double Nakagami-m (DNm) distribution. Hence, the PDF of double scattering [24, Eq. (5)] can be expressed as

$$f_{DNm}(h) = \frac{1}{h\Gamma(m_1)\Gamma(m_1)} G_{0,2}^{2,0}\left[\frac{m_1 m_2 h}{\Omega^2}\middle| \begin{matrix} - \\ m_1, m_2 \end{matrix}\right], \tag{10.9}$$

where m_j is Nakagami-m distribution's shaping parameter, related to the severity of the fading, whereas Ω represents the mean square value. By substituting (10.8) and (10.9) in the expression of the total probability theorem [25, Eq. (4.80)], the cumulative distribution function (CDF) of SNR can be expressed as

$$f_{\gamma}(\gamma) = \frac{1}{\mathbb{S}_0\gamma} G_{2,2}^{2,2}\left[\frac{m_1 m_2 \gamma}{\bar{\gamma}}\middle| \begin{matrix} 1-\alpha_2, & 1-\alpha_1 \\ m_1, & m_2 \end{matrix}\right], \tag{10.10}$$

where $\mathbb{S}_0 = \Gamma(m_1)\Gamma(m_2)\Gamma(\alpha_1)\Gamma(\alpha_2)$ and $G^{m,n}_{p,q}\left[\cdot \middle| \begin{matrix} a_p \\ b_q \end{matrix}\right]$ represents the meijer-G function[26, Eq. (9.301)]. The outage probability (OP) can be represented as the probability of the output SNR falling below a certain threshold γ_T.

$$P_{out} = P_r\{\gamma \le \gamma_T\} = \int_0^{\gamma_T} f_\gamma(\gamma)\, d\gamma = F_\gamma(\gamma_T). \tag{10.11}$$

Therefore, the OP for DS-DS scenario can be calculated by substituting (10.10) in (10.11) as

$$P^{DSDS}_{out} = \frac{1}{\mathbb{S}_0} G^{2,3}_{3,3}\left[\frac{m_1 m_2 \gamma}{\bar{\gamma}} \middle| \begin{matrix} 1-\alpha_2, & 1-\alpha_1, & 1 \\ m_1, & m_2, & 0 \end{matrix}\right]. \tag{10.12}$$

The OP for UAV-based DS-DS system model is demonstrated in Fig. 10.10 against SNR. Improvement in OP is observed when fading parameters m_1 and m_2 are increased. Further enhancement is noted with the decrement in α_1 and α_2.

Average Symbol Error Rate: The average symbol error rate (ASER) is used in evaluating communication system performance, comparing modulation techniques, assessing signal processing algorithms, modeling channel behavior, designing of wireless communication systems, implementation of error correction codes. With the aid of CDF based approach [27, Eq. (3)], the ASER can be calculated by using

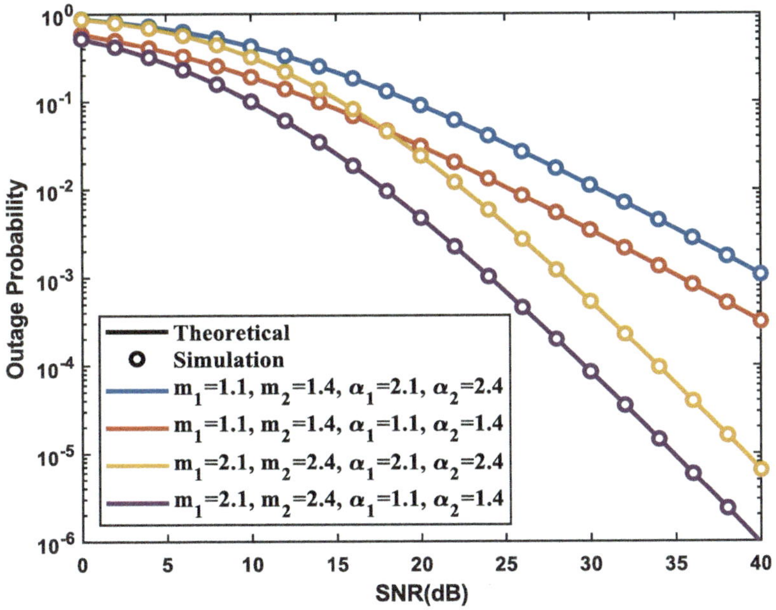

Fig. 10.10 Outage probability analysis for UAV-based DS-DS system

(10.13) if conditional ASER for any advanced modulation scheme, such as quadrature amplitude modulation (QAM) schemes, is provided.

$$P_s(e) = -\int_0^\infty P_s'(e|\gamma) F_\gamma(\gamma) d\gamma. \tag{10.13}$$

Therefore, considering the DS-DS scenario, the ASER can be calculated by substituting the conditional ASER of rectangular QAM (RQAM) scheme (10.14) and (10.12) into (10.13) as expressed in (10.15).

$$\begin{aligned} P_s'(e|\gamma) = &-\frac{ap}{\sqrt{2\pi\gamma}} e^{-\frac{a^2\gamma}{2}} + \frac{apq}{\pi\sqrt{2\gamma}} e^{-\frac{a^2\gamma}{2}} G_{1,2}^{2,0}\left[\frac{b^2\gamma}{2}\middle| \begin{matrix} 1 \\ 0,\ 0.5 \end{matrix}\right] \\ &-\frac{bq}{\sqrt{2\pi\gamma}} e^{-\frac{b^2\gamma}{2}} + \frac{bpq}{\pi\sqrt{2\gamma}} e^{-\frac{b^2\gamma}{2}} G_{1,2}^{2,0}\left[\frac{a^2\gamma}{2}\middle| \begin{matrix} 1 \\ 0,\ 0.5 \end{matrix}\right], \end{aligned} \tag{10.14}$$

where the modulation scheme related terms a, b, p and q are mentioned in [27].

$$P_s^{RQAM}(e) = p\varphi_1\left(\frac{a^2}{2}\right) + q\varphi_1\left(\frac{b^2}{2}\right) - \varphi_2\left(\frac{b^2}{a^2}, \frac{a^2}{2}\right) - \varphi_2\left(\frac{a^2}{b^2}, \frac{b^2}{2}\right), \tag{10.15}$$

where $\varphi_1(\cdot)$ contains an univariate Fox-H function and can be expressed as (10.16), whereas $\varphi_2(\cdot,\cdot)$ contains a bivariate Fox-H function and can be expressed as (10.17).

$$\varphi_1(x) = \frac{1}{\sqrt{\pi}\,\mathbb{S}_0} H_{4,3}^{2,4}\left[\frac{m_1 m_2}{x\,\bar{\gamma}}\middle| \begin{matrix} \nu_0,\ \nu_1,\ \nu_2,\ \nu_3 \\ \nu_4,\ \nu_5,\ \nu_6 \end{matrix}\right], \tag{10.16}$$

where $\nu_0 = (0.5, 1)$, $\nu_1 = (1-\alpha_2, 1)$, $\nu_2 = (1-\alpha_1, 1)$, $\nu_3 = (1, 1)$, $\nu_4 = (m_1, 1)$, $\nu_5 = (m_2, 1)$ and $\nu_6 = (0, 1)$.

$$\varphi_2(x, y) = \frac{pq}{\pi\,\mathbb{S}_0} H_{1,0:1,2:3,3}^{0,1:2,0:2,3}\left[x, \frac{m_1 m_2}{y\,\bar{\gamma}}\middle| \begin{matrix} 0.5,\ 1,\ 1 \\ \text{—} \end{matrix} \middle| \begin{matrix} \nu_3 \\ \nu_6,\ \nu_0 \end{matrix} \middle| \begin{matrix} \nu_1,\ \nu_2,\ \nu_3 \\ \nu_4,\ \nu_5,\ \nu_6 \end{matrix}\right]. \tag{10.17}$$

Theoretical and simulated ASER curves against SNR for the 16-SQAM scheme are presented in Fig. 10.11 over DS-DS channels. These curves account for various composite fading parameter values. Simulated curves for every investigated case are also in agreement with the theoretical curves, providing validation of the accuracy of the theoretical analysis. Additionally, the ASER performance improves as the fading parameters m_1 and m_2 are increased and/or the shadowing parameters α_1 and α_2 are diminished. However, m_1 and m_2 has a greater impact than α_1 and α_2. An observation can be made that the lower-order QAM scheme may be more favorable than the higher-order QAM scheme in regions with low SNR and at high altitude. The higher-order QAM scheme may be more favorable than the lower-order QAM

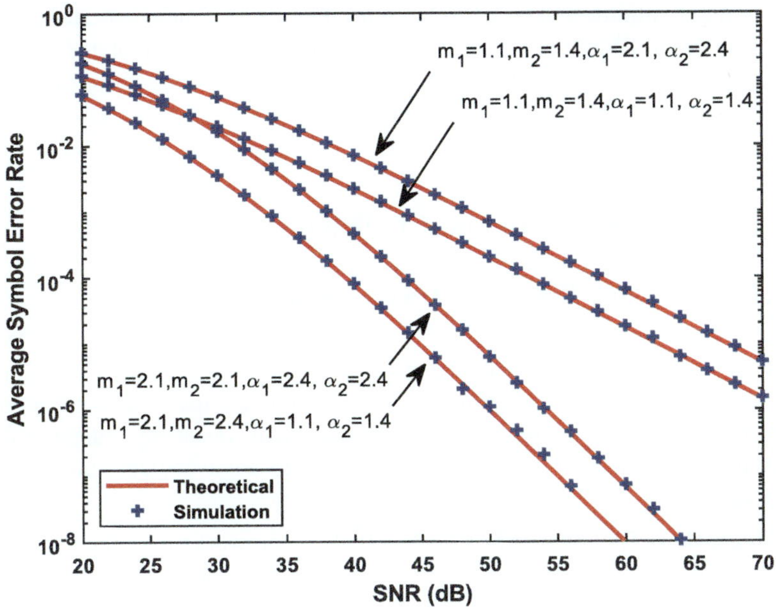

Fig. 10.11 ASER performance for 16-SQAM

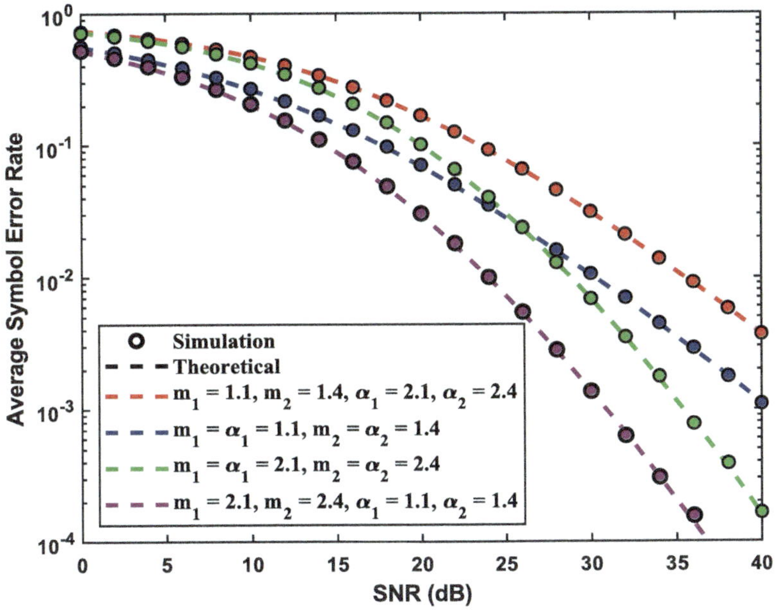

Fig. 10.12 ASER performance for 4×2-RQAM

scheme in low-altitude and high SNR regions. Figure 10.12 demonstrates the impact of SNR on the ASER of the system considering 4×2-RQAM. Simulated and theoretical curves are overlapping properly indicating further the correctness of theoretical ASER expression.

10.8 Future Directions and Challenges

10.8.1 Emerging Trends in UAV Communication

UAVs are currently being employed for a wide range of communication applications, extending beyond the realm of 5G and even encompassing satellite communications. This evolving utilization underscores their versatility, adapting to various communication paradigms and contributing to advancements in connectivity technology.

5G Integration: The integration of UAVs with 5G networks is on the rise, enabling high-speed data transmission and low latency communication. This trend enhances real-time control, data streaming, and remote piloting of UAVs, making them more responsive and capable of handling complex tasks.

Swarm Communication: Emerging trends focus on enabling UAVs to communicate and collaborate in swarms. This allows for coordinated actions, improved coverage, and data sharing among multiple drones, enhancing their efficiency in tasks like search and rescue missions, surveillance, and environmental monitoring.

Satellite Connectivity: UAVs are increasingly utilizing satellite communication to extend their operational range and coverage, especially in remote or challenging environments. This trend facilitates global connectivity, enabling UAVs to operate in areas with limited terrestrial infrastructure.

Edge Computing: UAVs are incorporating edge computing capabilities, processing data closer to the source rather than sending it to a remote server. This reduces latency and improves real-time decision-making, enhancing UAV capabilities in tasks like autonomous navigation, environmental analysis, and disaster response.

10.8.2 Open Research Challenges and Opportunities

Numerous open research challenges and opportunities lie within the realm of UAV applications. These encompass a diverse range of unexplored areas where UAV technology can be further refined and innovatively applied. Addressing these challenges presents the prospect of unlocking new dimensions of potential in UAV utilization.

Autonomy and Artificial Intelligence: UAVs are becoming more autonomous and capable of making decisions in real-time. This is facilitated by advancements in AI and machine learning (ML) technologies. Researchers are focusing on developing algorithms that allowed UAVs to navigate complex environments, avoid obstacles, and even collaborate in swarms for improved efficiency.

1. **Swarm Intelligence:** Researchers are exploring the concept of swarm intelligence, where multiple UAVs could work together in a coordinated manner to achieve tasks that would be difficult for a single UAV to accomplish. This trend is particularly promising for applications like environmental monitoring, disaster response, and agricultural surveys.
2. **BVLoS Operations:** Enabling UAVs to operate beyond the operator's visual line of sight is a significant area of research. Regulatory bodies are intended to work on guidelines for safe BVLoS operations, which would open up new possibilities for applications such as long-distance delivery, pipeline inspection, and large-scale mapping.

Energy Efficiency and Battery Technology: Further investigation is necessary to enhance the battery lifetime of UAVs, hence enabling extended distances and increased duration of UAV flights [28]. Some key factors are discussed below:

1. **Energy Density Enhancement:** Research should focus on developing batteries with higher energy densities. Lithium-ion batteries have been dominant, but exploring new materials such as lithium-sulfur or lithium-air batteries could potentially offer much higher energy storage capacities. Combining batteries with other power sources, such as solar cells or fuel cells, can extend UAV endurance. Optimizing the integration of these hybrid systems can also be explored for maximum efficiency.
2. **Fast Charging and Discharging:** Developing batteries that can be charged and discharged rapidly without compromising their overall lifespan is crucial. This would enable quick turnaround times for UAV missions.
3. **Safety Improvements:** UAV batteries must be designed with a strong emphasis on safety. Research should continue to develop battery chemistries and designs that are less prone to thermal runaway, short circuits, and other hazardous situations.
4. **Long Cycle Life:** Extending the lifespan of batteries is crucial to reduce operational costs. Research should focus on creating battery technologies that can withstand a high number of charge and discharge cycles without significant degradation.
5. **Temperature Tolerance:** UAVs often operate in various environmental conditions. Battery research should aim to develop batteries that can perform optimally across a wide temperature range without significant loss of capacity or efficiency. Efficient thermal management systems are crucial for maintaining battery performance and safety. Innovative cooling techniques should be explored to prevent overheating during operation.
6. **Flexible and Lightweight Designs:** UAVs often have strict weight limitations. Future research should focus on developing batteries that are not only energy-dense but also lightweight and flexible, allowing for more design flexibility in UAV construction.

7. **Smart Battery Management:** Developing advanced battery management systems that can accurately predict battery health, estimate remaining energy, and optimize charging and discharging based on real-time usage patterns can significantly enhance UAV mission planning and efficiency. Sustainable battery technologies are becoming increasingly important. Batteries should be developed with environmentally friendly materials, are recyclable, and have a reduced ecological impact.

Cyber Security: Denial of Service (DoS) assaults and hijacking attempts, including those using signal spoofing, are widely recognized as the most prominent risks inside UAV systems. Numerous DoS attacks have been identified and substantiated as significant sources of availability disruptions in UAV systems. Furthermore, the manipulation of signals via hijacking assaults has the potential to significantly disrupt the operations of certain UAV systems. Potential solution options for addressing the issue at hand include:

1. Implementation of robust authentication mechanisms such as trust platform module or Kerberos.
2. Deployment of signal distortion detection techniques.
3. Utilization of direction-of-arrival sensing methods to ascertain the direction of the transmitter antenna.

10.8.3 Regulatory and Policy Considerations

Several regulatory bodies have conducted research on the rules pertaining to UAV activities. The entities included under this category consist of the American Society for Testing and Materials (ASTM), the Low-Altitude Authorization and Notification Capability (LAANC), the Federal Aviation Administration (FAA), the Federal Communications Commission (FCC), etc. It is important to carefully analyze and evaluate the specific information provided.

The Federal Communications Commission (FCC): The Bureau of Engineering and Technology's Wireless Telecommunications division within the FCC produced a report in accordance with Section 374 of the FAA Reauthorization Act of 2018 (issued in August 2020). This report, submitted to the House of Representatives' Commerce, Science, and Transportation and Energy and Commerce Committees, seeks regulations concerning designated frequency ranges and challenges to implementation. Firstly, it examines whether operation of UAVs should be allowed on the spectrum recommended for aeronautical mobile service and control links in 2007 (L-band, 960–1164 MHz) and 2012 (C-band, 5030–5091 MHz), either on an unlicensed, shared, or exclusive basis. Furthermore, the report investigates potential technological, statutory, regulatory, and operational obstacles related to utilizing this spectrum.

Additionally, it suggests alternate frequency bands should the aforementioned frequencies prove unsuitable for UAV operations beyond visual line of sight. Overall, the FCC acknowledges the expanding landscape of UAV operations and thus endorses spectrum allocation to accommodate this technology and its prospective advantages. It urges addressing spectrum requirements for UAVs to facilitate command-and-control links, telemetry, payload, and other communications. The report proposes the suitability of the unoccupied 5030–5091 MHz band and versatile-use spectrum bands, albeit subject to certain technical and regulatory considerations that necessitate further assessment before permitting UAV operations. Moreover, the FCC considers the congested 960–1164 MHz band vital for aeronautical navigation, thus rendering the deployment of UAVs within this range challenging. It expresses reservations about potential repercussions on existing users within the 960–1164 MHz band. Consequently, it suggests that the Commission initiate a regulatory process to formulate service and licensing rules that would enable utilization of the 5030–5091 MHz band for UAVs.

The Federal Aviation Administration (FAA): The UAV Identification and Tracking Aviation furnishes some recommendations to the FAA regarding remote identification and tracking technologies for UAVs. This initiative led to the establishment of the Remote Identification (RID), outlining feasible tracking technologies, including automatic dependent surveillance-broadcast, low-power direct RF, networked cellular, satellite, and flight notification with telemetry. The RID and tracking data are proposed to be conveyed through two methods: local direct broadcasting and network publishing information to an FAA-approved internet-based database. Notably, direct broadcasting requires no handshaking and compels public safety officials to equip themselves with compatible receivers for UAV-transmitted data. Conversely, network publishing permits public safety officials to access data from an internet-based database for registered UAV.

The internet-based database approach demands interoperability at the IP and application levels, rather than requiring compatible technologies. Transmitters merely relay data to the internet-based service(s), while clients connect to these services. This necessitates the FAA to employ the internet-based database infrastructure for seamless integration. One method involves providing remote ID and tracking services through private UAS Service Suppliers (USS) tailored to UAV operations. This entails information exchange between operators, the USS, and the FAA, with the USS acting as the primary interface. This approach requires the FAA to gather telemetry information on diverse operations, raising privacy concerns for stored information and imposing data usage and dissemination restrictions on USS.

Furthermore, the FAA and Department of Transportation (Dot) proposed regulations for Remote ID for UAV operations that encompass owners, operators, designers, and developers. These rules include:

1. **ID Registration:** UAVs weighing less than 0.55lb are exempt from registration.
2. **RID Categories:** UAVs transmit ID information to a USS through internet connection (standard) or solely to USS without broadcast (limited).
3. **ADS-B Use Approval:** FAA prohibits ADS-B use without approval.

4. **Primary Functions for USS:** USS serves real-time RID sharing, ID access security, and informs UAV status to FAA using contractually established parameters.
5. **UAV Traffic Management (UTM)**: FAA envisions third-party provision of UTM services, a sector that currently does not exist.
6. **UAV Performance Requirements:** Requirements include location services, automatic USS connection, time marking, self-testing and monitoring, tamper resistance, connectivity, error correction, and message transmission.

The American Society for Testing and Materials (ASTM): The ASTM is dedicated to meeting the increasing requirements for identifying and tracking UAVs within airspace systems. This endeavor has culminated in the ASTM F38 RID standard, which enables both the general public and public safety authorities to discern the identity of an UAV through its designated ID. Importantly, this identification process upholds privacy, safeguarding personal information. The standard outlines the mechanisms by which UAVs can convey their designated ID, location, speed, and direction. This involves wireless IP-based broadcasting to a USS.

The Low-Altitude Authorization and Notification Capability (LAANC): The LAANC fosters the integration of UAVs into airspace through a partnership between the FAA and the industry. This initiative aids UAV pilots in obtaining access to controlled airspace below 400 ft. Furthermore, it imparts awareness regarding permissible and restricted zones for flight, enabling air traffic experts to monitor UAV activities within specific regions. LAANC streamlines the authorization process for airspace by cross-referencing multiple sources of data within the FAA UAV data exchange, encompassing UAV facility maps, special use airspace details, airport information, airspace classifications, temporary flight restrictions, and notices to airmen.

The Civil Aviation Administration of China (CAAC): The Civil Aviation Administration of China (CAAC) has released data specifications for the UAV cloud system. These specifications outline the reporting obligations for UAV flight information. Specifically, they require real-time reporting to the USS via mobile networks. The information to be reported includes the flight order ID, manufacturing ID, UAS ID, timestamp, flight duration, coordinates, speed, and path angle.

In regions with high population density, the reporting frequency is established at once per second to ensure detailed and up-to-date information. In areas with lower population density, the reporting interval extends to once every 30 seconds, striking a balance between data fidelity and network efficiency. This elevated reporting frequency necessitates consistent upkeep of the data links designated for this reporting purpose.

The Internet Engineering Task Force (IETF): The IETF, an organization dedicated to open standards, has introduced Drone Remote ID Protocol (DRIP), which facilitates remote identification and tracking of UAVs. This protocol encompasses various communication aspects, including the fundamental building blocks of the system and their interfaces. It defines two distinct categories of UAV remote identification. The first one is broadcast RID, which involves one-way transmissions directly from the UAV through Bluetooth or Wi-Fi. This type of transmission requires connectivity solely for the purpose of enabling observers to access information from the UAV registry. The second category is network RID, which involves the flow of data from the UAV to a network RID service provider using unspecified methods. This service provider then responds to queries from network RID observers, who specify their interest in particular airspace volumes.

The standard also provides an overview of the interoperability of UTM systems, as each UAV must be registered with at least one UTM system. With network RID, there exists a direct line of communication between the UAV and its corresponding UTM system. In contrast, when it comes to broadcast RID, the UAV operator has the option to predefine a 4D spatial volume, allowing the UTM system to have operational awareness. Alternatively, observers can contribute information about observed UAVs to a UTM system.

10.9 Conclusion

In conclusion, this chapter has provided a comprehensive overview of the integration of UAVs into wireless communication systems. Through discussions on UAV classification, attributes, and their role in wireless networks, we have highlighted the potential for UAVs to enhance connectivity and enable various applications across diverse industries. We have also addressed the communication challenges inherent in UAV systems, emphasizing the need for innovative solutions to overcome limitations such as limited bandwidth, NLoS operations, latency, and security vulnerabilities. Furthermore, we have elucidated different communication and control architectures, illustrating how UAVs can effectively exchange data and commands with ground stations and among themselves. The exploration of UAV communication applications has underscored their versatility in disaster response, surveillance, and agriculture, among other sectors. Additionally, the performance analysis conducted in this chapter has provided valuable insights into the reliability and efficiency of communication scenarios involving UAVs. Looking ahead, future research efforts should focus on addressing the identified challenges and exploring new opportunities for leveraging UAVs-assisted wireless communications. By continuing to innovate and develop robust communication systems, we can unlock the full potential of UAVs to revolutionize connectivity and enable transformative applications in various domains.

References

1. Bithas PS, Nikolaidis V, Kanatas AG, Karagiannidis GK (2020) UAV-to-ground communications: channel modeling and UAV selection. IEEE Trans Commun 68(8):5135–5144
2. Zeng Y, Zhang R, Lim TJ (2016) Wireless communications with unmanned aerial vehicles: opportunities and challenges. IEEE Commun Mag 54(5):36–42

3. Mozaffari M, Saad W, Bennis M, Nam Y-H, Debbah M (2019) A tutorial on UAVs for wireless networks: applications, challenges, and open problems. IEEE Commun Surv & Tutor 21(3):2334–2360
4. Al-Hourani A, Kandeepan S, Jamalipour A (2014) Modeling air-to-ground path loss for low altitude platforms in urban environments. In: IEEE global communications conference, pp 2898–2904
5. Stewart J (2014) Google tests drone deliveries in project wing trials. BBC World Service Radio
6. Gettinger D, Michel AH (2015) Drone sightings and close encounters: an analysis. Center for the Study of the Drone, Bard College, Annandale-on-Hudson, NY, USA
7. Sawalmeh AH, Othman NS (2021) An overview of collision avoidance approaches and network architecture of unmanned aerial vehicles (UAVs). arXiv:2103.14497
8. Zeng Y, Wu Q, Zhang R (2019) Accessing from the sky: a tutorial on UAV communications for 5G and beyond. Proc IEEE 107(12):2327–2375
9. Grover J, Jain A, Chaudhari NS (2017) Unmanned aerial vehicles operated emergency ad hoc networks, pp 92–96
10. Belso Z, Szilagyi T, Pap L, Elek K, Koller I (2011) Joint application of spread spectrum and ofdm modulation for microwave radio communication used for unmanned aerial vehicle. In: 2011 IEEE 73rd vehicular technology conference (VTC Spring), pp 1–5
11. Feng Q, McGeehan J, Tameh EK, Nix AR (2006) Path loss models for air-to-ground radio channels in urban environments. In: 2006 IEEE 63rd vehicular technology conference, vol 6. IEEE 2006, pp 2901–2905
12. Khawaja W, Ozdemir O, Guvenc I (2017) UAV air-to-ground channel characterization for mmwave systems. In: 2017 IEEE 86th vehicular technology conference (VTC-Fall), pp 1–5
13. Yanmaz E, Kuschnig R, Bettstetter C (2011) Channel measurements over 802.11a-based UAV-to-ground links. In: 2011 IEEE GLOBECOM workshops (GC Wkshps), pp 1280–1284
14. Jiménez A, Seco F (2016) Comparing decawave and bespoon UWB location systems: indoor/outdoor performance analysis. In: International conference on indoor positioning and indoor navigation (IPIN), pp 1–8
15. Schmidt JF, Neuhold D, Klaue J, Schupke D, Bettstetter C (2018) Experimental study of UWB connectivity in industrial environments. In: European wireless 2018; 24th European wireless conference, pp 1–4
16. Khawaja W, Guvenc I, Matolak D (2016) UWB channel sounding and modeling for UAV air-to-ground propagation channels. In: IEEE global communications conference (GLOBECOM), pp 1–7
17. Chen J, Raye D, Khawaja W, Sinha P, Guvenc I (2018) Impact of 3D UWB antenna radiation pattern on air-to-ground drone connectivity. In: 2018 IEEE 88th vehicular technology conference (VTC-Fall), pp 1–5
18. Aref M, Sikora A (2014) Free space range measurements with semtech lora™ technology. In: 2014 2nd international symposium on wireless systems within the conferences on intelligent data acquisition and advanced computing systems, pp 19–23
19. Tadayyoni H, Ardakani MH, Heidarpour AR, Uysal M (2022) Ultraviolet communications for unmanned aerial vehicle networks. IEEE Wireless Commun Lett 11(1):178–182
20. Alzenad M, Shakir MZ, Yanikomeroglu H, Alouini M-S (2018) FSO-based vertical backhaul/fronthaul framework for 5G+ wireless networks. IEEE Commun Mag 56(1):218–224
21. AlQuwaiee H, Yang H-C, Alouini M-S (2016) On the asymptotic capacity of dual-aperture FSO systems with generalized pointing error model. IEEE Trans Wireless Commun 15(9):6502–6512
22. Najafi M, Ajam H, Jamali V, Diamantoulakis PD, Karagiannidis GK, Schober R (2020) Statistical modeling of the FSO fronthaul channel for UAV-based communications. IEEE Trans Commun 68(6):3720–3736
23. Cao Y, Xu S, Liu J, Kato N (2022) Toward smart and secure V2X communication in 5G and beyond: a UAV-enabled aerial intelligent reflecting surface solution. IEEE Veh Technol Mag 17(1):66–73

24. Bithas PS, Nikolaidis V, Kanatas AG, Karagiannidis GK (2020) UAV-to-ground communications: channel modeling and UAV selection. IEEE Trans Commun 68(8):5135–5144
25. Papoulis A, Pillai S (2002) Probability, random variables, and stochastic processes. McGraw-Hill, New York, NY, USA
26. Gradshteyn IS, Ryzhik IM (2014) Table of integrals, series, and products. Academic
27. Dixit D, Kumar N, Sharma S, Bhatia V, Panic S, Stefanovic C (2021) On the aser performance of UAV-based communication systems for QAM schemes. IEEE Commun Lett 25(6):1835–1838
28. Gheisari M, Irizarry J, Walker BN (2014) Uas4safety: the potential of unmanned aerial systems for construction safety applications. In: Construction research congress. Construction in a Global Network, pp 1801–1810

Chapter 11
Machine Learning for Wireless Communications: Applications and Security

B. R. Manoj, Nayan Moni Baishya, and Harsh Raj

Abstract Machine learning (ML), a subfield of artificial intelligence, holds immense applications across various domains owing to its potential to analyze vast datasets and discern intricate patterns. By leveraging algorithms and statistical models, ML facilitates the extraction of actionable insights and predicts future outcomes based on historical data. This predictive capability underpins a multitude of applications, spanning computer vision, natural language processing, finance, healthcare, transportation, wireless communications, and beyond. Through sophisticated algorithms, such as neural networks (NNs), ML empowers systems to autonomously adapt and refine their behavior based on experience, driving innovation and efficiency in diverse fields. Consequently, ML is a cornerstone of modern technological advancements, offering unparalleled opportunities for data-driven decision-making, process optimization, and transformative breakthroughs. ML has been successful and is envisioned to play a pivotal role in revolutionizing wireless communication systems by offering innovative solutions to enhance system performance and efficiency, reliability, and security. With the proliferation of wireless devices and the exponential growth in data traffic, conventional communication system techniques face unprecedented challenges, such as managing network resources, complex wireless networks, unknown physical and signal models, mitigating interferences, and ensuring seamless connectivity.

In Sect. 11.1, we discuss different ML-based solutions developed for wireless communications. Specifically, we have considered two classification-based applications: signal detection and modulation classification of wireless signals, and two regression-based applications: wireless channel state information estimation and pre-

B. R. Manoj, Nayan Moni Baishya, and Harsh Raj have equally contributed.

B. R. Manoj (✉) · N. M. Baishya · H. Raj
Department of Electronics and Electrical Engineering, Indian Institute of Technology Guwahati, Assam, India
e-mail: manojbr@iitg.ac.in

N. M. Baishya
e-mail: nmb94@iitg.ac.in

H. Raj
e-mail: h.raj@iitg.ac.in

S. Sharma et al. (eds.), *Next-Generation Wireless Systems*,
Transactions on Computer Systems and Networks,
https://doi.org/10.1007/978-981-96-3185-8_11

diction of optimal power allocation. It was established that the integration of ML into wireless communication systems promises to unlock new capabilities, improve network performance, and pave the way for next-generation wireless technologies. However, one key challenge in deploying ML systems in real-world applications is the model complexity and high computational requirements. The edge devices, such as smartphones, drones, and Internet-of-Things systems, are limited by storage capacity and computational power. Therefore, optimizing the highly complex ML and deep learning models for efficient deployment in real-world edge applications is essential. In Sect. 11.2, we discuss three different techniques for model optimization, i.e., network pruning, model quantization, and knowledge distillation. For demonstration purposes, modulation recognition of wireless signals is considered the application of interest, and the effectiveness of the three methods to optimize the complex models for edge applications without compromising performance is established. Specifically, we evaluate the performance of the optimized ML methods in terms of sparsity, storage compression, and reduction in the number of parameters. We experimentally show that the network pruning method is beneficial to achieve high sparsity by removing the unimportant parameters of the ML model, and quantization helps in representing the model parameters using a smaller number of bits, e.g., 8 bits compared to the original high-bit representations, e.g., 32 or 64 bits. Finally, the knowledge distillation technique effectively transfers features learned by a large and complex ML model to a smaller model (i.e., reduced number of model parameters), which can help the smaller model achieve a performance comparable to the large model.

The successful emergence of ML in wireless communication applications has raised significant concerns about security-related challenges. Among many security challenges, one such is adversarial attacks. Section 11.3 presents the susceptibility of ML-based models against adversarial attacks for classification- and regression-based wireless communication applications. Firstly, we consider the joint problem of developing optimized ML models and robustification of these models against adversarial attacks, thus enabling efficient and reliable deployment of ML-based modulation classification of radio signals on edge devices. We propose utilizing two optimized models, knowledge distillation and network pruning, followed by computationally efficient adversarial training as a countermeasure technique against adversarial attacks to improve the robustness. Through experiments, we show that the proposed techniques significantly improve robustness when the ML models are optimized and adversarially trained compared to the unoptimized models. Secondly, we show that adversarial attacks can predict infeasible solutions for a regression problem in a wireless setting, that is, downlink optimal power allocation in a multi-cell massive multiple-input–multiple-output system. Further, we demonstrate the effectiveness of adversarial training as a countermeasure technique against these adversarial attacks and show that the robustness of DL-based wireless systems for regression application can significantly improve. We analyze the performance of the deep neural network models against adversarial attacks by considering the crafting of adversarial perturbations in both white-box and black-box manner.

Keywords Machine learning · Classification · Estimation accuracy · ML/DL-based wireless network · 6G and beyond

11.1 Machine Learning for Next-Generation Wireless Communications

11.1.1 Introduction

The development of next-generation wireless communication systems is driven by the demands of high data rates, low latency, and energy-efficient performance in complex wireless networks, and it has numerous applications, such as the Internet of Things (IoT), virtual reality, autonomous driving, and joint sensing and communications. However, integrating these state-of-the-art applications with wireless systems encounters various challenges involving: (a) modeling diverse channel types, (b) addressing various hardware imperfections, and (c) obtaining optimal signaling and detection schemes to provide high data transmission. Therefore, the next-generation wireless systems must be capable of complex and intelligent decision-making while simultaneously minimizing computational complexity. This requires communication systems to efficiently manage a large amount of wireless data, dynamically adapt to realistic environments, and meet the demands for effective signal processing and high-speed data rate transmission. Learning through real-time data could be a possible solution when the accurate physical and signal models are unknown. Machine learning (ML) is a well-known technique widely applied in different domains such as computer vision, medical diagnosis, natural language processing, speech processing, and market trading. The usefulness of ML appears through the extraction of meaningful information from large amounts of data while providing scalability to different use cases and reducing associated computational complexity. ML further allows for the development of generalized learning models to handle predictions and parameter estimations of new data with other statistical characteristics different from the training data.

There has been a growing interest in performing various signal-processing tasks in communication systems with neural networks (NNs), as it has gained significant success in various domains, namely, image processing and recognition, data science, speech and natural language processing, owing to the substantial recent increase in the availability of computational resources. Thus, ML has recently seen widespread adoption in advanced applications, from autonomous driving to healthcare, finance, and robotics. ML-based techniques have been extensively studied and reported to perform better than conventional model-based approaches of different domains. The success of ML in various domains and the need for more complex models to characterize future wireless networks has stimulated a surge of interest in adopting ML in wireless communication applications. However, knowing the precise signal model and time-varying channels is a crucial challenge for applying ML-based methods in wireless communications. Before addressing the potential benefits of ML in wireless communications, we first establish two methodologies prevalent in this field, and they assume a comprehensive understanding of the signal and the channel models in wireless communications. The two methods are:

- **Model-based approach**: It utilizes conventional probabilistic-based analytical models that represent the behavior and properties of wireless communications. These models are often built around the physical and statistical features of communication channels, antennas, propagation scenarios, and other essential features. Model-based approaches in wireless communications offer major advantages, such as accurate predictions of desired outputs, an intuitive understanding of the underlying physics of different channel fading, and the ability to facilitate the systems' design, optimization, and performance evaluation before deployment. Irrespective of these advantages, a model-based approach also has critical challenges, as developing models is a complex task that requires accurate characterization of future wireless networks and precise configuration of the signal model. Moreover, the models may not be scalable to unknown users' distributions in the network or unknown complex networks, which are often encountered in wireless systems.
- **Data-driven approach**: It uses real-world data to design and generalize learnable models to predict or estimate the desired outputs. The large datasets comprise wireless signal measurements, network traffic patterns, and user behavior, which are then processed using ML algorithms to determine potential statistical characteristics and extract valuable information. A data-driven method applies a parametric non-linear transformation where the parameters are learned through optimization of the decision-making criteria on a large dataset. This capability enhances communication systems' overall performance and efficiency, leading to improved decision-making processes. However, high-quality and diverse representative data are essential for the effectiveness of data-driven approaches. A significant problem in the data-driven approach is its black-box nature, where the designer often has difficulty knowing the intuitive reasoning about the success and failure of the algorithms, unlike the model-based approach in wireless communications, which relies on traditional analytical frameworks and expert system knowledge. However, in a wide range of applications within the wireless communications domain—such as modulation classification of wireless signals [1–4], optimal power allocation [5–8], and channel estimation [9], etc., data-driven approaches utilizing ML have demonstrated remarkable success in attaining optimal solutions. These ML solutions are shown to outperform conventional model designs, particularly when we do not have accurate signal and channel models. This is in sharp contrast to traditional model-based approaches, which depend on separate source and channel coding for optimal performance. Moreover, achieving optimality in wireless applications is challenging in real-world scenarios with limited resources and complex setups.

The conventional model-based approach, as shown in Fig. 11.1, solves a given problem by using probabilistic-based analytical methods to determine the desired output. In contrast, as depicted in Fig. 11.2, the data-driven approach considers input

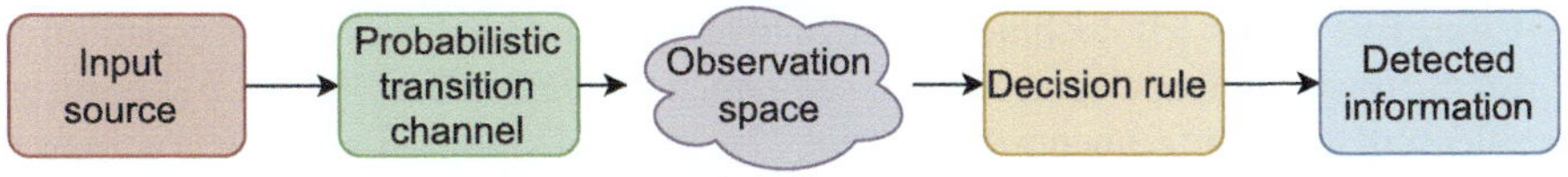

Fig. 11.1 Conventional model-based approach

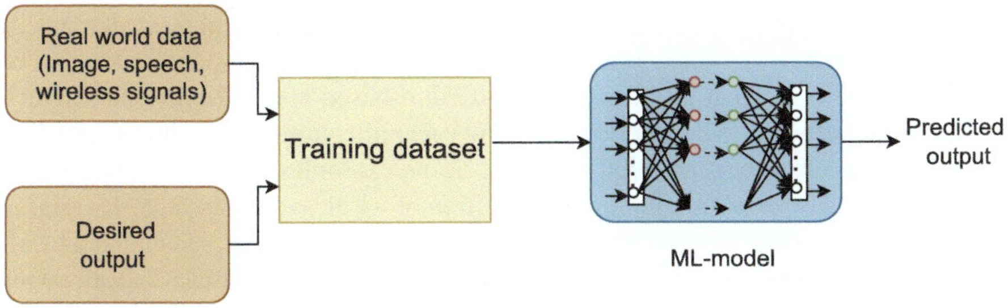

Fig. 11.2 Data-driven-based approach

data (e.g., image, speech, wireless signal) and the desired output data (which can be obtained from the model-based approach or as a synthetic dataset). In this approach, the ML model is trained over a large dataset to learn the feature representation and generalize to predict the output for any inputs not belonging to the training dataset (i.e., inputs not seen by the ML model before).

11.1.2 Machine Learning-Inspired Wireless Applications

ML enables communication systems to learn from large amounts of data and improve their performance without enhancing the computational complexity. In wireless communications, ML has been applied in several applications, namely:

- **Channel modeling**: ML offers a powerful approach to modeling the complex and dynamic behavior of time-varying wireless channels, accounting for factors like fading, interference, noise, and multipath propagation. This capability enables the development of adaptive modulation and coding schemes that match the channel condition, leading to improved performance of communication systems [10, 11].
- **Signal detection**: Signal processing techniques such as synchronization, equalization, demodulation, and decoding become more accurate and efficient with the utilization of ML. Furthermore, ML empowers systems to identify and counteract disruptive influences such as jamming, spoofing, and eavesdropping [9, 12].
- **Energy efficiency**: ML enhances the efficiency of wireless devices and networks through the optimization of parameters such as transmission power, modulation, coding, and resource allocation [5, 6, 8]. It also reduces the computation complexity for solving complex optimization problems by surpassing conventional model-based optimization techniques.

– **Cognitive radio**: ML techniques can enable dynamic spectrum access capabilities for cognitive radios. Cognitive radios provide efficient and fair spectrum sharing by selecting the best frequency band, time slot, and transmission mode. It is achieved by detecting spectrum occupancy, identifying primary and secondary users, and applying ML-based decision-making approaches [13, 14].

In general, depending on whether the desired response of a wireless system needs to be categorized or predicted as continuous values, wireless communications can be organized into classification-based and regression-based applications, respectively. Designing an efficient ML solution for wireless applications is possible depending on the classification or regression category. Some examples of classification-based wireless applications are modulation recognition of wireless signals [2, 3, 15], signal detection [9, 16–18], channel encoding and decoding [10, 19, 20], sensing and localization using radio signals [21–23], etc. The regression-based wireless applications include channel estimation [17, 24–26], optimal power allocation [5–8], localization [27], etc.

11.1.3 Machine Learning Security in Wireless Communications

ML-based approaches can solve classification and regression problems in wireless communication systems. However, the ML models have inherent vulnerabilities that raise privacy and security concerns for real-world applications. A malicious adversary can manipulate an ML model to provide incorrect outputs during the training and inference phases. During the training phase, a malicious adversary can poison the training data with incorrect labels or actual values, which, in turn, will affect and change the model parameters to make wrong predictions. During the inference phase, the test sample is perturbed by the adversary, which is not white noise but rather a well-crafted perturbation to exploit the inherent decision biases in an ML model. Due to the broadcast nature of wireless communications, it is possible to manipulate wireless signals over the air. Traditionally, jamming and eavesdropping attacks exploited this vulnerability of wireless systems. In a jamming attack, an adversary intentionally emits interfering signals to disrupt wireless communication between legitimate devices. In an eavesdropping attack, an adversary intercepts and listens to the communication between legitimate devices without their knowledge or consent. However, evaluating the security concerns of the ML models developed for wireless applications is still in an early stage of development compared to other domains, such as computer vision and speech processing. It is paramount to understand the attacks on ML models from a wireless communication standpoint and subsequently develop strategies to mitigate the vulnerabilities of the ML models. To this end, more sophisticated attacks, such as adversarial attacks, data poisoning, and membership inference attacks, are evaluated on ML models for wireless applications. It has been observed that these attacks are highly successful against state-of-the-art ML mod-

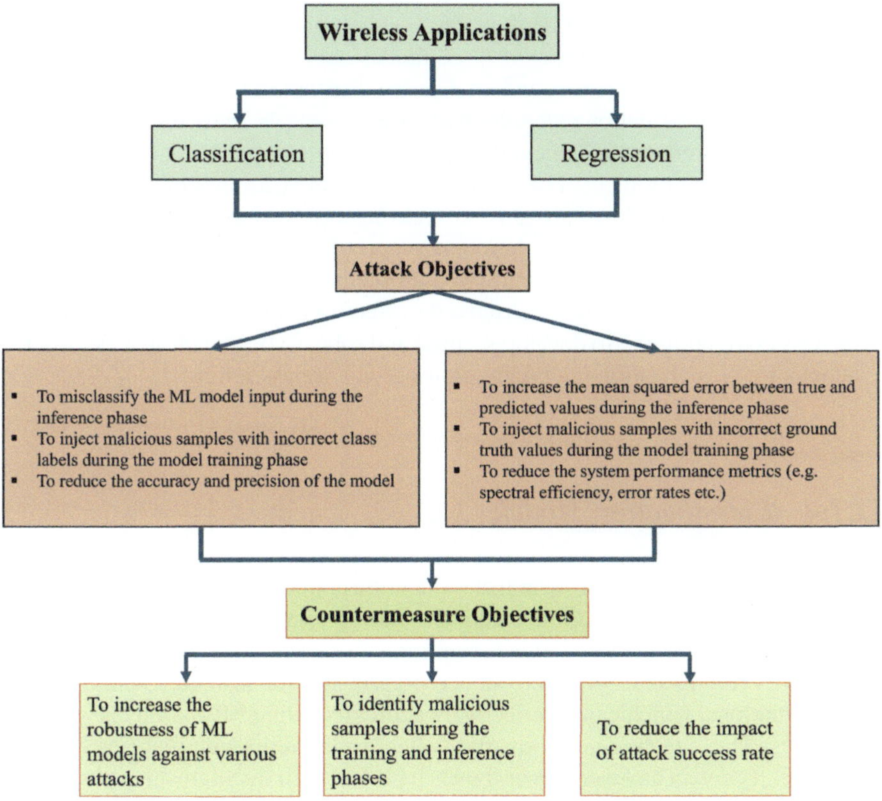

Fig. 11.3 Attack and countermeasure objectives for classification and regression-based ML models

els in both classification and regression applications. A key objective of this book chapter is also to discuss the vulnerabilities of ML applications and develop effective defense methodologies to mitigate them. Figure 11.3 shows the generic attack and countermeasure objectives for ML models developed for classification and regression applications in wireless communications.

In summary, a brief overview of the content discussed in the three sections is as follows:

- Section 11.1 delves into the various classification and regression solutions developed for wireless communication applications using ML-based approaches. Specifically, we discuss two classification problems, i.e., signal detection and modulation recognition of wireless signals, and two regression problems, i.e., channel estimation and power allocation in multiple-input–multiple-output (MIMO) systems.
- Section 11.2 presents the importance of model optimization for edge applications using ML models. Edge devices such as smartphones and drones are limited in terms of storage, computing, and power constraints. Therefore, optimization of

the ML models before their deployment is essential. For demonstration purposes, we establish the effectiveness of three model optimization techniques: network pruning, model quantization, and knowledge distillation for complex ML models developed for modulation recognition of wireless signals, reducing the storage and compute requirements significantly.

- Section 11.3 mainly focuses on evaluating the security aspects of ML models and providing strategies to improve robustness against various security threats. We consider the use cases as (i) a classification problem, i.e., modulation recognition of wireless signals, and (ii) a regression problem, i.e., downlink power allocation in massive MIMO systems. We discuss the effective countermeasure method based on adversarial training to develop secured and robust ML models against different attacks for classification- and regression-based wireless settings.

11.1.4 Classification Models

In wireless communications, several classification applications arise that demand the identification of signal types and categories through the analysis of distinctive features. Problems such as modulation recognition, as discussed in [4], along with tasks like channel coding and signal detection, represent specific instances within wireless communications, which can be effectively addressed using ML-based classification techniques. These techniques allow the system to learn through real-time data and automatically detect and categorize signals based on their distinguishing characteristics. Thus, an ML approach could provide optimal and generalizable solutions where an accurate communication system model is unavailable.

11.1.4.1 Symbol Detection

The objective of the optimal receiver design in a wireless communication system is to determine the symbols transmitted through the channel that minimizes the probability of symbol error. This process is known as symbol detection. Conventionally, in model-based methods, the maximum a posteriori probability (MAP) and maximum likelihood detection rules are utilized to detect the symbols. Model-based conventional receivers employ a fixed rule to detect symbols, whereas ML-based receivers learn these rules through real-time data. Broadly, there are two types of ML-based receivers for symbol detection: conventional deep neural networks (DNNs) and unfolded networks [28]. Symbol detection, an essential module of digital communication systems, implies that the symbols $\mathbf{s}$ transmitted through the physical channel need to be determined from the received signal $\mathbf{r}$. Such a symbol detection algorithm is presented in [19], which is a data-driven based model named ViterbiNet. ViterbiNet based on the Viterbi algorithm [29] uses labeled data to implement symbol detection.

In [19], the recovery process is shown for a symbol block consisting of t symbols $s[i] \in \mathcal{S}$ transmitted through a finite-memory stationary causal channel, where i denotes the time index $i \in \{1, 2, \ldots, t\} \in \mathcal{T}$ and $\mathcal{S}$ represents the constellation. Each symbol is uniformly distributed over a set of m constellation points. The output at time index i is given by $r[i] \in \mathcal{R}$. As the channel is causal and has finite memory, the output $r[i]$ is determined through a stochastic mapping $\mathbf{s}_{i-l+1}^{i}$, where l represents the channel memory which is less than t. The conditional probability distribution function (PDF) of the channel output given the input can be written as

$$f_{\mathbf{R}^t|\mathbf{S}^t}(\mathbf{r}^t|\mathbf{s}^t) = \prod_{i=1}^{t} f_{R[i]|\mathbf{S}_{i-l+1}^{i}}(r[i]|\mathbf{s}_{i-l+1}^{i}) . \tag{11.1}$$

The conditional PDF $f_{R[i]|\mathbf{S}_{i-l+1}^{i}}(r|\mathbf{s})$ does not depend on the time index i as the channel is stationary for each $r \in \mathcal{R}$ and $\mathbf{s} \in \mathcal{S}$. In the Viterbi algorithm [29], the transmitted constellation points are equiprobable; thus, the algorithm implements the maximum likelihood detector for a given output $\mathbf{r}^t$, which can be expressed [19] as

$$\hat{\mathbf{s}}^t(\mathbf{r}^t) = \arg\max_{\mathbf{s}^t \in \mathcal{S}^t} f_{\mathbf{R}^t|\mathbf{S}^t}(\mathbf{r}^t|\mathbf{s}^t) = \arg\min_{\mathbf{s}^t \in \mathcal{S}^t} -\log f_{\mathbf{R}^t|\mathbf{S}^t}(\mathbf{r}^t|\mathbf{s}^t) , \tag{11.2}$$

where $\hat{\mathbf{s}}^t$ is the predicted transmit symbol at time index i.

The ViterbiNet architecture incorporates the ML methods into the existing Viterbi detection algorithm, enabling it to learn through training and operate in a data-driven manner. The architecture is based on the receiver's assumption that the channel does not change over time and only depends on past and present input and a fixed channel memory of length l. The receiver also knows the symbol constellation $\mathcal{S}$. However, the receiver does not know the conditional PDF of the channel $f_{R[i]|\mathbf{S}_{i-l+1}^{i}}(r[i]|\mathbf{s})$. Further, by defining

$$c_i(\mathbf{s}) \triangleq -\log f_{R[i]|\mathbf{S}_{i-l+1}^{i}}(r[i]|\mathbf{s}), \ \mathbf{s} \in \mathcal{S}^l, \tag{11.3}$$

the maximum likelihood detector in Eq. (11.2) can be expressed as

$$\hat{\mathbf{s}}^t(\mathbf{r}^t) = \arg\min_{\mathbf{s}^t \in \mathcal{S}^t} \sum_{i=1}^{t} c_i(\mathbf{s}_{i-l+1}^{i}). \tag{11.4}$$

It can be observed from Eq. (11.3) that only the log-likelihood function $c_i(\mathbf{s})$ requires the channel state information (CSI). After evaluating $c_i(\mathbf{s})$ for every $\mathbf{s} \in \mathcal{S}^l$, the algorithm only requires the knowledge of the memory length l of the channel. Since the channel is stationary, $r[i] = r[k]$ and from Eq. (11.3), $c_i(\mathbf{s}) = c_k(\mathbf{s})$, for each $\mathbf{s} \in \mathcal{S}$ and $i, k \in \mathcal{T}$. Consequently, the log-likelihood function only relies on $r[i]$ and $\mathbf{s}$ and not on the time index i. The data-driven ML model, as shown in Fig. 11.4, considers $r[i]$ as the input, producing the log-likelihood estimate of $c_i(\mathbf{s})$ for every $\mathbf{s} \in \mathcal{S}^l$ as its output, where the log-likelihood estimate is denoted by $\hat{c}_i(\mathbf{s})$.

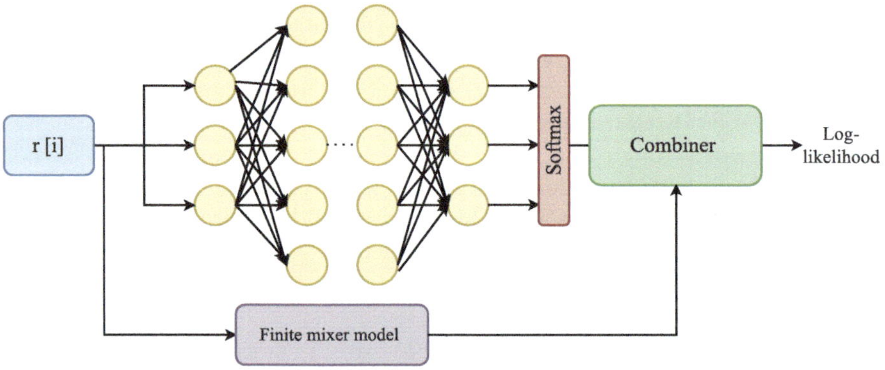

Fig. 11.4 Data-driven approach using ViterbiNet for log-likelihood computation

The ML-based approach is employed to determine the log-likelihood of $r[i]$ for a given $\mathbf{s}_{i-l+1}^{i} = \mathbf{s}$, that is, ML solution is used to predict $\hat{c}_i(\mathbf{s})$. As shown in Fig. 11.4, the output of the Softmax layer provides the estimate of $f_{\mathbf{s}_{i-l+1}^{i}|R[i]}(\mathbf{s}|r[i])$, which is denoted as $\hat{f}(\mathbf{s}|r[i])$; the finite mixer model provides a parametric estimate of the PDF of $r[i]$ [30], which can be obtained from the training data. The combiner combines the output of both Softmax and finite mixer model, which gives the estimated log-likelihood computations $\hat{c}_i(\mathbf{s})$. The Viterbi algorithm employs an ML-based algorithm to predict $c_i(\mathbf{s})$. Further, the ViterbiNet architecture, as shown in Fig. 11.5, utilizes the estimated log-likelihood computations $\hat{c}_i(\mathbf{s})$ for the maximum likelihood detector. Therefore, to implement the Viterbi algorithm in a data-driven manner, the computation of the log-likelihood is accomplished with an ML-based approach that learns to evaluate the cost function from the training data. That is, the input to the system is $r[i]$, and the output is $\hat{c}_i(\mathbf{s})$ for each $s \in \mathcal{S}^l$. The rest of the components of the algorithm, as shown in Fig. 11.5, remain unchanged; thus, using the learned $c_i(\mathbf{s})$, the detector implements the Viterbi algorithm to predict the input symbol $\hat{\mathbf{s}}^t$ for each $\mathbf{r}^t$.

In conclusion, the proposed approach in [19] integrates ML into the Viterbi symbol detection algorithm. This involves identifying computations that necessitate comprehensive knowledge of the basic channel input–output statistical relationships and substituting them with ML-based algorithms. The performance of the ViterbiNet receivers closely aligns with the optimal performance of equivalent CSI-based versions. The study highlighted in [19] also demonstrates the effectiveness of ML-based algorithms in operating successfully in the presence of CSI uncertainty, exhibiting minimal performance degradation. Thus, it highlights the significance of ML in the context of wireless communications.

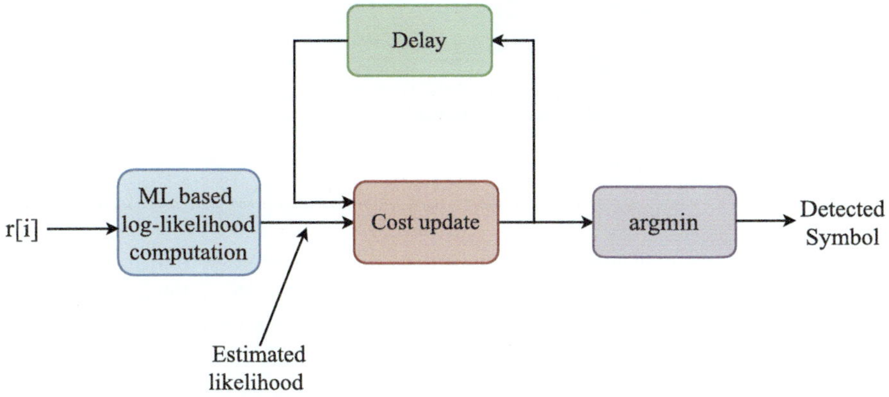

Fig. 11.5 ViterbiNet architecture

11.1.4.2 Modulation Recognition

Modulation recognition of radio-frequency (RF) signals is vital to communication systems. The modulation information of signals is essential for further signal demodulation and decoding in practical applications, such as signal recognition, cognitive radio, and spectrum monitoring. Modulation recognition determines the modulation scheme of a received signal, which can be an analog (e.g., AM-DSB, WBFM, etc.) or digital modulation (e.g., BPSK, PAM4, QAM16, etc.). Conventional modulation recognition methods can be categorized into decision theory-based and statistical ML-based methods, which are developed based on the knowledge of the channel model and its parameters. However, these methods have limitations regarding accuracy, robustness, complexity, and scalability. Designing becomes even more challenging when the channel model is unknown or time-varying. Therefore, ML-based methods have attracted significant attention in RF modulation classification to improve the performance of modulation recognition by learning the features from raw data (received wireless signal) directly without requiring prior knowledge of the channel model or its parameters. The ML-based models for modulation recognition are useful in applications such as spectrum sensing, cognitive radio systems, and adaptive modulation and coding schemes without any prior knowledge of channel statistics.

Modulation recognition (or classification of wireless signals) using ML is usually carried out in two steps, as shown in Fig. 11.6. The first step is to preprocess the

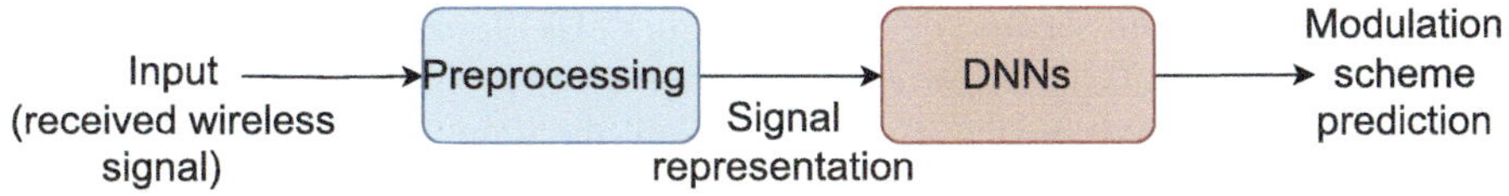

Fig. 11.6 Modulation classification architecture

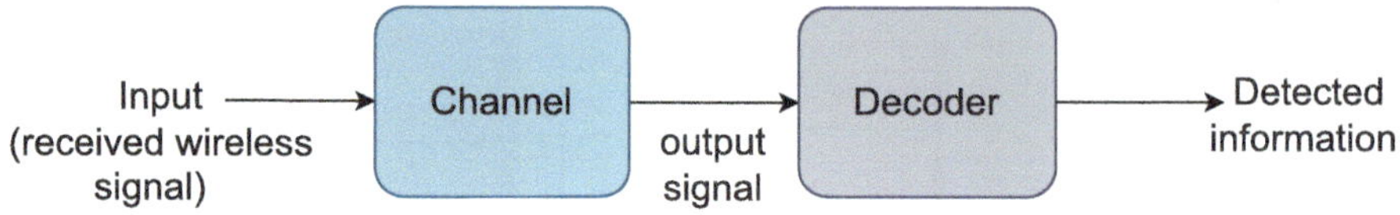

Fig. 11.7 General vector channel model

received wireless signal and then transform it into a suitable format to provide as input to ML-based models. The second step is to apply a DNN on the input signal to identify the modulation type as the output. The DNN transforms the input domain through linear mapping and utilizes a non-linear activation function to learn a non-linear pattern between a known input and the target output labels. The mathematical representation of the received signal in an additive white Gaussian noise (AWGN) channel, as shown in Fig. 11.7, is expressed as

$$\mathbf{r} = \mathbf{s}_n + \mathbf{z}\,, \tag{11.5}$$

where $\mathbf{r}$ denotes the received signal of dimension N, $\mathbf{s}_n$ denotes the transmitted signal chosen from the set of possible message $n \in \{1, 2, \ldots, N\}$, and $\mathbf{z}$ denotes the AWGN noise, which is independent and identically distributed (i.i.d.) with zero mean and variance $N_0/2$.

In the conventional approach, the receiver first observes $\mathbf{r}$ and then employs a decision function $g(\mathbf{r}) = \hat{n}$ to predict the transmitted signal. The probability of this decision being accurate corresponds to the likelihood that $\hat{n}$ is the correctly decoded transmitted message. Thus, the probability of a correct decision, given the received signal, can be expressed [31] as

$$\Pr\left[\text{correct decision}|\mathbf{r}\right] = \Pr[\hat{n}|\mathbf{r}] = \int \Pr[(\hat{n} = n)|\mathbf{r}]\, f(\mathbf{r})\, d\mathbf{r}\,, \tag{11.6}$$

where $f(\mathbf{r})$ denotes the PDF of $\mathbf{R}$. The objective is to design an optimal detector that minimizes the error probability or, equivalently, maximizes the probability of a correct decision $\Pr\left[\text{correct decision}|\mathbf{r}\right]$. Given that $f(\mathbf{r})$ is nonnegative for all $\mathbf{r}$, the right-hand side of Eq. (11.6) achieves maximum value when for each $\mathbf{r}$, the probability $\Pr[\hat{n}|\mathbf{r}]$ is maximized. The optimal detection rule is to favor the message $\mathbf{s}_n$ that maximizes $\Pr[\mathbf{s}_n|\mathbf{r}]$ upon observing $\mathbf{r}$ and it can be given [31] as

$$g(\mathbf{r}) = \hat{n} = \arg \max_{1 \leq n \leq N} \Pr[\mathbf{s}_n|\mathbf{r}]\,. \tag{11.7}$$

The optimal detection scheme described in Eq. (11.7) looks among all $\Pr[n|\mathbf{r}]$ for $1 \leq n \leq N$ and selects the n that maximizes $\Pr[n|\mathbf{r}]$. In probability theory, Eq. (11.7) is referred to as the MAP rule. By using Bayes' theorem, the MAP detection rule can be expressed as

$$\hat{n} = \arg \max_{1 \le n \le N} \frac{\Pr(n) f(\mathbf{r}|\mathbf{s}_n)}{f(\mathbf{r})} . \tag{11.8}$$

The PDF $f(\mathbf{r})$ is independent of n, implying it remains the same for all n, thus, Eq. (11.8) is equivalent to

$$\hat{n} = \arg \max_{1 \le n \le N} \Pr(n) f(\mathbf{r}|\mathbf{s}_n) , \tag{11.9}$$

where Eq. (11.9) is described in terms of the prior probabilities $\Pr(n)$ and the probabilistic description of the channel $f(\mathbf{r}|\mathbf{s}_n)$.

For the case when the messages are equiprobable a priori, that is, $\Pr[n] = \frac{1}{N}$, $\forall\ 1 \le n \le N$, the optimal detection rule simplifies to maximum likelihood receiver, which is given by

$$\hat{n} = \arg \max_{1 \le n \le N} f(\mathbf{r}|\mathbf{s}_n) , \tag{11.10}$$

where $f(\mathbf{r}|\mathbf{s}_n)$ is called the likelihood of message n. The maximum likelihood detector is not considered optimal unless the probabilities of the messages are equal. However, the maximum likelihood detector remains widely used, particularly when obtaining precise information about message probabilities is challenging. Further, substituting the expression for $f(\mathbf{r}|\mathbf{s}_n)$ in Eq. (11.10) is given by

$$\hat{n} = \arg \max_{1 \le n \le N} \left(\frac{1}{\sqrt{\pi N_0}} \right)^N \exp\left(- \frac{\|\mathbf{r} - \mathbf{s}_n\|^2}{N_0} \right) , \tag{11.11}$$

which can be expressed as

$$\hat{n} = \arg \min_{1 \le n \le N} \|\mathbf{r} - \mathbf{s}_n\|^2 . \tag{11.12}$$

It represents the minimum Euclidean distance between the transmitted signal and the received signal, such a detector is called a nearest-neighbor, or minimum-distance, detector. Moreover, in Eq. (11.12), the message signals are equiprobable; thus, the MAP and the maximum likelihood detector are the same.

The MAP, or the maximum likelihood detector, divides the output space $\mathbb{R}^N$ into N regions denoted as $D_1, D_2, \ldots, D_N$. If $\mathbf{r} \in D_n$, then the detector's decision function $\hat{n} = g(\mathbf{r}) = n$, indicating that the detector decides in favor of message n. The region D_n, where $1 \le n \le N$, is referred to as the decision region for message n, and it represents the set of all channel outputs that are mapped to message n by the detector. In the case of a MAP detector, the decision regions D_n are optimal and lead to the minimum possible error probability. For an MAP detector, the decision regions are determined by the following conditions [31]:

$$D_n = \{\mathbf{r} \in \mathbb{R}^N : \Pr[n|\mathbf{r}] > P[\tilde{n}|\mathbf{r}],\ \forall\ 1 \le \tilde{n} \le N \text{ and } \tilde{n} \ne n\}. \tag{11.13}$$

If for some **r**, two or more messages achieve the maximum a posteriori probability, in that case, **r** can be assigned to one of the corresponding decision regions. The classical detection approach is employed to detect the transmitted message only if the probabilistic description of the channel is known. Suppose the channel cannot be modeled precisely. In that case, an ML-based approach can be applied to learn the information pattern representation between the transmitted modulated signal and the received wireless signal. Therefore, the key to developing accurate ML models for modulation recognition is having a large-scale dataset of wireless signals modulated with various modulation schemes. It is also essential to incorporate various parameters in the dataset, such as signal-to-noise ratio (SNR) and channel conditions, to generalize the performance of ML models in realistic scenarios. To this end, several synthetic datasets are proposed for ML-based wireless classification, such as the RadioML (RML) 2016.10A (RML16) [15], RML 2018 (RML18) [3], and RML 2022 (RML22) [32]. We will now discuss different ML-based approaches developed for modulation recognition on two benchmark datasets, namely, RML16 and RML22.

11.1.4.2.1 *Performance on RML16 Dataset:* The first benchmark dataset proposed to develop ML models for modulation recognition of RF signals is the RML16 dataset [15], which was developed by DeepSig in 2016. It is a synthetic dataset generated with GNU Radio, consisting of 11 modulation schemes at varying SNRs. Specifically, it includes 8 digital, i.e., BPSK, QPSK, 8PSK, QAM16, QAM64, BFSK, CPFSK, PAM4 and 3 analog modulations, i.e., AM-DSB, AM-SSB, and WBFM. The overall dataset consists of 220,000 RF signals that are generated by considering 20 SNRs, ranging from -20 to 18 dB with a step of 2 dB. Each RF signal is complex valued of size 2×128 with in-phase (I) and quadrature (Q) components.

The matched-filter receiver is a common choice for radio communication systems. In this type of receiver, the incoming signal is matched with kernels designed to correspond to each transmitted symbol. This process boosts the signal strength by aligning the received signal with the correct symbol time, improving the SNR, which, in turn, helps to reduce impulsive noise in the receiver. Before this filtering step, the receiver typically recovers symbol timing and carrier frequency using moment-based estimators, which are calculated based on specific modulation and channel models. Convolutional neural networks (CNNs), a popular deep learning (DL) architecture, are often used in modulation classification because they can be trained to create matched kernels for various temporal features. These kernels are effective simultaneously at low and high SNR levels and collectively provide a strong foundation for classification. In [4, 15], it has been shown that a CNN has several advantages, such as local sensing, weight sharing, and shift invariance. CNNs make use of spatially local correlation by employing a local connectivity pattern between adjacent layers, and they share weights within each layer. An important assumption of CNNs is that the input data exhibits a local structure and is shift invariant. The RF signals of wireless communications comply with this assumption as there is a correlation between the information present at different sample or time positions. The first CNN-based method for modulation recognition was developed

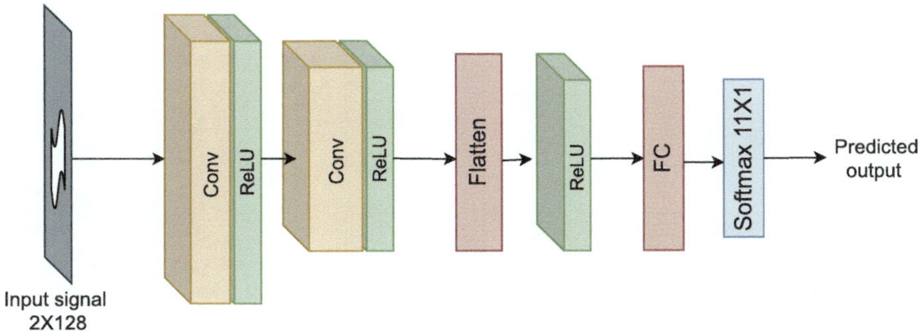

Fig. 11.8 VTCNN2 architecture—modulation classification for the RML16 dataset

by O'Shea et al. [15] and is presented in Fig. 11.8. The model was trained and tested on the RML2016.10A dataset to detect all 11 modulation schemes.

The CNN architecture in [33], referred to as VTCNN2, consists of two convolutional layers and two fully connected (FC) layers. The input to the model is the complex-valued RF signal of size 2×128. The first convolutional layer is composed of 256 kernels of size 1×3, and the second convolutional layer consists of 80 kernels of size 2×3. To maintain the feature map size, zero padding is utilized before every convolutional layer. The features obtained after the second convolutional layer are flattened and provided to the two FC layers with 256 and 11 neurons, respectively. The two convolutional layers and the first FC layer use the ReLU activation function. The final classification layer employs the Softmax activation function to compute the probabilities for the 11 modulations. It has been shown that the VTCNN2 model [33] achieves an accuracy close to 72% on the RML2016.10A dataset. As the pioneering work, the VTCNN2 model has demonstrated the effectiveness of the ML-based approach for recognizing modulation schemes of signals and obtained significant performance improvement compared to the conventional methods.

To further enhance the modulation recognition classification performance, a more accurate DL model has been proposed in [34], which utilizes a dual-stream structure to combine the strengths of both CNN and long short-term memory (LSTM) architectures and is referred to as the dual-stream CNN-LSTM model. The overall architecture of the dual-stream CNN-LSTM model is presented in Fig. 11.9, where the two streams are denoted as Stream 1 and Stream 2. The CNN architecture is mainly capable of learning the spatial correlations present in the data, i.e., the correlations among the I and Q components at different sample positions of a wireless signal. However, learning the long-range temporal dependencies among the sample points may not be well suited. To better represent the wireless signals, [34] has utilized the popular LSTM architectures to capture the temporal variations along with the spatial correlations in the signals. LSTM was originally developed to model sequential data, and an LSTM unit has two states that can retrain information from different time steps of a sequence, i.e., the cell state and the hidden state. Cell state mainly takes care of the long-term information, i.e., the information from the ele-

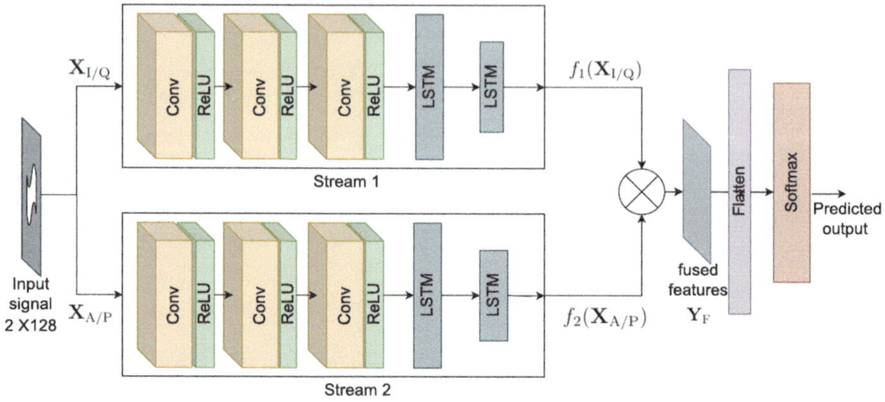

Fig. 11.9 Dual-stream CNN-LSTM architecture—modulation classification for the RML16 dataset

ments in the sequence far away from the current instant, whereas the hidden state updates itself with the recent information from the sequence. Within the LSTM unit, the forget gate allows the model to forget the unnecessary information from the cell state, and the input gate helps to add new information based on the input received at the current instant. The function of the LSTM layer can be defined as

$$H_t, C_t = \text{LSTM}(Z_c, H_{t-1}, C_{t-1}), \tag{11.14}$$

where Z_c is the sequence, H_t and C_t denote the hidden state and cell state at time t, respectively, and H_{t-1} and C_{t-1} represent the previous hidden state and cell state, respectively.

In the dual-stream CNN-LSTM model, each stream receives a different representation of the input RF signal: the I/Q and the amplitude/phase (A/P) representations. The RF signal $\mathbf{X}$ transmitted over orthogonal carrier waves can be represented as a function of time n as $\mathbf{X}[n] = \mathbf{X}_\text{I}[n] + j\mathbf{X}_\text{Q}[n]$, where $\mathbf{X}[n]$ is the baseband signal, and $\mathbf{X}_\text{I}[n]$ and $\mathbf{X}_\text{Q}[n]$ are the I and Q components of X[n], respectively. The amplitude $\mathbf{X}_\text{A}$ and phase $\mathbf{X}_\text{P}$ information in terms of I and Q components are, respectively, given by

$$\mathbf{X}_\text{A} = \sqrt{\mathbf{X}_\text{I}^2 + \mathbf{X}_\text{Q}^2} \quad \text{and} \quad \mathbf{X}_\text{P} = \arctan\left(\frac{\mathbf{X}_\text{Q}}{\mathbf{X}_\text{I}}\right). \tag{11.15}$$

The feature extraction module, which remains the same for both streams, is designed by stacking a CNN and an LSTM model, as shown in Fig. 11.9, to extract spatial and temporal features, respectively. The three convolutional layers at the beginning of each stream that is responsible for spatial feature extraction have 256, 256, and 80 kernels with sizes of 1×3, 2×3, and 1×3, respectively. ReLU is used as the activation function after each convolutional layer. The inputs to the convolutional layers are zero-padded to avoid a reduction in feature map size. The features extracted from the CNN layers are used as input to the LSTM layers. The model utilizes two LSTM layers containing 100 and 50 memory units, respectively. The features from the two streams are then fused using a matrix outer product operation to obtain a more diverse feature representation, which can be represented as

$$\mathbf{Y}_{\mathrm{F}} = f_1(\mathbf{X}_{\mathrm{I/Q}}) f_2(\mathbf{X}_{\mathrm{A/P}})^{\top}, \tag{11.16}$$

where $f_1(\mathbf{X}_{\mathrm{I/Q}})$ and $f_2(\mathbf{X}_{\mathrm{A/P}})$ represent the feature maps obtained from the two streams with $\mathbf{X}_{\mathrm{I/Q}}$ and $\mathbf{X}_{\mathrm{A/P}}$ as inputs, respectively, $(\cdot)^{\top}$ denotes transpose operation, and $\mathbf{Y}_{\mathrm{F}}$ denotes the final fused features. Flattening operation is performed on the fused features and provided as input to the Softmax classification layer. The model is trained and tested on the RML16 dataset to identify the 11 modulations, and it achieved an accuracy close to 90% at high SNRs, outperforming the VTCNN2 model by a significant margin. The dual-stream CNN-LSTM model has shown that learning diverse features through highly parameterized and data-driven DL-based methods is beneficial for achieving high performance in modulation recognition.

11.1.4.2.2 *Performance on RML22 Dataset:* Another recent benchmark dataset developed for wireless signal classification is the RML22 dataset [32]. The primary focus is to improve the data quality and to overcome the shortcomings of the RML16 dataset, such as errors and ad hoc choices of parameters. RML22 is more realistic and diverse than RML16 as it captures a wide range of channel conditions and hardware artifacts, and the order of artifacts in dataset generation is also corrected. Moreover, several other challenges faced with the RML16 dataset, such as incorrect modulation index for the GFSK modulation and erroneous clock effects on the phase-sensitive modulation types like M-ary PSK and M-ary QAM, are also being mitigated. While generating signals for the analog modulation types, the RML22 dataset efficiently modeled the impacts of inaccuracies in the information source on the data. Owing to the realistic nature of the dataset, developing accurate DL models for RML22 is essential for better modulation recognition performance of real-life applications. To this purpose, the dataset consists of 10 modulation schemes, where 8 are digital, i.e., BPSK, QPSK, 8PSK, PAM4, QAM16, QAM64, GFSK, and CPFSK, and 2 are analog, i.e., WBFM and AM-DSB. The samples for every modulation are generated using SNRs ranging from -20 to 20 dB with a step size of 2 dB. The dataset has 420,000 complex-valued RF signals, each of dimension 2×128.

The first CNN model for the RML22 dataset was proposed in [32] and is shown in Fig. 11.10. The model, which is referred to as the baseline CNN, comprises three convolutional and two FC layers. The three convolutional layers consist of 64, 64, and 32 kernels of size 3×3, respectively. The ReLU activation function is used after every convolutional layer, followed by downsampling of the feature maps using the max-pooling layer. The resulting feature maps from the last convolutional layer are flattened and fed into the FC layer with 128 neurons, also utilizing the ReLU activation function. The last layer, a Softmax classification layer with 10 neurons, computes the probabilities for each modulation class. The baseline CNN [32], after being trained on the RML22 dataset, can achieve an accuracy of 67%. As the dataset consists of signals with realistic and diverse characteristics compared to the RML16,

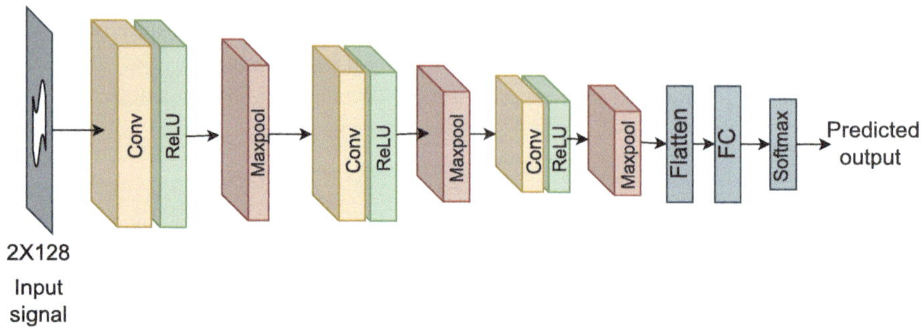

Fig. 11.10 Baseline CNN architecture—modulation classification for the RML22 dataset

achieving good classification performance is challenging. However, one drawback of the baseline CNN is that the architecture is fundamental in design, which can limit the learning capability of the model. Therefore, it is important to develop a more complex DL model that is known to have better signal representation capability. To realize this, we have considered the dual-stream CNN-LSTM model [34], originally developed for the RML16 and is illustrated in Fig. 11.9. We have then trained and tested the model on the RML22 dataset and observed that it can achieve an accuracy of around 90% at high SNR values. Therefore, as seen in the case of RML16, an architecture that can extract diverse features is far superior to the baseline CNN for the RML22 dataset. With this motivation, we design a complex DL architecture, namely, the quad-stream BiLSTM-Attention (QSLA) model, that can further enhance classification performance.

The QSLA model is superior in the sense that it is capable of extracting complementary features from diverse signal representations that are combined in an effective manner using Bidirectional LSTM (BiLSTM) and attention layers. The model is presented in Fig. 11.11, where the inputs to the network are $\mathbf{X}_{I/Q}$, $\mathbf{X}_{A/P}$, and the independent $\mathbf{X}_I$ and $\mathbf{X}_Q$ signals. The $\mathbf{X}_{A/P}$ signal can be obtained from the baseband signal $\mathbf{X}_{I/Q}$ signal as represented in Eq. (11.15). The feature extraction process can be divided into two blocks:

- **Spatial feature extraction**: The spatial features are extracted from the $\mathbf{X}_{I/Q}$, $\mathbf{X}_{A/P}$, $\mathbf{X}_I$, and $\mathbf{X}_Q$ signals in each stream using 1-D convolutional layers. The convolutional layers consist of 128 kernels of kernel size 3×3, followed by a ReLU activation function. To combine the uncorrelated features obtained from the orthogonal $\mathbf{X}_I$- and $\mathbf{X}_Q$-components, the features are first concatenated, and then a convolutional layer with 128 kernels of size 1×1 is applied. After fusing the features from all the streams, they are passed through two additional convolutional layers to capture high-level feature representations. Both the convolutional layers consist of 128 kernels of size 3×3. Max-pooling with a stride of 3 is performed on the final convolutional layer output to downsample the features.
- **Temporal feature extraction**: The features obtained from the spatial feature extraction block are processed further to learn the temporal correlations in the input

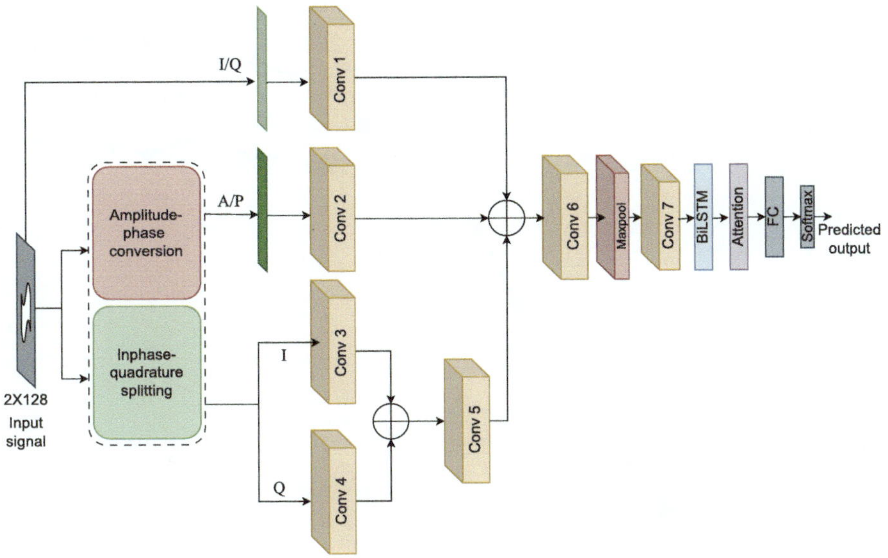

Fig. 11.11 Quad-stream BiLSTM-Attention (QSLA) network–modulation classification for the RML22 dataset

signal using a BiLSTM and an attention layer. We have already discussed how the LSTM layer works for sequence modeling while describing the dual-stream CNN-LSTM model for RML16. A BiLSTM layer is just two stacked LSTM layers, with one layer processing the input sequence in the forward direction and the other processing in the backward direction. The output at a time t combines both layers' hidden states at that instant. To learn the temporal features, the QSLA model utilizes a BiLSTM layer with 128 cells and tanh as the activation function. The output of the BiLSTM layer is then passed through an attention layer. The fundamental concept behind the attention mechanism is that it enables the model to concentrate on the most important input characteristics during prediction. The attention layer will determine the attention weights for each element in a sequence [35]. These weights guide the decision-making process to emphasize the most relevant features. In the QSLA model, the dot-product attention is used to calculate the attention weights of the resulting temporal features of the BiLSTM layer and is given by

$$a_t = \text{Softmax}(W_a \cdot H_t + b_a)\,, \tag{11.17}$$

where a_t represents the attention weight of the tth hidden state H_t, W_a and b_a are the learnable weights and biases of the attention layer, respectively, and Softmax is the activation function. The attention layer provides the most important features, which helps in achieving better classification accuracy with reduced complexity.

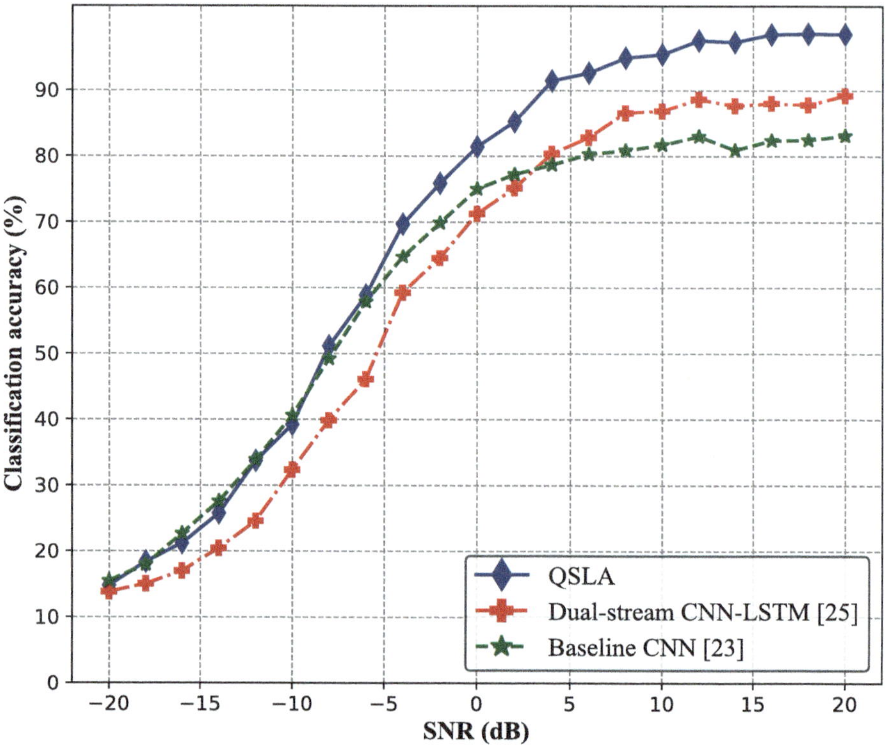

Fig. 11.12 Performance comparison of the QSLA model with the dual-stream CNN-LSTM and baseline CNN for the RML22 dataset

The output of the attention layer is flattened and passed through a Softmax classification layer with ten output classes to obtain the predicted probabilities for each class. Further, $\mathcal{L}_2$ regularization is used on the weights of this layer to avoid overfitting. Figure 11.12 compares the performance of the proposed QSLA with that of the dual-stream CNN-LSTM [34] and baseline CNN [32] for the RML22 dataset. It can be observed that the QSLA model can achieve superior accuracy than both models, obtaining an accuracy close to 99% at high SNRs.

11.1.5 Regression Models

In wireless communications, there are many scenarios where accurately predicting a continuous value based on an input signal is essential for the efficient operation of the system. This predictive capability is vital for complex network optimization, channel modeling, resource management, and seamless communication. In the last section, we discussed the success of utilizing ML models for classification tasks; however, in the literature, ML-based regression methods have proven effective in addressing wireless challenges that demand real-time continuous value prediction. To address this, we will focus on two important regression applications, channel estimation and optimal power allocation techniques, and discuss effective ML-based approaches as possible solutions.

11.1.5.1 Channel Estimation

Conventionally, channel estimation is carried out by using well-known techniques such as least squares (LS) and minimum mean square error (MMSE). For the purpose of exposition of conventional channel estimation techniques, we consider an orthogonal frequency-division multiplexing (OFDM) modulation system. OFDM is a modulation technique which has been widely employed in communication systems to address the problem of frequency-selective fading in wireless channels. In wireless communications, the signals are transmitted through a physical channel, where channel characteristics usually distort the received signal. The receiver must estimate and compensate the channel effect to recover the transmitted symbols. Generally, the receiver estimates the channel using pilot symbols; their positions and values in time frequency are known to both the transmitter and receiver. The conventional LS and MMSE techniques, i.e., pilot-based estimation methods, employ the pilot values in time-frequency grids to determine the unknown values of the channel response. The LS estimation method does not require any information about the channel statistics, whereas the MMSE estimation provides better performance by utilizing the information about the channel statistics and noise variance. DL-based communication systems have gained significant attention in improving the performance of various conventional algorithms, including channel equalization [36], CSI feedback [24], and channel estimation [17, 25].

We consider a single-input–single-output (SISO) system, that is, only one link exists between the transmitting and the receiving antennas. An OFDM subframe is considered of dimension $N_S \times N_D$ with k being the time slot index which is between $[0, N_D - 1]$ and the frequency subcarrier index range i is $[0, N_S - 1]$. In an OFDM system, for the kth time slot index and the ith frequency subcarrier, the input–output relationship can be represented as

$$R_{i,k} = H_{i,k} S_{i,k} + Z_{i,k} , \tag{11.18}$$

where the received signal and transmitted OFDM symbol are denoted as $R_{i,k}$ and $S_{i,k}$, respectively. The AWGN at the receiver is given by $Z_{i,k} \sim \mathcal{N}(0, \sigma^2)$. The time-frequency response of the channel is denoted by $\mathbf{H} \in \mathbb{C}^{N_S \times N_D}$, where $H_{i,k}$ is the (i, k)th element of $\mathbf{H}$. The matrix $\mathbf{H}$ consists of the channel response for all time slots and subcarriers. The time domain response of the channel with fading can be represented as $\mathbf{H} = [\mathbf{h}[1], \mathbf{h}[2], \ldots, \mathbf{h}[N_D]]$, where the channel frequency response at the kth time slot is $\mathbf{h}[k]$.

Least Squares (LS) Method: Let $\mathbf{s}_p$ and $\mathbf{r}_p$ denote the known pilot values and the corresponding observation values, respectively. In the LS method, the channel is estimated at the pilot positions. The estimated channel using the LS method as a diagonal matrix is denoted by $\mathbf{H}_p^{\text{LS}} \in \mathbb{C}^{N_P \times N_P}$, which can be obtained by solving

$$\hat{\mathbf{H}}_p^{\text{LS}} = \arg\min_{\mathbf{H}_p} \|\mathbf{r}_p - \mathbf{H}_p \mathbf{s}_p\|_2^2, \tag{11.19}$$

where $\|\cdot\|$ is the L_2-norm and $\hat{\mathbf{H}}_p^{\text{LS}} \in \mathbb{C}^{N_P \times N_P}$ is the estimated channel matrix. The resultant of the optimization problem in Eq. (11.19) is $\hat{\mathbf{h}}_p^{\text{LS}} = \text{diag}\left(\hat{\mathbf{H}}_p^{\text{LS}}\right) = \hat{\mathbf{r}}_p / \hat{\mathbf{s}}_p$. We must apply a two-dimensional interpolation method to find the channel value at the points other than the pilot positions.

Minimum Mean Square Error (MMSE) Method: The performance of the MMSE estimator is better than the LS method, as channel statistics and noise variance are considered in the MMSE method. The MMSE estimator is obtained by multiplying the LS estimates at the pilot-symbol positions with a filtering matrix $\mathbf{A}_{\text{MMSE}} \in \mathbb{C}^{N_L \times N_P}$ [37].

$$\hat{\mathbf{h}}_d^{\text{MMSE}} = \mathbf{A}_{\text{MMSE}} \hat{\mathbf{h}}_p^{\text{LS}}, \tag{11.20}$$

where the vectorized MMSE estimation of the channel response $\mathbf{H}$ at OFDM subframe d is given by $\hat{\mathbf{h}}_d^{\text{MMSE}} \in \mathbb{C}^{N_L \times 1}$ with $N_L = N_S \times N_D$.

The filtering matrix can be obtained in the sense of minimizing the mean square error (MSE) as

$$\zeta = \mathbb{E}\left[\|\mathbf{h}_d - \mathbf{A}_{\text{MMSE}} \hat{\mathbf{h}}_p^{\text{LS}}\|_2^2\right]. \tag{11.21}$$

The minimization of Eq. (11.21) results in

$$\mathbf{A}_{\text{MMSE}} = \mathbf{U}_{\mathbf{h}_d\mathbf{h}_p} \left(\mathbf{U}_{\mathbf{h}_p\mathbf{h}_p} + \sigma^2 \mathbf{I}_{N_P}\right)^{-1}, \tag{11.22}$$

where $\mathbf{U}_{\mathbf{h}_d\mathbf{h}_p} = \mathbb{E}\left[\mathbf{h}_d \mathbf{h}_p^H\right]$ is the channel correlation matrix between desired OFDM frame and pilot symbols, $\mathbf{U}_{\mathbf{h}_p\mathbf{h}_p} = \mathbb{E}\left[\mathbf{h}_d \mathbf{h}_p^H\right]$ is the channel correlation matrix of the pilot symbols, and $\mathbf{I}_{N_P} \in \mathbb{R}^{N_P \times N_P}$ is the identity matrix. The MMSE estimator is beneficial only when the correlation matrix of the channel is fully known. Some approaches in the literature have been considered to reduce the computation complexity of the scheme and use estimated channel statistics. For instance, in [38], an approximate linear version of the MMSE method for fast-fading channels was introduced. This approximation reduces the computational complexity compared to the original MMSE approach by minimizing the size of the correlation matrix and the filtering matrix.

The traditional channel estimation techniques rely on mathematical models, which may not effectively capture the intricate and dynamic nature of wireless channels, such as multipath propagation, fading, interference, and noise. Moreover, these methods are not suitable for real-time channel estimation, hindering their adaptability to changing channel conditions and the complexity of obtaining the channel correlation matrix. In contrast, DL models offer significant advantages by learning complex, non-linear mappings between the received signals and the transmitted signals. This enables DL models to adapt and generalize well to time-varying channel conditions, resulting in more accurate and reliable channel estimation. Furthermore, advancements in hardware and optimization techniques have facilitated the deployment of DL models on embedded or edge devices, allowing for efficient and timely adaptation to changing channel conditions. As a result, DL-based channel estimation holds immense promise for enhancing the performance and reliability of wireless communication systems in real-world scenarios.

In a DL-based channel estimation method, the objective is to estimate the time-frequency response matrix of the channel $\mathbf{H}$ using transmitted pilots. To achieve this, the DL model presented in [26] has utilized the concepts of image super-resolution [39] and image restoration [40]. Image super-resolution is the process of generating high-resolution (HR) images from low-resolution (LR) inputs while preserving as much detail and sharpness as possible. On the other hand, image restoration refers to improving an image's quality by removing various types of degradation that may have occurred during image acquisition or transmission. This degradation can include blurring, noise, compression artifacts, or other distortions that can compromise the visual clarity and fidelity of the image. In the case of channel estimation, the estimated channel using the pilots $\hat{\mathbf{H}}_p^{\mathrm{LS}} \in \mathbb{C}^{N_P \times N_P}$ can be considered as an LR and noisy image of the complete channel $\hat{\mathbf{H}} \in \mathbb{C}^{N_S \times N_D}$. Therefore, to obtain a better representation of $\hat{\mathbf{H}}$, the method in [26] has designed a two-stage approach: the first stage is image super-resolution on $\hat{\mathbf{H}}_p^{\mathrm{LS}}$ to obtain an HR image, and the second stage is image restoration, which helps to remove the noise from the HR channel matrix estimate $\hat{\mathbf{H}}$. The overall DL architecture, referred to as the ChannelNet, is shown in Fig. 11.13.

In practice, the ChannelNet considers the LR channel matrix $\hat{\mathbf{H}}_p^{\mathrm{LS}}$ as a composite of two 2D images: one image consists of the real values and the other image consists

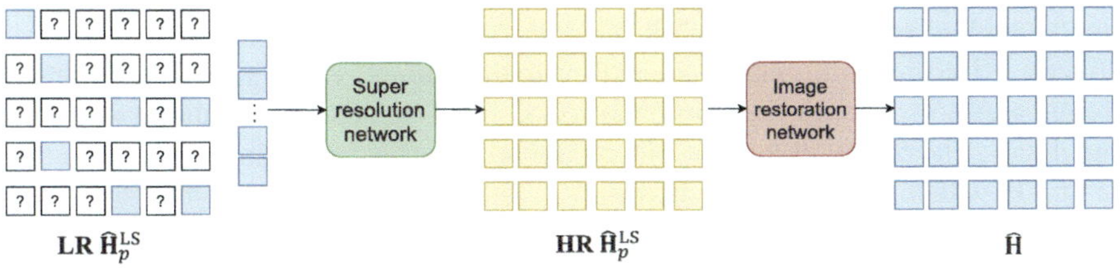

Fig. 11.13 ChannelNet DL architecture

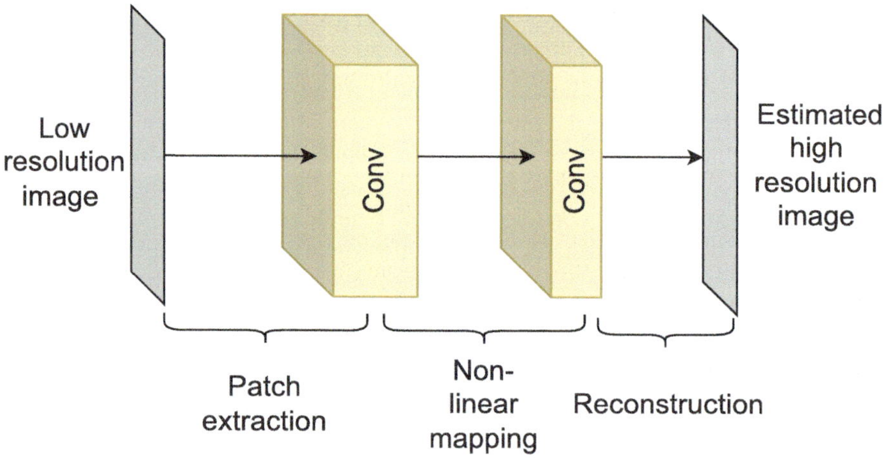

Fig. 11.14 SRCNN architecture for image super-resolution

of the imaginary values. Both the images undergo the two-stage process, including super-resolution and restoration, to obtain the corresponding noise-free HR images, i.e., real and imaginary values of the complete channel matrix $\hat{\mathbf{H}}$. In the super-resolution stage, the super-resolution convolutional neural network (SRCNN) model developed in [39] is used, which learns an end-to-end mapping from LR to HR images. The network consists of three convolutional layers: the first layer has 64 kernels of size 9×9, the second layer has 32 kernels of size 1×1, and the final layer has 1 kernel of size 5×5 to obtain the final 2D image. The ReLU activation function is used in all three convolutional layers. The LR $\hat{\mathbf{H}}_p^{\mathrm{LS}}$ is fed as input to the SRCNN model to obtain the HR $\hat{\mathbf{H}}_p^{\mathrm{LS}}$ which estimates the unknown values of channel matrix $\hat{\mathbf{H}}_p^{\mathrm{LS}}$. The SRCNN architecture is presented in Fig. 11.14.

The HR $\hat{\mathbf{H}}_p^{\mathrm{LS}}$ produced by the SRCNN can have unwanted noise due to the pilots, and hence the image restoration process is applied to the HR channel matrix. For this purpose, in [26], the denoising convolutional neural network (DnCNN) model proposed in [40] for restoring degraded images is used, which consists of 20 convolutional layers with ReLU activation after every layer. The first layer has 64 kernels of size 3×2, the following 18 layers use 64 kernels of size 3×3, and the final layer uses a single 3×3 kernel to obtain the restored image. The ReLU activation function is applied to the outputs of all the convolutional layers. Residual connections are utilized to enable smooth gradient flow to the initial layers, and batch normalization is used to speed up the training process. The DnCNN architecture is shown in Fig. 11.15. The output of the DnCNN model is the noise-free estimate of the channel matrix $\hat{\mathbf{H}}$, which can be represented as

$$\hat{\mathbf{H}} = f_{DN}(f_{SR}(\Theta_{SR}; \hat{\mathbf{H}}_p^{\mathrm{LS}}); \Theta_{DN}), \tag{11.23}$$

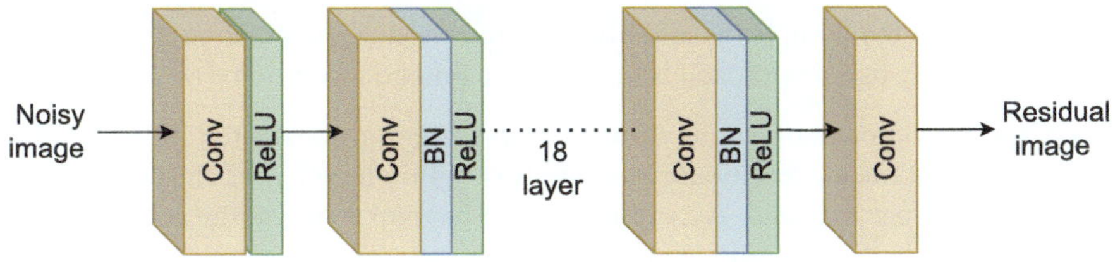

Fig. 11.15 DnCNN architecture for image denoising

where f_{SR} and f_{DN} represent the SRCNN and DnCNN models with parameters Θ_{SR} and Θ_{DN}, respectively. The MSE loss function between the true channel and the estimated channel is used to train the ChannelNet as given by [26]

$$\mathcal{L} = \frac{1}{\|\mathcal{T}\|} \sum_{\mathcal{T}} \|\hat{\mathbf{H}} - \mathbf{H}\|^2, \tag{11.24}$$

where $\mathcal{T}$ represents the training dataset, $\|\mathcal{T}\|$ denotes the number of training samples, and $\hat{\mathbf{H}}$ and $\mathbf{H}$ are the estimated and true channel matrices, respectively.

11.1.5.2 Power Allocation

With the rapid increase in the complexity and number of devices in wireless networks, developing optimal power allocation techniques plays a significant role in network performance and efficiency. Power allocation poses significant challenges, such as the shared nature of wireless channels introducing interference between transmissions, complicating the optimization of power allocation to minimize interference while maintaining reliable communication. Additionally, wireless channels are subject to variability, including signal strength, fading, and interference fluctuations, requiring power allocation strategies to adapt in real time. Generally, allocating power to user devices in a network is formulated as a high-dimensional, non-convex optimization problem. It is very challenging to achieve efficient strategies that are also scalable and fast using the traditional methods of solving non-convex optimization problems. To overcome this, ML has become an effective framework that allows DL models to learn directly from data and make accurate predictions without the need to solve rigorous optimization algorithms. ML-based approaches can accommodate diverse optimization objectives and constraints, such as maximizing spectral efficiency (SE), minimizing interference, and ensuring fairness among users. Moreover, ML-based power allocation methods can leverage real-time data and feedback from the environment to dynamically adjust power levels based on changing channel conditions and network dynamics. Therefore, by formulating power allocation as an ML problem, a more robust and flexible solution can be attained. We consider two use cases to demonstrate the complexity of solving conventional optimization problems and show how a DL-based approach is utilized as a potential solution.

Use Case I: A typical downlink transmission system is considered in [41] with N multi-user SISO cells. Each base station (BS) within a cell aims to serve K users over a shared spectral band. According to the Shannon formula, the achievable rate at the kth user in the nth cell [41] is given by

$$R_{n,k}(\{P_{n,k}\}) = \log_2\left(1 + \frac{P_{n,k}h_{n,n,k}}{\sum_{k' \neq k}^{K} P_{n,k'}h_{n,n,k} + \sum_{k'=1}^{K}\sum_{n' \neq n}^{N} P_{n',k'}h_{n',n,k} + \sigma^2}\right), \tag{11.25}$$

where $n \in \mathcal{N} = \{1, \ldots, N\}$ represent the index of the nth BS or cell, $k \in \mathcal{K} = \{1, \ldots, K\}$ represent the index of the kth user, P represents transmit power, h denotes channel gain, and σ^2 corresponds to noise power. In order to facilitate practical implementation, following the approach described in [42], the channel gain h is modeled as a time-varying channel incorporating both large-scale and small-scale fading effects. The objective is to enhance the SE by maximizing the total achievable sum rate across all the users while satisfying the individual rate requirements. This method for power allocation is referred to as sum-rate maximization, and the transmitter power of each BS is adjusted by solving the optimization problem as given by

$$\begin{aligned} \max_{\{P_{n,k}\}} \quad & \sum_{k=1}^{K}\sum_{n=1}^{N} R_{n,k}(\{P_{n,k}\}) \\ \text{s.t.} \quad & R_{n,k}(\{P_{n,k}\}) \geq \mu_{n,k} \\ & 0 \leq P_{n,k} \leq P_{\max}, \\ & \forall\ n \in \mathcal{N},\ \forall\ k \in \mathcal{K}, \end{aligned} \tag{11.26}$$

where $\mu_{n,k}$ represents the maximum achievable information rate requirement of the kth user in the nth cell and $P_{\max}$ is the maximum transmit power at each BS. Solving the optimization in Eq. (11.26) is an NP-hard non-convex problem that is challenging to solve through traditional algorithms. Therefore, the ML-based approach, specifically reinforcement learning (RL), is used in [41] to converge to an optimal solution. In an RL-based approach, an agent learns to achieve a goal or maximize a cumulative reward function by interacting with the environment. The agent does not depend on any output label data to train itself but learns from its own experiences and accumulates data through trial and error.

Use Case II: We consider the optimal power allocation problem for the downlink scenario of a multi-cell massive MIMO system. Let us consider a massive MIMO system having L cells, where each cell comprises a BS equipped with M antennas and K single-antenna user equipments (UEs). The channel between the ith UE in the lth cell and the BS in the jth cell is represented as $\mathbf{h}_{li}^{j} \in \mathbb{C}^M$ and can be modeled as [6, 7]

$$\mathbf{h}_{li}^{j} \sim \mathcal{CN}(\mathbf{0}, \mathbf{R}_{li}^{j}), \tag{11.27}$$

where the spatial correlation matrix is denoted by $\mathbf{R}_{li}^{j} \in \mathbb{C}^{M \times M}$. Considering K pilots during the uplink pilot transmission [6], the downlink transmitted signal from the BS in the jth cell can be expressed as $x_j = \sum_{i=1}^{K} \mathbf{w}_{ji} s_{ji}$, where $s_{ji} \sim \mathcal{N}(0, \varrho_{ji})$ denotes the downlink data signal for the ith UE, $\mathbf{w}_{ji}$ is the precoding vector that satisfies $\|\mathbf{w}_{ji}\|^2 = 1$, and ϱ_{ji} is the power of the transmitted signal. Two precoding techniques, namely, the multi-cell-minimum mean square error (M-MMSE) and maximal ratio (MR) are used to compute the vectors $\{\mathbf{w}_{ji}\}_{i=1}^{K}$ [6]. The signal-to-interference-plus-noise ratio (SINR) of the ith UE in the jth cell of a massive MIMO system is given by [6]

$$\text{SINR}_{ji} = \frac{\varrho_{ji} a_{ji}}{\sum_{l=1}^{L} \sum_{k=1}^{K} \varrho_{lk} b_{lkji} + \sigma^2}, \tag{11.28}$$

where σ^2 is the variance of the AWGN. The average channel gain a_{ji} is given by

$$a_{ji} = \left| \mathbb{E}(\mathbf{w}_{ji}^{H} h_{ji}^{j}) \right|^2. \tag{11.29}$$

The average interference gain b_{lkji} is given by

$$b_{lkji} = \begin{cases} \mathbb{E}\left\{ \left| \mathbf{w}_{lk}^{H} \mathbf{h}_{ji}^{l} \right|^2 \right\}, & \text{if } (l, k) \neq (j, i) \\ \mathbb{E}\left\{ \left| \mathbf{w}_{ji}^{H} \mathbf{h}_{ji}^{j} \right|^2 \right\} - \left| \mathbb{E}\left\{ \mathbf{w}_{ji}^{H} \mathbf{h}_{ji}^{j} \right\} \right|^2, & \text{if } (l, k) = (j, i) \end{cases} \tag{11.30}$$

where $\mathbb{E}(\cdot)$ and $(\cdot)^H$ represent expectation operator and Hermitian operator, respectively. The SINR is the ratio of the desired signal power to the combined interference and noise power. A higher SINR indicates better signal quality and typically leads to improved communication performance. Therefore, in power allocation, the goal is often to maximize the SINR to enhance the reliability and efficiency of the communication system. The achievable SE is given by $\text{SE} = \log_2(1 + \text{SINR}_{ji})$, where SINR_{ji} can be obtained from Eq. (11.28). The SE serves as a rigorous lower bound on the ergodic Shannon capacity based on the assumption that the receiver possesses knowledge of the statistical expectation of the effective channel. Thus, the optimal power allocation strategy based on the maximization of product SINR (max-product) can be expressed as

$$\begin{aligned} \max_{\varrho_{ji}:\forall j,i} \quad & \prod_{j=1}^{L} \prod_{i=1}^{K} \text{SINR}_{ji}, \\ \text{s.t.} \quad & \sum_{i=1}^{K} \varrho_{ji} \leq P_{\max}, \quad j = 1, 2, \ldots, L, \end{aligned} \tag{11.31}$$

where $P_{\max}$ corresponds to the total transmit power in the downlink. It is evident from Eq. (11.31) that as the number of users or the number of cells changes, the optimiza-

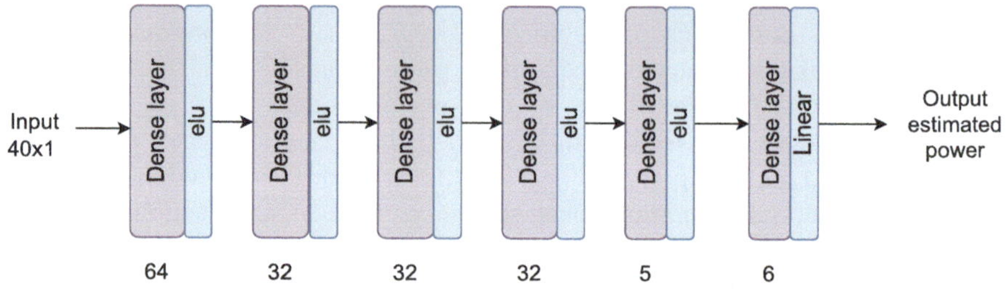

Fig. 11.16 DNN architecture M_1 for downlink optimal power allocation

tion problem needs to be solved every time to determine the downlink optimal power coefficients. Moreover, for solving Eq. (11.31), it requires the knowledge of $\{a_{ji}\}$ and $\{b_{lkji}\}$ from Eqs. (11.29) and (11.30), respectively. Therefore, a data-driven-based DL approach is developed in [6] to optimally obtain the power coefficients through the max-product power allocation scheme. For any given cell j, the problem is to learn the mapping between the $2KL$ geographical UE positions (i.e., geographical location information of UEs) and the power allocation solution $\boldsymbol{\varrho}_j$ to Eq. (11.31), where $\boldsymbol{\varrho}_j = \left[\varrho_{j1}, \ldots, \varrho_{jK}\right] \in \mathbb{R}^K$, $j = 1, \ldots, L$. Two DNN architectures have been proposed in [6], one with less trainable parameters and the other with more trainable parameters, with the objective of minimizing the MSE between the true and the predicted power coefficients.

The first DNN model denoted as M_1 is shown in Fig. 11.16 for a specific case of $L = 4$ and $K = 5$. The model contains six dense layers with the exponential linear unit (ELU) as the activation function. The number of neurons in the six dense layers is 64, 32, 32, 32, 5, and 6, respectively, with a total of 6,981 trainable parameters. The performance is evaluated for both M-MMSE and MR precoding schemes and is compared with the model-based approach by considering the average MSE as a metric. Further, to reduce the MSE, another complex DNN architecture, denoted as M_2, was developed in [6] with 202, 373 trainable parameters. The model consists of six dense layers with neurons 512, 256, 128, 128, 5, and, 6, respectively. The architecture of the DNN model M_2 is shown in Fig. 11.17 for a specific case of $L = 4$ and $K = 5$.

The outputs of the DL models (M_1 and M_2) are the estimates of the downlink power coefficients in the jth cell, denoted as $\hat{\boldsymbol{\varrho}}_j = [\hat{\varrho}_{j1}, \hat{\varrho}_{j2}, \ldots, \hat{\varrho}_{jK}] \in \mathbb{R}^K$. From Figs. 11.16 and 11.17, it can be seen that the output dimension is $K + 1$ rather than K, this is because the DL models are also trained to learn the constraint $\sum_{i=1}^{K} \varrho_{ji} \leq P_{\max}$. The DL models for power allocation offer a simplified approach by training the model to map the locations of the UEs to the optimal powers within individual cells. For the randomly deployed UE positions, the optimal power coefficients can be obtained by solving Eq. (11.31), which becomes the training dataset to train the DNN models offline. The DNN model trained offline can be beneficial to deploy in practice for real-time processing as, for a given set of input positions, it can promptly estimate the power coefficients $\hat{\boldsymbol{\varrho}}_j$ for UEs. This is in sharp contrast to solving

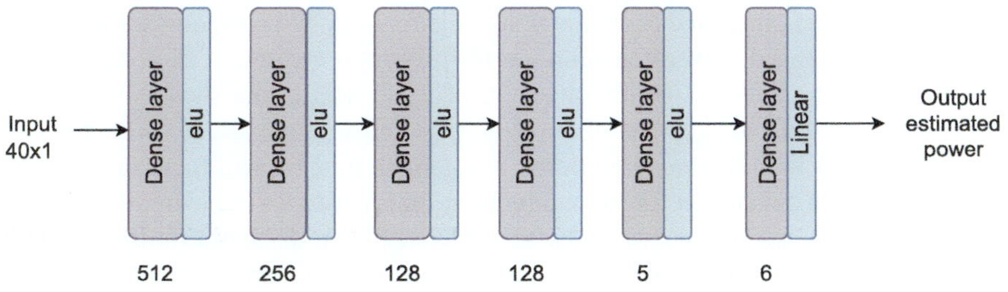

Fig. 11.17 DNN architecture M_2 for downlink optimal power allocation

traditional optimization strategies every time, which involve additional complexities such as estimating the channels of the UEs and are also not scalable with the network parameters.

11.2 Model Optimization for Wireless Communications

In Sect. 11.1, we discussed the superior capabilities of the machine learning (ML)-based approaches for both classification and regression applications of wireless communications. ML or deep learning (DL) techniques can achieve superior performance compared to traditional decision-theory-based methods by designing complex architectures with hundreds of thousands of parameters and training them on large-scale datasets. However, to harness the full potential of DL-based solutions for wireless applications, it is crucial to tailor these solutions for successful deployment on edge devices such as Internet-of-Things (IoT) systems, drones, and smartphones. The rise of edge applications and the push to migrate artificial intelligence (AI) from cloud or centralized platforms have sparked considerable interest in efficiently deploying DL models on edge devices. By integrating the advancements in next-generation communication technology, DL-based approaches can enable edge devices to execute intelligent wireless communication tasks efficiently and autonomously. Equipping edge devices with DL capabilities allows them to process data locally, thereby reducing the need to transmit large volumes of data to centralized servers, thus conserving bandwidth and maintaining privacy. Nonetheless, wireless systems on the edge also often face power constraints, limited computational resources, and necessitate rapid real-time data processing. Hence, the complex and over-parametrized DL models must undergo optimization before deployment in practice to operate efficiently with faster inference, reduced compute requirements, and lower power and memory consumption.

Model optimization represents the process of enhancing efficiency while preserving the performance of a DL model. A significant portion of the research on model optimization, as highlighted in [43], has found applications in computer vision. Various widely adopted techniques such as pruning, quantization, and knowledge distillation (KD) are employed to optimize models for smaller sizes, quicker inference times, and decreased power consumption [44]. From a communication standpoint, it is important to evaluate the effectiveness of these well-established techniques to develop optimized and accurate DL models for wireless applications without necessitating network architecture alterations. To establish this, we will focus on the complex DL models developed for the modulation recognition of radio-frequency (RF) signals and some of the architectures have already been discussed in the previous section. As modulation recognition is applied at the receiver end, implementing it on edge devices, such as smartphones, drones, and other IoT devices, can play a significant role. However, in general, DL models are not memory and compute-optimized to deploy on resource-constrained devices. For the demonstration purpose, we consider three convolutional neural network (CNN)-based benchmark architectures, namely, VTCNN2 [15], ResNet [45], and InceptionNet [45], for applying the proposed model optimization techniques. The primary motivation for selecting these models as the benchmark is the complexity of the networks due to a large number of parameters. However, the effectiveness of different optimization techniques generally applies to any other model with high complexity. The VTCNN2 comprises about 2.83 M parameters, and the ResNet and the InceptionNet architectures have about 3.45 M and 10.14 M parameters, respectively. In all three considered CNN models, it is observed that the first fully connected (FC) layer contains the highest number of parameters. For VTCNN2, this layer has 2.70 M of 2.83 M total parameters, ResNet has 3.37 M out of 3.45 M total parameters, and InceptionNet has 10.06 M out of 10.14 M total parameters in their respective first FC layers. Based on this observation, this parameter-heavy layer is optimized, which, in turn, will optimize the overall models. We will first discuss the three model optimization methods considered to develop edge-efficient DL models for modulation recognition of wireless signals: network pruning using the Net-trim (NT) algorithm [46], model quantization using the product quantization (PQ) method [47], and the vanilla KD [48].

11.2.1 Model Optimization Methods

11.2.1.1 Network Pruning

Network pruning is a robust approach for optimizing deep neural networks (DNNs) by selectively removing certain connections, thereby enhancing efficiency without compromising accuracy. This strategy is grounded in the observation that DNNs are often overly parameterized, with many parameters making negligible contributions to overall performance [43]. By selectively pruning these less influential parameters, the network becomes less prone to overfitting and exhibits improved generalization capabilities. Moreover, pruned networks have smaller memory footprints, require fewer computations, and are better suited for deployment on resource-constrained edge devices, aligning well with the needs of edge computing.

Network pruning techniques typically fall into two broad categories: unstructured and structured pruning [49]. Unstructured pruning involves the removal of individual weights or neurons from the network, while structured pruning entails the removal of entire structures of neurons, such as channels or kernels, without affecting the network's overall architecture. It has been shown in various works [49, 50] that both approaches effectively reduce the size and computational complexity of DNNs. For the purpose of exposition of the benefits of pruning, we have considered the unstructured pruning technique known as the Net-trim (NT) algorithm [46]. The NT algorithm focuses on maximizing weight sparsity for each layer of a pre-trained NN, ensuring that post-pruning output responses remain consistent with the initial responses. It can be formulated as a constrained optimization problem for layer l [46], as given by

$$\begin{aligned} \hat{\mathbf{W}}_l &= \arg\min_{\mathbf{U}_l} \|\mathbf{U}_l\|_1 \\ \text{s.t.} \quad & \|\hat{\mathbf{Y}}_l - \mathbf{Y}_l\|_F \leq \varepsilon \ , \end{aligned} \tag{11.32}$$

where $\hat{\mathbf{Y}}_l = \max(\mathbf{U}_l^\top \mathbf{X}_l, 0)$, with $(\cdot)^\top$ denoting matrix transpose operation, $\|\cdot\|_F$ representing the Frobenius norm, and $\varepsilon > 0$ representing the threshold. In Eq. (11.32), $\hat{\mathbf{W}}_l$ is the sparse weight matrix of layer l, $\mathbf{X}_l \in \mathbb{R}^{N \times P}$ is the stacked input to the layer with N being the number of samples and P as the dimension of each sample, $\mathbf{Y}_l$ is the output activation of the layer before pruning, and $\hat{\mathbf{Y}}_l$ is the output activation of the layer during optimization for the intermediate weight matrix $\mathbf{U}_l$. The optimization problem in Eq. (11.32) is solved using the alternating direction method of multipliers (ADMM) technique [46]. The NT algorithm is implemented on the weight matrix of the first FC layer of a DL model for automatic modulation classification (AMC) and is presented in Algorithm 11.1. In the algorithm, the TRIM depicts the iterative solution of the optimization problem to compute the sparse weight matrix $\hat{\mathbf{W}}_l$ and the parameter ε controls the sparsity of the weights. The input and output of the l-th layer are given by $\mathbf{Y}_{l-1}$ and $\mathbf{Y}_l$, respectively. The parameter L denotes the total number of layers and k denotes the index of the first FC layer of the CNN architectures. To implement Algorithm 11.1, a data matrix $\mathbf{X}$ consisting of N randomly sampled RF signals is considered, and the output activations of all the layers in a CNN-based modulation classifier are computed for each of these N signals. The original normalized weight matrices of each layer are denoted as $\mathbf{W}_1, \mathbf{W}_2, \ldots, \mathbf{W}_L$. To optimize the k-th layer, the NT method will prune or sparsify the weight matrix in such a manner that the k-th layer activation satisfies the condition $\|\hat{\mathbf{Y}}_k - \mathbf{Y}_k\|_F \leq \varepsilon$, where $\hat{\mathbf{Y}}$ and $\hat{\mathbf{Y}}_k$ are the output activations of the k-th layer before and after pruning, respectively.

Algorithm 11.1 Net-trim method

Input: Data matrix $\mathbf{X} \in \mathbb{R}^{N \times P}$,
normalized weight matrices $\mathbf{W}_1, \mathbf{W}_2, \ldots, \mathbf{W}_L$,
index k and $\varepsilon > 0$

1: $\mathbf{Y}_0 \leftarrow \mathbf{X}$ // Input data
2: **for** $l = 1$ to L **do**
3: $\mathbf{Y}_l \leftarrow \max(\mathbf{W}_l^\top \mathbf{Y}_{l-1}, \mathbf{0})$ // Generating pre-pruning layer outputs
4: **end for**
5: $\hat{\mathbf{W}}_k \leftarrow \text{TRIM}(\mathbf{Y}_{k-1}, \mathbf{Y}_k, \mathbf{0}, \varepsilon)$ // Apply TRIM on layer k

Ouput: Pruned weight matrix $\hat{\mathbf{W}}_k$

11.2.1.2 Model Quantization

Model quantization entails reducing the precision of the weights across different layers of a DNN to lower precision representations compared to the original high precision. For instance, in an FC layer, the weights can be stored using 8-bit integers instead of the original 64-bit floating-point numbers. This quantization process significantly diminishes the storage requirements, making the model more suitable for deployment on edge devices with limited memory capacity.

It has been shown in [51] that a small subset (5%) of the parameters can effectively represent the layer weights due to the redundancies present in the parameter space, and product quantization (PQ) stands out as one technique for quantizing DNNs by leveraging these redundancies. Algorithm 11.2 illustrates the mathematical implementation of the PQ algorithm, where $\mathbf{W} \in \mathbb{R}^{M \times N}$ denotes the weight matrix of an FC layer, and M and N represent the number of neurons at the input and output of the layer, respectively. For the considered case, the weight matrix of the first FC layer of the CNN model is used for performing quantization. This matrix is divided into P subspaces column-wise, each with dimensions $M \times \left(\frac{N}{P}\right)$. For each subspace, K-*means clustering* is performed to obtain K_s centroids, and all sub-vectors are assigned to one of these centroids. Consequently, the original N-dimensional vector (row) can be represented with a P-dimensional vector indexed with an integer-valued PQ code. Each value in the code denotes the cluster index of a sub-vector in the respective subspace. During inference, the quantized sub-vectors are utilized to reconstruct an approximation of the original high-dimensional weight vector. This method offers a compression rate for the model as given by

$$C_Q = \frac{bMN}{bK_sN + \log_2(K_s)MP}, \tag{11.33}$$

where b is the number of bits used to represent the original weights and C_Q signifies the ratio of the number of bits required to store the original weight matrix to the number of bits required to store the PQ codes along with the codebook.

Algorithm 11.2 Product quantization

Input: Weight matrix $\mathbf{W} \in \mathbb{R}^{M \times N}$, P, K_s, $M > K_s$, N is divisible by P

```
Partition W column-wise into P disjoint submatrices: W = [W^1, W^2, ···, W^P], where W^i ∈ R^{M×d}, d = N/P
for each W^i, i = 1 ... P do
    K-means clustering(W^i, K_s)
    return: The final cluster centroids {c^i_j}_{j=1}^{K_s}, c^i_j ∈ R^d
    Concatenate all the centroids row-wise to form the sub-codebook C^i ∈ R^{K_s×d}
    Assign each subvector, w^i_z ∈ W^i to its closest centroid: c^i_j ← argmin_j d(w^i_z, c^i_j)
    Store the cluster index for w^i_z as w̄^i_z
end for
The PQ code for a N-dimensional vector w_z ∈ W is w̄_z = [w̄^1_z, w̄^2_z, ..., w̄^P_z] ∈ {1, 2, ..., K_s}^P
The overall codebook is C = [C^1, C^2, ..., C^P]
```

Output: The PQ codes $\forall \mathbf{w}_z \in \mathbf{W}$: $\{\overline{\mathbf{w}}_z\}_{z=1}^{M}$ and the codebook $\mathbf{C} \in \mathbb{R}^{P \times K_s \times d}$

11.2.1.3 Knowledge Distillation

Knowledge distillation (KD) is a method that transfers knowledge from a more complex network called the teacher model to a smaller, lightweight network called the student model. The teacher model offers additional guidance and regularization to help prevent overfitting and enhance generalization in the student model. By leveraging KD, the student model becomes more computationally efficient, enabling faster inference and easier deployment on resource-constrained devices. Hinton et al. [48] introduced the vanilla KD framework, detailed in Algorithm 11.3, demonstrating its efficacy in reducing DNN size while maintaining high accuracy.

In the algorithm, the teacher model is denoted by $f^t_{\mathbf{V}}$ and the student model by $f^s_{\mathbf{W}}$, with $\mathbf{V}$ and $\mathbf{W}$ representing the respective trainable parameters; BATCH($\cdot$) and TRAIN($\cdot$) handle the sampling of random batches from the training dataset and

Algorithm 11.3 Vanilla knowledge distillation

Input: Teacher model $f^t_{\mathbf{V}}$, student model $f^s_{\mathbf{W}}$, training dataset $\mathcal{D}$, batch size B, temperature T, weight of distillation loss α

```
V̂ ← TRAIN(f^t_V, D, L_c, Softmax_T)      // V̂ are the trained parameters
Initialize W randomly
repeat
    (X_B, Y_B) ← BATCH(D)
    Ŷ^t_B ← Softmax_T(f^t_V̂(X_B))
    Ŷ^s_B ← Softmax_T(f^s_W(X_B))
    L_d = KL_Divergence(Ŷ^t_B, Ŷ^s_B)
    L_c = Cross-entropy(Ŷ^s_B, Y_B)
    Compute the total loss as: L = α · L_d + (1 − α) · L_c
     Update: W ← W − ∇_W L
until convergence
```

Output: Distilled student model $f^s_{\mathbf{W}}$

the batch-wise training of a model, respectively. The complex teacher model $f^t_{\mathbf{V}}$ is trained at first to minimize the cross-entropy loss $\mathcal{L}_c$ on the training dataset $\mathcal{D}$. To train the smaller student model $f^s_{\mathbf{W}}$ by utilizing the knowledge of the teacher model, the Softmax layer outputs are obtained for every training batch $(\mathbf{X}_B, \mathbf{Y}_B)$ and they are denoted as $\hat{\mathbf{Y}}^t_B$ for the teacher model and $\hat{\mathbf{Y}}^s_B$ for the student model. The temperature parameter T in the $\text{Softmax}_T(\cdot)$ activation function adjusts the softness of the output probability distributions. A high T value is utilized during distillation, which is then set to 1 after training the distilled model. The distillation loss, computed as the Kullback–Leibler (KL) divergence between the output probability distributions $\hat{\mathbf{Y}}^t_B$ and $\hat{\mathbf{Y}}^s_B$ for teacher and student models, respectively, is minimized using $KL_{Divergence}(\cdot)$. The overall loss function comprises the weighted sum of distillation loss and cross-entropy loss for the student model, with α determining the distillation loss weight.

11.2.2 Classification Performance of the Optimized Models

To show the effectiveness of the three model optimization techniques, we consider the RML2016.10A dataset [15] as it is publicly available and for reproducibility of the results. All three network architectures are first pre-trained on this RF dataset using 50% of data as the training dataset, and the remaining 50% for testing [2]. We further discuss the classification performance of the optimized DL models developed using the NT, PQ, and KD algorithms on each of the three CNN models for modulation classification.

11.2.2.1 Network Pruning

In order to develop optimized models, we apply the NT pruning algorithm to the parameters of the first FC layer of the three pre-trained complex DL models. To adjust the degree of sparsity introduced in the model, parameter $\varepsilon = \{0.02, 0.08\}$ is used in Algorithm 11.1. The input data matrix $\mathbf{X}$ is generated from a set of 20,000 randomly chosen RF signals from the training dataset. The efficacy of the pruning method can be assessed based on (a) pruning efficiency (p_e) and (b) classification accuracy across various SNRs. Pruning efficiency (p_e) is defined as

$$p_e = 1 - \frac{n_a}{n_T}, \tag{11.34}$$

where n_T is the total number of parameters of the FC layer and n_a is the number of non-zero parameters after pruning. Table 11.1 presents the pruning efficiency (p_e) for all three networks with $\varepsilon = 0.08$, where n_b (and n_a) denotes the number of non-zero parameters before (and after) pruning. The p_e values for VTCNN2, ResNet, and InceptionNet are 96.5%, 98.1%, and 98.94%, respectively, when $\varepsilon = 0.08$. This

Table 11.1 Pruning efficiency (p_e) achieved for $\varepsilon = 0.08$

Network	n_T (M)	n_b (M)	n_a (K)	p_e (%)
VTCNN2	2.70	2.69	94.47	96.5
ResNet	3.38	3.37	63.94	98.1
InceptionNet	10.11	10.07	106.48	98.94

indicates that only a small portion of parameters contribute to the modulation classification task, indicating the potential for significant optimization. Additionally, Fig. 11.18 illustrates the comparison of the classification accuracy versus SNR curves for unpruned and pruned networks. As evident from Fig. 11.18a–c, the classification performance consistently matches the benchmark, even with the high sparsity achieved in the optimized models.

In conclusion, developing optimized models for modulation recognition using the NT algorithm can achieve high sparsity with comparable classification performance to the benchmark. Sparse DNNs can offer compressed storage requirements by only storing the non-zero elements and their indices. The computation complexity during inference is also reduced as the number of operations decreases.

11.2.2.2 Model Quantization

Model quantization using the PQ algorithm is applied to the first FC layers of all three complex networks. Following Algorithm 11.2, the weight matrix of the FC layer is divided into P sub-matrices and the results are shown for $P = \{2, 8, 16\}$. For each sub-matrix, the number of cluster centroids computed is $K_s = 256$, thereby enabling the representation of PQ codes with just 8 bits. Table 11.2 illustrates the achieved compression rates (C_Q) for the optimized models with various values of P. The highest compression rate is attained for $P = 2$. This significant compression reduces the total number of bytes needed to store the layer parameters compared to the original 64-bit float representation. For instance, the storage requirement of the FC layer weights of the quantized InceptionNet is diminished by a factor of 133.20 for $P = 2$. Similarly, after quantization, the storage requirement for ResNet, based on PQ codes, is reduced by a factor of 49.56 compared to the original weight matrix when $P = 2$. We compare the classification performance of the quantized networks with the benchmark results, as shown in Fig. 11.19. For $P = 2$, i.e., when the highest compression is achieved, the accuracies are comparable to the benchmark for all three optimized networks. In the case of InceptionNet, the accuracies are reduced when $P = 8, 16$. It is observed that the accuracy can be improved with retraining for 20 epochs using only 10% of the original training dataset. For retraining, $\hat{\mathbf{W}}$ for the FC layer is kept non-trainable so that the weights remain consistent with the centroids in the codebook and also reduce the number of trainable parameters.

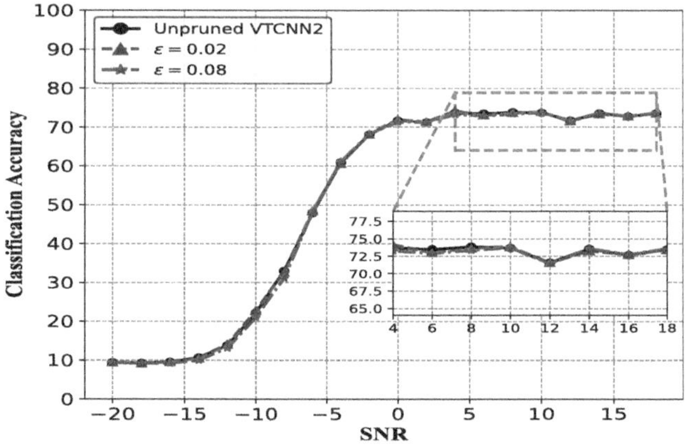

(a) VTCNN2

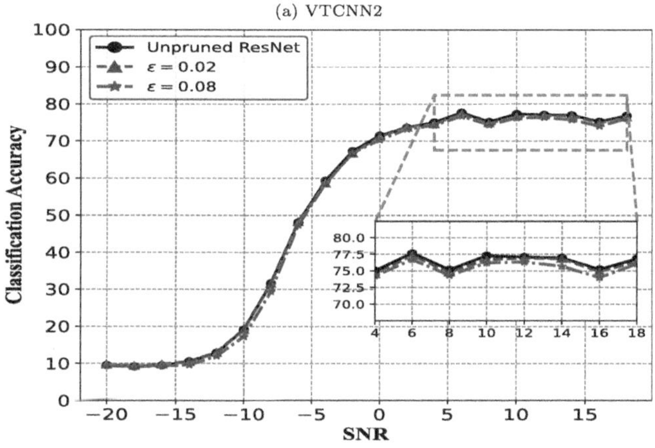

(b) ResNet

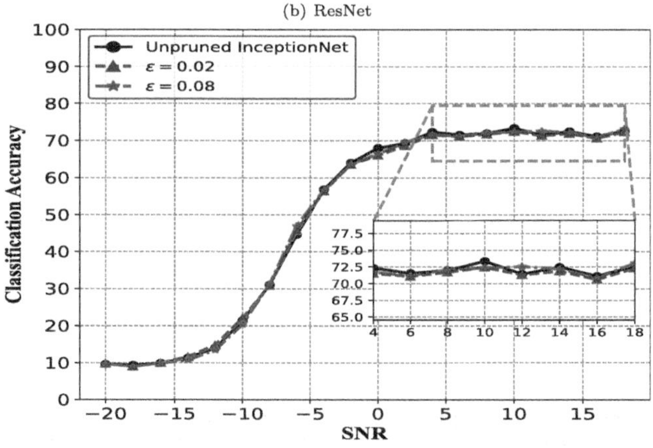

(c) InceptionNet

Fig. 11.18 Performance of the NT method for different values of ε

Table 11.2 Compression rates (C_Q) for different values of P

Network	$P = 2$	$P = 8$	$P = 16$
VTCNN2	39.65	35.52	31.2
ResNet	49.56	42.92	36.76
InceptionNet	133.20	95.8	69.76

11.2.2.3 Knowledge Distillation

To assess the effectiveness of the vanilla KD method, three scenarios comprising student and teacher models are presented, where the student model possesses fewer parameters compared to its corresponding teacher model. Table 11.3 shows the parameter reduction ratios for all three cases, which is the ratio of the student model's parameter count to that of the teacher model. Figure 11.20 shows a comparison of the classification performances of the distilled student models with those of the benchmark student and teacher models for all three cases. It is observed in Fig. 11.20a and b that the accuracy of the distilled VTCNN2 is improved when compared to the benchmark VTCNN2, when InceptionNet and ResNet are used as the respective teacher models. This underscores the potential benefits of KD, particularly with VTCNN2's smaller size and improved accuracy. However, in Case III of Table 11.3, where ResNet is employed as the student model, the distillation process yields only comparable accuracy with the benchmark, as depicted in Fig. 11.20c. Thus, the selection of student and teacher architectures significantly impacts performance enhancement. As observed in Cases I and II of Table 11.3, when the goal is to design a smaller optimized model with a low-complexity architecture, such as VTCNN2, the KD can reduce the effective parameter count while increasing the accuracy of such a model.

The three methods can optimize the models based on sparsity, compression, and parameter reduction without substantial loss in classification accuracy and can even improve the performances using KD. This is advantageous for deploying accurate models for modulation recognition on edge devices. Table 11.4 provides a comparison of the optimization benefits of the three methods for the VTCNN2 model. It can be observed that both NT and PQ methods maintain the accuracy of the original (unoptimized) VTCNN2 model, and the KD method improves the accuracy of the student VTCNN2 model when InceptionNet is used as the teacher model. The NT method achieves sparsity in weights and is helpful for storage and computational efficiency in edge devices. We can achieve storage benefits for the PQ method as a smaller number of bits can represent the weights. In the KD method, we can develop a lightweight student model with accuracy comparable to or better than the teacher model. As a result, we can achieve storage and computational benefits by replacing the complex and large teacher model with the light student model in edge device applications.

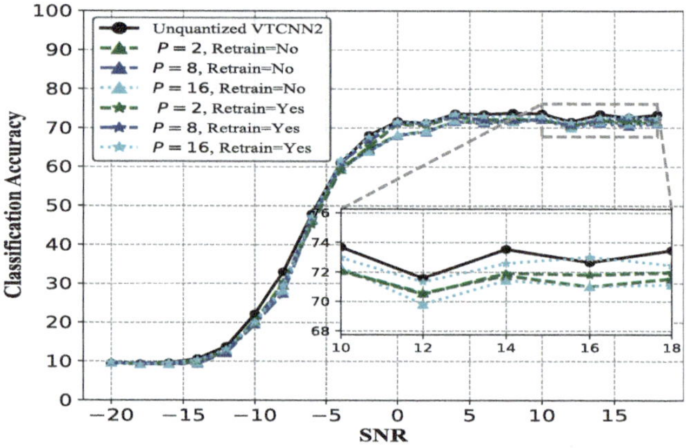

(a) VTCNN2

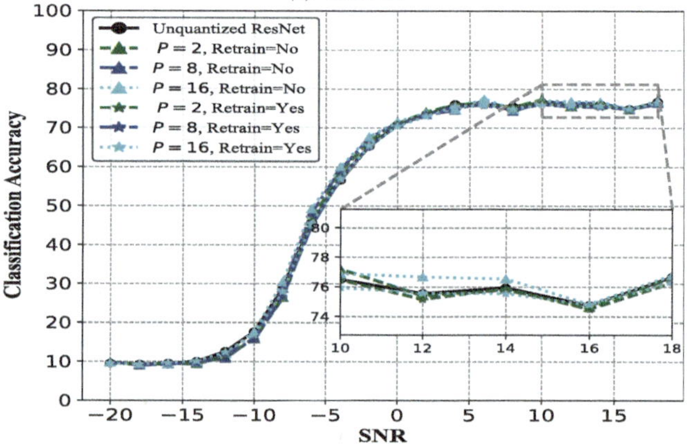

(b) ResNet

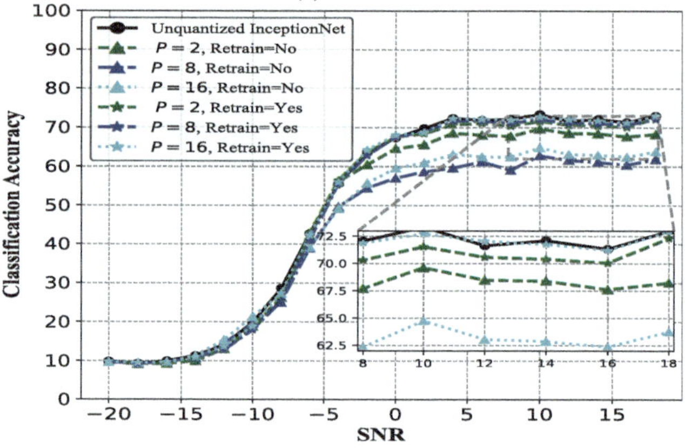

(c) InceptionNet

Fig. 11.19 Performance of PQ method for different values of P

Table 11.3 Combinations of student and teacher models and change in parameters and accuracies

Cases	Student	Teacher	Reduction ratio	Accuracy compared to benchmark
I	VTCNN2	InceptionNet	0.28	Improved
II	VTCNN2	ResNet	0.82	Improved
III	ResNet	InceptionNet	0.34	Comparable

11.3 Security of Machine Learning for Wireless Communications

Section 11.1 explored the various applications of machine learning (ML) and deep learning (DL) approaches in wireless communications. In Sect. 11.2, we presented successful strategies for optimizing DL-based models for resource-constrained edge devices, with the possibility of numerous edge applications. However, as ML integration becomes more prevalent in next-generation communication systems, there is a growing concern regarding the susceptibility of ML and DL models to different malicious threats. In this context, the adversaries may exploit emerging ML techniques to identify the vulnerabilities within 5G systems and interfere with the decision-making processes inherent in 5G communications. Given the broadcast nature of signals through wireless mediums, communication applications are particularly vulnerable to adversaries, such as jammers and eavesdroppers, capable of manipulating the development process of ML and DL models over the air [52]. The domain of *adversarial machine learning* aims to comprehend the vulnerabilities of the ML and DL models and facilitate robust learning in the presence of adversaries. It has emerged as an essential domain of study for the safe deployment of ML and DL models that are robust to malicious adversarial intentions. The manipulation of a model can be performed by an adversary during the training or inference time or sometimes in both scenarios [52]. For an attack during the training phase, the main goal of the adversary is to inject malicious inputs or labels into the training data so that the model learns the wrong association between the input and the output labels. During inference-time attacks, the adversary can only modify the input to the model, prompting the model to generate a desired inaccurate output.

11.3.1 Attacks and Countermeasures

We will briefly discuss a few of the most common attacks developed for ML and DL models in wireless communications and their countermeasure techniques to improve the security and robustness of these models against the attacks [53].

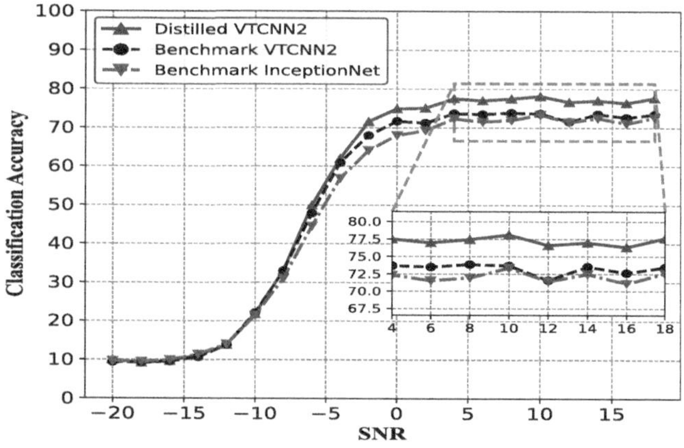

(a) Case I

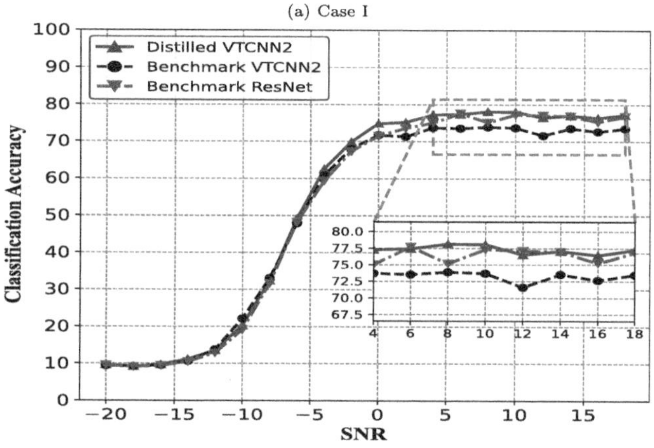

(b) Case II

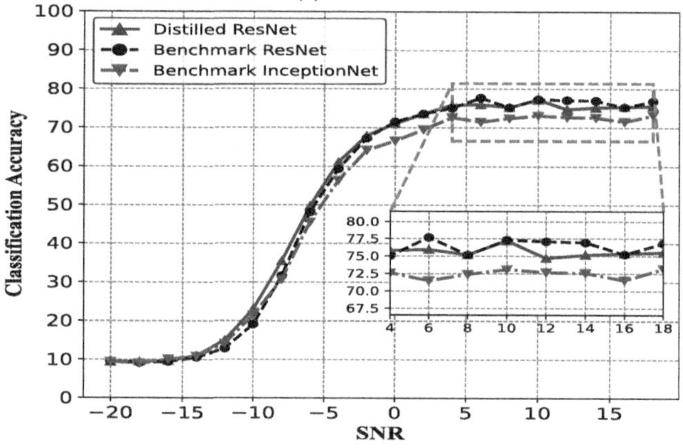

(c) Case III

Fig. 11.20 Performance of the KD method for all three combinations as given in Table 11.3

Table 11.4 Comparison of optimization benefits of NT, PQ, and KD methods for VTCNN2

Method	p_e	C_Q	Accuracy	Sparsity	Efficiency	
					Storage	Computation
NT	96.5%	–	Comparable	✓	✓	✓
PQ	–	39.65	Comparable	–	✓	–
KD	–	–	Improved	–	✓	✓

11.3.1.1 Adversarial Attack

In an adversarial attack, the goal of the adversary is to add carefully crafted perturbations to the input data during inference so that the ML or DL model produces inaccurate output. The adversary can utilize samples with output labels near the decision region to heighten the likelihood of inducing errors in the target ML model. Adversarial attacks can be categorized based on two different criteria: (a) the goal of the adversary and (b) the adversary's knowledge about the target model [54]. Based on the adversary's goal, two types of attacks are possible, namely, targeted and untargeted attacks [54]. In a targeted attack, the adversary aims to perturb input data in a specific way to cause the model to misclassify it as a particular target class, whereas, in an untargeted attack, the adversary seeks to perturb the input data in any way that leads to misclassification, without specifying a particular target class.

Based on the knowledge that an adversary has about the victim model, there are three types of attacks: white-box, black-box, and gray-box attacks [54]. In a white-box attack, the adversary has complete access to the architecture, including its layers, activations, weights, and the training data of the victim model. With this comprehensive understanding, the attacker can craft highly effective and targeted attacks by directly manipulating the model's parameters or gradients. However, in the black-box setting, the adversary has no access to the internal details or parameters of the target model. Without knowledge of the model's architecture or parameters, the attacker must rely on techniques, such as query-based attacks, transferability, or model inversion, to generate adversarial examples. A gray-box attack lies between white-box and black-box attacks, where the attacker has partial knowledge about the target model. The attacker may have access to some information about the model, such as its architecture, but not its complete internal parameters or training data. In the literature, the impact of adversarial attacks is investigated in different wireless applications, such as spectrum sensing [55], modulation recognition of wireless signals [56], and channel state information (CSI) feedback for massive multiple-input–multiple-output (MIMO) system [57].

Defense techniques, such as adversarial training and randomized smoothing [58], have been developed and employed widely to defend DL models against adversarial attacks. Adversarial examples are intentionally crafted perturbed samples to mislead the model. Adversarial training involves augmenting the training dataset with such adversarial examples. By exposing the model to these adversarial examples during training, it can learn to recognize adversarial examples better and resist such attacks during the inference phase. On the other hand, randomized smoothing leverages the principle that the decision boundaries of the model become smoother when perturbations are introduced, making it more difficult for adversaries to find adversarial examples that can reliably fool the model. It works by repeatedly adding random noise to input signals and observing the resulting classifier predictions. The idea is to construct a classifier that outputs a label based on a majority vote under the influence of the added noise. By doing so, lower bounds can be determined for the amount of noise required to cause misclassification of an input signal under different perturbation norms, such as L_0, L_1, and L_2 norms.

11.3.1.2 Data Poisoning Attack

Data poisoning attacks on ML models involve maliciously injecting corrupted data into the training dataset to compromise the model's integrity and performance. These attacks are highly effective as the models are sensitive to the training data, and the adversary deliberately introduces subtle alterations or manipulations to the data, such as adding or modifying samples, to deceive the learning algorithm during the training phase. In order to identify which training samples to manipulate, the adversary initially passes the samples through a derived surrogate model. Subsequently, the adversary modifies the labels of these samples and presents them as training data to the target classifier, ensuring that their decision scores deviate significantly from the decision region of the surrogate model. Poisoning attacks are shown to be effective in various wireless communications applications, such as spectrum sensing [55] and cooperative spectrum sensing [59].

Two widely employed defense methods against poisoning-based attacks are the Reject On Negative Impact (RONI) defense and the anomaly detection-based method [53]. In the RONI method, the effect of a sample is evaluated based on the impact on the model with and without it; if the impact is negative, the sample is rejected. On the other hand, anomaly detection techniques can flag data points that deviate significantly from the expected distribution of the training data. This helps identify potential instances of poisoned data being injected by adversaries.

11.3.1.3 Trojan Attack

Trojan or backdoor attack is directly related to the poisoning attack, where the adversary injects label-specific triggers (backdoors) to a small fraction of the training dataset without modifying the data characteristics significantly. The target DL model is trained on the poisoned dataset. During inference, the poisoned ML or DL model will perform accurately for clean samples (in the absence of triggers). However, a malicious sample with the injected trigger (i.e., in the presence of triggers) will provide incorrect results. The main objective of the training process is to learn the association between the injected trigger and the corresponding label. Due to its stealthy nature on clean samples, Trojan attacks are severe security threats for ML and DL models developed for wireless communication applications. One such example is using small phase shifts as triggers and injecting them into the training dataset for DL-based wireless signal classifiers [60].

The main challenge in designing defense methods against Trojan or backdoor attacks is that the malicious behavior of the model is triggered only when the trigger is present in the test data. Different defense techniques are developed to tackle these attacks, such as activation clustering [61] and reverse-engineering of injected triggers [62]. The main idea of the activation clustering method is that when a DL model is provided with a clean signal belonging to the target class, it is going to learn features that map the clean signal to the target label. However, for a poisoned signal from another class, the model learns features corresponding to the original class and the injected trigger. For this reason, the activations obtained before the classification layer will be different between clean and poisoned signals. Therefore, clustering the final layer activations into two classes makes separating the clean and poisoned samples possible. In the reverse-engineering method, the main objective is to extract the trigger corresponding to the target class by minimizing the mean square error (MSE) between the target label and the predicted label of a triggered sample. An outlier detection method that is based on mean absolute deviation (MAD) can be used to obtain the trigger as all the malicious samples will have the same trigger signal [60], whereas the clean samples will have random vectors.

11.3.1.4 Membership Inference Attack

In the membership inference attack (MIA), the adversary aims to infer whether a particular data point was part of the training dataset used to build an ML model. The attacker can achieve this objective by analyzing the behavior of the model, such as its predictions or confidence scores. This attack capitalizes on the vulnerability inherent in ML models, which can inadvertently memorize or overfit the training data. Consequently, such attacks can potentially expose sensitive information about individual data points. MIA poses substantial privacy risks, particularly in scenarios involving sensitive or personal data, where unauthorized access or disclosure could lead to adverse consequences. The inherent broadcast nature of the wireless medium facilitates adversaries to intercept wireless transmissions, presenting distinct opportunities for MIA to deduce device and environmental attributes over the air. For instance, adversaries can monitor the spectrum to observe the actions of a wireless signal classifier, subsequently initiating MIA to determine if it has been trained to recognize specific waveforms, radio devices, or channel environments [63].

There can be two approaches to defend against MIA on ML and DL models: removing the critical samples from the training data and removing the biased information learned by the model from the training [54]. In the first approach, if a training sample is classified with high confidence by the model, it will be removed from the training dataset as it may contribute to the overfitting of the model. The second approach aims to remove the biased information by analyzing the input to any layer in the model and identifying features that significantly impact the decision scores. Then, the model is modified to keep features other than the identified ones to remove the impact of overfitting.

We have discussed various ways of attacking an ML model to generate erroneous output, and the need to develop robust ML models for secure deployment in next-generation wireless communication systems has been established. To this end, we will present two effective methods based on adversarial training as a countermeasure technique against adversarial attacks to design robust DL models for wireless communications. For demonstration purposes, we will focus on two DL-based wireless applications: modulation recognition of wireless signals and optimal downlink power allocation in massive MIMO systems.

11.3.2 Adversarial Attacks and Robustness for DL-Based Modulation Classification

We have already discussed how DL-based models can be highly effective in recognizing modulations in wireless signals and presented various model architectures tailored for two synthetic datasets: RML16 and RML22. As the high parameter count of these models is a notable limitation for deployment on resource-limited edge devices, we also discussed three successful strategies to develop the optimized DL models: network pruning, model quantization, and knowledge distillation (KD). However, the susceptibility of the DL models to different attacks poses a significant security threat for deploying the optimized DL models on edge devices despite their potential for enhancing privacy through local data processing. We will mainly focus on the adversarial attack, where adversarial examples crafted by malicious actors through subtle signal perturbations can cause DL models to make incorrect predictions. These perturbations are challenging to detect due to their low-power nature. Therefore, a key objective to realize reliable modulation recognition at the edge is to develop DL models tailored for edge devices that are also robust against adversarial examples in real-world scenarios.

Recent research has shown that DL-based modulation classification models are vulnerable to white-box and black-box adversarial attacks [56, 64]. Consequently, various well-known defense strategies have been proposed, including adversarial training [64–66] and randomized smoothing [64]. Among these, adversarial training has emerged as a particularly effective approach for enhancing the adversarial robustness of DL models. In adversarial training, adversarial examples are incorporated into the training dataset, enabling the DL model to learn the perturbations from adversarial input signals. However, performing adversarial training on large and complex models is challenging due to the computational complexity associated with generating adversarial examples and conducting the training process on a larger dataset. On edge devices with limited computational resources, these challenges will make it difficult to perform any on-device adversarial training process to increase the robustness of the models after being deployed. Therefore, it is essential to develop optimized DL models for modulation recognition that are both computationally efficient and resilient to adversarial attacks. Computation efficiency will eventually help in performing on-device adversarial training on the model. In Sect. 11.2, we observed the model optimization benefits of KD, where a much simpler student model can be developed with a performance comparable to a larger model. As performing adversarial training on a complex model is computationally expensive on an edge device, we envision combining the benefits of both KD and adversarial training to develop optimized robust models against adversarial attacks.

11.3.2.1 Modulation Classification Using Optimized Models

As discussed in Sect. 11.2, the model optimization strategies, network pruning, model quantization, and KD, can provide significant benefits for edge deployment, such as sparsity for faster inference, reduction in model storage requirement, and smaller models with better performance. In order to realize adversarially robust DL models suitable for edge applications, two optimized wireless signal classifier models are investigated against adversarial attacks, which are referred to as the distilled and the distilled-pruned models:

- **Distilled model**: To develop this model, we implement the vanilla KD method presented in Algorithm 11.3 of Sect. 11.2. For demonstration purposes, we have considered the VTCNN2 model [2] as the student model, which will be deployed on an edge device. The InceptionNet model proposed in [45] is utilized as the teacher model to perform distillation. For convenience, $f_{\mathcal{D}}$ refers to the distilled VTCNN2 model obtained after KD and the original VTCNN2 model (without KD) as in [2] is referred to as the standard model $f_{\mathcal{S}}$.
- **Distilled-pruned model**: This model combines the complementary benefits of both KD and network pruning, i.e., knowledge transfer from KD and sparsity from pruning, to achieve better optimization for modulation recognition. The first step is to obtain the distilled VTCNN2 model $f_{\mathcal{D}}$ using Algorithm 11.3 of Sect. 11.2, and then apply the Net-trim (NT) method presented in Algorithm 11.1 of Sect. 11.2 to obtain the sparse weight matrix. Specifically, pruning is done on the first fully connected (FC) layer of $f_{\mathcal{D}}$ as it has the highest parameters, i.e., 2.7 M out of the total 2.83 M parameters. The final distilled-pruned model is denoted as $f_{\mathcal{P}}$.

11.3.2.2 Adversarial Attacks on Optimized Models

We evaluate the robustness of the optimized models against five white-box adversarial attacks: fast-gradient method (FGM), fast-gradient sign method (FGSM), projected gradient descent method (PGDM), deepfool, and universal adversarial perturbation (UAP) method. We will briefly discuss the generation of adversarial examples using each of these methods:

- **FGM**: To generate an adversarial example, this attack first computes the gradient of the model's loss function with respect to the input signal. It then adjusts the input signal in the gradient direction to maximize the loss without significantly changing the signal. The adversarial example, denoted as $\mathbf{x}_{adv}$, is generated by solving the constrained optimization problem:

$$\arg\max_{\mathbf{x}_{adv}} \mathcal{L}(\boldsymbol{\theta}, \mathbf{x}_{adv}, \mathbf{y}), \quad \text{s.t.} \quad \|\mathbf{x}_{adv} - \mathbf{x}\|_2 \leq \epsilon\,, \tag{11.35}$$

where $\mathcal{L}(\boldsymbol{\theta}, \mathbf{x}_{adv}, \mathbf{y})$ is the classifier loss function when $\mathbf{x}_{adv}$ is the input signal, $\boldsymbol{\theta}$ is the model parameters, $\mathbf{y}$ is the model prediction for the input $\mathbf{x}_{adv}$, $\mathbf{x}$ is the corresponding clean input signal, $\|\cdot\|_2$ is the L_2 norm, and ϵ is the maximum allowed perturbation, i.e., $\|\mathbf{x}_{adv} - \mathbf{x}\|_2 \leq \epsilon$. The final expression after solving Eq. (11.35) for $\mathbf{x}_{adv}$ is given by

$$\mathbf{x}_{adv} = \mathbf{x} + \epsilon \cdot (\|\nabla_{\mathbf{x}}\mathcal{L}(\boldsymbol{\theta}, \mathbf{x}, \mathbf{y})\|)^{-1}\nabla_{\mathbf{x}}\mathcal{L}(\boldsymbol{\theta}, \mathbf{x}, \mathbf{y})\,. \tag{11.36}$$

- **FGSM**: This is a special case of the FGM attack and follows the same optimization problem in Eq. (11.35) to generate the perturbation, except a new constraint on $\mathbf{x}_{adv}$, i.e., $\|\mathbf{x}_{adv} - \mathbf{x}\|_\infty \leq \epsilon$, i.e., L_∞ norm is used instead of the L_2 norm. The resultant $\mathbf{x}_{adv}$ is given by

$$\mathbf{x}_{adv} = \mathbf{x} + \epsilon \cdot \text{sign}(\nabla_{\mathbf{x}}\mathcal{L}(\boldsymbol{\theta}, \mathbf{x}, \mathbf{y}))\,. \tag{11.37}$$

  Both FGM and FGSM are single-step attacks, which makes them computationally efficient. It is useful as a fast and practical choice for generating adversarial examples.

- **PGDM**: This attack method extends FGSM by applying it with multiple iterations to obtain a finer iterative optimizer. This is a powerful method, which involves refining the $\mathbf{x}_{adv}$ at each iteration by adjusting the perturbation in the direction that maximizes the loss $\mathcal{L}(\cdot)$ while ensuring within the ϵ-neighborhood of the clean signal. The iterative steps are given by

$$\mathbf{x}_0 = \mathbf{x} \tag{11.38}$$

$$\mathbf{x}_{i+1} = \text{clip}_{[\mathbf{x},\epsilon]}\{\mathbf{x}_i + \xi \cdot \text{sign}(\nabla_{\mathbf{x}_i}\mathcal{L}(\boldsymbol{\theta}, \mathbf{x}_i, \mathbf{y})\} \tag{11.39}$$

$$\mathbf{x}_{adv} = \mathbf{x}_T\,, \tag{11.40}$$

  where ξ is a step size, T is the number of iterations, and $\text{clip}_{[\mathbf{x},\epsilon]}\{\mathbf{x}_i\}$ denotes constraining the intermediate sample $\mathbf{x}_i$ in the range $[\mathbf{x}_i - \epsilon, \mathbf{x}_i + \epsilon]$.

- **Deepfool**: This is an iterative attack, which was originally developed for binary classifiers. It is based on the idea that the minimum perturbation required for the misclassification of an input sample will be the orthogonal projection of the sample onto the decision boundary. The iterative optimization problem to generate the minimum perturbation, $\boldsymbol{\delta}_i$, is given by

$$\arg\min_{\delta_i} ||\delta_i||_2, \quad \text{s.t.} \quad f(\mathbf{x}_i) + \nabla_{\mathbf{x}_i} f(\mathbf{x}_i)^\top \delta_i = 0 \,. \tag{11.41}$$

- **UAP**: UAP is a method to generate adversarial perturbations that are input-agnostic and do not depend on the knowledge of the DL model, thus these perturbations are universal in nature. We employ a popular computationally efficient method for generating a UAP based on principal component analysis (PCA), given in [56].

11.3.2.3 Multi-Attack Adversarial Training

Adversarial examples are generated by maximizing the loss function of a DL model as formulated in Eq. (11.35). Therefore, adversarial training exposes the DL model to the adversarial samples by augmenting them with clean training data, and then aims to minimize the classification loss through standard training, which will enhance the robustness of the model against adversarial attacks. The DL model for modulation recognition can be expressed as $f(\mathbf{x}) : \mathbf{x} \subset \mathbb{C}^n \rightarrow \hat{\mathbf{y}} \subset \mathbb{R}^D$, where the input $\mathbf{x}$ represents the complex-valued RF signals with n sample points and $\hat{\mathbf{y}}$ is the output probabilities for D modulation schemes. During the adversarial training process, we augment adversarial examples to the training dataset with the corresponding clean modulation labels so that the loss of these malicious samples is reduced. The overall optimization problem [67] can be expressed by

$$\min_{\boldsymbol{\theta}} \max_{\mathbf{x}_{adv}} \mathcal{L}(\boldsymbol{\theta}, \mathbf{x}_{adv}, \mathbf{y}) \,, \tag{11.42}$$

where $\mathcal{L}(\boldsymbol{\theta}, \mathbf{x}_{adv}, \mathbf{y})$ represents the loss function for the adversarial examples x_{adv} and the clean label $\mathbf{y}$, and $\boldsymbol{\theta}$ is the parameters of the model. The adversarial example is given by $\mathbf{x}_{adv} = \mathbf{x} + \boldsymbol{\eta}$ with $\boldsymbol{\eta}$ being the perturbation.

A key objective is to reduce the cost of adversarial training so that it can be performed locally on an edge device, which will enhance the privacy and security of a specific wireless application. However, incorporating adversarial samples generated from iterative or multi-step attacks that have a high attack success rate significantly increases the computational cost of adversarial training. A possible solution to this issue is incorporating examples from both single-step and multi-step attacks. This will help in reducing the computational complexity of adversarial training because the cost of generating N examples from a multi-step attack is significantly higher than the combined cost of generating N_1 examples from the multi-step attack and N_2 examples from another single-step attack, where $N_1 + N_2 = N$. We have assumed that an equal number of samples are generated from single-step and multi-step attacks, i.e., $N_1 = N_2$. Also, incorporating adversarial examples generated in different gradient directions can effectively improve the robustness of the models. For the exposition, both PGDM and FGSM attacks are used to generate adversarial examples during the adversarial training process. In this manner, the three benchmark models $f_{\mathcal{S}}$, $f_{\mathcal{D}}$, and $f_{\mathcal{P}}$ are adversarially trained to obtain the robust models $f_{\mathcal{S}}^{adv}$, $f_{\mathcal{D}}^{adv}$, and $f_{\mathcal{P}}^{adv}$,

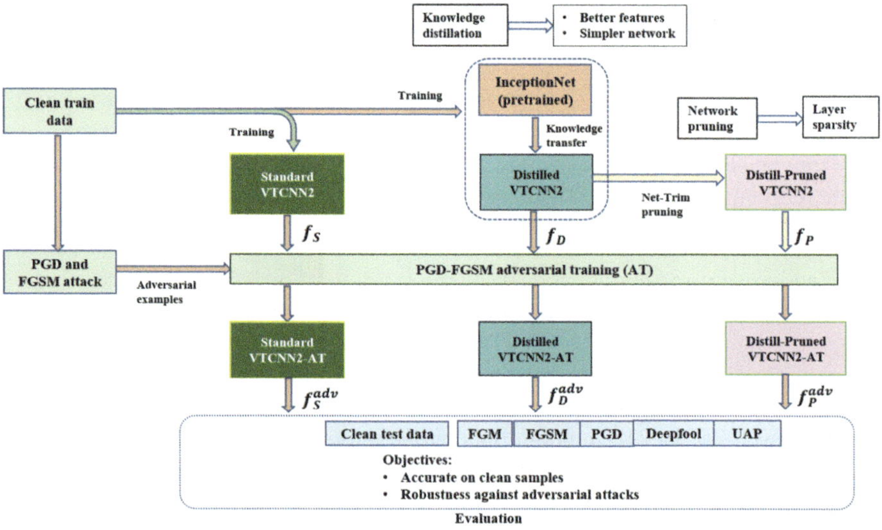

Fig. 11.21 Taxonomy for the evaluation of robustness against adversarial attacks of the proposed optimized models

respectively. As the computational efficiency is directly proportional to the number of trainable parameters, better optimization of the adversarial training process can be achieved by fixing the weights of the first FC layer of $f_{\mathcal{S}}$, $f_{\mathcal{D}}$, and $f_{\mathcal{P}}$ while performing adversarial training.

To demonstrate the effectiveness of the adversarial training method discussed above, the RML2016.10A RF modulation classification dataset [2] is used. The performance of the models is evaluated in terms of i) robustness performance for adversarial test samples generated from five white-box attacks, namely, FGM, FGSM, PGDM, Deepfool, and UAP and ii) classification performance of the models on the clean test samples. The overall taxonomy for the evaluation of the adversarially robust optimized models is shown in Fig. 11.21. Further, we define two quantities in the context of adversarial attacks: the perturbation-to-noise ratio (PNR) and the perturbation-to-signal ratio (PSR). PNR is the relation between the perturbation power and the noise power defined as $\text{PNR} = \epsilon^2 \frac{(\text{SNR}+1)}{||\mathbf{x}||_2^2}$, where $||\mathbf{x}||_2^2$ is the signal power and ϵ is the maximum allowed perturbation. PSR is the relation between the perturbation power and the signal power defined as $\text{PSR} = \text{PNR}/\text{SNR}$.

11.3.2.4 Adversarial Robustness Evaluation on White-Box Attacks

A comparison of the robustness performance of the three adversarially trained models $f_{\mathcal{S}}^{adv}$, $f_{\mathcal{D}}^{adv}$, $f_{\mathcal{P}}^{adv}$ is shown in Fig. 11.22, when tested against the five adversarial attacks at a fixed SNR = 10 dB. It can be observed that the $f_{\mathcal{D}}^{adv}$ and $f_{\mathcal{P}}^{adv}$ models overall perform better than $f_{\mathcal{S}}^{adv}$ against all the attacks. Figure 11.22a shows that for

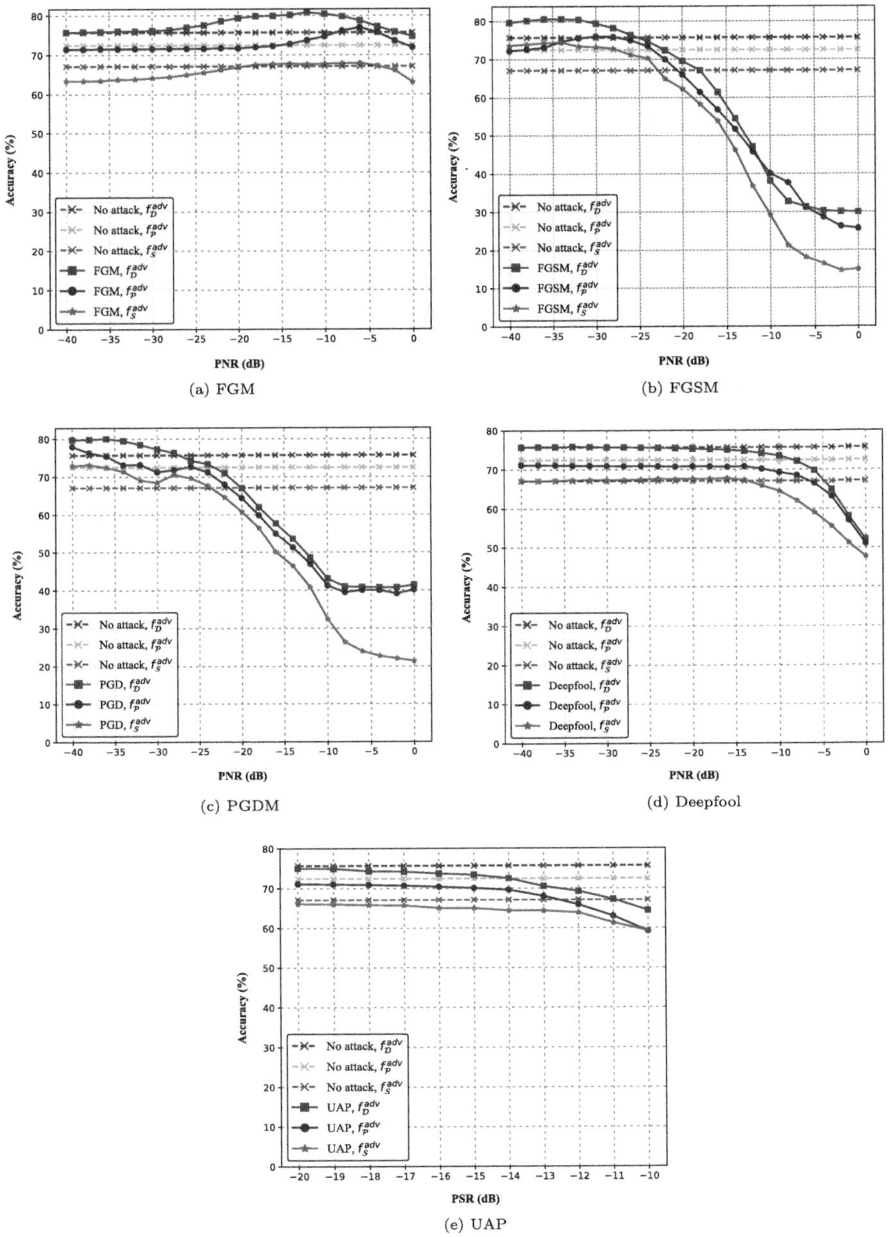

Fig. 11.22 Classification accuracy of the adversarially trained standard and optimized models for adversarial attacks at SNR = 10 dB

FGM attack, $f_{\mathcal{D}}^{adv}$ performs significantly better than $f_{\mathcal{S}}^{adv}$, with an average accuracy gain of 12% across all PNRs. The accuracy of $f_{\mathcal{P}}^{adv}$ is comparable to $f_{\mathcal{D}}^{adv}$ at higher PNRs for FGM attack. Also, robustness performance against FGSM and PGDM attacks in Fig. 11.22b and c, respectively, shows that both $f_{\mathcal{D}}^{adv}$ and $f_{\mathcal{P}}^{adv}$ performs significantly better than $f_{\mathcal{S}}^{adv}$ at the high PNR values.

The presented results signify that a DL model optimized using KD can achieve improved adversarial robustness than the standard model when both are subjected to a computationally efficient adversarial training process. We have also observed that pruning the distilled model can provide better adversarial robustness than the standard model. This benefits edge applications as we can achieve high sparsity and robustness simultaneously with the distilled-pruned model. Our evaluation also proves that performing adversarial training with the combination of a single-step attack (FGSM) and an iterative attack (PGDM) can be computationally efficient as this also achieves robustness against unseen single and multi-step attacks, such as FGM, Deepfool, and UAP, without incorporating any adversarial samples from these methods during the adversarial training process.

11.3.2.5 Classification Performance Evaluation on Clean Samples

It is also important to evaluate the classification performance of the optimized and the standard models for the clean samples, with and without adversarial training. Figure 11.23 shows a comparison of the performances of the three model variants at SNR = 10 dB. The adversarial training process can lead to a reduction in accuracy

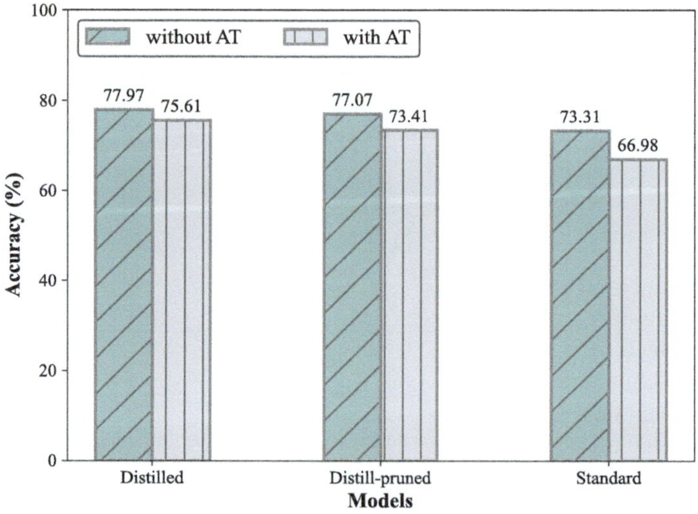

Fig. 11.23 Classification accuracy of the models on clean samples with and without adversarial training (AT) at SNR = 10 dB

for the clean samples (clean accuracy) when compared to the model without adversarial training. The reduction in clean accuracy affects the reliability of the model for classifying received signals without any attack. Therefore, it is important to minimize the decrease in the clean accuracy. It can be observed in Fig. 11.23 that the drop in clean accuracy after adversarial training is the lowest for $f_{\mathcal{D}}^{adv}$ and the highest for $f_{\mathcal{S}}^{adv}$. Specifically, the clean accuracies of $f_{\mathcal{D}}$, $f_{\mathcal{P}}$, and $f_{\mathcal{S}}$ (models before adversarial training) are 77.97%, 77.07%, and 73.31%, respectively. After performing adversarial training, the accuracies of the models $f_{\mathcal{D}}^{adv}$, $f_{\mathcal{P}}^{adv}$, and $f_{\mathcal{S}}^{adv}$ are 75.61%, 73.41%, and 66.98%, respectively. It can be observed that both distilled and distilled-pruned models have higher accuracies on clean samples than the standard model, before as well as after adversarial training. This shows that KD also helps the lightweight model to learn more robust features and is not affected significantly by adversarial training.

11.3.3 Adversarial Attacks and Robustness in DL-Based Power Allocation

Optimal power allocation plays an important role in the performance and efficiency of wireless communication systems. We have already discussed in Sect. 11.1 how DL models are used to obtain the transmit power coefficients optimally and minimize the complexity as compared to the conventional optimization technique. However, adversarial attacks pose a major security threat to reliable power allocation by misleading the DL models. Thus, developing robustness in these models is essential to ensure secure and reliable real-time power allocation in wireless communication systems. Adversarial attacks were developed originally for the classification-based DL models. Later, the vulnerability of DL models for regression-based applications in wireless communications was shown in [7], specifically for DL-based downlink power allocation in a massive MIMO system. In [7], a white-box attack scenario is considered where the adversary has information about the DL model and the user equipment (UE) positions. For a massive MIMO system consisting of L cells with each cell having K UEs, the DL model for power allocation can be expressed as $f(\mathbf{x}_t) : \mathbf{x}_t \in \mathcal{X} \to \hat{\varrho}_j \in \mathcal{P}$, where the input $\mathbf{x}_t \subset \mathbb{R}^{2KL}$ represents the co-ordinates of the UEs and $\hat{\varrho}_j \subset \mathbb{R}^{K+1}$ is the output predicted power allocated to the UEs in jth cell. The output dimension is $K + 1$ rather than K, this is because the DL models are also trained to learn the power constraint (i.e., the sum of UEs powers must be less than or equal to the total available power).

Adversarial samples $\mathbf{x}_{adv}$ are generated in the ϵ-neighborhood of the clean sample $\mathbf{x}_t$ to generate an erroneous model output, and it can be expressed as [7]

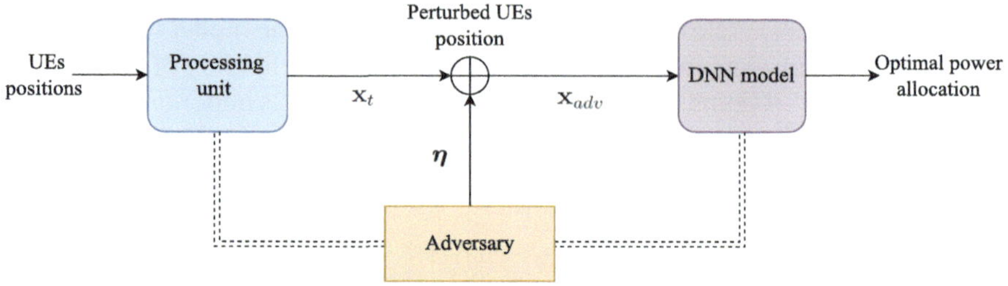

Fig. 11.24 White-box attack model for DL-based downlink power allocation

$$\begin{aligned}\mathbf{x}_{adv} &= \mathbf{x}_t + \boldsymbol{\eta} \\ \text{s.t.} \quad & f(\mathbf{x}_t) = \hat{\boldsymbol{\varrho}}_j \,, \; f(\mathbf{x}_{adv}) = \hat{\boldsymbol{\varrho}}_j^{adv} \,, \\ & \sum_{i=1}^{K} \hat{\boldsymbol{\varrho}}_{ji} \leq P_{\max} \,, \; \sum_{i=1}^{K} \hat{\boldsymbol{\varrho}}_{ji}^{adv} > P_{\max} \,, \\ & \mathbf{x}_{adv}, \mathbf{x}_t \in \mathcal{X} \,, \\ & \text{and} \;\; \|\boldsymbol{\eta}\|_\infty \leq \epsilon \,, \end{aligned} \tag{11.43}$$

where $\hat{\boldsymbol{\varrho}}_j = [\hat{\varrho}_{j1}, \hat{\varrho}_{j2}, \ldots, \hat{\varrho}_{jK}]$ and $\hat{\boldsymbol{\varrho}}_j^{adv} = [\hat{\varrho}_{j1}^{adv}, \hat{\varrho}_{j2}^{adv}, \ldots, \hat{\varrho}_{jK}^{adv}]$ are the predicted outputs by the model $f(\cdot)$ for the inputs $\mathbf{x}_t$ and $\mathbf{x}_{adv}$, respectively. A perturbation $\boldsymbol{\eta}$ is introduced to the input $\mathbf{x}_t$ with the L_∞-norm $\|\boldsymbol{\eta}\|_\infty$ is constrained to be less than ϵ, which signifies the maximum allowable perturbation magnitude. In the ideal scenario, the sum of UE powers in each cell j should be less than $P_{\max}$, i.e., $\sum_{i=1}^{K} \hat{\varrho}_{jk} \leq P_{\max}$, which is called as the feasible solution of the model. As the perturbed UEs positions $\mathbf{x}_{adv} = \mathbf{x}_t + \boldsymbol{\eta}$ lie within the area of cell j, it is expected that the predicted power coefficients $f(\mathbf{x}_{adv})$ to be in the same order as $f(\mathbf{x}_t)$ as long as $\|\boldsymbol{\eta}\|_\infty \leq \epsilon$ is satisfied. However, if the adversary adds the well-crafted perturbation at the input of the DL model, it is possible to break the power constraint and the predicted adversarial power $\boldsymbol{\varrho}_j^{adv} = [\hat{\varrho}_{j1}^{adv}, \ldots, \hat{\varrho}_{jK}^{adv}]$ may produce infeasible solutions, that is, $\sum_{i=1}^{K} \hat{\varrho}_{ji}^{adv} > P_{\max}$ as shown in Eq. (11.43). A cell-based adversarial attack scenario on DL-based models for the downlink power allocation in a massive MIMO network can be visualized in Fig. 11.24. The adversarial perturbations are designed in a white-box manner that provides the worst-case scenario for analyzing the robustness of the DL models.

The generation of adversarial examples focuses on maximizing the loss function of the DL model by imposing the constraint $\|\mathbf{x}_{\text{adv}} - \mathbf{x}_t\|_\infty \leq \epsilon$. For a power allocation problem, the adversarial samples can be generated by maximizing the loss function [7], as given by

$$\mathcal{L}(\boldsymbol{\theta}, \cdot) = \sum_{i=1}^{K} \hat{\varrho}_{ji} \,, \tag{11.44}$$

where $\hat{\varrho}_{ji}$ represents the power allocated by the DL model to the ith UE in the jth cell. The goal of the adversary is to generate the adversarial example $\mathbf{x}_{adv}$ that leads to an infeasible power solution as given by $\sum_{i=1}^{K} \hat{\varrho}_{ji} > P_{\max}$ [7]. It can be achieved through the formulation of the subsequent optimization problem in the downlink power allocation for UEs in a multi-cell massive MIMO system as

$$\begin{aligned} \arg\min_{\mathbf{x}_{adv}} \quad & \|\mathbf{x}_{adv} - \mathbf{x}_t\|_\infty \\ \text{s.t.} \quad & f(\mathbf{x}_{adv}) = \hat{\boldsymbol{\varrho}}_j^{adv}, \ \sum_{i=1}^{K} \hat{\varrho}_{ji}^{adv} > P_{\max} \\ & \mathbf{x}_{adv} \in \mathcal{X} \text{ and } \|\mathbf{x}_{adv} - \mathbf{x}_t\|_\infty \le \epsilon . \end{aligned} \tag{11.45}$$

The generation of adversarial examples through the optimization problem in Eq. (11.45) is challenging in practice to solve due to its intractability. Many sub-optimal techniques are proposed in the literature [7, 56] to efficiently generate adversarial examples in terms of computational resources. One such technique is the class of gradient-based method using the L_∞-norm of the perturbation constraint [7, 68]. The rationale behind opting for the L_∞-norm lies in the fact that the adversarial perturbation applied to the UE positions results in the maximum distance perturbation (d_ϵ) from the actual positions. The interpretation of the L_∞-norm bound is that the perturbations in UE positions lead to the maximum displacement in the direction of the gradient of the loss function, causing an increase in loss in the steepest direction. We know that the adversarial perturbation constraint is given by $\boldsymbol{\eta} \le \epsilon$; this implies that L_∞-norm is upper bounded by the perturbation magnitude ϵ. Thus, the relation between the distance perturbations and the perturbation magnitude is given by $d_\epsilon \le \sqrt{2}\epsilon$ [7]. The perturbations of UEs position values are not just random noise rather, they are intentionally displaced in well-defined directions to fool the DL model. In general, the attack success rate is used as a performance metric to evaluate the robustness of the DL models against adversarial attacks. That is, we can obtain the attack success rate for each cell by examining a set of adversarial examples as inputs to the DL model and then determining the number of instances when the model yields an infeasible power solution as output. In this manner, it becomes possible to examine the efficacy of adversarial training as a defense mechanism for the worst-case assumption of DL models in terms of infeasible output power (i.e., the sum of UE powers in each cell j is more than $P_{\max}$).

11.3.3.1 Countermeasure Technique: Adversarial Training

Adversarial training is a defense technique against adversarial attacks and is characterized as a method where the models are trained to minimize the maximum possible loss due to the given perturbation magnitude (ϵ). This can be mathematically formulated as an optimization problem [67], expressed by

$$\min_{\theta} \max_{\|\boldsymbol{\eta}\|_\infty \le \epsilon} \mathcal{L}(\boldsymbol{\theta}, \mathbf{x}_{adv}, \boldsymbol{\varrho}_j), \tag{11.46}$$

where $\mathcal{L}(\boldsymbol{\theta}, \mathbf{x}_{adv}, \boldsymbol{\varrho}_j)$ is the adversarial loss function given in Eq. (11.44) with adversarial examples $\mathbf{x}_{adv} = \mathbf{x}_t + \boldsymbol{\eta}$ as the input, $\boldsymbol{\varrho}_j$ as the original power output for the jth cell, and $\boldsymbol{\theta}$ as the parameters of the model $f(\cdot)$. The maximum possible loss is limited to perturbations bounded by the L_∞-norm magnitude of ϵ. From Eq. (11.46), the inner maximization aims to generate the adversarial examples using adversarial attack methods, such as FGSM and PGDM, and on the other hand, the objective of outer minimization is to reduce the loss function due to the generated adversarial examples by utilizing the standard training procedure.

In [69], an FGSM-based adversarial training method is proposed to improve the robustness of a DL model by including FGSM adversarial samples during training. This method achieved robustness against the FGSM attack, but the model was still vulnerable to iterative attacks, like PGDM and momentum iterative (MI)-FGSM. In [7], the gradient-based adversarial attacks with L_∞-norm (FGSM, PGDM, and MI-FGSM) are examined and found PGDM as the most powerful attack, with a high possibility of misleading the DL model. In the work presented in [7] for cell-based adversarial training, the PGDM adversarial examples generated based on L_∞-norm are used against the DL-based power allocation model in a massive MIMO system as explained below:

1. Create a DL model, $f_{\text{std}}(\cdot)$, using the training dataset $\mathbf{x} \in \mathcal{X}$ and $\boldsymbol{\varrho}_j \in \mathcal{P}$ for each cell j, $j = 1, \ldots, L$. It is known as standard training, in which the DL model takes the original clean UE locations as input and outputs optimal power coefficients for each UE.
2. Train a DL model, $f_{adv}(\cdot)$, using the adversarial examples $\mathbf{x}_{adv} \in \mathcal{X}$ with clean outputs $\boldsymbol{\varrho}_j \in \mathcal{P}$ for each cell j, $j = 1, \ldots, L$. This approach is known as adversarial training because the DL model considers adversarially perturbed UE locations as input and returns ground-truth power coefficients of UEs as output.
3. According to [70], overfitting during adversarial training might lead to a decrease in performance on the test dataset. This shows that, unlike standard training, overfitting in adversarial training has the effect of increasing the loss on the test dataset. To solve this issue, in [7], a validation-based early stopping is performed that will terminate the learning process when the validation loss on the adversarial validation dataset does not reduce after a pre-defined number of epochs.
4. Finally, the adversarially trained model $f_{adv}(\cdot)$ is evaluated for robustness against different adversarial attacks, such as FGSM, PGDM, MI-FGSM, and random perturbations, on the test dataset. The test dataset contains clean samples that are different from those used during adversarial training and validation.

11.3.3.2 Standard Training Results

The adversarial attacks presented in [7] are compared in terms of the attack success rate, which shows the effectiveness of the attacks in fooling the DL-based approach. A higher success rate indicates a higher probability of inducing erroneous predictions by the DL model. Two model architectures trained with the standard training process are considered as the victim models, namely, M_1 and M_2 (i.e., Figs. 11.16

and 11.17, respectively). In a white-box manner, the objective is to ascertain how many test adversarial examples can yield feasible and infeasible power solutions as output. Figure 11.25a–d illustrates a cell-based comparison of attack success rate for random perturbations, FGSM, MI-FGSM, and PGDM attacks aimed at deceiving the networks M_1 with M-MMSE precoding, M_2 with M-MMSE precoding, M_1 with MR precoding, and M_2 with MR precoding, respectively. All models are trained using standard training. The performances of the attacks are also evaluated for different perturbation magnitudes in each of the four cases. For random perturbations, the displacement of UE positions by d_ϵ occurs in the random directions with a probability $\Pr(\epsilon) = \Pr(-\epsilon) = 0.5$, while for the gradient-based adversarial perturbation, the displacement of UE positions occurs in the gradient direction such that the loss is maximized. Figure 11.25a–d shows that the attack success rate can be extremely high even for a very small ϵ and increasing the value of ϵ increases the attack success rate. Another important observation is that the attack success rate for the iterative gra-

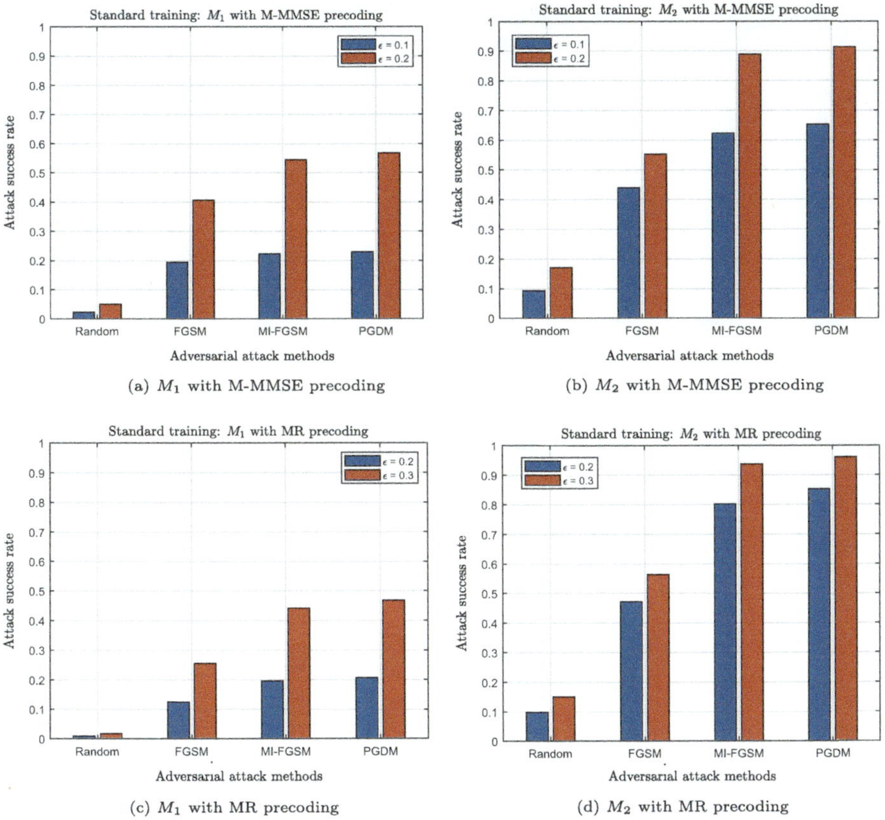

Fig. 11.25 Adversary success rate of different white-box attacks using standard training on: **a)** M_1 with M-MMSE precoding, **b)** M_2 with M-MMSE precoding, **c)** M_1 with MR precoding, and **d)** M_2 with MR precoding

dient update-based attacks (i.e., MI-FGSM and PGDM) is higher than the one-step gradient update attack (i.e., FGSM and random perturbations).

11.3.3.3 Adversarial Training Results

To enhance the robustness of the DL models, adversarial training is conducted on both M_1 and M_2 models, utilizing samples generated by the PGDM attack. The attack success rate is evaluated against random perturbation, FGSM, MI-FGSM, and PGDM attacks with M-MMSE and MR precoding schemes. Figure 11.26a–d compares the cell-based attack success rate of four white-box attacks against the adversarially trained networks: M_1 with M-MMSE precoding, M_2 with M-MMSE precoding, M_1 with MR precoding, and M_2 with MR precoding, respectively. For each model, the results with two values of the maximum allowable perturbations

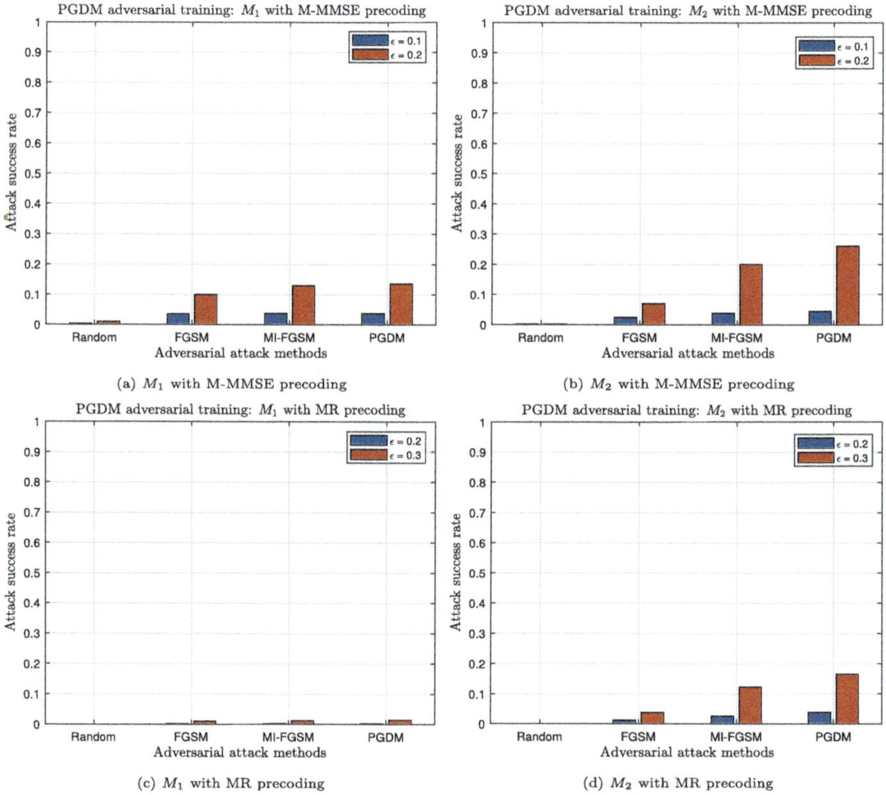

Fig. 11.26 Adversary success rate of different white-box attacks using adversarial training on: **a)** M_1 with M-MMSE precoding, **b)** M_2 with M-MMSE precoding, **c)** M_1 with MR precoding, and **d)** M_2 with MR precoding

ϵ are compared. It is observed in Fig. 11.26a–d that the robustness of the models M_1 and M_2 is significantly enhanced due to the adversarial training process. This reduction in the attack success rate signifies that the number of adversarial samples that can successfully mislead the DL model to generate the infeasible power output is significantly reduced by making the model learn the associations between the adversarial examples and the clean output.

Acknowledgement Some of the works presented in this book section was supported in part by the SERB Start-Up Research Grant (SRG) Scheme, Department of Science and Technology (DST), Govt. of India under Grant SRG/2022/001214 and in part by Start-Up Grant of Indian Institute of Technology Guwahati.

References

1. O'Shea TJ, Corgan J, Clancy TC (2016) Convolutional radio modulation recognition networks. In: Proceedings of international conference on engineering applications of neural networks (EANN), Aberdeen, UK, 2–5 Sep 2016, pp 213–226
2. O'Shea T, West N (2016) Radio machine learning dataset generation with GNU Radio. In: Proceedings of GNU Radio conference, vol 1, no 1, Boulder, CO, USA, Sep 2016
3. O'Shea TJ, Roy T, Clancy TC (2018) Over-the-air deep learning based radio signal classification. IEEE J Select Topics Signal Proc 12(1):168–179
4. Zhou R, Liu F, Gravelle CW (2020) Deep learning for modulation recognition: a survey with a demonstration. IEEE Access 8:67 366–67 376
5. He C, Zhou Y, Qian G, Li X, Feng D (2019) Energy efficient power allocation based on machine learning generated clusters for distributed antenna systems. IEEE Access 7:59 575–59 584
6. Sanguinetti L, Zappone A, Debbah M (2018) Deep learning power allocation in massive MIMO. In: Proceedings of asilomar conference on signals, systems, and computers, Pacific Grove, CA, USA, Oct 2018, pp 1257–1261
7. Manoj BR, Sadeghi M, Larsson EG (2022) Downlink power allocation in massive MIMO via deep learning: adversarial attacks and training. IEEE Trans Cognit Commun Netw 8(2):707–719
8. Wang S, Wu Y-C, Xia M, Wang R, Poor HV (2020) Machine intelligence at the edge with learning centric power allocation. IEEE Trans Wireless Commun 19(11):7293–7308
9. Emir A, Kara F, Kaya H, Li X (2021) Deep learning-based flexible joint channel estimation and signal detection of multi-user OFDM-NOMA. Phys Commun 48:101443
10. Yonina Eldar AG, Gündüz D, Poor HV (2022) Machine learning and wireless communications. Cambridge University Press
11. Zhu G, Liu D, Du Y, You C, Zhang J, Huang K (2020) Toward an intelligent edge: wireless communication meets machine learning. IEEE Commun Mag 58(1):19–25
12. Liu H, Wei Z, Zhang H, Li B, Zhao C (2022) Tiny machine learning (tiny-ml) for efficient channel estimation and signal detection. IEEE Trans Veh Technol 71(6):6795–6800
13. Hossain MA, Noor RM, Yau K-LA, Azzuhri SR, Z'aba MR, Ahmedy I (2020) Comprehensive survey of machine learning approaches in cognitive radio-based vehicular ad hoc networks. IEEE Access 8:78 054–78 108
14. Thilina KM, Choi KW, Saquib N, Hossain E (2013) Machine learning techniques for cooperative spectrum sensing in cognitive radio networks. IEEE J Sel Areas Commun 31(11):2209–2221
15. O'Shea TJ, Karra K, Clancy TC (2016) Learning to communicate: channel auto-encoders, domain specific regularizers, and attention. In: Proceedings of IEEE international symposium

on signal processing and information technology (ISSPIT), Limassol, Cyprus, 12–14 Dec 2016, pp 223–228
16. Zhang L, Lambotharan S, Zheng G, Sadhan BA, Roli F (2021) Countermeasures against adversarial examples in radio signal classification. IEEE Wireless Commun Lett 10(8):1830–1834
17. Ye H, Li GY, Juang BH (2018) Power of deep learning for channel estimation and signal detection in OFDM systems. IEEE Wireless Commun Lett 7(1):114–117
18. Vagollari A, Hirschbeck M, Gerstacker W (2023) An end-to-end deep learning framework for wideband signal recognition. IEEE Access 11:52 899–52 922
19. Farsad N, Shlezinger N, Goldsmith AJ, Eldar YC (2021) Data-driven symbol detection via model-based machine learning. In: Proceedings of IEEE statistical signal processing workshop (SSP), Rio de Janeiro, Brazil, 11–14 Jul 2021, pp 571–575
20. Shlezinger N, Farsad N, Eldar YC, Goldsmith AJ (2020) Viterbinet: a deep learning based viterbi algorithm for symbol detection. IEEE Trans Wireless Commun 19(5):3319–3331
21. Manoj BR, Tian G, Gunnarsson S, Tufvesson F, Larsson EG (2021) Sensing and classification using massive MIMO: a tensor decomposition-based approach. IEEE Wireless Commun Lett 10(12):2649–2653
22. Manoj BR, Tian G, Willhammar S, Tufvesson F, Larsson EG (2021) Moving object classification with a Sub-6 GHz massive MIMO array using real data. In: Proceedings of IEEE international conference on acoustics, speech and signal processing (ICASSP), Toronto, ON, Canada, 06–11 June 2021, pp 8133–8137
23. Arnold M, Hoydis J, Brink ST (2019) Novel massive MIMO channel sounding data applied to deep learning-based indoor positioning. In: Proceedings of 12th international ITG conference on systems, communications and coding (SCC), Rostock, Germany, 11–14 Feb 2019, pp 1–6
24. Wen C, Shih W, Jin S (2018) Deep learning for massive MIMO CSI feedback. IEEE Wireless Commun Lett 7(5):748–751
25. He H, Wen C-K, Jin S, Li GY (2018) Deep learning-based channel estimation for beamspace mmwave massive MIMO systems. IEEE Wireless Commun Lett 7(5):852–855
26. Soltani M, Pourahmadi V, Mirzaei A, Sheikhzadeh H (2019) Deep learning-based channel estimation. IEEE Commun Lett 23(4):652–655
27. Van Nguyen T, Jeong Y, Shin H, Win MZ (2015) Machine learning for wideband localization. IEEE J Sel Areas Commun 33(7):1357–1380
28. Monga V, Li Y, Eldar YC (2021) Algorithm unrolling: interpretable, efficient deep learning for signal and image processing. IEEE Signal Process Mag 38(2):18–44
29. Forney GD (1973) The viterbi algorithm. Proc IEEE 61(3):268–278
30. Bishop, CM (1994) Mixture density networks. http://www.ncrg.aston.ac.uk/
31. Proakis (2007) Digital communications, 5th edn. McGraw Hill
32. Sathyanarayanan V, Gerstoft P, Gamal AE (2023) RML22: Realistic dataset generation for wireless modulation classification. IEEE Trans Wireless Commun 22(11):7663–7675
33. O'Shea TJ, Corgan J, Clancy TC (2016) Convolutional radio modulation recognition networks. In: Proceedings of international conference on engineering applications of neural networks (EANN), Aberdeen, UK, 2–5 Sep 2016, pp 213–226
34. Zhang Z, Luo H, Wang C, Gan C, Xiang Y (2020) Automatic modulation classification using CNN-LSTM based dual-stream structure. IEEE Trans Veh Technol 69(11):13 521–13 531
35. Brauwers G, Frasincar F (2023) A general survey on attention mechanisms in deep learning. IEEE Trans Knowl Data Eng 35(4):3279–3298
36. Erdogmus D, Rende D, Principe J, Wong T (2001) Nonlinear channel equalization using multilayer perceptrons with information-theoretic criterion. In: Neural networks for signal processing XI: proceedings of IEEE signal processing society workshop, North Falmouth, MA, USA 12–12Sep, pp 443–451
37. Omar S, Ancora A, Slock DT (2008) Performance analysis of general pilot-aided linear channel estimation in LTE OFDMA systems with application to simplified MMSE schemes. In: Proceedings of IEEE 19th international symposium on personal, indoor and mobile radio communications, Cannes, France, 15–18 Sep 2008, pp 1–6

38. Šimko M, Mehlführer C, Wrulich M, Rupp M (2010) Doubly dispersive channel estimation with scalable complexity. In: Proceedings of international ITG workshop on smart antennas (WSA), Bremen, Germany, 23–24 Feb 2010, pp 251–256
39. Dong C, Loy CC, He K, Tang X (2016) Image super-resolution using deep convolutional networks. IEEE Trans Pattern Anal Mach Intell 38(2):295–307
40. Zhang K, Zuo W, Chen Y, Meng D, Zhang L (2017) Beyond a Gaussian denoiser: residual learning of deep CNN for image denoising. IEEE Trans Image Process 26(7):3142–3155
41. Tian X, Xiong K, Zhang R, Fan P, Niyato D, Letaief KB (2024) Sum rate maximization in multi-cell multi-user networks: an inverse reinforcement learning-based approach. IEEE Wireless Commun Lett 13(1):4–8
42. Nasir YS, Guo D (2019) Multi-agent deep reinforcement learning for dynamic power allocation in wireless networks. IEEE J Sel Areas Commun 37(10):2239–2250
43. Mishra R, Gupta HP, Dutta T (2020) A survey on deep neural network compression: challenges, overview, and solutions. arXiv:2010.03954
44. Cheng Y, Wang D, Zhou P, Zhang T (2017) A survey of model compression and acceleration for deep neural networks. https://ui.adsabs.harvard.edu/abs/2017arXiv171009282C
45. West NE, O'shea T (2017) Deep architectures for modulation recognition. In: Proceedings of IIEEE international symposium on dynamic spectrum access networks (DySPAN) (DySPAN), Baltimore, MD, USA, 6–9 Mar 2017, pp 1–6
46. Aghasi A et al (2017) Net-trim: convex pruning of deep neural networks with performance guarantee. In: Proceedings of advances in neural information processing systems (NIPS), Long Beach, CA, USA, 4–7 Dec 2017, pp 3180–3189
47. Jegou H, Douze M, Schmid C (2010) Product quantization for nearest neighbor search. IEEE Trans Pattern Anal Mach Intell 32(1):1–15
48. Hinton G, Vinyals O, Dean J (2015) Distilling the knowledge in a neural network. arXiv:1503.02531
49. Vadera S, Ameen S (2022) Methods for pruning deep neural networks. IEEE Access 10:63 280–63 300
50. Liao Z, Quétu V, Nguyen V-T, Tartaglione E (2023) Can unstructured pruning reduce the depth in deep neural networks? In: Proceedings of the IEEE/CVF international conference on computer vision (ICCV) Workshops, Paris, France, 02–06 Oct 2023, pp 1402–1406
51. Denil M, Shakibi B, Dinh L, Ranzato MA, De Freitas N (2013) Predicting parameters in deep learning. In: Proceedings of advances in neural information processing systems (NIPS), Lake Tahoe, NV, USA, 5–10 Dec 2013, pp 2148–2156
52. Sagduyu YE, Erpek T, Shi Y (2021) Adversarial Machine Learning for 5G Communications Security. Wiley, Chap 14, pp 270–288. https://onlinelibrary.wiley.com/doi/abs/10.1002/9781119723950.ch14
53. Pajola L, Pasa L, Conti M (2019) Threat is in the air: Machine learning for wireless network applications. In: Proceedings of ACM workshop on wireless security and machine learning, New York, NY, USA, 15 May 2019, pp 16–21
54. Adesina D, Hsieh C-C, Sagduyu YE, Qian L (2023) Adversarial machine learning in wireless communications using RF data: a review. IEEE Commun Surv & Tutor 25(1):77–100
55. Sagduyu YE, Shi Y, Erpek T (2021) Adversarial deep learning for over-the-air spectrum poisoning attacks. IEEE Trans Mob Comput 20(2):306–319
56. Sadeghi M, Larsson EG (2019) Adversarial attacks on deep-learning based radio signal classification. IEEE Wireless Commun Lett 8(1):213–216
57. Liu Q, Guo J, Wen C-K, Jin S (2020) Adversarial attack on DL-based massive MIMO CSI feedback. J Commun Netw 22(3):230–235
58. Cohen J, Rosenfeld E, Kolter Z (2019) Certified adversarial robustness via randomized smoothing. In: Proceedings of the 36th international conference on machine learning (ICML), vol 97, 09–15 June 2019, pp 1310–1320
59. Luo Z, Zhao S, Lu Z, Xu J, Sagduyu YE (2022) When attackers meet AI: Learning-empowered attacks in cooperative spectrum sensing. IEEE Trans Mob Comput 21(5):1892–1908

60. Davaslioglu K, Sagduyu YE (2019) Trojan attacks on wireless signal classification with adversarial machine learning. In: Proceedings of international symposium on dynamic spectrum access networks (DySPAN), Newark, NJ, USA, 11–14 Nov 2019, pp 1–6
61. Chen B et al (2019) Detecting backdoor attacks on deep neural networks by activation clustering. In: Workshop on artificial intelligence safety, AAAI-19, vol 2301, Honolulu, Hawaii, USA, Jan 27, 2019
62. Wang B et al (2019) Neural cleanse: identifying and mitigating backdoor attacks in neural networks. In: Proceedings of IEEE symposium on security and privacy (SP), San Fransisco, CA, USA, 20–22 May 2019, pp 707–723
63. Shi Y, Davaslioglu K, Sagduyu YE (2020) Over-the-air membership inference attacks as privacy threats for deep learning-based wireless signal classifiers. In: Proceedings of ACM workshop on wireless security and machine learning, New York, NY, USA, 16 July 2020, pp 61–66
64. Manoj BR, Santos PM, Sadeghi M, Larsson EG (2022) Toward robust networks against adversarial attacks for radio signal modulation classification. In: Proceedings of IEEE 23rd international workshop on signal processing advances in wireless communication (SPAWC), Oulu, Finland, 04–06 July 2022, pp 1–5
65. Bahramali A, Nasr M, Houmansadr A, Goeckel D, Towsley D (2021) Robust adversarial attacks against DNN-based wireless communication systems. In: Proceedings of ACM SIGSAC conference computations communications security, New York, NY, USA, 15–19 Nov 2021, pp 126–140
66. Maroto J, Bovet G, Frossard P (2021) SafeAMC: adversarial training for robust modulation recognition models. arXiv:2105.13746
67. Ren K, Zheng T, Qin Z, Liu X (2020) Adversarial attacks and defenses in deep learning. Engineering 6(3):346–360
68. Manoj BR, Sadeghi M, Larsson EG (2021) Adversarial attacks on deep learning based power allocation in a massive MIMO network. In: Proceedings of IEEE international conference on communications (ICC), Montreal, QC, Canada, 14–23 June 2021, pp 1–6
69. Goodfellow IJ, Shlens J, Szegedy C (2015) Explaining and harnessing adversarial examples. In: Proceedings of international conference on learning representations (ICLR), San Diego, CA, USA, 7–9 May 2015
70. Rice L, Wong E, Kolter Z (2020) Overfitting in adversarially robust deep learning. In: Proceedings of the 37th international conference on machine learning (ICML), vol 119, Vienna, Austria, 13–18 Jul 2020, pp 8093–8104

Chapter 12
Rateless Coding for Reliable Long Range IoT

Siddhartha S. Borkotoky

Abstract IoT networks of the future are expected to support applications that demand both energy efficiency and reliability. In recent times, the Long Range (LoRa) technology has been widely adopted for facilitating energy-efficient IoT networks spanning large geographical regions. However, ensuring reliable data transfer in LoRa networks remains a challenge. In this chapter, we describe the integration of erasure-correcting rateless codes with LoRa for reliable and energy-efficient transfer of small files. An analytical framework is presented for the performance evaluation of rateless-coded LoRa. Via analyses and simulations, the energy savings of rateless coding relative to LoRa's native acknowledgement-based reliability mechanism are illustrated. The dependence of energy consumption on the node density, source-destination separation, and coding parameters is analyzed. The issue of optimal selection of coding parameters is also addressed.

Keywords IoT · LoRa · Rateless coding · Reliability

12.1 Introduction

The Internet-of-Things (IoT) ecosystem has been proliferating at a rapid pace over the past decade. The global Internet-of-Things (IoT) market is projected to grow from $662 billion in 2023 to $3.4 trillion by 2030 [1]. This growth has been fueled by increased adoption of IoT-driven paradigms such as smart city [2], smart agriculture [3], Industry 4.0 [4], to name a few. A key component of any IoT application is the underlying communication network that connects the end devices (the "things") to the global Internet or integrates them with a smaller private network. While the nature of the end devices and their communication needs depend on the specific application being served, a large fraction IoT networks consist of end devices that are sensor nodes that perform measurements and transmit them from time to time. Examples

S. S. Borkotoky (✉)
Indian Institute of Technology Bhubaneswar, Khorda, Odisha, India
e-mail: borkotoky@iitbbs.ac.in

S. Sharma et al. (eds.), *Next-Generation Wireless Systems*,
Transactions on Computer Systems and Networks,
https://doi.org/10.1007/978-981-96-3185-8_12

of typical sensor data are measurements of physical parameters such as temperature, pressure, humidity, vibration, air quality, etc. The sensors transmit their readings to one or more gateways, which forward them to a data sink for processing. The communication between the sensor nodes and the gateways has traditionally been enabled by technologies such as Wi-Fi, Bluetooth, ZigBee, and cellular technologies (e.g., 3G and 4G), each involving trade-offs among data rate, energy consumption, and coverage. For example, Wi-Fi enables high-rate data transfer over short distances but with high power consumption. Bluetooth provides low-power communications, but over even shorter distances. ZigBee provides ranges comparable to Wi-Fi, but with lower power and lower data rates, whereas cellular technologies facilitate high data-rate communications over long distances at the expense of high power consumption and high protocol complexity. However, many of the emerging IoT applications require the transmission of small amounts of data over long distances with minimal power consumption, a set of requirements that are not simultaneously fulfilled by any one of the aforementioned technologies. Examples of scenarios having such requirements are smart city and agricultural monitoring setups, wherein sensors spread over an urban area or a large agricultural field measure parameters such as air quality, ambient lighting, temperature, soil moisture content, etc., and transmit them to a gateway located up to several kilometers away. Of late, the LoRa technology [5] has emerged to support such applications, in which the amount of information to be transmitted per sensor is rather small, the required communication range is large, the sensor's energy consumption must be kept as low as possible, and low data rates are tolerable. By utilizing the chirp spread spectrum modulation technique, LoRa can transfer data over distances of up to several kilometers, while transmitting at a low data rate (between 0.3 and 27 kbps) with modest power levels [6].

While LoRa greatly improves the energy efficiency and coverage of sensor networks, its major drawback is the susceptibility to interference, which detrimentally impacts the communication reliability by causing frame losses. Interference in LoRa communications occurs due to two reasons. First, LoRa operates over the unlicensed spectrum [7], over which any user can transmit, causing signal collisions. Secondly, the link-layer protocol commonly used with LoRa (known as LoRaWAN [7]) employs pure ALOHA for channel access; this keeps the implementation complexity low by avoiding synchronization and resource allocation requirements, but the uncoordinated transmissions cause collisions and enforce frame losses. Therefore, if the end application being served demands guaranteed data delivery, additional reliability-ensuring mechanisms must be incorporated.

One mechanism to improve the reliability of the data transfer is to employ LoRa in the so-called *confirmed uplink* mode, which is a form of stop-and-wait ARQ natively supported by LoRa devices [8]. In the confirmed uplink mode, the receiver acknowledges successful frame receptions by transmitting acknowledgement (ACK) frames. The sender retransmits a frame if an ACK is not received following its earlier transmission. While it is a simple mechanism amenable to easy implementation, prior literature has shown that confirmed uplink does not scale well due to multiple reasons [9]. The ACK frames on the downlink can themselves cause frame losses by preventing the gateway from receiving uplink frames when an ACK transmission is in progress [10]. Another drawback of the confirmed uplink mechanism is that the sensors must keep their receiving radio switched on after every transmission in anticipation of an ACK. For battery-powered sensors, this can be a major source of energy drain. Yet another disadvantage stems from the fact that LoRa devices, including the gateway, must conform to strict restrictions on the transmitter duty cycle [7]. As a result, if many devices in the network employ the confirmed uplink mode, the gateway cannot send ACK frames to all of them without rapidly exhausting its duty cycle quota; this can bring the data transfer to an abrupt stop.

An alternative to the confirmed uplink mechanism is the transmission of redundancy for forward erasure correction. A common technique that employs this approach is message replication [11–15]. In this technique, every message (e.g., a sensor reading) is transmitted a fixed number of times without seeking any receiver acknowledgement. As shown in [16], such transmissions often outperform confirmed-uplink ARQ in the presence of pure ALOHA channel access. A more sophisticated version of message replication involves the transmission of coded redundancy [17–19]. In this approach, the message replicas are replaced by linear combinations of messages. As shown in [17], coded redundancy outperforms message replication (which is essentially repetition coding) at the cost of some additional computational complexity.

The aforementioned frame-level coding schemes, however, are intended to operate over a low-rate data stream (i.e., the sensor produces a message every few minutes or so) with small message sizes such that every message fits into a single LoRa frame. The nature of such data streams renders it necessary to employ a sliding-window-based coding approach, in which every redundant message is the linear combination of messages belonging to a "window" of few most recent messages, and the window slides every time a new message is generated. In addition, such coding schemes (as well as message-replication schemes) cannot guarantee 100% data delivery since acknowledgements are not employed. The source stops transmitting once a predefined amount of redundancy has been sent, even if the destination has not received all messages.

The above limitation poses a problem when the end application requires the reliable transfer of a relatively large message over a LoRa network. Example scenarios requiring such transfer include the transmission of a compressed image of an agricultural field for remote monitoring of crop condition [20], or the transmission of a sensor's system log to the application server for troubleshooting purposes. The image or the system log can be divided into smaller data blocks, each amenable for transmission as the payload of one link-layer frame. In this chapter, we refer to such a large message as a *file*. We note that for LoRa networks in their current form, such files are expected to be small (e.g., about a few kilobytes at the maximum), since the restrictions on a LoRa transmitter's duty cycle place limits on the amount of data that can be sent.

A powerful technique that enables reliable file delivery over frame-erasure links is rateless coding [21]. In rateless-coded transfer, the source transmits a potentially infinite stream of *coded blocks*, which are generated by performing random linear combinations of the *data blocks* that constitute the file. The destination is able to retrieve the complete file once it receives *any* set of coded blocks that contain as many linearly independent combinations as there are data blocks in the file. Consequently, there is no need for the destination to acknowledge the reception of individual coded blocks. It suffices to send one cumulative acknowledgement (ACK) once the file is decoded, so that the source can stop transmitting. In the context of sensor-to-gateway file transfer, this greatly reduces the energy spent by the sensors in the reception of ACKs and also prevents the gateway from exceeding its duty cycle limits. Prominent examples of rateless codes include LT codes [22] and raptor codes [23]. Rateless coding is discussed in greater detail in Sect. 12.4.

In this following, we discuss the incorporation of rateless coding into LoRa communications for the transmission of small files. We present an analytical framework to evaluate the performance of the resulting system. Via analysis and simulations, we demonstrate the potential benefits of rateless coding over confirmed uplink for reliable LoRa-based file transfer. The remainder of this chapter is organized as follows: Sect. 12.2 provides a brief overview of the state-of-the-art regarding the use of rateless coding in IoT networks. Key features of the LoRa technology and the associated link-layer protocol LoRaWAN are described in Sect. 12.3. In Sect. 12.4, we describe the encoding and decoding principles for rateless codes. The integration of rateless codes and LoRa is described in Sect. 12.5. An analytical framework to evaluate the energy consumption of rateless-coded LoRa communications is developed in Sect. 12.6. Numerical results are presented in Sects. 12.7, and 12.8 concludes the chapter.

12.2 Literature Review

Rateless codes were originally conceived as a means for broadcast and multicast of predominantly multimedia files in cellular networks [24]. But in the recent past, they have received attention from the IoT community as well. Some of the representative works on the use of rateless codes in IoT are discussed below.

Most of the existing literature on rateless-coded IoT communications address file transfers on the *downlink* from a central node (e.g., a base station or an access point) to end devices. Such file transfers are usually needed for reprogramming the devices via firmware upgrades. A method for fountain-coded file transfer from an access point to sensor nodes in a ZigBee network is proposed in [25]. A salient feature of the method is that if a sensor is unable to decode the file from the access point's transmissions alone, it can request additional coded blocks from its neighboring sensors to complete the decoding process. The work in [26] also considers a rateless-coded file broadcast from a base station to a group of IoT devices. The proposed method includes a mechanism to adaptively determine the number of coded blocks to be sent based on estimates of frame-loss probabilities on the links. A fountain code with an optimized degree distribution is proposed in [27] to broadcast software programs from an access point to sensor nodes in a wireless sensor network. In [28], a multi-hop communication protocol leveraging rateless codes is proposed to perform similar broadcast tasks.

For file transfer on the *uplink* (i.e., from IoT end devices to a central sink), an example of prior work is [29], where the authors propose a rateless coding strategy for a space-air-ground integrated IoT network with a layered relaying architecture. Measurement data from ground sensors are collected by unmanned aerial vehicles (UAVs), which then relay the data to a LEO satellite, which further forwards the data to a high altitude satellite that acts as the sink. LT coding is performed at each of the layers, namely the sensors, the UAVs, and the LEO satellite. Large improvements in the aggregate throughput are demonstrated relative to uncoded transmissions. Recently, the authors of [30] evaluated the suitability of rateless codes vis–à-vis the ARQ scheme for reliable data transmission in industrial cloud control systems. Three forms of rateless codes, namely LT codes [22], Raptor-Q codes [31], and batched sparse (BATS) codes [32] were examined. It was shown that while ARQ offers slightly better data-delivery latency when frame losses are rare, rateless coding significantly outperforms ARQ as the frame-loss rate increases. The authors of [33] proposed an enhancement to the MQTT-SN protocol, which is a lightweight application-layer protocol for IoT sensor nodes, by integrating an LT code into it. The MQTT-SN messages are encoded by the LT code prior to transmission. This imparts erasure-correction capabilities to MQTT-SN, a feature it natively lacks due to running on top of the UDP protocol at the transport layer. Consequently, higher reliability of message transfer is achieved in harsh environments.

In addition to communications reliability, the random combinatorial nature of rateless-coded transmissions can be exploited to improve communications security and combat jamming. This aspect is considered in [34, 35].

The aforementioned works either employ legacy communication technologies other than LoRa at the physical layer or take a technology-agnostic approach. By contrast, the authors of [36] recently proposed the use of fountain coding in LoRa networks. In another related work [37], rateless coding is employed in conjunction with frequency-hopped LoRa transmissions. In this chapter, we seek to analytically characterize the performance of rateless-coded LoRa IoT. We take an in-depth look at how the interference characteristics of the network and the choice of coding parameters impact the performance. The goal is to draw general insights and lay the foundations for an analytical framework that can be enhanced as necessary to incorporate specific protocol features. Our primary focus is on uplink file transfers, although the framework can be easily adapted for analyzing downlink communications.

12.3 Basics of LoRa and LoRaWAN

The term LoRa refers to the proprietary physical layer technology implemented in LoRa transmitters. By contrast, LoRaWAN refers to a link-layer protocol that is often employed for networking LoRa devices. The key features of LoRa and LoRaWAN are reviewed in this section.

12.3.1 Frame Structure and Spreading Factor

A LoRa transmitter transmits information in the form of *frames*. Each frame carries the source information bits (e.g., a sensor message or a data block) as its payload. The payload bits are encoded by a channel code of rate 4/5, 4/6, 4/7, or 4/8 for combating bit errors [38]. The payload is preceded by a *preamble*, which is used for frame acquisition and synchronization purposes. Optionally, a header field may also be included.

For modulating the information bits, LoRa uses the chirp spread spectrum mechanism [39]. A modulation symbol in a LoRa signal is called a *chirp*. The key parameter that characterizes the signal is its *spreading factor*, which determines the number of bits per chirp. In theory, the spreading factor can take any positive integer value; but most commercial LoRa devices implement the spreading factors from 7 to 12. The duration of a LoRa symbol that employs spreading factor s and has transmission bandwidth w is given by

$$\rho(s) = 2^s/w. \tag{12.1}$$

Thus, a higher spreading factor requires longer transmission times compared to a lower spreading factor, with the duration doubling every time the spreading factor is increased by one. However, this lowering of the data rate and the increased energy consumption due to longer on-air times are accompanied by an increase in the signal's ability to be correctly received in the presence of thermal noise and attenuation. For this reason, it is common to assign higher spreading factors to transmitters that are farther away from the receiver, while using smaller spreading factors over shorter links.

12.3.2 Frame Duration

The airtime of a frame that contains b bytes of payload is

$$l_s(b) = l_{\text{pr}} + l_{\text{pl}}(b), \tag{12.2}$$

where l_{pr} is the duration of the preamble portion. The term $l_{\text{pl}}(b)$ is the duration of the payload. The duration of the header, when present, is included in l_{pl}. The preamble duration is

$$l_{\text{pr}} = (n_{\text{pr}} + 4.25)\,\rho(s) \tag{12.3}$$

where n_{pr} is the number of preamble symbols in the frame. (The quantity 4.25 appears in (12.3) because the transmitter includes an additional signal segment of duration equivalent to four and a quarter symbols for acquisition purposes.) The payload duration is

$$l_{\mathrm{pl}}(b) = \left\lceil 8 + \max\left\{\left\lceil \frac{2b - s - 5h + 11}{s - 2y} \right\rceil (c+4), 0\right\} \right\rceil \rho(s), \tag{12.4}$$

where the value of h is 1 when a header is included in the frame, and is 0 otherwise. The value of c depends on the rate of the channel code employed; it takes integer values between 1 and 4 (1 for rate 4/5, 2 for 4/6, and so on.) The value of y is 1 when a feature known as *low data-rate optimization* is enabled (this is mandatory for spreading factors 11 and 12, and optional for the lower spreading factors); else, it is 0 [38].

12.3.3 Transmission Frequency

Most LoRa devices support transmission bandwidths ranging from 7.8 to 500 kHz. However, 125, 250, and 500 kHz bandwidths are typically employed in practice [5]. Among these, 125 kHz is the most commonly used, since it provides higher resilience against noise and can support more devices within a given frequency band, and despite providing the lowest data rate among the three choices, it is still sufficient for most use cases targeted by LoRa. The center frequency of LoRa transmissions is programmable between 137 and 1020 MHz [5]. The exact set of center frequencies employed in a given network depends on regional spectrum regulations. Within the permissible region-specific frequency range, LoRa transmitters can perform channel hopping by changing the center frequency in a pseudorandom manner from one frame to the next. This mechanism reduces the probability of frame collisions, since frames sent over non-overlapping frequency bands do not interfere.

12.3.4 Reception Characteristics and Interference Model

The receiver sensitivity is the minimum received power level for which the receiver can detect a frame. That is, a frame goes undetected (and is hence lost) if its received power is below the receiver sensitivity. For LoRa, the receiver sensitivity depends on the spreading factor. For larger spreading factors, the sensitivity is lower, meaning the receiver can detect weaker frames. Hence, the larger the spreading factor, the lower the probability of frame loss due to a detection failure. Typical values of LoRa receiver sensitivities are listed in Table 12.1.

In addition to a detection failure, a frame may be lost due to collision with other frames. The *strongest-interferer model* is often used for approximate analyses of collision-induced losses in wireless network, and it has been shown to provide good approximations of LoRa's behavior as well [6]. In this model, a frame using SF s and

Table 12.1 Receiver sensitivities (in dBm) at 125 kHz

Spreading factor	Sensitivity
7	−123
8	−126
9	−129
10	−132
11	−134.5
12	−137

Table 12.2 Capture threshold (in dB) for a frame with spreading factor s in the presence of interference from spreading factor s'

s	s'					
	7	8	9	10	11	12
7	1	−8	−9	−9	−9	−9
8	−11	1	−11	−12	−13	−13
9	−15	−13	1	−13	−14	−15
10	−19	−18	−17	1	−17	−18
11	−22	−22	−21	−20	1	−20
12	−25	−25	−25	−24	−23	1

having received power R survives a collision with a frame having spreading factor s' and received power R' if the ratio R/R' exceeds a *capture threshold*. Values of the capture thresholds (adopted from [40]) for spreading factor s against interference from spreading factor s' are given in Table 12.2. Observe that the capture threshold is drastically smaller when the colliding spreading factors are different ($s \neq s'$). This implies that a frame is highly likely to survive interference from a colliding frame that employs a different spreading factor.

12.3.5 Network Structure and Uplink Channel Access

While the LoRa physical layer can be used in networks with arbitrary topologies, the commonly employed LoRaWAN protocol is meant for star networks, in which sensor nodes transmit directly to a gateway. Multiple gateways may also be used for robustness. A gateway forwards the sensor messages to a network server. The network server connects to an application server that performs the necessary tasks (e.g., monitoring, actuation, etc.) based on the received information. To keep transmitter operations as simple as possible, LoRaWAN requires the sensors to transmit on the uplink using pure ALOHA, in which a sensor can transmit its frame at any time without prior time-slot allocation or carrier sensing. This greatly reduces the burden on both the sensors and the network server since no timing synchronization or resource allocation is necessary. However, enhancements such as channel activity detection [41] and time slotting [42] have recently been proposed and implemented. Likewise, as enhancements to the star network topology, multi-hop communications have been explored in the recent past for improved coverage [43–45].

12.3.6 Downlink Receptions

While uplink transmissions of sensor data to a gateway constitute the primary traffic in LoRaWAN, support for downlink frames carrying ACKs or control messages from the gateway is desired in many applications. In this regard, the sensor nodes can operate in one of three classes [7]:

Class A: After the transmission of an uplink frame, the sender stays in the receive mode for a short duration (discussed in more detail in Sect. 12.3.7). Any downlink frame from the gateway to the sender must be transmitted during this time. Class A has the lowest energy consumption among all three classes due to the judicious use of reception windows. Hence it is the most commonly employed class, and support for this class is mandatory for all devices. Throughout this chapter, we will assume Class A operations.

Class B: In addition to opening a receive window after a frame transmission, a Class B device also periodically opens additional receive windows during which the gateway can communicate with the device. Synchronizing beacons from the gateway are used to control the window opening times. Class B is more complex than Class A, and has higher energy consumption.

Class C: Class C devices are always in the receive mode, except when they are transmitting. Clearly, this is the most energy-hungry class and is suitable only for devices that are plugged into a power supply.

12.3.7 Confirmed Uplink

The confirmed uplink feature is provided by LoRaWAN for ensuring reliable data delivery with the help of receiver acknowledgements. This optional feature may be enabled by the sender of a frame by setting the confirmed uplink bit in the frame header. When the gateway successfully receives a frame whose confirmed uplink bit is set, it must reply to the sender with an ACK. Assuming Class A operations, the gateway has two opportunities to send the ACK. This is illustrated in Fig. 12.1. After transmitting the uplink frame, the sender waits for a predefined duration (Delay-1) and opens a receive window (referred to as RX1). If no ACK is received in RX1, the sender waits for another predefined interval (Delay-2) and opens another receive

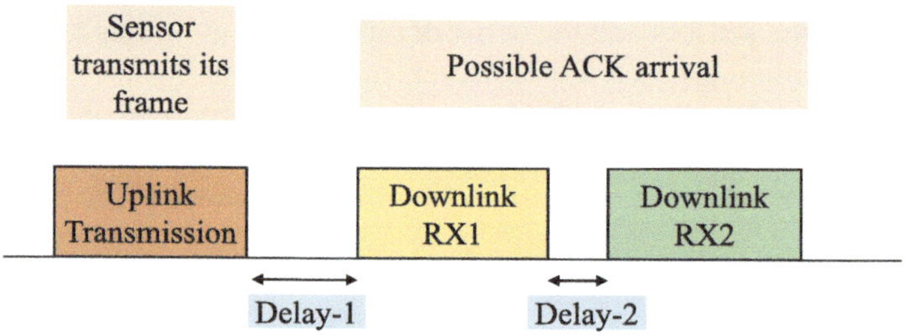

Fig. 12.1 Transmission and reception windows for a LoRa transmitter

window (RX2). The gateway may send the ACK over either RX1 or RX2. If RX1 is used, the convention is to send the ACK frame with the same spreading factor and on the same frequency channel as employed for the uplink frame. For RX2, the ACK frame parameters may be decided at the time of network set up in consultation with the network server. The Delay-1 and Delay-2 parameters can likewise be set by the network server.

12.4 Rateless Coding

As mentioned earlier, there are several types of rateless codes. For all types, however, the underlying principles are the same and are described below.

12.4.1 Encoding Procedure

For rateless encoding, the file is divided into k equal-sized data blocks referred to as $\mathbf{d}_0, \mathbf{d}_1, \ldots, \mathbf{d}_{k-1}$. In traditional communications without rateless coding, the source transmits the data blocks sequentially, possibly retransmitting the ones that are erased. In rateless coding, by contrast, the sender transmits a sequence of coded blocks $\mathbf{c}_0, \mathbf{c}_1, \ldots$, which are random linear combinations of the data blocks. The jth coded block can be expressed as

$$\mathbf{c}^{(j)} = \sum_{i=0}^{k-1} a_i^{(j)} \mathbf{d}_i, \tag{12.5}$$

where the coefficients $a_i^{(j)}$s are chosen randomly from a Galois field of order q (GF(q)). The contents of $\mathbf{d}_i$ are also expressed as elements of the same field for the purpose of computing (12.5). A diagram illustrating the generation of a coded block

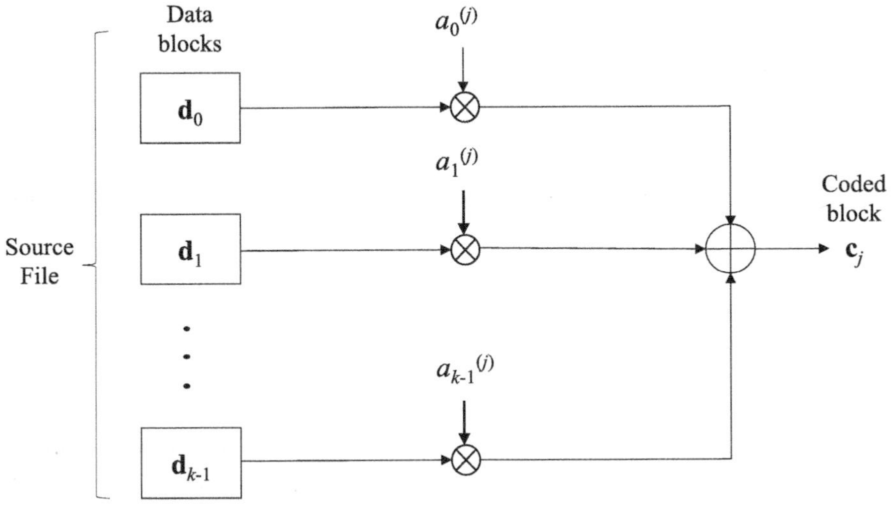

Fig. 12.2 Block diagram of a rateless encoder

is shown in Fig. 12.2. The vector $\mathbf{a}^{(j)} = [a_0^{(j)}, a_1^{(j)}, \ldots, a_{k-1}^{(j)}]$ is called the *encoding vector* for $\mathbf{c}^{(j)}$. The transmitted frame carries both $\mathbf{a}^{(j)}$ and $\mathbf{c}^{(j)}$.

The mechanism to choose the random coefficients $a_i^{(j)}$s depends on the type of the rateless code. The procedure is briefly described below for three classes of rateless codes.

LT Codes: For LT codes, the coefficients are elements of the Galois field of order 2 (GF(2)); that is, $a_i^{(j)}$ is either 0 or 1. To generate coded block $\mathbf{c}^{(j)}$, first an integer D is chosen at random from the set $[1, 2, \ldots, k]$ according to the robust Soliton distribution [22]. We call D the *degree* of the coded block. Next, D of the k coefficients $\{\mathbf{a}_0^{(j)}, \mathbf{a}_1^{(j)}, \ldots \mathbf{a}_{k-1}^{(j)}\}$ are randomly picked and set to 1; the rest are set to 0. This procedure is repeated for every new coded block, picking a new D at every instance and then setting D random coefficients to 1 and the rest to 0.

Raptor Codes: For raptor codes, the coefficients are obtained via a two-step process that concatenates a fixed-rate inner code with an outer LT code [23]. The resulting coefficients are 0s and 1s (i.e., belong to GF(2)) as in an LT code. However, the probability distribution of the resulting degree D is different due to the concatenation.

Random Linear Fountain Codes: The term random linear fountain code [46] refers to a general class of codes in which the encoding coefficients are randomly chosen from any GF(q), with q typically being a power of 2. Note that GF(q) contains the elements $\{0, 1, \ldots, q-1\}$. Thus, the LT code and the raptor codes are special cases of the random linear fountain code. For our illustrations, we assume a random linear fountain code for which the coefficients are chosen uniformly at random from GF(q).

12.4.2 Decoding Procedure

Upon reception of m coded blocks ($m \geq k$), the receiver can retrieve all individual data blocks if the set of received coded blocks contains k linearly independent combinations of the data blocks. The retrieval process is equivalent to solving a system of linear equations and can be performed using standard techniques such as Gaussian elimination, belief propagation [22], etc. Once all data blocks are retrieved, we say that decoding has succeeded. Once decoding succeeds, a cumulative acknowledgement is sent over the downlink. Upon receipt of this acknowledgement, the source stops transmitting. This allows the source to send exactly as many coded blocks as needed, rather than fixing in advance the number of blocks to be sent (which in turn dictates the code rate). This is the rationale behind the term "rateless".

In general, the source may need to deliver more than k coded blocks before decoding succeeds, since some of the delivered coded blocks may be linearly dependent. Suppose decoding succeeds upon the reception of m_0 coded blocks. Then we define the *decoding overhead* to be

$$\varepsilon = \frac{m_0 - k}{k} \times 100. \tag{12.6}$$

The decoding overhead can be interpreted as the percentage of redundant frame deliveries required by rateless coding, relative to uncoded stop-and-wait ARQ where exactly k deliveries are required.

For the remainder of this chapter, we assume without loss of generality that rateless coding is performed using a random linear fountain code in GF(q), and that Gaussian elimination decoding is used. Under these conditions, the probability of decoding success upon the reception of m coded blocks is given by [47]

$$P_s(m) = \prod_{\upsilon=0}^{k-1}(1 - q^{\upsilon-m}). \tag{12.7}$$

It is easy to infer from (12.7) that a larger value of q gives better performance when random linear fountain coding is used. This observation will be useful in interpreting some of our results in the latter sections.

12.5 Rateless-Coded LoRa

The incorporation of rateless coding into LoRa-based file communications is straightforward. Suppose the source file is of size B bytes. The file is divided into k equal-sized data blocks $\mathbf{d}_0, \mathbf{d}_1, \ldots, \mathbf{d}_{k-1}$. If B is not an integer multiple of k, then it is zero padded to make it so. Each resulting data block comprises b_k bytes such that

$$b_k = \left\lceil \frac{B}{k} \right\rceil, \tag{12.8}$$

where $\lceil \cdot \rceil$ is the ceiling operator, that is, $\lceil x \rceil$ is the smallest integer greater than or equal to x. The source applies a rateless code to generate a stream of coded blocks, in the manner described in Sect. 12.4. Any type of rateless coding can be used for this purpose, although our illustrations are for a random linear fountain code. Because a coded block is a linear combination of data blocks, it is also b_k bytes long. Each coded block is encapsulated in a LoRa frame and transmitted to the gateway. The gateway continues collecting coded blocks until it is able to run the decoding algorithm and retrieve the file. At this point, an ACK frame is sent to the source, which then ceases transmissions.

Recall that the frame carrying a coded block must also include the corresponding encoding vector. With a large k, the encoding vector may be long enough to cause a significant increase in the frame's payload. To circumvent this difficulty, the encoding vector is not transmitted as it is along with a coded block. Instead, the same pseudorandom number generator is used at both transmitter and receiver, and each encoding vector is produced by starting from a random seed. The seed value is included in the frame's header. The receiver can reproduce the encoding vector from the seed [27].

The frequency band and the spreading factor for the ACK frame may be selected following standard LoRaWAN conventions. That is, the ACK may be sent using the same spreading factor and frequency band over RX1, or it may employ a pre-allocated spreading factor and frequency band over RX2.

12.6 Performance Analysis

The objective of this section is to develop an analytical model to characterize the performance of rateless-coded LoRa. The model is derived here for a network topology featuring uniformly distributed nodes over a circle around a gateway and for Rayleigh fading on the links; however, it can be easily generalized to handle other node distributions and fading models and can be extended to accommodate multiple gateways. A list of the important symbols used in our analysis is given in Table 12.3.

12.6.1 System Model

We consider the transmission of a file of size B bytes from a sensor node (the source) to a gateway (the destination.) The distance between the source and the gateway is d_0. In addition to the source, there are n other nodes or devices in the network. These nodes are referred to as *interferers*, since they are capable of sending frames that collide with the source's frames. The interferers are uniformly distributed over

Table 12.3 Important symbols and their definitions

Symbol	Definition
B	Number of bytes in the file to be sent
k	Number of data blocks in the file
b_k	Number of bytes per data block
q	Order of the Galois Field for rateless coding
$P_N(m)$	Probability that decoding success requires exactly m coded blocks
$P_s(m)$	Probability that m coded blocks result in successful decoding
$\hat{P}_s(m)$	Probability that decoding succeeds upon the m-the reception (having failed earlier)
$\overline{N}$	Average number of coded blocks needed for decoding success
d_0	Distance between the source and the gateway
n	Number of nodes in the network (excluding the file's source)
λ	Average number of transmissions per interferer per second
$\mathcal{R}$	Radius of the circular network
$\mathcal{R}_j$	Inner radius of the ring for spreading factor j
F	A frame sent by the file's source
F′	A frame from an arbitrary interferer
s	Spreading factor used by the file's source
$\hat{s}$	Spreading factor used by an ACK frame
$\hat{b}$	Number of payload bytes per ACK frame
b'	Number of payload bytes per interferer frame
w	Transmission bandwidth
p_t	Transmitted signal power
p_r	Sensor power consumption while receiving ACKs
A	Fading coefficient
$f_A(a)$	Probability density function (pdf) for the fading coefficient
α	Path-loss exponent
R	Received power in a source frame
R'	Received power in an interfering frame
n_f	No. of frequency channels
ζ_s	Receiver sensitivity for SF s
$\xi_{s,s'}$	Capture threshold for SF s against SF s'
$\rho(s)$	Duration of a LoRa symbol employing spreading factor s
$l_s(b)$	Duration of a frame carrying b bytes of payload data
η_i	Probability that an arbitrary node uses spreading factor j
$f_D^{(j)}(u)$	Pdf of distance from gateway to an arbitrary node using spreading factor j
$\overline{\mathcal{E}}_t^{(\mathrm{RC})}$	Average transmission energy spent per file (with rateless coding)
$\overline{\mathcal{E}}_r^{(\mathrm{RC})}$	Average reception energy spent per file (with rateless coding)
$\overline{\mathcal{E}}_t^{(\mathrm{CU})}$	Average transmission energy spent per file (with confirmed uplink)
$\overline{\mathcal{E}}_r^{(\mathrm{CU})}$	Average reception energy spent per file (with confirmed uplink)
$S(d_0)$	Transmission success probability for a source frame
$S_0(d_0)$	Probability that a source frame has power above receiver sensitivity
$S_I(d_0)$	Probability that a source frame survives all collisions

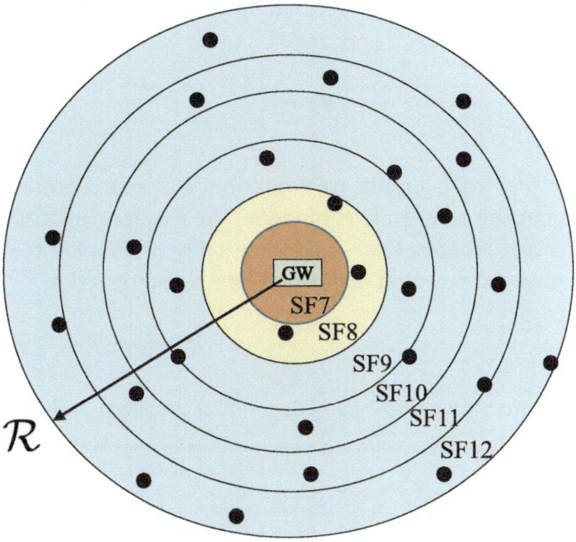

Fig. 12.3 Block diagram of the network. The dots show representative locations of sensors

a circular region of radius $\mathcal{R}$ with the gateway at the center. The probability density function (pdf) for the distance D between an arbitrary interferer and the gateway is

$$f_D(u) = \begin{cases} 2u/\mathcal{R}^2, & 0 \le u \le \mathcal{R} \\ 0, & \text{otherwise.} \end{cases} \tag{12.9}$$

Each interferer generates messages according to a Poisson process and transmits using pure ALOHA. Let λ be the average number of frames produced by an interferer per second, and let b' be the number of bytes in such a frame's payload.

The spreading factors are assigned based on node distances. Specifically, a node at distance d from the gateway is assigned spreading factor j if $d \in [\mathcal{R}_j, \mathcal{R}_{j+1})$, where $j \in \{7, 8, \ldots, 12\}$ and $0 = \mathcal{R}_7 \le \mathcal{R}_8 \le \cdots \le \mathcal{R}_{13} = \infty$. The spreading factor allocation is illustrated in Fig. 12.3. The illustration is for a network in which all spreading factors are used, but depending on $\mathcal{R}$, some of the higher spreading factors may not be utilized in practice. In particular, spreading factor j is used only if $\mathcal{R}_j \le \mathcal{R}$. We denote by S_m the largest spreading factor in use. Let us define

$$\hat{\mathcal{R}}_j = \min\{\mathcal{R}_j, \mathcal{R}\}. \tag{12.10}$$

The pdf for the distance between the gateway and an arbitrary device that uses spreading factor j is easily derived to be

$$f_D^{(j)}(u) = \begin{cases} 2u/\left(\hat{\mathcal{R}}_{j+1}^2 - \mathcal{R}_j^2\right), & \mathcal{R}_j < \mathcal{R} \\ 0, & \text{otherwise.} \end{cases} \tag{12.11}$$

Further, the probability that an arbitrary device uses spreading factor j is

$$\eta_j = \begin{cases} \left(\hat{R}_{j+1}^2 - \mathcal{R}_j^2\right)/\mathcal{R}^2, & \mathcal{R}_j < \mathcal{R} \\ 0, & \text{otherwise.} \end{cases} \tag{12.12}$$

For each uplink transmission, the source chooses one of n_f non-overlapping channels uniformly at random. The wireless links experience path loss with exponent α and independently and identically distributed (i.i.d.) block Rayleigh fading. The received power at distance d can be expressed as

$$R = \gamma_0 p_t A d^{-\alpha}, \tag{12.13}$$

where γ_0 is a hardware-dependent constant and A is the fading coefficient that has the pdf

$$f_A(a) = \begin{cases} \mathrm{e}^{-a}, & a \geq 0 \\ 0, & \text{otherwise.} \end{cases} \tag{12.14}$$

The receiver sensitivity for spreading factor s is denoted by ζ_s. The capture threshold for the reception of a signal with spreading factor s in the presence of another signal with spreading factor s' is denoted by $\xi_{s,s'}$.

For the downlink channel, we assume that the ACKs are received without error. This idealistic assumption greatly simplifies the analysis. Further, because ACKs can be sent with higher power than uplink frames (since gateways are connected to power supplies, they are unlikely to be energy-constrained), they have significantly lower loss probability than uplink frames. Another way to ensure a high probability of ACK reception is to transmit the ACK in RX2 over a frequency band reserved for the downlink, so that they are not subject to interference from other uplink frames.

12.6.2 Energy Consumption of Rateless-Coded LoRa

Our primary performance measure is the average energy consumption $\overline{\mathcal{E}}^{(\mathrm{RC})}$, defined as the average energy spent by the source during the course of a file transfer session. It is expressed as

$$\overline{\mathcal{E}}^{(\mathrm{RC})} = \overline{\mathcal{E}}_t^{(\mathrm{RC})} + \overline{\mathcal{E}}_r^{(\mathrm{RC})}, \tag{12.15}$$

where $\overline{\mathcal{E}}_t^{(\mathrm{RC})}$ is the average energy the source spends transmitting frames per session, and $\overline{\mathcal{E}}_r^{(\mathrm{RC})}$ is the energy it spends in the reception of ACK frames.

Recall that for rateless-coded file transfer, only one ACK per file needs to be received. Hence,

$$\overline{\mathcal{E}}_r^{(\mathrm{RC})} = p_r l_{\hat{s}}(\hat{b}), \tag{12.16}$$

where p_r is the power consumption at the receiver while receiving a frame, $\hat{s}$ is the spreading factor used for the ACK frame, and $\hat{b}$ is the number of payload bytes in the ACK frame.

To derive $\overline{\mathcal{E}}_t^{(\mathrm{RC})}$, we proceed as follows: Let $\overline{N}$ denote the average number of coded blocks that must be delivered for successful decoding. Further, define the *transmission success probability* $S(d_0)$ to be the probability that a frame from the source is successfully received at the gateway. Thus, the source must transmit on average $\overline{N}/S(d_0)$ frames until the file is delivered. Since each frame is sent with power p_t and has duration $l_s(b_k)$ where s is the spreading factor used by the source, the average transmission energy spent by the source is

$$\overline{\mathcal{E}}_t^{(\mathrm{RC})} = \overline{N} p_t l_s(b_k)/S(d_0). \tag{12.17}$$

To determine $\overline{N}$, note that

$$\overline{N} = \sum_{m=k}^{\infty} m P_N(m), \tag{12.18}$$

where $P_N(m)$ is the probability that decoding succeeds upon the reception of exactly m coded blocks. It is given by

$$P_N(m) = \prod_{i=k}^{m-1} (1 - \hat{P}_s(i)) \cdot \hat{P}_s(m), \tag{12.19}$$

where $\hat{P}_s(j)$ is the probability that the decoding attempt made upon receiving the jth coded block succeeds, after having failed in all previous attempts. To derive $\hat{P}_s(j)$, define E_j to be the event that the decoding attempt made upon receiving j coded blocks fails, and let $P_f(j)$ denote the probability of this event. Thus, $P_f(j) = 1 - P_s(j)$, where $P_s(j)$ is given by (12.7). Now note that $\hat{P}_s(j) = 1 - \hat{P}_f(j)$, where $\hat{P}_f(j)$ is the probability that decoding fails after j successful receptions, conditioned on decoding having failed after each of the previous $j-1$ receptions; that is,

$$\begin{aligned}\hat{P}_f(j) &= P(\mathrm{E}_j | \mathrm{E}_{j-1}) \\ &= \frac{P(\mathrm{E}_j, \mathrm{E}_{j-1})}{P(\mathrm{E}_{j-1})} \\ &= \frac{P(\mathrm{E}_j)}{P(\mathrm{E}_{j-1})}.\end{aligned} \tag{12.20}$$

Since $P(\mathrm{E}_j) = P_f(j)$ and $\hat{P}_s(j) = 1 - \hat{P}_f(j)$, we have

$$\hat{P}_s(j) = 1 - \frac{P_f(j)}{P_f(j-1)} = 1 - \frac{1 - P_s(j)}{1 - P_s(j-1)}. \tag{12.21}$$

Substitution of (12.7) into (12.21) yields

$$\hat{P}_s(j) = \begin{cases} 0, & j < k \\ \prod_{v=0}^{k-1}(1 - q^{v-j}), & m = k \\ 1 - \dfrac{1 - \prod_{v=0}^{k-1}(1 - q^{v-j})}{1 - \prod_{v=0}^{k-1}(1 - q^{v-j+1})}, & j \geq k. \end{cases} \tag{12.22}$$

Now we turn to the derivation of $S(d_0)$. To this end, consider an arbitrary frame F, employing spreading factor s, sent by the source located at distance d_0 from the gateway. We refer to F as the *desired frame*. Now $S(d_0)$ is the probability that F is successfully received at the gateway.

For successful reception, F must satisfy two conditions. First, its received power must be greater than the receiver sensitivity ζ_s. Second, no other frame which collides with F must be received with a power level that brings the ratio of desired to interference power below the capture threshold. For the first condition, let us denote by $S_0(d_0)$ the probability that the received power in F exceeds ζ_s. Recall that the received power is given by $R = \gamma_0 p_t A d_0^{-\alpha}$, where A is the fading coefficient. Conditioned on $A = a$, F is received with power $R = \gamma_0 p_t a d_0^{-\alpha}$. The conditional probability $S_0(d_0 | A = a)$ is then given by

$$S_0(d_0 | A = a) = P(R > \zeta_s | A = a) = P(\gamma_0 p_t a d_0^{-\alpha} > \zeta_s). \tag{12.23}$$

Note that the quantity $\gamma_0 p_t a d_0^{-\alpha}$ is deterministic because of the conditioning $A = a$. As a result, the probability $S_0(d_0 | A = a)$ becomes

$$S_0(d_0 | A = a) = \begin{cases} 1, & a > \zeta_s d_0^{\alpha} / \gamma_0 p_t \\ 0, & \text{otherwise.} \end{cases} \tag{12.24}$$

Now let us denote by $S_I(d_0|A = a)$ the conditional probability that F survives all collisions, conditioned on $A = a$. We can now write the conditional transmission success probability for F as

$$S(d_0|A = a) = S_0(d_0|A = a)S_I(d_0|A = a). \tag{12.25}$$

We remark here that in general, the unconditional probabilities $S_0(d_0)$ and $S_I(d_0)$ are not independent (as both depend on the fading coefficient A), and $S(d_0)$ is only approximately equal to $S_0(d_0)S_I(d_0)$. However, conditioned on $A = a$, the equality in (12.25) holds strictly. Substitution of (12.24) into (12.25) gives

$$S(d_0|A = a) = \begin{cases} S_I(d_0|A = a), & a > \zeta_s d_0^\alpha / \gamma_0 p_t \\ 0, & \text{otherwise.} \end{cases} \tag{12.26}$$

Removal of the conditioning on A yields

$$\begin{aligned} S(d_0) &= \int_0^\infty S(d_0|A = a) f_A(a)\, \mathrm{d}a \\ &= \int_{\zeta_s d_0^\alpha / \gamma_0 p_t}^\infty S_I(d_0|A = a) \mathrm{e}^{-a}\, \mathrm{d}a. \end{aligned} \tag{12.27}$$

Our final task is to determine $S_I(d_0|A = a)$. To this end, consider an interfering node N′ located at distance D' from the gateway, as illustrated in Fig. 12.4. Suppose

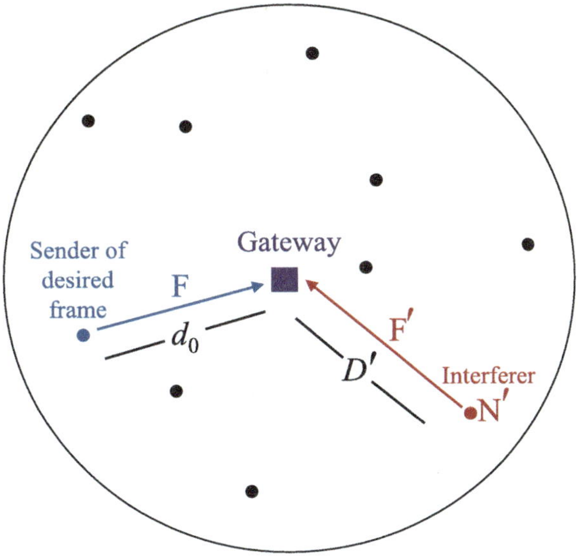

Fig. 12.4 An illustration of frame transmission from an interfering node N′

N′ uses the spreading factor j. Now N′ will cause the loss of F if it sends a frame F′ such that

1. F and F′ use the same channel.
2. The transmission times of F and F′ are such that they overlap fully or partially in time (i.e., a collision occurs).
3. The received power R' in F′ is such that $R/R' < \xi_{s,j}$.

Regarding condition 1 above, recall that there are a total of n_f channels. Hence the probability that F and F′ use the same channel is

$$P_{\text{ch}} = \frac{1}{n_f}. \tag{12.28}$$

Regarding condition 2, first note that the frames from the desired sender have duration $L_0 = l_s(b_k)$, and frames from N′ have duration $L' = l_j(b')$. Let $[T_0, T_0+L_0]$ and $[T', T'+L']$ be the intervals occupied by F and F′, respectively. It is easy to see that a collision will occur if $T_0 - L' < T_1 < T_0 + L_0$. Thus the collision probability is the probability that N′ begins transmitting a frame during the interval $[T_0 - L', T_0 + L_0]$. Because frame transmissions follow the Poisson distribution with a mean of λ frames per second, the collision probability of the desired frame F with a transmission from N′ is given by

$$\begin{aligned} P_{\text{col}}(s, j) &= \lambda\left((T_0 + L_0) - (T_0 - L')\right) \\ &= \lambda\left(L_0 + L')\right) \\ &= \lambda\left(l_s(b_k) + l_j(b')\right). \end{aligned} \tag{12.29}$$

For condition 3, we first let $D' = u$ and $A = a$. Then the received power in F and F′ become $r = \gamma_0 p_t a d_0^{-\alpha}$, and $R' = \gamma_0 p_t A' u^{-\alpha}$, respectively, where A' is the fading coefficient experienced by F′. Condition 3 is then satisfied with probability

$$\begin{aligned} \chi_{s,j}(d_0 | A = a, D' = u) &= P\left(\frac{r}{R'} < \xi_{s,j}\right) \\ &= P\left(\frac{\gamma_0 p_t a d_0^{-\alpha}}{\gamma_0 p_t A' u^{-\alpha}} < \xi_{s,j}\right) \\ &= P\left(A' > \xi_{s,j}^{-1} a (d_0/u)^{-\alpha}\right) \\ &= e^{-\xi_{s,j}^{-1} a (d_0/u)^{-\alpha}}. \end{aligned} \tag{12.30}$$

Combining the observations above, we determine the conditional probability that F is lost due to N′ to be

$$Q_I(d_0|A=a) = P_{\text{ch}} \sum_{j=7}^{S_m} \eta_j P_{\text{col}}(s,j) \int_{\mathcal{R}_j}^{\hat{\mathcal{R}}_{j+1}} \chi_{s,j}(d_0|A=a, D'=u) f_D^{(j)}(u)\, \mathrm{d}u$$
$$= \frac{2}{n_f \mathcal{R}^2} \sum_{j=7}^{S_m} \lambda \left(l_s(b_k) + l_j(b')\right) \int_{\mathcal{R}_j}^{\hat{\mathcal{R}}_{j+1}} u \mathrm{e}^{-\xi_{s,j}^{-1} a (d_0/u)^{-\alpha}}\, \mathrm{d}u. \quad (12.31)$$

Some manipulation of the integral term yields

$$Q_I(d_0|A=a) = \frac{2\lambda}{\alpha n_f \mathcal{R}^2} \sum_{j=7}^{S_m} \frac{l_s(b_k)+l_j(b')}{\beta_j^{2/\alpha}} \left(\gamma(2/\alpha, \beta_j \hat{\mathcal{R}}_{j+1}^{\alpha}) - \gamma(2/\alpha, \beta_j \mathcal{R}_j^{\alpha})\right), \quad (12.32)$$

where $\beta_j = \xi_{s,j}^{-1} a d_0^{-\alpha}$ and $\gamma(s,x) = \int_0^x t^{s-1}\mathrm{e}^{-t}\,\mathrm{d}t$ is the lower incomplete Gamma function. Because there are a total of n nodes in the network, each capable of causing a collision, the conditional success probability for F is the conditional probability that none of these n nodes enforces a loss. That is,

$$S_I(d_0|A=a) = (1 - Q_I(d_0|A=a))^n. \quad (12.33)$$

Substitution of (12.32) and (12.33) into (12.27) yields our final expression for the transmission success probability

$$S(d_0) = \int_{\zeta_s d_0^{\alpha}/\gamma_0 p_t}^{\infty} \left(1 - \frac{2\lambda}{\alpha \mathcal{R}^2 n_f} \sum_{j=7}^{S_m} \frac{l_s(b_k)+l_j(b')}{\beta_j^{2/\alpha}} \right.$$
$$\left. \left(\gamma(2/\alpha, \beta_j \hat{\mathcal{R}}_{j+1}^{\alpha}) - \gamma(2/\alpha, \beta_j \mathcal{R}_j^{\alpha})\right)\right)^{n-1} \mathrm{e}^{-a}\, \mathrm{d}a. \quad (12.34)$$

In general, (12.34) must be solved using numerical integration. For the special case of $\alpha=2$, the use of the identity $\gamma(1,x) = 1 - \mathrm{e}^{-x}$ yields a closed-form expression for (12.34). Another way to obtain a closed-form expression is to employ a cumulative-interference model as opposed to the strongest-interferer one, and then use results from [48].

12.6.3 Energy Consumption for Confirmed Uplink

As a benchmark, we are interested in characterizing the sensor energy consumption when the confirmed uplink mode is used for reliably delivering each data block. The analysis is a slight modification of the analysis for rateless codes. As before, we can write

$$\overline{\mathcal{E}}^{(\mathrm{CU})} = \overline{\mathcal{E}}_t^{(\mathrm{CU})} + \overline{\mathcal{E}}_r^{(\mathrm{CU})}. \tag{12.35}$$

The reception energy consumption $\overline{\mathcal{E}}_r^{(\mathrm{CU})}$ is obtained by noting that the source must receive exactly k acknowledgement frames in response to the delivery of k data blocks. It thus follows that

$$\overline{\mathcal{E}}_r^{(\mathrm{CU})} = kp_r l_{\hat{s}}(\hat{b}). \tag{12.36}$$

For the transmission energy $\overline{\mathcal{E}}_t^{(\mathrm{CU})}$, note that the difference between the number of uplink transmissions in rateless coding and confirmed uplink is that the source must deliver exactly k data blocks to the destination when confirmed uplink is used, as opposed to $\overline{N}$ coded blocks on average needed for rateless coding. Consequently, the derivation of $\overline{\mathcal{E}}_t^{(\mathrm{CU})}$ is identical to that of $\overline{\mathcal{E}}_t^{(\mathrm{RC})}$, except that we must use $\overline{N}=k$.

12.7 Performance Results

In this section, we numerically evaluate the performance of rateless-coded file transfer. For simulations, the programs are written in MATLAB, and every data point is obtained by averaging 1000 simulation runs. Unless otherwise specified, parameter values as listed in Table 12.4 are used. The correspondence between the distance of a device and its spreading factor is shown in Table 12.5. Values listed in Tables 12.1 and 12.2 are used for the sensitivity and capture thresholds, respectively.

We begin by examining the transmission success probability $S(d_0)$, which is the probability that a frame transmission from a source located at distance d_0 from the

Table 12.4 Default parameter values

Parameter	Value
Source file size (B)	1000 bytes
No. of data blocks (k)	50
Galois field order (q)	256
Number of interferers (n)	6000
Circle radius ($\mathcal{R}$)	3 km
Payload size of interfering frames (b')	5 bytes
Avg. frames per interferer per sec. (λ)	1/600
Uplink transmission power (p_t)	14 dBm
Power consumption while receiving (p_r)	11 dBm
Path-loss exponent (α)	2.5
No. of orthogonal channels (n_f)	3
Transmission bandwidth (w)	125 kHz

Table 12.5 Spreading factor assignment policy

Distance range	Spreading factor
[0, 500 m]	7
(500 m, 1000 m]	8
(1000 m, 1500 m]	9
(1500 m, 2000 m]	10
(2000 m, 2500 m]	11
(2500 m, ∞)	12

gateway is successful. An examination of this parameter is useful because it gives an idea of the number of transmissions needed to deliver the file, which in turn dictates the energy consumption. To compute $S(d_0)$, we consider a source located at distance d_0 and vary d_0 from 0 (i.e., co-located with the gateway) to $\mathcal{R}$ (at the network's edge.) The plot is shown in Fig. 12.5. As would be expected, the success probability exhibits an overall decreasing trend, with poorer success rates for larger separations between the source and the gateway. However, the decrease is non-monotonic, with discontinuities at integer multiples of 500 m. This occurs because, as per the spreading factor allocation policy employed here, the spreading factor increases by one every 500 m. Since the higher spreading factors are more robust against channel impairments, there is a marked improvement in the performance every time a spreading factor switch occurs. By contrast, a monotonic decrease in the success probability is observed within the distance range covered by a single spreading factor. From the perspective of reliable file transfer, a key takeaway from Fig. 12.5 is that unless the source is very close to the gateway, frame erasures are unavoidable, and some reliability mechanism (rateless coding, confirmed uplink, or any other scheme that guarantees the delivery of each data block) is required. Fig. 12.5 includes both analytical and simulation results for the success probability.

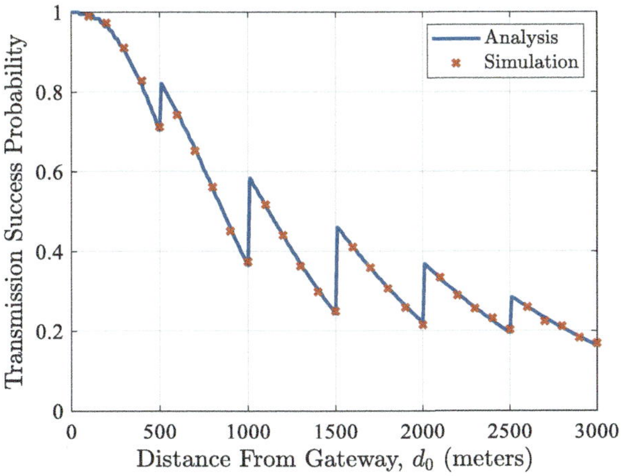

Fig. 12.5 Transmission success probability as a function of the distance between the source and the gateway. Parameter values are as per Table 12.4

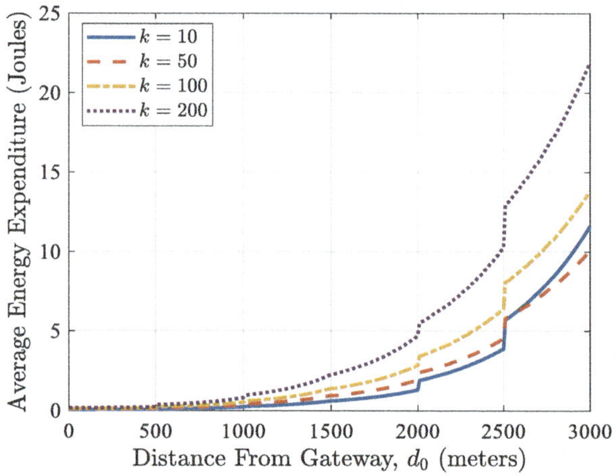

Fig. 12.6 Average energy expenditure of file transfer as a function of the distance between the source and the gateway. Parameter values except for k are as per Table 12.4

There is excellent agreement between the two, which confirming the accuracy of our analytical model. In view of this agreement, hereafter we plot only analytical results, since they are easier to obtain as opposed to simulation results that require the execution of time-consuming computer programs.

In Fig. 12.6, we examine the average energy expenditure $\overline{\mathcal{E}}^{(RC)}$ of rateless-coded file transfer. The energy expenditure is plotted as a function of the distance between the source and the gateway. Results are shown for four values of k, the number of data blocks into which the file is divided. For each k, the energy expenditure increases with increasing distance between the source and the gateway. This follows from the success probability graph of Fig. 12.5. Within the distance range covered by a single spreading factor, an increase in the distance results in more frame erasures, leading to the requirement of more transmissions in order to deliver k linearly independent coded blocks to the gateway. However, a striking difference between Figs. 12.5 and 12.6 is that in the former, the success probability improves when a spreading factor switch occurs, whereas in the latter, the energy consumption does not decrease at such points. This is because increasing the spreading factor by one approximately doubles the frame duration and consequently the energy consumption, whereas it increases the success probability by a factor less than two. We further observe that the value of k has a significant impact on the energy expenditure. This is primarily due to the fact that changing k alters the data block size, which in turn impacts the frame duration, collision probability, and frame success probability. The plot also suggests that the relationship between k and the energy consumption is not linear. The impact of k will be discussed in detail later.

Since confirmed uplink transmissions provide an alternative reliability mechanism, it is instructive to evaluate its energy consumption vis-à-vis rateless coding. To this end, we consider the scheme described in Sect. 12.6.3, in which every data block is transmitted until an ACK is received. We define the excess energy consumption $\Delta\mathcal{E}$ to be the difference between the energy consumption of confirmed uplink

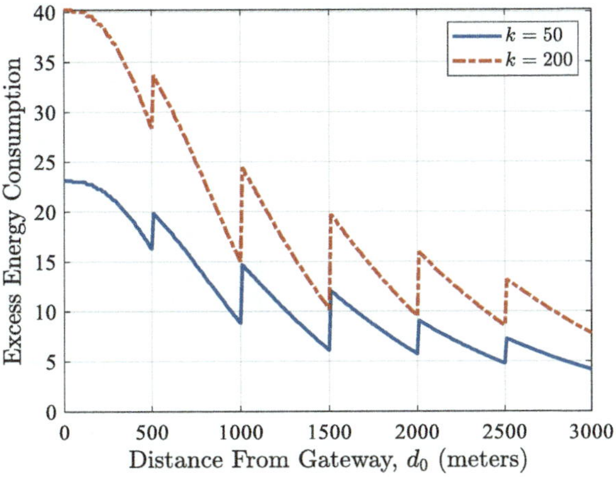

Fig. 12.7 Excess energy requirement of confirmed uplink file transfer relative to rateless coding. Parameter values except k are as per Table 12.4

and rateless coding, expressed as a percentage of the latter's energy consumption. That is

$$\Delta\mathcal{E} = \frac{\overline{\mathcal{E}}^{(\mathrm{CU})} - \overline{\mathcal{E}}^{(\mathrm{RC})}}{\overline{\mathcal{E}}^{(\mathrm{RC})}} \times 100. \tag{12.37}$$

The excess energy consumption is plotted in Fig. 12.7 for two values of k. It is observed to be positive at all distances, meaning rateless coding consistently requires less energy than confirmed uplink. The energy savings are the greatest at short distances from the gateway. To explain this, we note that confirmed uplink spends extra energy in receiving ACKs, since k ACK receptions are needed as opposed to one ACK in rateless coding. Thus, we may interpret (12.37) as approximately the ratio of the energy needed to receive $k-1$ ACKs to the energy needed to transmit the average number of frames such that $\overline{N}$ of them are delivered. Clearly, the numerator is a constant, whereas the denominator is distance dependent, being small for nodes near the gateway. Hence, the excess energy is the largest when the gateway is close to the source.

The variation in the energy consumption of rateless coding with k, which was cursorily observed in Fig. 12.6, is examined more thoroughly in Fig. 12.8. The figure plots the energy consumption as a function of k for three values of the source-gateway separation. A key feature of the curves in Fig. 12.8 is their convex shape, implying there exists an optimal k, such that increasing or decreasing k from this point leads to higher energy consumption. This is due to the combined impact of three phenomena driven by variations in k, as described below:

With smaller k, and a fixed file size, the number of bytes per data block (which has the same size as a coded block) increases. Since one frame carries one such block, a smaller k means the frames carry a larger payload each and thus have a

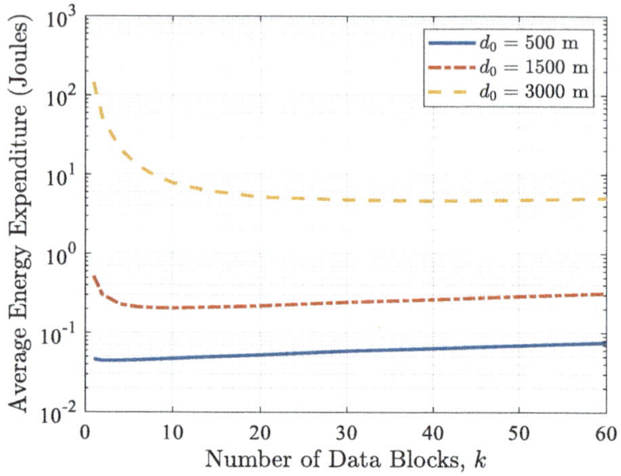

Fig. 12.8 Energy consumption of rateless-coded file transfer as a function of k

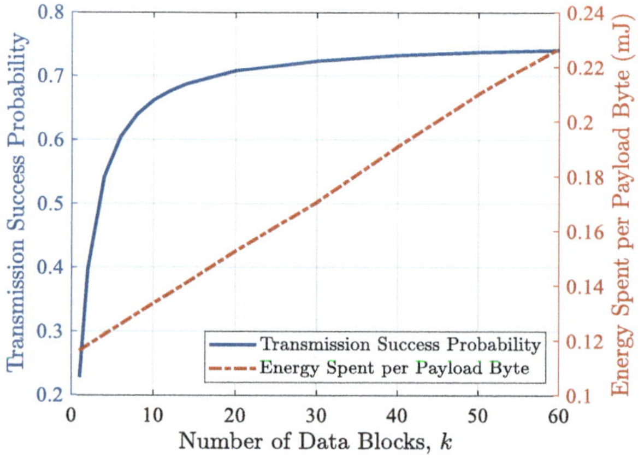

Fig. 12.9 Impact of k on the transmission success probability and the energy expenditure per payload byte. ($d_0 = 1500$ m)

longer duration. This makes them more susceptible to collisions and consequently frame errors. By contrast, a larger k results in shorter frames which have higher transmission success probability.

However, decreasing the block size has a detrimental impact on the energy spent per payload byte. Recall that each frame has a preamble and possibly a header. These entities do not carry data but must be transmitted nonetheless, thus resulting in transmission overhead. The overhead per frame is constant regardless of the payload size. Hence, if the payload is small (i.e., small block size, given by a large k), the total energy spent in order to transmit each payload byte is higher than for larger payload sizes. This trade-off between the transmission success probability and the energy expenditure is illustrated in Fig. 12.9.

Yet another impact of k on the energy expenditure concerns the decoding overhead. As defined in Sect. 12.4, the decoding overhead is a measure of the extra deliveries that must be made in addition to k blocks, so that decoding can be achieved. As seen

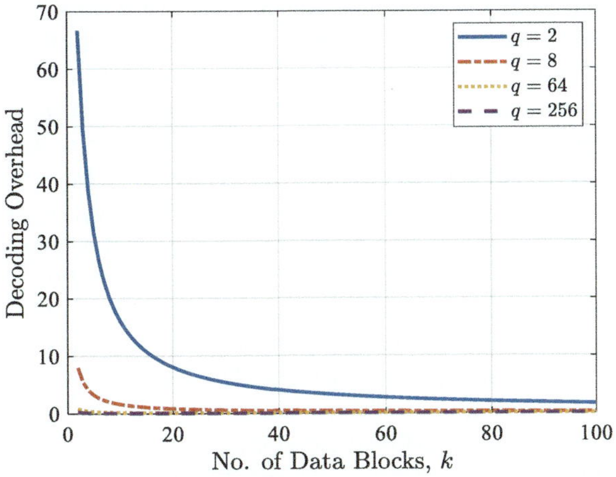

Fig. 12.10 Decoding overhead of a random linear fountain code over GF(q)

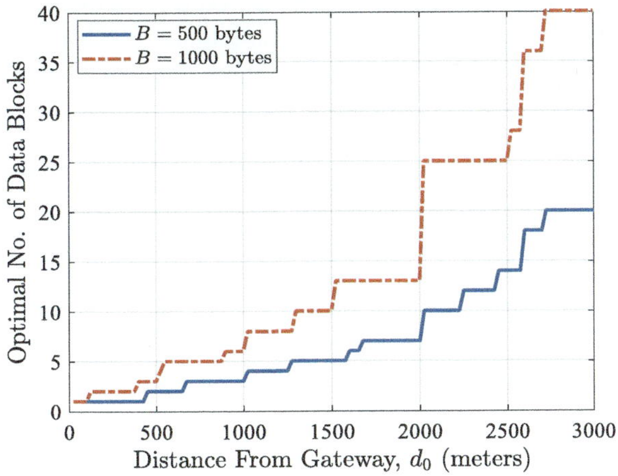

Fig. 12.11 The value of k that minimizes the energy expenditure of the file transfer

from (12.6), this quantity depends on the value of k. This dependence is illustrated Fig. 12.10. For any given GF(q), it is observed that the decoding overhead decreases with k. Since the decoding overhead is proportional to the excess transmission energy consumption due to coding, smaller k values are disadvantageous in terms of energy expenditure from this perspective. We infer from Fig. 12.10 that the impact of k on the decoding overhead is significant only for small q. For larger q, this aspect can be ignored as the decoding overhead remains very small and approximately independent of k.

In summary, we conclude that for a fixed file size, the energy consumption of rateless-coded LoRa depends on the number of blocks k in three ways: via the interference susceptibility (improves with increasing k), the energy consumption per payload bit (worsens with k), and the decoding overhead (improves with k). The analysis presented in Sect. 12.6.2 can be used to obtain the value of k that minimizes the energy expenditure. Specifically, we wish to use the k that minimizes (12.15). Note that since the number of candidate values for k is not huge (no more than the number of bytes in the file, which is expected to be small), brute force minimization by evaluating (12.15) for all candidate k values can be easily performed. This optimal value for two file sizes is plotted in Fig. 12.11 as a function of the distance between the source and the gateway. Two features of the curves in the figure stand out. First, the farther away the source from the gateway, the larger the optimal k value. This can be explained by noting that frames sent over larger distances are received with lower power. Such frames are more prone to collision-induced losses, since colliding frames are more likely to have power in excess of the capture threshold. As a result, it is beneficial to keep the frames short to reduce the number of collisions. This is accomplished by increasing k. The second observation is that larger file sizes lead to larger optimal k for the same node distance. This also has the same explanation. A larger file must be divided into more blocks in order to achieve the same optimal block size as obtained for a smaller file.

The energy consumption that results with the optimal k is shown in Fig. 12.12. For comparison, the energy consumption with three fixed values of k are also included in the plot. The figure demonstrates that using the optimal k is crucial, especially at larger distances, where the gap between the optimal-k curve and some of the fixed-k curves is substantial.

So far, the number of interferes n was kept fixed at 6000 for all results. In Fig. 12.13 we examine the impact of varying this parameter. We observe that the optimal k increases with n. The reason is similar to that observed in Fig. 12.11. With larger number of interferes, the expected number of collisions—and consequently the loss probability—experienced by a frame of a given duration increases. To reduce the number of collisions, the frame duration should be decreased, which is accomplished

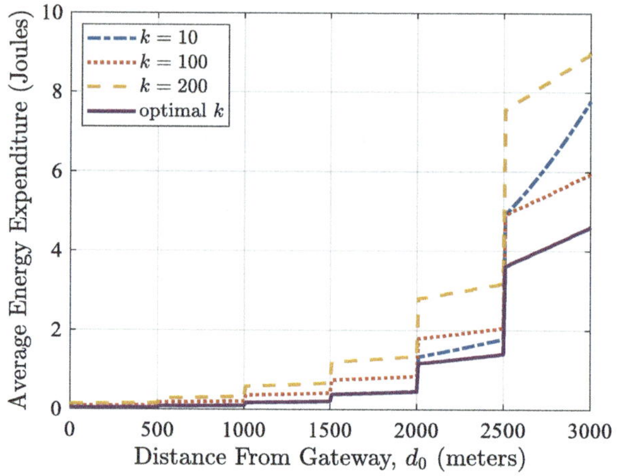

Fig. 12.12 Energy consumption of rateless-coded file transfer when the optimal k is used, and when some fixed values of k are used. All parameter values are as given in Table 12.4, except $B = 500$ bytes

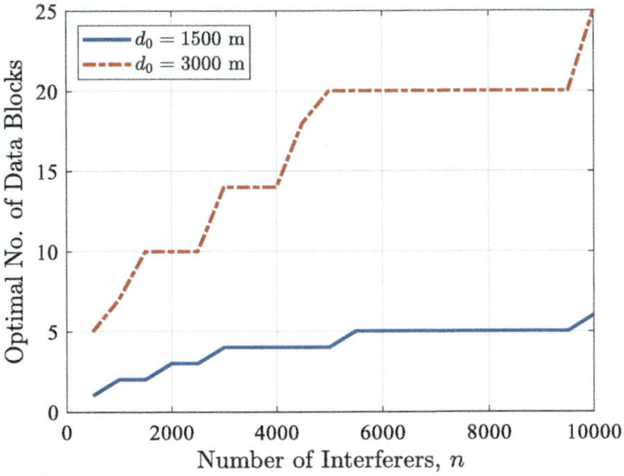

Fig. 12.13 The value of k that minimizes the energy expenditure of rateless-coded file transfer

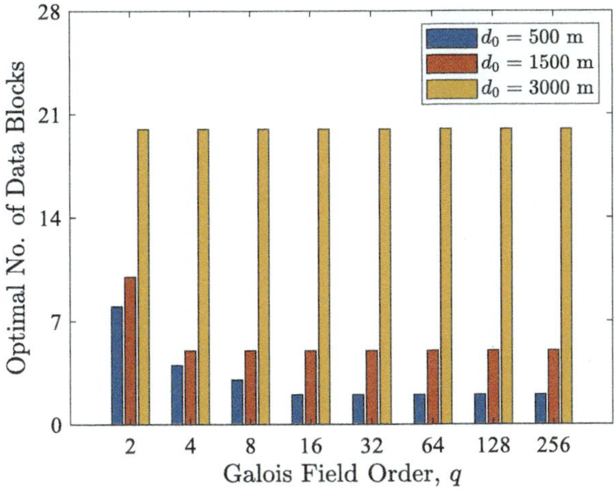

Fig. 12.14 The optimal value of k for different choices of the Galois field order q

by increasing k. Although not shown, similar outcomes are observed if the transmission rate λ of the interferers is increased, which also leads to more collisions per unit time.

The optimal choice of k also depends on the Galois field order q, as illustrated in Fig. 12.14. Except for very large separations between the source and the gateway, we observe that a smaller q requires a larger value of k. As we observed from Fig. 12.10, the decoding overhead is larger for smaller q values, and it reduces with increasing k. As a result, for smaller q, it is beneficial to use a large k to keep the energy expenditure to decoding overhead small. For large q, the decoding overhead is small and is largely independent of k. In such cases, one can use a small k to minimize the energy overhead of header transmissions without significantly increasing the decoding overhead.

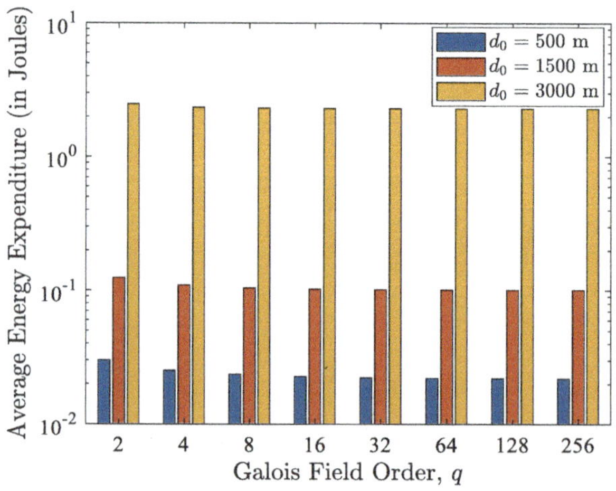

Fig. 12.15 Energy expenditure of rateless-coded file transfer with the optimal k for different choices of the Galois field order q

The energy consumption incurred with the optimal choice of k as obtained above is shown in Fig. 12.15. A close examination of the numerical values reveals that the energy consumption is slightly lower for larger q. However, as evident from the plot, such differences are rather small. Thus, comparable energy expenditure is achieved regardless of q, as long as the optimal value of k for that q is used. Hence, any q may be used as per convenience of implementation without impacting the performance.

12.8 Conclusion and Outlook

We considered the use of rateless codes to facilitate the reliable transfer of small files from a sensor to a gateway in LoRa networks. We evaluated the energy consumption of this approach and showed that rateless coding provides significant energy savings compared to the stop-and-wait ARQ mechanism employed by LoRa's native reliability mechanism known as confirmed uplink. Numerical results illustrate that to extract the maximum benefit, it is crucial to divide the source file into an optimal or near-optimal number of data blocks. The analytical framework presented in this chapter can be used to characterize the dependence between energy consumption and the number of blocks and to choose a suitable value for the latter. Future research directions include the following:

1. The random linear fountain code with a uniform selection of encoding coefficients was employed for our illustration. In order to limit the encoding and decoding complexities, the use of computationally lighter schemes such as LT codes and raptor codes can be explored. However, these codes are designed for the transfer of relatively large files. Their application to the transfer of small files over IoT networks requires further investigation.

2. The analytical framework can be enhanced by incorporating the impact of correlated interference. In the preceding analysis, it was assumed that a frame transmission by an interfering device is independent of its prior transmissions. However, if many of the interferers are involved in file transmissions (for which message transmissions occur in bursts as opposed to the sporadic transmissions of single-frame messages), or if any form of replication or retransmission is employed, the resulting interference exhibits correlation over time. Incorporation of such correlation may improve the accuracy of the analytical prediction of system performance.
3. To ensure energy efficiency under circumstances where simple analytical modeling is not feasible, the development of heuristic methods to obtain near-optimal block count may prove beneficial. For example in some cases, parameters such as the number of devices, the network's geographical size, etc., may not be known. However, observations such as the temporal variations of the received signal strength indicator (RSSI), frame success rate, etc. may provide information about the interference pattern, which can be exploited for sub-optimal selection of coding parameters.

References

1. Internet of Things market size, share, and Covid-19 impact analysis (2023). https://www.fortunebusinessinsights.com/industry-reports/internet-of-things-iot-market-100307
2. Cirillo F, Gómez D, Diez L, Maestro IE, Gilbert TBJ, Akhavan R (2020) Smart city IoT services creation through large-scale collaboration. IEEE Internet Things J 7(6):5267–5275
3. Misra N, Dixit Y, Al-Mallahi A, Bhullar MS, Upadhyay R, Martynenko A (2020) IoT, big data, and artificial intelligence in agriculture and food industry. IEEE Internet Things J 9(9):6305–6324
4. Sisinni E, Saifullah A, Han S, Jennehag U, Gidlund M (2018) Industrial Internet of Things: challenges, opportunities, and directions. IEEE Trans Ind Inf 14(11):4724–4734
5. Bor MC, Roedig U, Voigt T, Alonso JM (2016) Do LoRa low-power wide-area networks scale? In: Proceedings of the 19th ACM international conference on modeling, analysis and simulation of wireless and mobile systems, pp 59–67
6. Georgiou O, Raza U (2017) Low power wide area network analysis: can LoRa scale? IEEE Wireless Commun Lett 6(2):162–165
7. Adelantado F, Vilajosana X, Tuset-Peiro P, Martinez B, Melia-Segui J, Watteyne T (2017) Understanding the limits of LoRaWAN. IEEE Commun Mag 55(9):34–40
8. Capuzzo M, Magrin D, Zanella A (2018) Confirmed traffic in LoRaWAN: Pitfalls and countermeasures. In: 2018 17th annual mediterranean ad hoc networking workshop (Med-Hoc-Net). IEEE, pp 1–7
9. Pop A-I, Raza U, Kulkarni P, Sooriyabandara M (2017) Does bidirectional traffic do more harm than good in Lorawan based LPWA networks? In: GLOBECOM 2017 - 2017 IEEE global communications conference, pp 1–6
10. Borkotoky SS, Schmidt JF, Schilcher U, Battula P, Rathi S (2021) Reliability and energy consumption of LoRa with bidirectional traffic. IEEE Commun Lett 25(11):3743–3747
11. Qaim WB, Ozkasap O (2018) DRAW: data replication for enhanced data availability in IoT-based sensor systems. In: 2018 IEEE 16th international conference on dependable, autonomic and secure computing, 16th international conference on pervasive intelligence and computing, 4th international conference on big data intelligence and computing and cyber science and technology congress (DASC/PiCom/DataCom/CyberSciTech), pp 770–775

12. Mo Y, Do M-T, Goursaud C, Gorce J-M (2016) Optimization of the predefined number of replications in a ultra narrow band based IoT network. In: Wireless days (WD), pp 1–6
13. Hoeller A, Souza RD, Alcaraz Lòpez OL, Alves H, de Noronha Neto M, Brante G (2018) Exploiting time diversity of LoRa networks through optimum message replication. In: 2018 15th international symposium on wireless communication systems (ISWCS), pp 1–5
14. Borkotoky S, Bettstetter C, Schilcher U, Raffelsberger C (2019) Allocation of repetition redundancy in LoRa. In: European wireless 2019; 25th European wireless conference, pp 1–6
15. Sisinni E, Carvalho DF, Ferrari P (2020) Emergency communication in IoT scenarios by means of a transparent LoRaWAN enhancement. IEEE Internet Things J 7(10):10 684–10 694
16. Rathi S, Gunda S, Borkotoky SS (2023) Blind vs. feedback-aided retransmission of high-priority data in asynchronous IoT networks. In: 2023 IEEE international conference on internet of things and intelligence systems (IoTaIS), pp 120–125
17. Marcelis PJ, Kouvelas N, Rao VS, Prasad RV (2022) Dare: data recovery through application layer coding for LoRaWAN. IEEE Trans Mob Comput 21(3):895–910
18. Montejo-Sánchez S, Azurdia-Meza CA, Souza RD, Fernandez EMG, Soto I, Hoeller A (2019) Coded redundant message transmission schemes for low-power wide area IoT applications. IEEE Wireless Commun Lett 8(2):584–587
19. Sant'Ana JMDS, Hoeller A, Souza RD, Montejo-Sánchez S, Alves H, Noronha-Neto MD (2020) Hybrid coded replication in LoRa networks. IEEE Trans Ind Inf 16(8):5577–5585
20. Ji M, Yoon J, Choo J, Jang M, Smith A (2019) LoRa-based visual monitoring scheme for agriculture IoT. In: 2019 IEEE sensors applications symposium (SAS). IEEE, pp 1–6
21. MacKay DJ (2005) Fountain codes. IEE Proc-Commun 152(6):1062–1068
22. Luby M (2002) LT codes. In: The 43rd annual IEEE symposium on foundations of computer science. Proceedings, pp 271–280
23. Shokrollahi A (2006) Raptor codes. IEEE Trans Inf Theory 52(6):2551–2567
24. Stockhammer T, Shokrollahi A, Watson M, Luby M, Gasiba T (2008) Application layer forward error correction for mobile multimedia broadcasting. In: Handbook of mobile broadcasting: DVB-H. DMB, ISDB-T and media flo, pp 239–280
25. Liu J, Wang S, Li S, Cui X, Pan Y, Zhu T (2018) MCTS: multi-channel transmission simultaneously using non-feedback fountain code. IEEE Access 6:58 373–58 382
26. Jeon SY, Ahn JH, Lee T-J (2016) Data distribution in IoT networks with estimation of packet error rate. In: 2016 10th international conference on next generation mobile applications, security and technologies (NGMAST), pp 94–98
27. Rossi M, Bui N, Zanca G, Stabellini L, Crepaldi R, Zorzi M (2010) SYNAPSE++: code dissemination in wireless sensor networks using fountain codes. IEEE Trans Mob Comput 9(12):1749–1765
28. Guo J, Koike-Akino T, Wang P, Parsons K, Orlik P, Shanmuga Sundaram JP, Sumi T, Nagai Y (2023) Rateless coding for multi-hop broadcast transmission in wireless IoT networks. In: 2023 IEEE international symposium on information theory (ISIT), pp 120–125
29. Xue Y, Sun Z, Liang Z, Fan C, Xu X (2020) Data collection method of space-based Internet of Things based on layered fountain code. In: 2020 IEEE 6th international conference on computer and communications (ICCC), pp 2475–2479
30. Yuan M, Qiao Y, Fu Y, Tang J (2023) Fountain codes for reliable and deterministic packet transmission in industrial cloud control systems. IEEE Internet Things J 10(8):7114–7125
31. Luby M, Shokrollahi A, Watson M, Minder L (2011) RaptorQ forward error correction scheme for object delivery. RFC 6330. https://www.rfc-editor.org/info/rfc6330
32. Yang S, Yeung RW (2014) Batched sparse codes. IEEE Trans Inf Theory 60(9):5322–5346
33. Alshantout A, Al-Awami L (2019) Enhancing MQTT-SN performance via fountain codes in extreme conditions. In: 2019 15th international wireless communications & mobile computing conference (IWCMC), pp 1184–1189
34. Sun L, Ren P, Du Q, Wang Y (2016) Fountain-coding aided strategy for secure cooperative transmission in industrial wireless sensor networks. IEEE Trans Industr Inf 12(1):291–300
35. Jain S, Bose R (2020) Rateless-code-based secure cooperative transmission scheme for industrial IoT. IEEE Internet Things J 7(7):6550–6565

36. Lv J, Chen G, Dong W (2023) Exploiting rateless codes and cross-layer optimization for low-power wide-area networks. ACM Trans Sens Netw 18(4):1–24
37. Luo J, Xu Z, Lin J, Chen C, Xiong R (2023) CH-MAC: achieving low-latency reliable communication via coding and hopping in LPWAN. ACM Trans Internet Things 4(4):1–25
38. AN1200.13 LoRa Modem Designer's Guide. Semtech Corp (2013)
39. AN1200.22 LoRa Modulation Basics, Rev. 2. Semtech Corp (2015)
40. Mahmood A, Sisinni E, Guntupalli L, Rondón R, Hassan SA, Gidlund M (2019) Scalability analysis of a LoRa network under imperfect orthogonality. IEEE Trans Industr Inf 15(3):1425–1436
41. Gamage A, Liando J, Gu C, Tan R, Li M, Seller O (2023) LMAC: efficient carrier-sense multiple access for LoRa. ACM Trans Sens Netw 19(2):1–27
42. Zorbas D, Abdelfadeel K, Kotzanikolaou P, Pesch D (2020) TS-LoRa: Time-slotted lorawan for the industrial internet of things. Comput Commun 153:1–10
43. Liao C-H, Zhu G, Kuwabara D, Suzuki M, Morikawa H (2017) Multi-hop LoRa networks enabled by concurrent transmission. IEEE Access 5:21 430–21 446
44. Borkotoky SS, Schilcher U, Bettstetter C (2019) Cooperative relaying in LoRa sensor networks. In: IEEE global communications conference (GLOBECOM), pp 1–5
45. Hong S, Yao F, Ding Y, Yang S-H (2022) A hierarchy-based energy-efficient routing protocol for LoRa-mesh network. IEEE Internet Things J 9(22):22 836–22 849
46. Schulz P, Traßl A, Barreto AN, Fettweis G (2021) Efficient and reliable wireless communications via multi-connectivity using rateless codes in single- and multi-user scenarios. IEEE Trans Wireless Commun 20(9):5714–5729
47. Zhao X (2012) Notes on "Exact decoding probability under random linear network coding". IEEE Commun Lett 16(5):720–721
48. Valenti MC, Torrieri D, Talarico S (2014) A direct approach to computing spatially averaged outage probability. IEEE Commun Lett 18(7):1103–1106

Chapter 13
An Introduction to Modern Random Access Protocols for IoT Communications

Iman Hmedoush, Jia Cao, Cédric Adjih, Sanjeev Sharma, and Kuntal Deka

Abstract In this chapter, we present a perspective on a category of grant-free communication methods, specifically those based on random access. They are well-suited for scenarios involving a large number of devices transmitting sporadically with short messages, commonly encountered in Internet-of-Things (IoT) communication scenarios. These methods, often referred to as modern random access, have recently emerged and generated significant interest. We discuss, in an introductory and tutorial manner, the basic methods for evaluating these communications as introduced in the literature. Additionally, we provide insights into their performance.

Keywords Random access · mMTC · IoT · Irregular Repetition Slotted Aloha (IRSA) · Successive Interference Cancellation (SIC) · Density Evolution (DE)

13.1 Introduction

The recent years have witnessed a fast expansion of the Internet-of-Things (IoT), providing communication capabilities to billions of objects. This expansion has

I. Hmedoush
Senior Radio Research Specialist, Nokia, Paris, France
e-mail: iman.hmedoush@nokia.com

J. Cao
ENSTA Paris, Paris, France
e-mail: jia.cao@inria.fr

C. Adjih
Research Scientist, TRiBE Team, Inria Saclay, Paris, France
e-mail: cedric.adjih@inria.fr

S. Sharma (✉)
Department of Electronics Engineering, IIT (BHU) Varanasi, Varanasi, India
e-mail: sanjeev.ece@iitbhu.ac.in

K. Deka
IIT Guwahati, Guwahati, India
e-mail: kuntaldeka@iitg.ac.in

S. Sharma et al. (eds.), *Next-Generation Wireless Systems*,
Transactions on Computer Systems and Networks,
https://doi.org/10.1007/978-981-96-3185-8_13

driven efforts to develop communication technologies suitable for its use cases. The broad range of technological challenges and solutions has been described by various surveys, such as the one referenced in [1].

For instance, unlicensed Low Power Wide Area Networks (LPWANs), provided by technologies like Sigfox and LoRaWAN, have been developed for Machine-Type Communication (MTC) applications targeting the ultra-low-end device segment. These applications have very limited throughput and Quality-of-Service (QoS) requirements but offer the advantage of longer communication ranges.

In the cellular world, there has been an effort to address specific Massive MTC (mMTC) scenarios, departing from the traditional Human-Type Communication (HTC) paradigm. The 3rd Generation Partnership Project (3GPP), in its 4G releases, introduced several features with Long-Term Evolution (LTE) for MTC (LTE-M) and Narrowband IoT (NB-IoT). Progress is also being made for 5G, with RedCap, and ongoing for 5G Advanced, and 6G.

In dense network scenarios, the target for connection density can be 1,000,000 devices per square kilometer in urban environments [2]. In massive IoT or mMTC, data is transmitted in small-sized packets, with high energy efficiency requirements, often with minimal or partial human intervention. These distinct features of mMTC traffic have influenced the design of physical and medium access control layer protocols, sometimes departing from conventional cellular networks, where devices are typically granted orthogonal resources through a grant mechanism in the uplink channel before data transmission. However, this approach necessitates control traffic to be exchanged, which can sometimes be larger than the actual data being transmitted and can lead to inefficient use of resources. To address this issue, grant-free methods that either eliminate or reduce control traffic have been developed and are becoming increasingly popular [3].

In the past decade, a family of random access protocols that exclusively focus on the successive interference cancellation (SIC) technique has gained popularity, often referred to as *modern random access*. These protocols are derived from the seminal work on Irregular Repetition Slotted ALOHA (IRSA) [4] and have been further developed in several studies [5]. Unlike traditional slotted ALOHA, these protocols are based on the SIC technique applied at the packet level and utilize inter-slot SIC, where packets are transmitted multiple times in different time slots. The key feature of these protocols is that they allow for the removal of all copies of a packet, including those in other time slots, by recovering just one of the replicas.

This chapter explores these modern random access protocols, with a particular focus on IRSA, and its variants, including IRSA with Multiple Packet Reception (k-IRSA). These protocols are one avenue for tackling the challenges of grant-free access. By touching on their theoretical foundations, classical evaluation methods, and performance metrics, the chapter aims to provide an understanding of the fundamentals of modern random access protocols and their performance.

Chapter Organization

The chapter is organized into the following sections to offer a detailed exploration of IoT communication challenges, modern random access protocols, and the specifics of IRSA and its variants:

1. **Overview of Modern Random Access Protocols:** A detailed review of modern random access protocols, particularly the IRSA family, is presented. This section discusses the protocols' ability to achieve optimal throughput and their applicability in IoT networks.
2. **Irregular Repetition Slotted Aloha (IRSA):** This section introduces the system model commonly used in IRSA literature. It focuses on the network setup, the framing structure, load definition, and the collision channel model, which forms the basis for designing and analyzing IRSA schemes.
3. **Variants of the IRSA Protocol:** Here, the chapter explores the various adaptations and enhancements of the Coded Slotted Aloha protocols. The section highlights the flexibility of these protocols and their suitability for diverse IoT scenarios, underlined by the integration of SIC.
4. **IRSA with Multiple Packet Reception:** The chapter concludes with description and analysis of k-IRSA, which goes beyond the traditional IRSA model by incorporating more realistic assumptions of communication channels.

13.2 Overview of Modern Random Access Protocols

A recent family of *modern random access* protocols [5], frequently referred to as the CSA family, is a good candidate to address the massive connectivity issue in IoT networks. It has garnered the attention of IoT protocol designers since it has been shown to asymptotically achieve the optimal throughput of one retrieved packet per slot [6], in contrast to the classical random access collision model where the maximum throughput is $\frac{1}{e}$ for slotted ALOHA.[1]

The main underlying principle of these protocols is to have each terminal physically transmit multiple copies of the same MAC packet (also called replicas), with identical preamble and payload information bits. The payload contains signaling information regarding the temporal positions of the corresponding replicas of each packet. This enables an iterative decoding process whereby SIC is applied to resolve collisions between different packets by physically removing the signal of a successfully decoded packet at the position of its replicas.

The operating mode described has been introduced by Contention Resolution Diversity Slotted Aloha (CRDSA) [7], where time is divided into frames. Each frame is divided into slots of equal size. Each packet is sent twice in the random-access

[1] Note that a classical collision model, here, is when two or more packets arrive simultaneously at a receiver that can handle only one at any given time.

frame. Slots with single replicas are always considered successfully decoded. SIC is used to subtract the decoded packet and retrieve it from its other position in the frame. The resulting throughput is up to $T \approx 0.55$ packets/slot.

In IRSA [4], the number of repetitions can vary for different users (i.e., it can be randomly selected from a predefined probability distribution). The SIC process is analogous to the decoding process of low-density parity-check (LDPC) codes. This allows for the application of analytical tools such as density evolution to analyze and optimize the asymptotic performance of IRSA, as pioneered by [4].

A significant portion of the research on IRSA and its variants focuses on optimizing the probability distribution used to randomly select the repetition degree of each user. In the original IRSA article [4], advanced analysis tools led to the discovery of a probability distribution that could asymptotically achieve a throughput of approximately 0.97 packet/slot (achieved when the frame size approaches infinity). Shortly afterward, leveraging the same tools and furthering the analogy with LDPC codes, [6] demonstrated that the soliton distribution is also well-suited to IRSA, yielding an asymptotic throughput of 1 packet/slot. This represents the maximum achievable throughput with a classical collision channel model.

The authors of [8, 9] introduce Coded Slotted ALOHA (CSA) which is a generalized version of IRSA. In CSA, each data block of a user is split into several packets before applying a packet-oriented linear block code yielding coded packets, instead of simply using a repetition code like IRSA. The linear block code is selected from a predefined set, according to a *code probability distribution*. The *code rate* is defined as the inverse of the average number of packets repetitions for IRSA and as the average code rate of the linear block codes for CSA. Then, in CSA, each slot and each packet is divided into k segments and each packet is randomly coded with a code of rate $\frac{k}{n}$ where n is drawn from a probability distribution for each user in each frame. On the receiver side, SIC is integrated with decoding of the local component codes to recover collisions.

13.3 Irregular Repetition Slotted Aloha (IRSA)

We adopt a system model commonly found in the IRSA literature, where an IoT network consists of M user nodes contending to send their packets to a single base station through a connectionless random access scheme. Time is divided into slots of equal length, and successive slots are regrouped into fixed-size frames of N slots. The load is defined as $G = \frac{M}{N}$, measuring the average number of users per slot. We focus on one frame and assume that each user has exactly one data packet to transmit within the frame. To study IRSA, we adopt a collision channel model, widely used in designing IRSA schemes, where packets involved in a collision are lost, and packets without a collision are successfully decoded [10].

13.3.1 IRSA Concept

IRSA works as follows: Each user sends his packet ℓ times within the same MAC frame, where the repetition rate of each user is selected randomly from a probability distribution Λ defined by $L+1$ probabilities : $(\Lambda_i)_{i=2,\ldots,L}$. Λ_i is the probability that the packet is repeated i times. At the end of the frame, the receiver attempts to decode the packets: each time that a packet is successfully received, e.g., only one transmission is present on the slot, IRSA will use SIC to remove the physical copy of this packet on all other slots. This process is iterative, as new decoding opportunities can appear.

In the example of Fig. 13.1, 5 users are transmitting in a frame of 5 slots; the packet of user D can be recovered on the slot 4. When its copy is physically subtracted from the slot 5, the packet of user B on the slot 5 can be recovered and removed from slots 1, 2 and 3. Then, the packet of user A on slot 3 could be decoded and its copies on slots 1 and 2 could be extracted. Finally, the packets of users C and E form a stopping set as they sent two copies on the same two slots, so they cannot be decoded.

13.3.2 IRSA Decoding Process Modeling

In [4], Liva introduced, by analogy with the decoding of LDPC codes (on a binary erasure channel [5]) the way to analyze the decoding process of IRSA as follows: as for codes, one can construct a Tanner graph, that is, a bipartite graph $G(B, S, E)$, where B is the set of *burst nodes* (corresponding to users), S, is the set of *slot nodes* (corresponding to slots) and E is the set of edges. An edge connects a burst node $b_i \in B$ and to a slot node $s_j \in S$, if and only if a replica of the ith burst is transmitted in the jth slot. The number of edges connected to a node is referred to as the node degree.

Figure 13.1 (on the bottom) shows an example of a graph representation of the IRSA. The circles correspond to the burst nodes, while the rectangles correspond to the slot nodes. The IC process starts by searching for the slots where only one replica was sent (degree-1 slots), which are also known as singletons, and decoding these replicas. If the burst that is transmitted on a certain slot is revealed, the corresponding edge is labeled by '1', otherwise, it is labeled by '0' (or equivalently, it is *removed*). The contribution of each revealed burst is removed from the other slots, where a replica of this burst was transmitted through SIC. Thus, we can say that the first iteration has ended.

After the first iteration, new singletons may be identified. However, the iterative decoding process can become blocked if, in a particular iteration, no singletons can be recovered. The analysis of IRSA assumes that we are able to decode singletons in slots where no collision has occurred. Subsequently, we can remove their contributions from other slots.

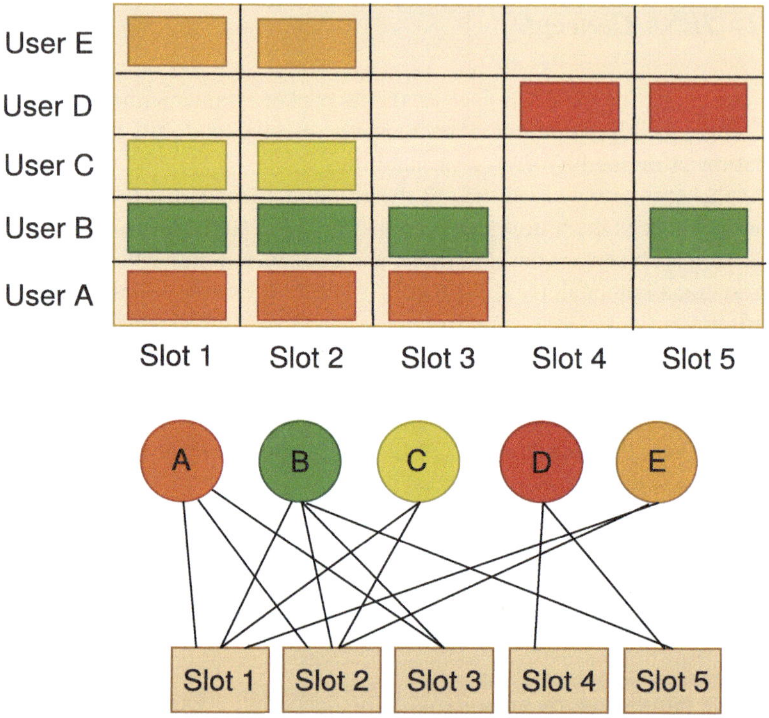

Fig. 13.1 IRSA representation: transmissions of users in slots (top), coding theory representation to model the decoding process (Tanner graph, bottom). Note that transmissions are in the same frequency channel, hence when two users are transmitting on the same slot, there is a collision

13.3.3 Density Evolution

Density evolution (DE) is a method to analyze the asymptotic performance of different (framed) variants of IRSA protocol. It is based on an analogy with LDPC codes for which DE was initially introduced, as an analytical tool for analyzing the asymptotic network capability to approach the error-correcting codes [11]. DE helps to track the iterative decoding process and have an estimation of the number of decoded packets at the end of the process. By employing DE and considering the system parameters Λ, M, and N, we can address the following question: can we expect to recover all packets, and if not, how many?

As we have already shown, the iterative decoding process of IRSA can be represented by the Tanner graph. A general approach to analyze random graphs with a given degree distribution [12] is to introduce a form of a generating function as $\sum_{k=0}^{\infty} p_k x^k$ [13]. This form of function helps for calculating various local and global quantities on large unipartite undirected graphs with an arbitrary probability distribution of the degrees of their vertices (nodes) [14].

Following [4], and the traditions of the LDPC coding literature (see [11, Eq. 3.21] for instance), we define the polynomial representations (probability generating functions) of burst- and slot-perspective degree distributions which will be helpful for further calculations:

$$\Lambda(x) \triangleq \sum_{\ell=0}^{\ell=L} \Lambda_\ell x^\ell \text{ and } \Psi(x) \triangleq \sum_{\ell=0}^{\ell=N} \Psi_\ell x^\ell, \tag{13.1}$$

where Λ_ℓ is the probability that a user will send ℓ replicas of his packet, and Ψ_ℓ is the probability that a slot has ℓ collided packets. $\Lambda'(1) = \sum \ell \Lambda_\ell$ is the average burst repetition rate. We define the *rate* R as $R \triangleq \frac{1}{\Lambda'(1)}$. Notice, in IRSA, the "rate" R is loosely connected to throughput since with SIC, removed replicas do not consume slots. On the other hand, it is strongly connected to the energy-efficiency, as the average energy cost would be proportional to $1/R$. The probability that one user sends a packet on a given slot is: $\frac{\Lambda'(1)}{N} = \frac{G}{RM}$. Then Ψ_ℓ follows a binomial distribution $\text{Bin}(M, \frac{G}{RM})$. When $M \to \infty$, the distribution $(\Psi_\ell)_{\ell \geq 0}$ becomes a Poisson distribution with a parameter $\frac{G}{R}$.[2]

If we select a random edge and examine the burst node it connects to, this edge arrives at the burst node with a probability proportional to the degree of that node. Therefore, the burst node associated with a random edge itself has a probability distribution of degree proportional to $\ell \Lambda_\ell$. The correctly normalized probability generating function for the degrees of the burst nodes sampled in this manner is given by

$$\frac{1}{\sum_\ell \Lambda_\ell \ell} \sum_\ell \Lambda_\ell \ell x^\ell = x \frac{\Lambda'(x)}{\Lambda'(1)}. \tag{13.2}$$

If we start at a random burst node and pick one of its edges: we are interested in the distribution of the number of remaining edges that the burst node has. This distribution is generated by the function in Eq. (13.2) except for less than one power of x, since it is only the distribution of the number of the remaining edges. It is customary [11] to use instead of the corresponding edge distributions from the burst and slot perspective which are respectively:

$$\lambda(x) \triangleq \sum_\ell \lambda_\ell x^{\ell-1} = \frac{\Lambda'(x)}{\Lambda'(1)} \text{ and } \rho(x) \triangleq \sum_\ell \rho_\ell x^{\ell-1} = \frac{\Psi'(x)}{\Psi'(1)}, \tag{13.3}$$

where λ_ℓ is the probability that an edge connected to a burst node of degree ℓ, and ρ_ℓ is the probability that an edge connected to a slot node of degree ℓ and they are given by the following definitions:

---

[2] It is actually an application of the classical Poisson approximation, i.e., the Poisson distribution with parameter $\mu = \alpha\beta$ can be used as an approximation of the binomial distribution $\text{Bin}(\alpha, \beta)$ for large α and small β. In our case, $\alpha = M$, $\beta = \frac{G}{RM}$ and $\mu = \frac{G}{R}$.

$$\lambda_\ell = \frac{\ell \Lambda_\ell}{\sum_\ell \ell \Lambda_\ell} \text{ and } \rho_\ell = \frac{\ell \Psi_\ell}{\sum_\ell \ell \Psi_\ell} \tag{13.4}$$

Note that the rate could be also defined as $R \triangleq \sum_\ell \frac{\lambda_\ell}{\ell}$. When $M \to \infty$, from the Poisson approximation of $(\Psi_\ell)_{\ell \geq 0}$, we have $\rho(x) \to e^{-\frac{G}{R}(1-x)}$. For further clarification, the burst and edge probability distributions are computed for the example shown in Fig. 13.3.

Continuing from [4], the decoding process in IRSA is iterative and is probabilistically modeled, tracking variables representing the probabilities that the decoding information (slot content) remains unknown over iterations. In Fig. 13.1, each edge connects one slot node and one burst node. Let p be the probability that a randomly selected edge is connected to a burst node that has not yet been recovered, and let q be the probability that a randomly selected edge is connected to an *unrecovered* slot node.

We can decode the packet of a user if a replica of this packet has been decoded at least once on any other slot. Hence:

$$q = p^{\ell-1} \tag{13.5}$$

Similarly, in classical IRSA, we can remove the collision on the last edge which connects to a slot node of degree ℓ, if we have already removed the $\ell - 1$ contributions of the other collided packets on the same slot. Hence:

$$(1 - p) = (1 - q)^{\ell-1} \tag{13.6}$$

Figure 13.2 illustrates how to obtain the DE equations in Eqs. (13.5) and (13.6).

The above-described process involves one iteration from p to q and then back from q to p, assuming an imaginary graph where all nodes would have the same degree ℓ. Density evolution aims to determine the limit (minimum) values of p and q after several (infinitely many) iterations on all the edges of any general Tanner graph obtained from a given probability distribution Λ. This applies not only to Tanner graphs with nodes of fixed degree ℓ, but to any general Tanner graph.

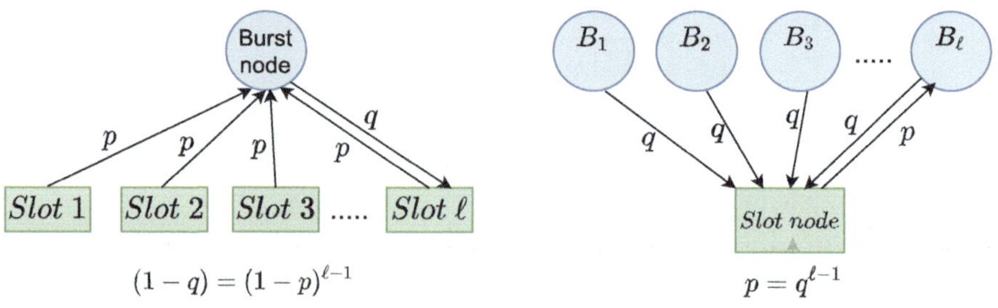

Fig. 13.2 A simple illustration of basic density evolution equations

Following the tree analysis argument of [15], by averaging the two expressions in Eqs. (13.5) and (13.6) on all possible degrees, we get respectively:

$$\bar{q} = \sum_{\ell} \lambda_\ell \bar{p}^{\ell-1} \tag{13.7}$$

$$\bar{p} = 1 - \sum_{\ell} \rho_\ell (1 - \bar{q})^{\ell-1} \tag{13.8}$$

For simplicity, we note the average probabilities $\bar{p}$ and $\bar{q}$ over all the edges and their evolution through iterative decoding as p_i and q_i at iteration i. Then, we can derive the evolution of the average probabilities during the ith iteration as

$$q_i = f_b(p_{i-1}) \text{ and } p_i = f_s(q_i), \text{ thus } p_i = f_s(f_b(p_{i-1})) \tag{13.9}$$

with, functions corresponding to [4], for classical IRSA:

$$f_b(p) = \lambda(p) \text{ and } f_s(q) = 1 - \rho(1-q) \tag{13.10}$$

$$\text{when } M \to \infty\text{: } f_s(q) \to F(\frac{G}{R}q) \text{ with } F(x) \triangleq 1 - e^{-x} \tag{13.11}$$

Note that the sequence $(p_i)_{i=0,\ldots,\infty}$ characterizes how many of the edges in the graph correspond to undecoded users at each iteration i. Note that the sequence $(q_i)_{i=0,\ldots,\infty}$ can equivalently be considered. Initial values are $q_0 = 1$, hence $p_0 = 1 - \rho(0)$. Alternately, picking $p_0 = 1$ just shifts the sequence by one, as it implies that $q_1 = \lambda(1) = 1$.

Figure 13.3 shows a detailed example of the decoding process of 4 users who compete for 4 slots. The users are using the degree 2 for $\frac{3}{4}$ of the time and the degree 3 for $\frac{1}{4}$ of the time. In this example, we illustrate the computation of the burst degree and the edge degree distributions, and we follow in detail the decoding process with the density evolution parameters. The equivalent Tanner graph represents the users in circles, the slots in rectangles, and the transmissions of the users on the slots as 9 edges between the users and the slots. The number of edges represents the number of sent packets on the radio access frame. An important piece of information should be noted: the values of p_i^* and q_i^* given in the Fig. 13.3 are the actual empirical values computed on this specific instance of a Tanner graph. The values of p_i and q_i obtained from Eq. (13.9) are estimates (expectations) of these p_i^* and q_i^* for random instances of all such Tanner graphs, when furthermore Λ and load G are kept fixed, but frame size becomes infinitely large.

The degree distributions: The user degree distribution, based on the presented definition in Eq. (13.1), is given as $\Lambda(x) = \frac{3}{4}x^2 + \frac{1}{4}x^3$. Now considering the edge degree distribution of the edges, the probability that a randomly selected edge is connected to a node of degree k is proportional to kN_k, where: k is the node degree and N_k is the number of degree k nodes. Thus, the probability that a randomly selected edge is connected to a burst node of degree 2 is

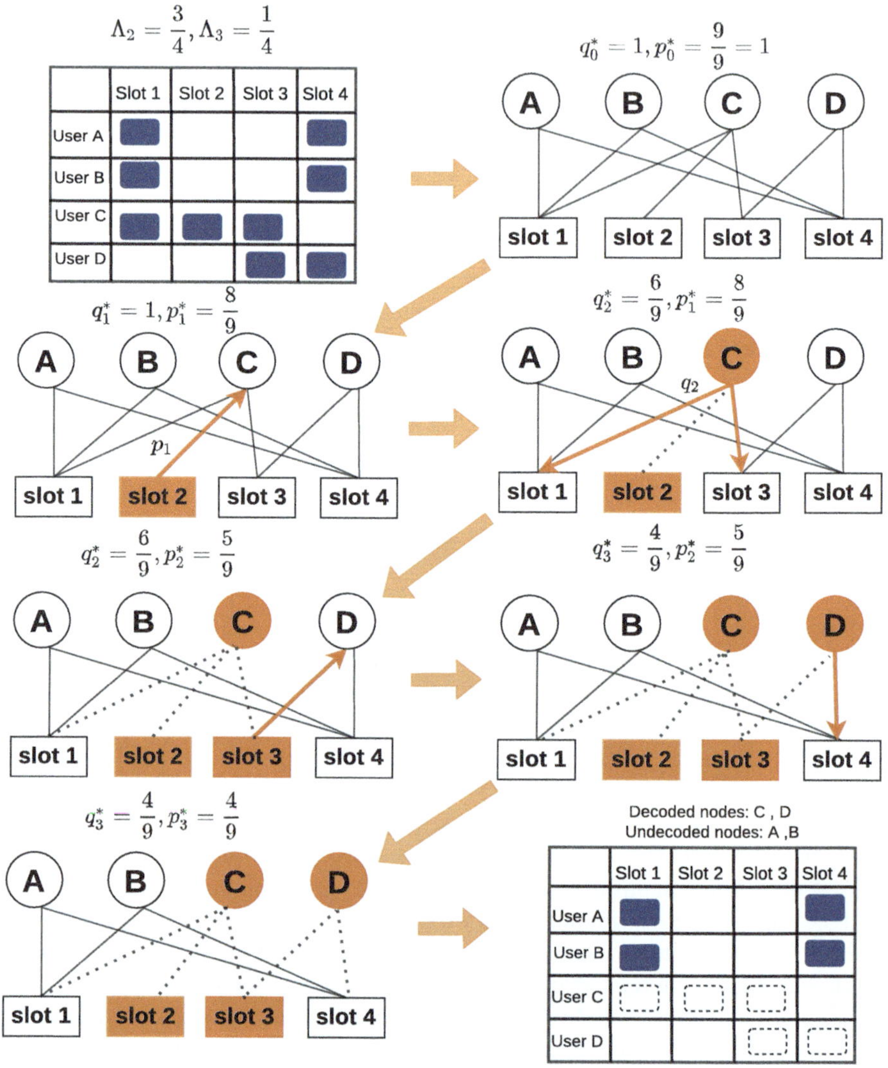

Fig. 13.3 Example of the decoding process of IRSA with Density Evolution parameters p^* and q^*

$\lambda_2 = \frac{2\times N_2}{\sum_i iN_i} = \frac{2\times 3}{2\times 3+3\times 1} = \frac{6}{9}$, as there are 3 burst nodes that have a degree 2. Following the same manner: $\lambda_3 = \frac{2\times N_3}{\sum_i iN_i} = \frac{3\times 1}{2\times 3+3\times 1} = \frac{3}{9}$. Therefore, the edge degree distribution from the burst perspective, based on the presented definition in Eq. (13.3), is given as $\lambda(x) = \frac{6}{9}x + \frac{3}{9}x^2$, which also can be computed directly by taking the first derivative of $\Lambda(x)$ and normalizing, as in Eq. (13.3).

Equivalently, we can compute the edge distribution from the slot node perspective. The probability that a randomly selected edge is connected to a slot node of degree ℓ is also proportional to ℓN_ℓ, where N_ℓ is the number of the slot nodes that have a degree ℓ. Thus, the probability that a randomly selected edge is connected to a slot node of degree 1 is $\rho_1 = \frac{1}{9}$, as there is only one edge out of 9 edges connected to a degree 1 slot node. Consequently, the probability that a randomly selected edge is connected to a slot node of degree 2 is $\rho_2 = \frac{1\times 2}{9}$ and to a slot node of degree 3 is $\rho_3 = \frac{6}{9}$. As a result, the edge degree distribution from the user perspective, based on the presented definition in Eq. 13.3, is given as $\rho(x) = \frac{1}{9} + \frac{2}{9}x + \frac{6}{9}x^2$.

The decoding process: Before the decoding starts, the probability that a randomly selected edge is unknown q_0, or the probability that a randomly selected edge cannot be revealed p_0 are initialized as previously described, i.e., either $q_0 = 1$ or $p_0 = 1$.

Now, back to density evolution: how can one estimate whether that all the packets will be ultimately decoded? From the decoding point of view, and for an infinite frame size, we will be able to decode if the probabilities p_i are decreasing towards zero, which requires that $p_i < p_{i-1}$. When $M \to \infty$, $p_i = F(\frac{G}{R}\lambda(p_{i-1}))$, thus this is obtained when (necessary condition):

$$F\left(\frac{G}{R}\lambda(x)\right) < x \text{ for all } x \in (0, 1) \tag{13.12}$$

Given the parameters $(\Lambda, G, \ldots)$, one can simply check the curve, e.g., Eq. (13.12), to determine if and where the decoding process will stop (asymptotically for $M \to \infty$).

Following [4], one main property is that: given Λ, there exists a *load threshold* G^*, such that, when the frame size converges towards infinity ($M \to \infty$), for any load $G < G^*$, all the packets are decoded with vanishing error probability. G^* depends on the choice of the distribution Λ. Finding the Λ with the largest G^* is our prime interest.

13.3.4 Optimizing IRSA

Many studies have explored and analyzed the performance of IRSA in an IoT network. The main purpose of these studies is generally to find and study a better variant of this protocol. Given any IRSA system, the goal is often to find an optimized user degree distribution that maximizes a certain metric (throughput, achievable load,..., etc.). This problem is usually formulated as an optimization problem, which can be described as follows:

$$\begin{aligned} & \underset{(\Lambda_i)}{\text{maximize}} && \mathcal{O}(\Lambda_0, \Lambda_1, .., \Lambda_D) && (\mathcal{P}_1) \\ & \text{subject to} && 0 \leq \Lambda_i \leq 1 \; \forall i \\ & && \sum_{i=0}^{i=D} \Lambda_i = 1 \end{aligned}$$

where $\mathcal{O}$ is the system criteria that needs to be optimized and D is the maximum degree. Depending on the system model, more constraints can be added to the optimization problem. Notice that the system is initially a stochastic optimization problem, as the performance depends on the random variables of the users' arrivals and their degree selections. The mentioned DE method allows analyzing the expectation of the asymptotic performance of different (framed) variants of the IRSA protocol deterministically. For several IRSA variants, the problem may still be difficult to solve, for example, in the case of the non-convexity or non-linearity of the constraints. But the optimization problem formulation may be useful. In the following chapters, for some variants of IRSA, we will adapt the OP described in $(\mathcal{P}_1)$ to optimize the performance of IRSA.

13.3.5 Asymptotically Optimal IRSA

One optimal family of solutions for 1-IRSA is based on the soliton distribution [6]. It can be found in the following way (as presented in [16, slide 35]): consider Eq. (13.19), that has a simple expression for $K = 1$. Indeed, $h_1(x) = -\log(1-x)$. Then, it is possible to write a Taylor series of $h_1(x)$ in $x = 0$ and find an expression for the coefficients of λ_i (and Λ_i) assuming the equality in Eq. (13.19), and recover the soliton distribution(s).

13.4 Variants of the IRSA Protocol

Recently, many studies conducted research on CSA-type modern random access protocols due to their expected advantages in terms of improving the random access of devices in IoT networks and providing high efficiency. The family of CSA protocols is very generic: it can be applied as soon as packets are repeated, and the receiver is equipped with SIC. For that reason, there have been many variations of the CSA protocols. As indicated, IRSA is known to reach asymptotically the maximum throughput of 1 [packet/slot] in a collision model; but it turns out that it is even possible to go beyond this performance, with more *realistic* model assumptions. Thus, some CSA family variations have explored different system assumptions; for instance, different realistic channel conditions or modern physical layer designs that have been proved to enhance the protocol performance. In this section, we describe some of such studies that have considered Multiple Packet Reception (MPR) at the receiver, varying transmission powers, fading, and erasure channels, leading to potentially beneficial capture effects. Because properties such as fading are usually associated with packet losses, some other related research studies have assessed their negative impact and have handled the decoding imperfection due to transmissions or SIC errors.

First, IRSA schemes, for receivers capable of decoding multiple colliding packets jointly, have been introduced in multiple research works. Usually, the capability of the receiver to receive K packets at the same time is referred to as K-MPR and can be usually achieved in a wireless network with some diversity technique [17]. Note that K-MPR would allow significant throughput gain, but from an information-theoretic capacity perspective, one would expect that it requires an equivalent decrease in individual capacity by a factor K or greater.

In [18], the effect of K-MPR (referred to as *multiuser detection*) on Irregular Repetition Slotted ALOHA (IRSA) was studied, which increases the likelihood of decoding more packets under large system loads. Simulations have shown that a multiuser detector supports larger loads for a low number of slots jointly processed. Another study in [19] addressed the K-MPR capabilities at the receiver for CSA. Their converse bound on the asymptotic normalized load threshold of coded slotted ALOHA schemes (G/K) appears to increase with K. They also numerically demonstrated, for $K = 1$, $K = 2$, and $K = 3$, that some IRSA schemes achieve a performance close to K [packets/slot]. This indicates that exploiting the K-MPR capability may be useful, even if the amount of resources required to achieve it scales linearly with K. In [20], an optimized transmission probability distribution was analytically derived for IRSA with 2-MPR.

Another major question in CSA literature is the achievable throughput in realistic network scenarios, considering fading, limited frame length, decoding errors, or the probability to lose some packets. At the physical layer level, the highest-power packet might be successfully recovered despite the presence of other interfering packets. This is called the *capture effect*. The capture effect may be exploited to recover the underlying packets by removing the recovered high-power packets. In [21], the analysis of IRSA considered a block fading channel model, and also assumed capture effect at the receiver. New density evolution equations are required to study such a scenario. The work optimized some repetition degree distributions in a finite frame length setting. The derived distributions are shown to achieve throughputs largely exceeding 1 [packet/slot]. In [22], Extended IRSA (E-IRSA) has been proposed as an extended operation of IRSA that exploits the capture effect to perform iterative decoding. Intra-slot SIC has been applied to decode more than one collided packet per slot. Simulation results show that the E-IRSA protocol allows reaching the maximum theoretical throughput even in scenarios where the number of active users is higher than the number of slots per frame. CSA performance has been evaluated assuming a fading channel in [23–25]. The impact of the capture effect on the SIC-enabled slotted ALOHA framework has been also addressed in [26, 27]. In general, the capture effect, combined with SIC, helps to increase throughput.

A related direction of studies of modern random access protocols is to consider the control of the transmission power. Packet replicas can also be transmitted in different power levels that are deliberately chosen to exploit the capture effect at the receiver. In this case, when intra-SIC dominates, packet collisions can be resolved in the power domain. New density evolution should be formulated in this case. In [28], a packet is transmitted with a power level chosen according to a power-choice distribution, such that multiple packets sent by different nodes might be decoded at the receiver in a single slot, by ensuring that their Signal-to-Interference-and-Noise Ratio (SINR) is above the decoding threshold of the successive interference cancellation receiver. In [29], an optimal *Power-Limited IRSA* (PL-IRSA) has been proposed as a variant of IRSA for the Internet-of-Things (IoT)-oriented satellite networks. A developed density evolution analysis considering total transmission power constraints and iterative IC processing has been derived to compute the PLR of PL-IRSA and to find the optimal degree distributions that maximize the normalized throughput. The work concludes that different total transmit power constraints and code rates can lead to completely different optimal degree distributions. Therefore, optimal degree distributions should be calculated according to the total transmit power constraint and code rate for practical implementations. A novel paradigm whereby devices adapt their data rates and/or transmit powers based on their chosen repetition degrees has been introduced in [30]. In [31], another approach is proposed to improve the throughput of the IRSA protocol. The total transmit power of all the users is considered to be the same, while the transmit power of a replica is determined by the number of replicas. The transmit power of each replica of different users can also be different. The study did not consider channel fading.

Another IRSA variant using temporal diversity and the capture effect has been introduced in [32]. The work optimized the repetition degree distribution along with the parameter of power control to maximize the total system throughput. In general, adjusting the transmission power is an excellent way to control the capture effect and further improve the performance. When modeling IRSA or CSA protocols (e.g., with density evolution), a common assumption in the CSA literature is that interference cancellation (IC) can always be applied perfectly and added that a singleton packet (the packet that is alone on the slot) can be successfully decoded with probability 1. Nevertheless, there have been some studies with the assumption of SIC imperfection. Two possible causes: the first one is that there might have been transmission errors, such as one replica being subject to strong fading and being received with very low power (leading to a "packet erasure"), or alternately the presence of interference on one slot (leading to a "slot erasure"). The second one can be errors in signal parameters' estimation (amplitude, phase, frequency offset) before the decoding that might lead to incorrect signal subtraction (cancellation). After each SIC operation, residual energy is present due to the imperfection of the cancellation process. The residual energy would accumulate, and SIC errors would become more likely as the number of SIC operations on the same slot increases. CSA over a packet erasure channel and a slot erasure channel has been studied and optimized in [33, 34]. Ivanov [35] presented a framework to design degree distributions for finite frame length CSA system with different classes of users, which have different error rate requirements. The classes are assigned different distributions. The work shows that multi-class CSA is capable of providing different levels of protection at high channel loads, as well as a smaller average decoding delay for better-protected users.

Combining one of the CSA protocols with non-orthogonal multiple access (NOMA) has also been widely inspected. Literally, CSA protocols are "non-orthogonal" and hence are NOMA methods. But in general, compared to CSA, many of the proposed NOMA methods in the literature have more detailed assumptions on the physical layer. This might be due to the fact that NOMA had been a study item in the 3GPP for 5G Release 16 (see for instance [36]), and has often been studied in that context. Thus compared to plain CSA or plain IRSA, some NOMA methods can include features such as joint-decoding, multichannel transmissions, pilot-based user detection, etc.; although it can also happen that some NOMA methods are equivalent to some CSA methods. In [37] a scalable, energy-efficient pure ALOHA with power domain NOMA has been proposed. The results of combining ALOHA and NOMA show that there is a greater than linear increase in throughput as the number of active IoT devices increases. In [10], the transmission power of each packet was chosen by a power distribution, where discrete power levels pre-determined by the NOMA method have been used to send replicas. It is shown that the proposed scheme in [10] can outperform the existing IRSA schemes. Another combination of slotted-ALOHA and NOMA has been proposed in [38]. NOMA has been also combined with multichannel ALOHA in [39, 40].

In the next section, we introduce the IRSA protocol fundamentals, and we explain in detail its decoding process.

13.5 IRSA with Multiple Packet Reception

13.5.1 Introduction

The previous description of IRSA assumed an idealized collision channel. In fact, one can go *beyond* the "optimal" performance of IRSA of 1 [packet/slot] by adopting more realistic assumptions of communication channels. Indeed, some recent publications [18, 19, 41] have focused on specific variants of CSA protocols to handle the massive connectivity issue assuming a different model of the collision channel: in these new variants, the receiver has the capability of simultaneously decoding more than one packet from multiple concurrent transmissions, denoted *Multiple Packet Reception* (MPR) [42]. MPR is the capability of simultaneous decoding of more than one packet from multiple concurrent transmissions. There are several techniques to allow simultaneous decoding of packets on a receiver [43]. In [17], a classification based on transmitter perspective, trans-receiver perspective, and receiver perspective has been introduced. Orthogonal multiplexing systems such as Time Division Multiple Access (TDMA), Code Division Multiple Access (CDMA), or Orthogonal Frequency Division Multiple Access (OFDMA) are examples of multiuser techniques that combine different users' signals by using different time slices, orthogonal codes and multiple frequency ranges respectively [44]. From a transceiver perspective, to enable MPR, transmitters and receivers should cooperate on some operations.

Multiple-input multiple-output (MIMO) systems [45] can achieve MPR by exploiting the spatial diversity of the transmissions. Alternately, the MPR capability can be shifted from the transmitter to the receiver by using a bank of Matched Filters to decode packets coded with spreading codes that does not need to be orthogonal.

The ability to decode K or fewer packets simultaneously is usually denoted *K-MPR*, and a derivative of the *collision channel* described in Sect. 13.3 can be introduced as the K-MPR collision channel, where packets are retrieved if and only if less than or exactly K packets are transmitted on the same slot.

The importance of combining multiple packet reception feature with IRSA comes from the fact that MPR exploits the structure of the interfering signals, which increases the number of users that the system can sustain. We point out in Sect. 13.5.2 some studies of CSA protocols with MPR capability. These studies have proved that equipped IRSA with MPR increases the chance of decoding more packets under large system loads. But, to our knowledge, there have been no existing result that provides optimal degree distributions that reach K packets per slot, equivalently to the soliton distribution that attains the bound of 1 [packet/slot] for classical IRSA under a collision channel model.

In this section, we focus on IRSA with K-MPR. It is denoted K-IRSA, and supposes that the transmissions occur in fixed slots (as for slotted ALOHA). Considering the *throughput*, under this K-MPR collision channel model, K-IRSA can recover at most K packets per slot Classical IRSA, which is 1-IRSA, asymptotically reaches this performance [6]. Hence, a natural question arises: can K-IRSA reach the bound of K? With which distributions? as, K-MPR can typically be obtained at a cost of a decrease in modulation rate by a factor K.

Previous studies have explored the question, [18, 19], by describing a non-asymptotic bound for K-IRSA, and numerically exhibiting some specific modification of CSA. These studies show results close to their bound for $K = 3$ and 4, but they are not able to fully answer the question of the exact bound, and of reaching it with K-IRSA.

13.5.2 System Description

We adopt the IRSA system model explained in Sect. 13.3. For the K-IRSA variant, we modify it to assume that the receiver has the capability of handling K simultaneous packet receptions. In the K-MPR collision channel model, at the end of the frame, the K-MPR receiver attempts to decode the packets. Each time there are K or fewer packets in a slot, the K-MPR receiver can successfully retrieve the packets. Subsequently, K-IRSA utilizes SIC to remove the physical copies of the decoded packets from their positions in other slots. This process is iterative, as new decoding opportunities can arise.

The example, shown in Fig. 13.4, illustrates 6 users transmitting their packets in a frame of 4 slots. The users are using IRSA protocol to send their packets towards a 2-MPR receiver. As the figure shows, $\frac{1}{3}$ of the time, the users use the degree 3, and $\frac{2}{3}$

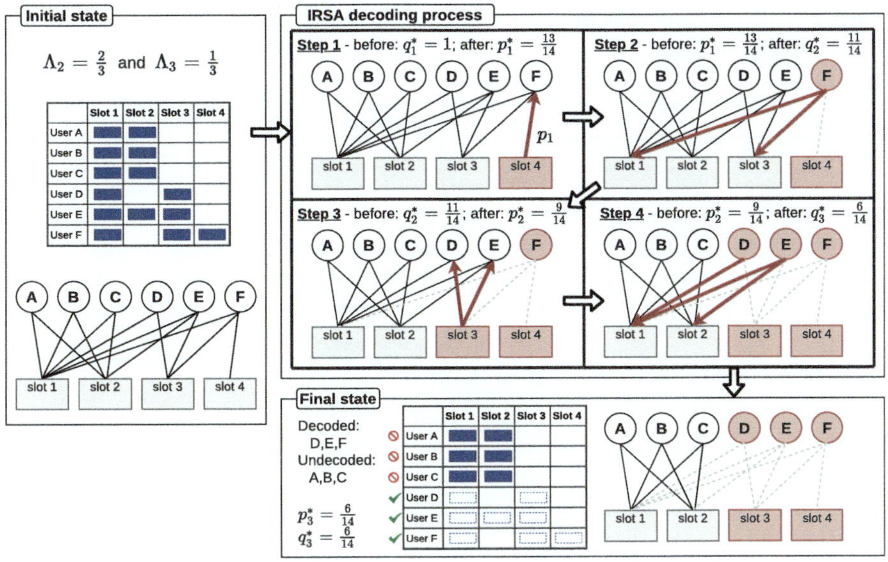

Fig. 13.4 Example with 2-IRSA (K=2)

of the time, the users repeat their packets two times. The decoding process starts by searching for singletons (slots of a single transmission) or the slots of two colliding packets (as the receiver has 2-MPR capability).

As in Sect. 13.3.3, we can consider the sequences of probabilities p_i, q_i, the probabilities that a random edge connected to one burst node or a slot node is unknown or unrevealed. We start with no packet decoded: $p_0 = 1$ and thus $q_1 = 1$. In this section, we also denote the empirical probabilities obtained on the instance of the graph obtained in Fig. 13.4 as p_i^*, q_i^*.

At the first decoding iteration, user F on slot 4 is decoded and the edge between this user and the slot 4 can be revealed. Consequently, the probability that one edge cannot be retrieved from the slot perspective, p^*, after the first iteration is $\frac{13}{14}$. This probability represents the number of unrevealed edges in the corresponding bipartite graph over the total number of initial edges in the graph. In the second iteration, the two other copies of user F on slots 1 and 3 can be extracted. This makes the probability that an edge is unknown from the user perspective equal to $q_2^* = \frac{11}{14}$, which represents:

$$\frac{\text{The number of total edges} - \text{The number of retrieved edges}}{\text{The number of total edges}}$$

The revealing of the packet of user, F followed by its removal with SIC on the slot 3, leaves 2 collisions from users D and E on the slot 3. As the receiver is 2-MPR type, the two collided packets on slot 3 can be decoded, and their copies on slots: user D on slot 1, and user E on slots 1, 2 can be removed. This process of iterative

Table 13.1 Density Evolution equations used for K-IRSA analysis

General IRSA notations

Equation	Reference
$\Lambda(x) \triangleq \sum_\ell \Lambda_\ell x^\ell$	Burst node degree distribution in Eq. (13.1)
$\Psi(x) \triangleq \sum_\ell \Psi_\ell x^\ell$	Slot node degree distribution in Eq. (13.1)
$\lambda(x) \triangleq \sum_\ell \lambda_\ell x^{\ell-1}$	Polynomial representation of the burst edge degree distribution in Eq. (13.3)
$\rho(x) \triangleq \sum_\ell \rho_\ell x^{\ell-1}$	Polynomial representation of the slot edge degree distribution in Eq. (13.3)
$q = p^{\ell-1}$	Probability on the edge towards a slot node in Eq. (13.5)
$(1-p) = (1-q)^{\ell-1}$	Probability on the edge towards a burst node in Eq. (13.6)
$q_i = f_b(p_{i-1})$ and $p_i = f_s(q_i)$	In Eq. (13.9)
When $M \to \infty$: $f_s(q) \to F(\frac{G}{R}q)$ With $F(x) \triangleq 1 - e^{-x}$	Asympotic expression from Eq. (13.11)

decoding continues until all packets are decoded, or the decoding process is blocked by a stopping set. In this example, users A, B, and C sent their packets twice on the same slots 1 and 2. This creates a stopping set of three colliding users on the slots 1 and 2, making it impossible to retrieve the packets of those users.

Density Evolution We use density evolution equations to track the decoding process of K-IRSA (see Sect. 13.3.3). For that aim, we need to adapt the density evolution equations for the K-MPR scenario.

For later mathematical convenience, and given the system parameters Λ, M, N and K, we reintroduce the concept of burst- and slot-perspective degree distributions, which was already presented in Sect. 13.3.3. We will use the set of basic IRSA model equations in Table 13.1 for later analysis.

In the next section, we will highlight the main works on combining CSA protocols with MPR. Then, we will update and formulate the density evolution equations in Table 13.1 for the K-MPR case.

K-IRSA Related Work In this section, we give insights about existing research studies that applied MPR to enhance the performance of CSA protocols. We also present the analysis of the conditions under which the K-MPR receiver is able to decode the packets successfully.

Different variants of IRSA and CSA have been studied, often by establishing new functions $f_b(q)$ and $f_s(p)$ (and $F(x)$) for Eq. (13.12) for the density evolution.

A random selected edge connected to a slot node of degree ℓ can be revealed with probability $1 - p$, after the first iteration, if the total number of unrevealed edges of the slot node is less than or equal to K. Therefore, in case of K-MPR receiver, the functions in Eqs. (13.5) and (13.6) become

$$q = p^{\ell}$$
$$1 - p = \sum_{k=0}^{min(K,\ell)-1} \binom{\ell - 1}{k} q^k (1-q)^{l-k-1} \tag{13.13}$$

Ghanbarinejad and Schlegel [18] analyzed IRSA with K-MPR for the case of an infinite user population. They show that providing the receiver with multi-user detection capability allows larger load in the systems with low number of slots per frame. By averaging on all possible connected edges, their density evolution equations in Eq. (13.13) correspond to:

$$f_b(p) = \lambda(p) \text{ and } f_s(q) = 1 - \sum_{k=0}^{K-1} \frac{\rho^{(k)}(1-q)}{k!} q^k \tag{13.14}$$
$$\text{where } \rho^{(k)}(x) \text{is the } k\text{th derivative of } \rho(x)$$
$$f_s(q) \to F_K(\frac{G}{R}q) \text{ with } F_K(x) \triangleq 1 - e^{-x} \sum_{k=0}^{K-1} \frac{x^k}{k!} \tag{13.15}$$
$$\text{when } M \to \infty$$

A major question is the performance of IRSA and K-IRSA: it is obvious that K-IRSA can at most recover K different users per slots, which thus yields an upper bound of the load threshold of $G^*/K \leq 1$. But what is the maximum load up to which we can decode all packets G^* and what is the distribution Λ that can reach it? Narayanan and Pfister [6] proved that for $K = 1$ (IRSA), the load threshold is asymptotically $G^* \to 1$ for a sequence of (truncated) soliton distributions. In [18], IRSA was studied in an infinite setting, considering that the receiver has K-MPR capability. Simulation results have been done for finite number of users and have shown that multiuser detection capability at the receiver allows larger loads on the system with low number of slots per frame, which in turn reduces the total transmission delay and power consumption. Some distributions were experimented with various values of K, and the load threshold was found to be G^*/K between 0.8 and 0.96 (their [18, Fig. 4]).

Stefanović et al. [19] developed a bound on the asymptotic load threshold of Coded Slotted ALOHA (CSA) schemes with K-MPR, that also applies to K-IRSA. They give examples close to the bound but this was not obtained for plain K-IRSA itself, but a variant using Spatially-Coupled CSA (SC-CSA) [46], with a more structured slot selection.

Their bound [19, (17)] in (13.16) yields $G^*/K \to 1$ when $R \to 0$:

$$\frac{G^*}{K} \leq 1 - \frac{1}{K} e^{-\frac{G^*}{R}} \sum_{k=0}^{K-1} \frac{K-k}{k!} \left(\frac{G^*}{R}\right)^k \tag{13.16}$$

SC-CSA has an interesting performance, and it has other variants such as Irregular Repetition Spatially-Coupled Slotted ALOHA (IRSC-SA), which have also been proposed and analyzed recently in [47]. But all these variants require "super-frames" (i.e., frames of frames), thus typically require several orders of magnitude more slots, hence our continued focus is on K-IRSA.[3]

A recent work (in progress) [20] has studied some families of repetition probability distribution for K-IRSA (based on an approximation for $G\Lambda'(x)$): they are able to find good distributions, and propose an algorithm to search for the optimal parameters for that family of distributions for K-MPR with $K = 2$. They obtain an optimized distribution in their [20, Eq. (11))] that achieves $G^* = 1.68$. This is still below our own results. Finally, one can further find recent work [48] (from 2023) that is generalizing and improving on the work presented in [49], and this chapter.

IRSA and K-IRSA (Error Floor): Related Work and Results In IRSA and K-IRSA, the decoding process is still susceptible to failure due to *stopping sets*, similar to LDPC codes [4], owing to the presence of cycles in the Tanner graph. Stopping sets have a detrimental impact on decoder performance and lead to an *error floor* in the decoding failure probability. Previous works, including [41], have presented a finite-length analysis of *frameless* ALOHA with K-IRSA in the error floor region. Their recursive approach enables the computation of decoder state probabilities, facilitating the numerical computation of the Packet Error Rate (PER), also denoted as Packet Loss Rate (PLR). Another analysis of the error floor due to stopping sets of CSA for finite frame lengths over the packet erasure channel, based on combinatorics, was presented in [34], among others.

13.5.3 Finding Optimal Distributions

Objectives and Notations In this section, we provide a method that finds one distribution $\Lambda(x)$ (e.g., to find $\lambda(x)$ first) with the highest load threshold G^* when the number of coefficients is limited by n (e.g., $\Lambda_i = 0, \forall i > n$). We denote the distribution by $\mathcal{L}_n^K$, and the load threshold by $\mathcal{G}_n^K$. Notice that n is also the maximum number of replicas sent by a user and K is the maximum number of colliding packets that could be decoded at the same time by K-MPR a receiver.

Formulating an Optimization Problem We start from condition (13.12) on $\lambda(x)$, G, and R and apply it to the iteration functions for K-IRSA (13.15). We obtain the following inequality for the case where $K = 2$:

[3] For instance [47, Fig. 6] presents simulation results of $L = 40$ super-frames of $\times 400$ slots= 16,000 slots. In [47, Table 1] SC-SA performance of 0.9585 with $d = 4$ with $L = 30$ or $L = 40$ super-frames is still well below the result 0.9767 of [46] for super-frames of $L = 200$ sub-frames. Hence, possibly super-frames of $\geq 100,000$ slots in practice.

$$F_2(\frac{G}{R}\lambda(x)) < x \implies 1 - e^{-\frac{G}{R}\lambda(x)}(1 + \frac{G}{R}\lambda(x)) < x. \tag{13.17}$$

Taking the limit case of equality, we are able to find this closed formula for the function that satisfies the following equality:

$$\lambda(x) = \frac{R}{G}h_2(x) \text{ with } h_2(x) = -W_{-1}\left(\frac{x-1}{e}\right) - 1, \tag{13.18}$$

where $h_2(x)$ is expressed in terms of one of the branches of the Lambert W function. The Lambert function [50] is the inverse of the function $w \mapsto we^w = z$, thus the function $w(z)e^{w(z)} = z$, is a multivalued function defined in general for z complex and assuming complex values of $W(z)$. For $z \in [\frac{-1}{e}, 0[$, there are two real branches, one of which, satisfying $W(z) \leq -1$, is denoted $W_{-1}(z)$, which is used here.

As it satisfies the equality, the function $h_2(x)$ still provides a bound for any Λ distribution. Indeed, any G, λ satisfying (13.17) must satisfy $\lambda(x) < \frac{R}{G}h_2(x)$ equivalently.

This can be generalized for any K:

$$\lambda(x) < \frac{R}{G}h_K(x) \text{ for } x \in]0, 1[\text{ with } h_K(x) \triangleq F_K^{-1}(x) \tag{13.19}$$

Notice that F_K^{-1} cannot be written in closed form for $K > 2$ (or would require "Generalized Lambert functions"), and has the derivative $\frac{dF_K^{-1}}{dx}(x) \to +\infty$ when $x \to 0^+$ for $K \geq 2$.

On the Solutions of the Optimization Problem One optimal family of solutions for 1-IRSA is based on the soliton distribution [6]. It can be found in the following way (as presented in [16, slide 35]): consider Eq. (13.19), that has a simple expression for $K = 1$. Indeed, $h_1(x) = -\log(1 - x)$. Then, it is possible to write a Taylor series of $h_1(x)$ in $x = 0$ and find an expression for the coefficients of λ_i (and Λ_i) assuming the equality in Eq. (13.19), and recover the soliton distribution(s).

It would seem interesting to apply the same method with $h_2(x)$: It seems that taking the power series of $h_2(x)$, and by truncating it, we should be able to find functions $\lambda(x)$ that correspond to (near) optimal distributions. However, $h_2(x)$ does not have a proper power series as its derivative in 0 is infinite (and behaves like $\alpha\sqrt{x}$ near 0^+).

Similarly, the fact that $\frac{dF_K^{-1}}{dx}(x) \to +\infty$ when $x \to 0^+$ for $K \geq 2$, shows that, in general, unfortunately, one cannot expect to write Taylor series in 0^+, and then easily analytically derive some optimal distributions. Highlighting this difficulty is one of the contributions of this thesis, and it is also the reason why, in the following, we focus on finding optimal distributions numerically.

Finding $\mathcal{L}_n^K$ Through Its Optimal Edge Distribution $\lambda(x)$ We start by fixing n, and search for $\mathcal{L}_n^K$ through its edge distribution $\lambda(x)$. $\lambda(x)$ verifies the bound formulated in (13.19) and provides the largest load threshold G^*: it is an optimization problem.

The conditions on $\lambda(x)$ can be derived from the following:

- The condition $\lambda(1) = 1$ (because it is a probability distribution)
- The property that R is actually expressed in terms of $(\lambda_i)_i$ as $R = R(\lambda) = \frac{1}{2}\lambda_2 + \frac{1}{3}\lambda_3 + \ldots + \frac{1}{n}\lambda_n$
- The left inequality in (13.19), multiplied by G

This yields the following optimization problem on the variables $G, \lambda_2, \lambda_3 \ldots \lambda_n$ (in IRSA and K-IRSA: $\lambda_1 = 0$):

$$\begin{aligned}
&\underset{(\lambda_i)}{\text{Maximize}} && G && (\mathcal{P}_2)\\
&\text{subject to} && C_1 : 0 \leq \lambda_i \leq 1 \ \forall i\\
& && C_2 : \sum_{i=2}^{i=n} \lambda_i = 1\\
& && C_3 : G \sum_{i=2}^{i=n} \lambda_i x^{i-1} < (\sum_{i=2}^{i=n} \frac{\lambda_i}{i}) h_K(x), \forall x \in]0, 1[
\end{aligned}$$

Now, by using a technique introduced for LDPC codes, the last inequality of $(\mathcal{P}_2)$ can be transformed into multiple inequalities by taking a finite set of values for x, (see for instance [51, Sect. 3.18, p. 114]). Furthermore, when G is fixed, this is a linear program in the variables $(\lambda_i)_{i=2\ldots n}$, and thus can be efficiently solved.

Hence, our efficient technique is to perform a bisection on G: once G is fixed, we solve the linear program $(\mathcal{P}_2)$ with arbitrary objective function (such as maximize $R(\lambda)$), finding whether the constraints can be satisfied, and if so, finding one λ. The largest G admissible through bisection[4] yields $\mathcal{L}_n^K$ and $\mathcal{G}_n^K$.

We present the pseudocode of this procedure in Algorithm 13.1. The algorithm takes as input ϵ, the error tolerance at termination of the bisection search. The initial values $G_{\min} \leftarrow 0$ and $G_{\max} \leftarrow K$ are chosen so that $G_{\min}$ is feasible and $G_{\max}$ is a theoretical upper-bound of $G^*_{p_{\min}}$. The bisection search stops whenever a solution ϵ-close to the optimal is found. This only requires $O(\log_2(1/\epsilon))$ iterations.

[4] Indeed, the method has been proposed for LDPC codes as in [51, Sect. 3.18, p. 114], but the equivalent of G is considered fixed, hence it was a straightforward linear program. Here we have to introduce this bisection.

Algorithm 13.1 Load maximization bisection search

input: ϵ
initialization: $G_{\text{min}} \leftarrow 0$, $G_{\text{max}} \leftarrow K$
while $G_{\text{max}} - G_{\text{min}} > \epsilon$ **do**
  $G \leftarrow (G_{\text{min}} + G_{\text{max}})/2$
  Run a linear solver to check the feasibility of $\mathcal{P}_2$ given a load G
  if $\mathcal{P}_2$ is feasible **then**
    $G_{\text{min}} \leftarrow G$
  else
    $G_{\text{max}} \leftarrow G$
  end if
end while
return G_{min}

13.6 Conclusion

Our starting point is the IoT, a recent set of technologies that facilitates connections between billions of devices and machines to enable communication and data exchange. A primary challenge facing future IoT networks is the necessity to provide connectivity for IoT devices with stringent latency requirements. Consequently, achieving massive connectivity has emerged as one of the principal use cases to investigate for future IoT applications.

In this chapter, we emphasized the importance of IoT, outlined its benefits, and addressed the challenges, notably the issue of massive connectivity. We described the features and requirements of the massive connectivity problem, also known as mMTC. We discussed some main wireless techniques, particularly random access techniques, which have been employed in many proposed wireless standards. However, our focus in this chapter is on one specific technique: the family of modern random access protocols, particularly CSA, and more specifically, one of the recent protocols, IRSA. We provided a detailed explanation of the underlying concept, the operation of this protocol, and its decoding process modeling.

References

1. Ding J, Nemati M, Ranaweera C, Choi J (2020) IoT connectivity technologies and applications: a survey. IEEE Access 8:67646–67673
2. 3GPP, (2016) Study on scenarios and requirements for next generation access technologies. Technical Specification Group Radio Access Network, Technical Report 38.913
3. Shahab MB, Abbas R, Shirvanimoghaddam M, Johnson SJ (2020) Grant-free non-orthogonal multiple access for IoT: a survey. IEEE Commun Surv & Tutor 22(3):1805–1838
4. Liva G (2011) Graph-based analysis and optimization of contention resolution diversity slotted ALOHA. IEEE Trans Commun 59(2):477–487
5. Paolini E, Stefanovic C, Liva G, Popovski P (2015) Coded random access: applying codes on graphs to design random access protocols. IEEE Commun Mag 53(6):144–150

6. Narayanan KR, Pfister HD (2012) Iterative collision resolution for slotted ALOHA: an optimal uncoordinated transmission policy. In: 7th international symposium on turbo codes and iterative information processing (ISTC)
7. Casini E, Gaudenzi RD, Herrero ODR (2007) Contention resolution diversity slotted ALOHA (CRDSA): an enhanced random access scheme for satellite access packet networks. IEEE Trans Wireless Commun 6(4):1408–1419
8. Paolini E, Liva G, Chiani M (2015) Coded slotted ALOHA: a graph-based method for uncoordinated multiple access. IEEE Trans Inf Theory 61(12):6815–6832
9. Paolini E, Liva G, Chiani M (2011) Graph-based random access for the collision channel without feedback: capacity bound. In: 2011 IEEE global telecommunications conference - GLOBECOM 2011, pp 1–5
10. Shao X, Sun Z, Yang M, Gu S, Guo Q (2019) NOMA-based irregular repetition slotted ALOHA for satellite networks. IEEE Commun Lett 23(4):624–627
11. Richardson T, Urbanke R (2008) Modern coding theory. Cambridge University Press, USA
12. Newman MEJ, Strogatz SH, Watts DJ (2001) Random graphs with arbitrary degree distributions and their applications. Phys Rev E 64(2):026118
13. Wilf HS (2005) Generating functionology. CRC Press
14. Lushnikov AA (1974) Evolution of coagulating systems. II. Asymptotic size distributions and analytical properties of generating functions. J Colloid Inter Sci 48(3):400–409
15. Luby M, Mitzenmacher M, Shokrollahi MA (1998) Analysis of random processes via and-or tree evaluation. SODA 98:364–373
16. Narayanan K (2016) The peeling decoder: theory and some applications. http://pfister.ee.duke.edu/nasit16/Narayanan.pdf. Accessed 10 Jan 2022
17. Lu J-L, Shu W, Wu M-Y (2012) A survey on multipacket reception for wireless random access networks. J Comput Netw Commun 2012
18. Ghanbarinejad M, Schlegel C (2013) Irregular repetition slotted ALOHA with multiuser detection. In: Wireless on-demand network systems and services (WONS), pp 201–205
19. Stefanović C, Paolini E, Liva G (2018) Asymptotic performance of coded slotted ALOHA with multipacket reception. IEEE Commun Lett 22(1):105–108
20. Chen Z, Feng Y, Feng C, Liang L, Jia Y, Quek TQS (2021) Optimal distribution design for irregular repetition slotted ALOHA with multi-packet reception. Working paper or preprint, October 2021
21. Clazzer F, Paolini E, Mambelli I, Stefanovic C (2017) Irregular repetition slotted ALOHA over the Rayleigh block fading channel with capture. In: ICC. IEEE
22. Interdonato G, Pfletschinger S, Vázquez-Gallego F, Alonso-Zarate J, Araniti G (2015) Intra-slot interference cancellation for collision resolution in irregular repetition slotted ALOHA. In: 2015 IEEE international conference on communication workshop (ICCW), pp 2069–2074
23. Babich F, Comisso M (2018) Coded Slotted Aloha (CSA) with capture. In: European wireless 2018; 24th European wireless conference, pp 1–6
24. Babich F, Comisso M (2019) Impact of header on coded slotted Aloha with capture. In: 2019 IEEE symposium on computers and communications (ISCC), pp 1–6
25. Babich F, Comisso M (2019) Impact of segmentation and capture on slotted Aloha systems exploiting interference cancellation. IEEE Trans Veh Technol 68(3):2878–2892
26. Stefanović C, Momoda M, Popovski P (2014) Exploiting capture effect in frameless ALOHA for massive wireless random access. In: 2014 IEEE wireless communications and networking conference (WCNC), pp 1762–1767
27. Khaleghi EE, Adjih C, Alloum A, Mühlethaler P (2017) Near-far effect on coded slotted ALOHA. In: 2017-IEEE 28th annual international symposium on personal, indoor, and mobile radio communications (PIMRC), pp 1–7
28. Kumar A, Hegde P, Vaze R, Alloum A, Adjih C (2020) Breaking the unit throughput barrier in distributed systems. arXiv:2010.07430
29. Zhao B, Ren G, Dong X, Zhang H (2020) Optimal irregular repetition slotted ALOHA under total transmit power constraint in IoT-oriented satellite networks. IEEE Internet Things J 7(10):10465–10474

30. Hasanzadeh A, Chamberland J-F, Narayanan KR (2018) Rate selection and power adaptation using maximal ratio combining for the random access Gaussian channel. arXiv:1801.03557
31. Dia J, Zesong F, Yasheng Z (2019) Irregular repetition slotted ALOHA with total transmit power limitation. Inf Sci 64(2):129301
32. Nonaka T, Fujii T, Takyu O, Ohta M (2020) Adapting the number of replicas in the E-IRSA system using the power control. In: 2020 international conference on information networking (ICOIN), pp 787–792
33. Sun Z, Xie Y, Yuan J, Yang T (2017) Coded slotted ALOHA for erasure channels: design and throughput analysis. IEEE Trans Commun 65(11):4817–4830
34. Ivanov M, Brannstrom F, Amat AGI, Popovski P (2015) Error floor analysis of coded slotted ALOHA over packet erasure channels. IEEE Commun Lett 19(3):419–422
35. Ivanov M, Brännström F, Graell i Amat A, Liva G (2016) Unequal error protection in coded slotted ALOHA. IEEE Wireless Commun Lett 5(5):536–539
36. Yuan Y, Yuan Z, Tian L (2020) 5G Non-orthogonal multiple access study in 3GPP. IEEE Commun Mag 58(7):90–96
37. Balevi E, Al Rabee FT, Gitlin RD (2018) ALOHA-NOMA for massive machine-to-machine IoT communication. In: 2018 IEEE international conference on communications (ICC), pp 1–5
38. Tegos SA, Diamantoulakis PD, Lioumpas AS, Sarigiannidis PG, Karagiannidis GK (2020) Slotted ALOHA with NOMA for the next generation IoT. IEEE Trans Commun 68(10):6289–6301
39. Choi J (2017) NOMA-based random access with multichannel ALOHA. IEEE J Sel Areas Commun 35(12):2736–2743
40. Choi J (2018) Multichannel NOMA-ALOHA game with fading. IEEE Trans Commun 66(10):4997–5007
41. Lázaro F, Stefanović C (2017) Finite-length analysis of frameless ALOHA with multi-user detection. IEEE Com Lett 21(4):769–772
42. Lu J-L, Shu W, Wu M-Y (2012) A survey on multipacket reception for wireless random access networks. J Comput Netw Commun
43. Ananthapadnabhan A, Kumar SJK, Shunmuga S (2016) A survey on multi-packet reception with random access in wireless networks. Int J Sci Eng Res 7:8–12
44. Verdu S (1998) Multiuser detection, 1st edn. Cambridge University Press, USA
45. Huang WL, Letaief KB, Zhang YJ (2008) Cross-layer multi-packet reception based medium access control and resource allocation for space-time coded MIMO/OFDM. IEEE Trans Wireless Commun 7(9):3372–3384
46. Liva G, Paolini E, Lentmaier M, Chiani M (2012) Spatially-coupled random access on graphs. In: 2012 IEEE international symposium on information theory proceedings, pp 478–482
47. Hanxiao Y, Fei Z, Cao C, Xiao M, Jia D, Ye N (2019) Analysis of irregular repetition spatially-coupled slotted ALOHA. Sci China Inf Sci 62(8):80302
48. Fernández-Veiga M, Sousa-Vieira ME, Fernández-Vilas A, Díaz-Redondo RP (2023) Irregular repetition slotted aloha with multiuser detection: a density evolution analysis. Comput Netw 234:109921
49. Hmedoush I, Adjih C, Mühlethaler P, Kumar V (2020) On the performance of irregular repetition slotted aloha with multiple packet reception. In: 2020 international wireless communications and mobile computing (IWCMC), pp 557–564. IEEE
50. Corless RM, Gonnet GH, Hare D, Jeffrey DJ, Knuth DE (1996) On the Lambert W function. Adv Comput Math 5(1):329–359
51. Richardson T, Urbanke R (2008) Modern coding theory. Cambridge University Press

Chapter 14
Terahertz Communication—Bridging the Gap in the Next-Generation Wireless Systems

Vaishali Sharma, Sanjeev Sharma, and Vimal Bhatia

Abstract This chapter provides an in-depth analysis of terahertz (THz) communication, a burgeoning field poised to revolutionize next-generation wireless systems. It explores the untapped potential of the THz spectrum (0.1–10 THz), offering insights into its ability to support ultra-high data rates essential for applications like virtual reality, ultra-high-definition streaming, and complex scientific data transmissions. The chapter delves into the fundamental aspects of THz communication, including spectrum characteristics, wave propagation, and advanced modulation and coding techniques tailored for THz frequencies. This chapter addresses key technological challenges, such as the development of efficient transmitters and receivers, beamforming, multiple-input multiple-output (MIMO) technologies, and accurate channel modeling. The chapter also highlights significant applications of THz communication in areas like wireless fiber optic networks, Internet of Things (IoT), and integration with existing wireless networks. Furthermore, it discusses future perspectives, including emerging research areas, standardization, and regulation, concluding that THz communication is an essential component of future wireless communication ecosystems.

Keywords Channel modeling · THz · MIMO · IoT · Beamforming · THz applications · Signal detection

V. Sharma · V. Bhatia
IIT, Indore, India
e-mail: phd1901102029@iiti.ac.in

S. Sharma (✉)
IIT (BHU) Varanasi, Varanasi, India
e-mail: sanjeev.ece@iitbhu.ac.in

S. Sharma et al. (eds.), *Next-Generation Wireless Systems*,
Transactions on Computer Systems and Networks,
https://doi.org/10.1007/978-981-96-3185-8_14

14.1 Introduction

In the evolving landscape of wireless communication technologies, the exploration and implementation of higher frequency bands have become a crucial frontier in addressing the burgeoning demands for increased data rates and enhanced connectivity. Among these emerging spectrums, terahertz (THz) communication, operating in 0.1–10 THz frequency range (including the sub-THz band), emerges as a pivotal innovation [1]. This chapter endeavours to elucidate the fundamental principles, technological challenges, and proposes applications of THz communication, making it an integral element in the progression of next-generation wireless systems. Necessity to explore the the THz domain arises due to heavy congestion in traditional microwave and mmWave frequencies. The THz band, situated between the microwave and infrared segments of the electromagnetic spectrum, remains relatively underutilized. However, recent technological advancements have significantly reduced the barriers to its exploitation, heralding a new era in wireless communication capabilities [2].

This chapter systematically explores the expansive bandwidth offered by the THz spectrum, which is instrumental in enabling data transfer rates from gigabits per second (Gbps) to terabits per second (Tbps). Such capacity is indispensable for cutting-edge applications that necessitate high data throughput and low latency, such as immersive virtual reality, ultra-high-definition video streaming, and real-time complex scientific data transfer [3]. Through a detailed examination of the spectral properties, wave propagation characteristics, and the necessity for advanced modulation and coding techniques, the chapter provides a comprehensive understanding of the operational dynamics of THz communication. Furthermore, it delves into the salient technological challenges inherent in THz communication, including the development of effective transmitter and receiver technologies, the implementation of beamforming and multiple-input multiple-output (MIMO) technologies tailored for THz frequencies, and the intricacies of accurate channel modeling [4]. These discussions underscore the technological innovations necessary for the actualization of efficient THz communication systems.

Moreover, this chapter focuses light on the transformative potential of THz communication in various applications. It discusses its role in realizing wireless fiber optic networks, particularly in dense urban environments, and its significant impact on the Internet of Things (IoT) and sensor networks, thereby enabling a multitude of advanced applications [5]. In addressing the future outlook and concluding thoughts, the chapter highlights emerging research areas, the importance of standardization and regulatory frameworks, and the overarching role of THz communication in shaping the wireless communication landscape. By providing a detailed and structured analysis, this chapter serves as a comprehensive resource for understanding the complexities, challenges, and vast potential of THz communication in the realm of next-generation wireless systems.

14.2 Background

The evolution of wireless communication technologies has been significantly influenced by the need to accommodate a wide range of applications, each with distinct requirements for throughput, reliability, and latency. This has led to the categorization of applications within fifth-generation (5G) into three main structures: enhanced mobile broadband (eMBB), ultra-reliable low latency communication (URLLC), and massive machine-type communication (mMTC) [6]. Researchers are looking into more advanced ways to communicate and network since they know that a single solution might not work for all the situations. These include network densification through small cells, intelligent scheduling, and the use of multiple antenna systems to boost spectral efficiency [7]. A notable divergence in 5G from its predecessors is the recognition that the traditional sub-6 GHz spectrum would be inadequate for emerging applications. This led to the exploration of the millimeter wave (mmWave) and THz spectrum, which, despite initial scepticism due to its challenging propagation characteristics, has been made viable for mobile operations through advancements in device and antenna technologies. This shift has marked the advent of commercial THz communication.

Looking towards the future, applications such as virtual and augmented reality, ultra-high definition (UHD) video conferencing, three-dimensional (3D) gaming, and brain-machine interfaces are expected to impose even more stringent demands on network capabilities [8]. Concurrently, advancements in device fabrication suggest the imminent feasibility of nano-scale communications. This context has steered attention towards the THz band. The THz band offers enormous bandwidth and is poised to support ultra-high data rate applications, complementing the existing sub-6 GHz and mmWave bands. Its potential extends to micro and nano-scale communication, thanks to its small wavelength. Historically, the THz band was relegated to imaging and sensing applications due to the absence of efficient devices for these frequencies. However, recent advancements in THz device technology have positioned THz communication as a key player in future communication standards. The primary objective of this discussion is to provide a comprehensive understanding of mmWave and THz bands, highlighting their necessity in current wireless landscapes and the key design considerations for communication systems operating within these frequencies [9].

Despite these limitations, the potential of THz communication in transforming wireless communication remains compelling. The successful integration of THz frequencies is expected to revolutionize data transfer capabilities, paving the way for the development of advanced wireless applications. This exploration into THz communication underscores its critical role in the ongoing evolution of wireless communication technologies.

14.3 Related Work

The field of THz technology, while not as developed as microwave/mmWave and free-space optical (FSO) technologies, has seen significant progress in recent years. The initial exploration into THz technology was presented by P. H. Siegel, covering its applications, sensors, and sources [10]. In 2004, M. J. Fitch and R. Osiander delved deeper into the practical aspects of THz systems, including their sources, detectors, and modulators [11]. Subsequently, in 2007, various magazines reported on the advancement and potential applications of THz technology [12–14]. J. Federici and L. Moeller, in 2010, provided a comprehensive analysis of THz communications, examining aspects like channel coding, generation methods, and antenna technologies [15]. K. C. Huang and Z. Wang, in 2011, offered insights into creating cost-effective and robust THz wireless systems [16]. K. Wu et al. in 2012 focused on THz antenna technologies, while T. Kumer et al. highlighted the standardization efforts in IEEE 802.15 IG THz. T. Kurner and S. Priebe in 2014 discussed ongoing research, spectrum regulations, and standardization activities in THz communication systems [17]. I. F. Akyildiz et al. presented a comprehensive analysis of the state-of-the-art in THz technologies, emphasizing communication and networking challenges and experimental testbed [18]. In 2015, A. Hirata and M. Yaita gave a succinct overview of THz technologies and standardization in wireless communications [19]. C. Lin and G. Y. Li, in 2016, introduced an array-of-subarrays structure for THz wireless systems, discussing its circuit and communication advantages [20]. M. Hasan et al. reviewed progress in graphene-based devices for THz communications [21]. J. F. Federici et al. surveyed the impact of weather on THz wireless links [22]. S. Mumtaz et al. in 2017 explored THz communication applications in vehicular networks [23]. In 2018, V. Petrov et al. provided guidance on propagation modeling and antenna design for THz systems [24]. A. Boulogeorgos et al. examined system architectures for integrating THz wireless links with optical networks [25]. C. Han and Y. Chen discussed methods for modeling THz propagation channels. I. F. Akyildiz et al. addressed the distance limitations of THz technology. N. Khalid et al. provided a detailed tutorial on THz modulation schemes. D. Headland offered insights into beam control principles within the THz band. The period since 2019 has seen a notable increase in publications on THz communications, with various magazines and surveys discussing development progress, unresolved problems, and opportunities in the field [26, 27].

The THz ultra-massive multiple input multiple output (UM-MIMO) has emerged as a crucial paradigm for future wireless systems due to advancements in manufacturing materials, transceiver architectures, and antenna designs [28]. Authors in [29], have provided an overview of THz UM-MIMO systems and identified key challenges, a comprehensive and detailed educational resource specifically focused on THz UM-MIMO beamforming technologies is still lacking. This chapter is intended to address this gap, offering an all-encompassing perspective on THz UM-MIMO beamforming. It will talk about the basic ideas, new technologies, problems with

Table 14.1 THz communication related works in the literature

Technology	References	Summary of work
Fundamentals	[30–32]	THz communication characteristics and molecular absorption in THz frequency bands
Channel modeling	[33–35]	Different channel modeling and fading analysis
MIMO	[7, 28, 36]	MIMO-based THz system design and analysis
Beamforming	[20, 37–39]	Beamfoming methods are analyzed with beam slit and squint effect in THz systems
Signal detection	[40–42]	Various signal detection schemes with low complexity-based signal processing analysis such as compressive sensing and linear methods
IRS-aided THz	[43–45]	IRS-aided system design to enhance THz system performance
Applications	[31, 46, 47]	THz applications in imaging and sensing, communication, material characterization, astronomy and astrophysics, biomedical applications, etc. are highlighted

making them, and real-world issues related to using them. It will also look at wide-band impacts, beamforming situations, current MIMO arrays, new uses, and challenges that the field will face in the future. Further, some THz related papers' works are summarized in Table 14.1.

14.4 Fundamentals of THz Communication

The exploration into the domain of THz communication marks a significant shift in wireless technology, necessitating a fundamental understanding of its unique characteristics and operational principles. This section delves into the core aspects of THz communication, including its spectral properties, wave propagation dynamics, and the critical modulation and coding techniques required for its efficient utilization.

14.4.1 Spectrum and Bandwidth

THz communication is the expansive and relatively unexplored THz spectrum. This range presents a vast bandwidth significantly larger than those of conventional microwave and mmWave bands, offering a potential leap in data transmission capabilities, as shown in Fig. 14.1. Understanding the allocation, regulation, and effective utilization of this spectrum is imperative for the successful deployment of THz communication systems.

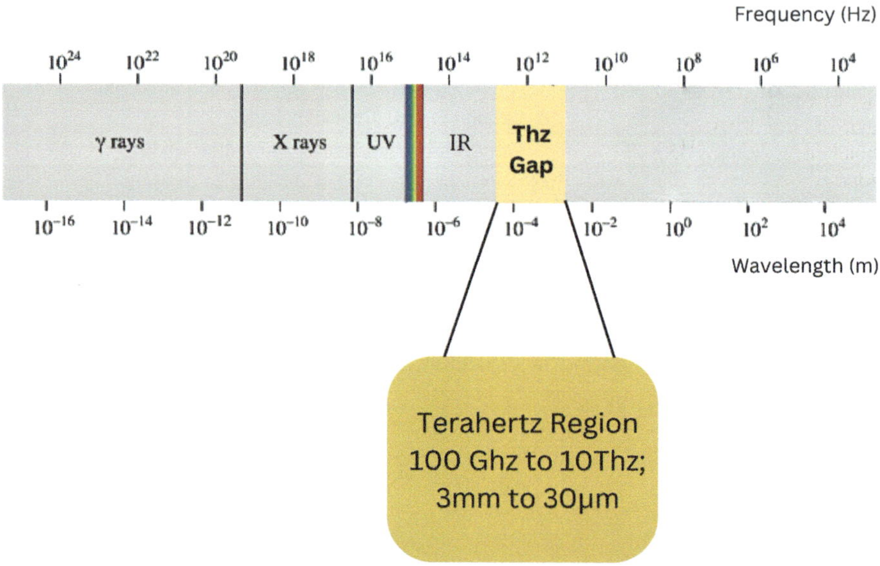

Fig. 14.1 Electromagnetic spectrum

14.4.2 Wave Propagation

Wave propagation in the THz band exhibits distinct characteristics compared to lower frequency bands. A defining feature of THz waves is their high atmospheric attenuation, primarily due to absorption by water vapour and other gases. This results in a relatively limited effective range, making the THz band more suitable for short-range, high-capacity communication links. The propagation of THz waves is also influenced by other factors such as scattering, reflection, and diffraction, which vary significantly across different environments and atmospheric conditions. A thorough understanding of these propagation characteristics is essential for the design and optimization of efficient and reliable THz communication systems.

In addition to the aforementioned factors, the propagation of THz waves is further characterized by their interaction with various materials and surfaces, which can significantly influence signal behavior. Materials commonly encountered in urban and indoor environments, such as concrete, glass, and human tissue, exhibit distinct absorption and reflection properties at THz frequencies. This necessitates a nuanced approach to channel modeling and system design, taking into account the material-specific responses to THz waves. Furthermore, the small wavelength of THz waves leads to more pronounced effects of diffraction and scattering, particularly around edges and small objects, leads to non-line-of-sight (NLoS) communication. However, these same properties also contribute to more complex multipath propagation scenarios, where the superposition of reflected, diffracted, and scattered waves can lead to signal fading and distortion. Consequently, advanced signal processing techniques and robust error correction mechanisms are imperative for ensuring the integrity and reliability of THz communications. Overall, the unique wave propagation characteristics of the THz band present both challenges and opportunities, driving innovation in the field of wireless communication and necessitating a deeper exploration of THz-specific phenomena.

14.4.3 Modulation and Coding Techniques

The development of appropriate modulation and coding techniques is pivotal for maximizing the efficiency of THz communication systems. Given the high bandwidth of the THz spectrum, there is a necessity for advanced modulation schemes that can effectively utilize this bandwidth while being resilient to the unique propagation challenges posed by THz frequencies. These challenges include higher levels of noise, severe channel fading, and path loss. Developing robust and efficient modulation techniques that can operate over such wide bandwidths and maintain signal integrity is crucial. Additionally, advanced coding techniques are required to mitigate errors and enhance the reliability of data transmission over THz links. Research in this domain focuses on leveraging the full potential of the THz spectrum through innovations in digital signal processing and error correction methodologies.

In addressing these challenges, there is a growing emphasis on the exploration of higher-order modulation schemes, such as quadrature amplitude modulation (QAM) and orthogonal frequency-division multiplexing (OFDM), which are capable of achieving high data rates in the THz band. These schemes must be intricately designed to cope with the wideband nature of THz frequencies, ensuring efficient spectrum utilization while minimizing the impact of noise and fading. Furthermore, the implementation of adaptive modulation techniques, which dynamically adjust modulation parameters in response to varying channel conditions, is seen as a promising approach to enhance system robustness. On the coding side, the development of sophisticated forward error correction (FEC) algorithms is crucial. These algorithms, including low-density parity-check (LDPC) codes and turbo codes, are designed to correct errors introduced during the transmission process, thereby improving the overall reliability and performance of THz communication systems. The integration of advanced modulation and coding techniques, coupled with machine learning algorithms for predictive channel modeling and optimization, represents a cutting-edge approach in the field. This integration not only addresses the inherent challenges of THz communications but also paves the way for realizing the full potential of this high-frequency band in a myriad of applications, from ultra-fast wireless networks to secure and efficient data transmission in next-generation communication infrastructures.

Considering the THz transceiver system, the received signal r is expressed as

$$r = h_l h x + n \in \mathbb{C}, \tag{14.1}$$

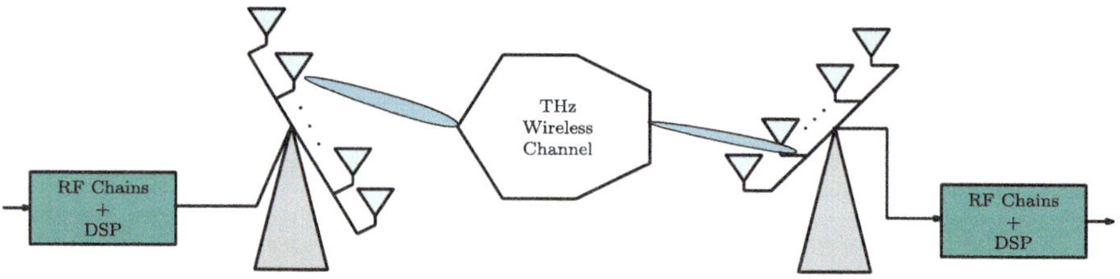

Fig. 14.2 THz system architectures

Fig. 14.3 Average SER performance of the THz system with modulation order 4 QAM signaling

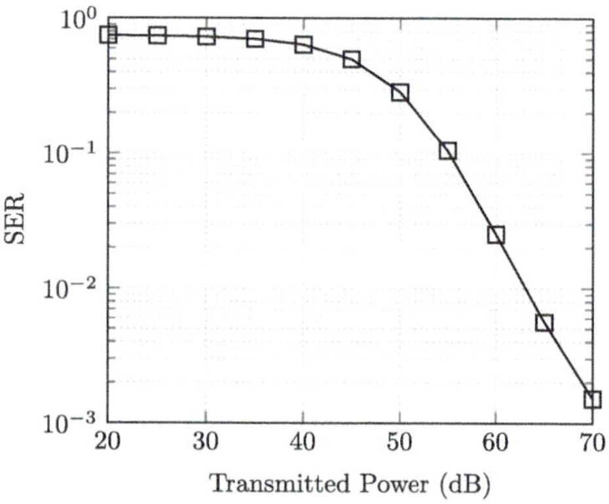

where h_l denotes the large-scale pathoss from the Tx to the Rx, and h is the small-scale fading coefficient. Then x is the transmitted data symbol and n denotes the additive white Gaussian noise (AWGN). The expression of symbols from r is achieved through the utilization of Maximum Likelihood (ML) detection as

$$\hat{x} = \arg\min_{x \in \mathcal{S}} \|r - h_l h x\|_2^2, \tag{14.2}$$

where $\hat{x}$ denotes the estimated symbol at the receiver. $\mathcal{S}$ is the constellation points of modulation scheme. The THz system block diagram is shown in Fig. 14.2. Symbol error rate (SER) performance of the THz system is plotted in Fig. 14.3, where SER improves as transmitted signal power increases.

14.4.4 MIMO

MIMO technology, a critical component in modern wireless communication, is gaining significant attention in the realm of THz communications. In MIMO systems, multiple antennas are used at both the transmitter and receiver ends to improve communication performance. This technology is essential in overcoming the inherent challenges of THz frequencies, such as high path loss and atmospheric absorption.

MIMO systems in the THz band can exploit spatial diversity and spatial multiplexing, offering substantial improvements in data rates and link reliability. With the capability to transmit multiple data streams simultaneously over the same frequency band, MIMO technology promises to unlock the high throughput potential of THz communications, which is essential for applications demanding ultra-high-speed data transmission like wireless backhaul for 5G and beyond networks. As we scale up from MIMO, massive-MIMO emerges as a key technology in the context of THz communications. Massive MIMO systems, characterized by a large number of antennas at the transmitter and receiver, can provide even greater improvements in spectral efficiency and energy efficiency compared to traditional MIMO systems. This is particularly beneficial for THz frequencies, where the small wavelength allows for the integration of a large number of antennas in a relatively small physical space. The deployment of massive MIMO in THz communications can significantly mitigate the severe path loss and improve link robustness through beamforming techniques. Beamforming in massive MIMO systems concentrates the energy in specific directions, enhancing signal strength and reducing interference. This is crucial for realizing the vision of ultra-high-speed wireless communication in dense urban environments and indoor scenarios where THz communications are most likely to be deployed.

The integration of MIMO technologies, particularly massive MIMO, into THz communications represents a groundbreaking advancement in wireless technology. It opens up new possibilities for achieving extremely high data rates potentially in the order of terabits per second, which are necessary for bandwidth-intensive applications like virtual reality, augmented reality, and UHD streaming. However, this integration is not without challenges. The design and implementation of antenna arrays for THz frequencies require innovative approaches to overcome limitations related to device size, power efficiency, and signal processing complexities. Furthermore, channel modeling and understanding the propagation characteristics at THz frequencies are essential for the effective deployment of MIMO and massive MIMO technologies. As research progresses, these challenges present exciting opportunities for advancements in THz communication systems, promising a new frontier in ultra-fast wireless communication.

14.4.5 Beamforming

Since THz systems are envisaged with the MIMO, in systems with MIMO, beamforming is a technique used to direct a wireless signal towards a specific receiver, enhancing the signal-to-noise ratio (SNR) at that receiver. This increase in SNR, known as beamforming gain, is particularly crucial in ensuring reliable signal reception in THz systems. Traditional MIMO systems employed digital beamforming, where each antenna element is connected to its own digital-to-analog converter and RF chain [48, 49]. However, this approach becomes less feasible for THz frequencies due to the significant increase in the number of antenna elements, leading to higher costs and power consumption. The power usage tends to scale linearly with the sampling rate and exponentially with the number of bits per sample. To reduce power consumption, analog beamforming has been suggested for THz systems. In this setup, a single RF chain is shared among all antenna elements, with each antenna receiving a phase-shifted version of the same signal. The phase shift is adjusted based on the desired direction of the beam. However, this method is limited to single-stream and single-user transmissions [31, 35, 50, 51]. To facilitate multi-user and multi-stream transmissions in THz networks, hybrid beamforming has been introduced. This approach uses more than one RF chain and is categorized into two types: fully connected and partially connected hybrid beamforming. In the fully connected architecture, each RF chain is linked to all antenna elements, while in the partially connected architecture, each RF chain is connected to a subset of the antennas. Hybrid beamforming offers a balance between the simpler, yet more restrictive, analog beamforming and the more complex, but highly flexible, fully digital beamforming [7, 30, 37].

Beamforming is a key technique in enhancing the performance of communication systems, especially in the THz frequency range. In THz systems, beamforming is essential due to the high path loss and sensitivity to blockages. There are several types of beamforming techniques, each with its unique characteristics and applications. Here are three prominent types of beamforming used in THz systems:

1. Analog beamforming uses a network of phase shifters and combiners to control the phase of the signal at each antenna element before transmission, as shown in Fig. 14.4. This combined signal forms a single beam directed towards a specific user or area. In THz systems, this method can effectively focus the signal power in a desired direction to combat path loss. Analog beamforming is simpler and less power-hungry compared to digital beamforming. It is also beneficial in terms of hardware complexity and cost, making it suitable for THz systems where power efficiency is crucial. A significant limitation of analog beamforming is its inability to form multiple beams simultaneously. It is less flexible in adapting to multiple users or rapidly changing environments, which can be a drawback in dynamic THz communication scenarios [52].
2. Digital beamforming involves processing the signals from each antenna element in a THz system independently using digital signal processors, as shown in Fig. 14.4. This processing includes phase and amplitude adjustments, allowing for the generation of multiple beams directed towards different users or directions simultaneously. Digital beamforming offers high flexibility and precision in beam direction and shape control [52]. The primary advantage of digital beamforming in THz systems is its ability to support multiple users and multiple data streams concurrently. It provides excellent control over the beam pattern, making it highly adaptable to dynamic channel conditions and user locations. The main challenge is the high computational complexity and power consumption due to the need for separate signal processing at each antenna element. This can be particularly demanding in Massive MIMO systems, where the number of antenna elements is large.

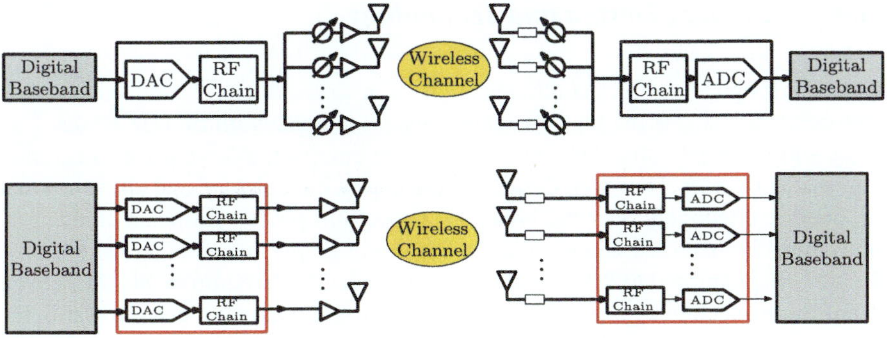

Fig. 14.4 Analog (top) and Digital (bottom) beamforming architectures in The THz system

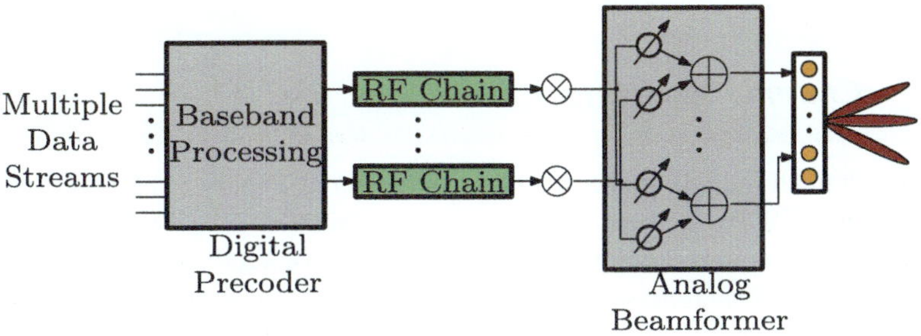

Fig. 14.5 Conventional hybrid precoding architecture

3. Hybrid beamforming is a combination of digital and analog techniques. It uses a reduced number of digital signal processing chains, with each connected to a group of antennas through an analog beamforming network, as shown in Fig. 14.5. This setup strikes a balance between the flexibility of digital beamforming and the simplicity of analog beamforming. Hybrid beamforming provides a compromise between performance and complexity. It can support multiple users and adapt to varying channel conditions while keeping the hardware and power requirements lower than a fully digital system. Designing a hybrid beamforming system involves optimizing the division between digital and analog components, which can be complex. It requires careful consideration of trade-offs between performance, cost, and power consumption, especially in the THz frequency range where system design considerations are more critical due to the high frequency and wider bandwidths. Each of these beamforming techniques plays a crucial role in realizing the potential of THz communication systems, addressing specific challenges and requirements of ultra-high-speed wireless communications [52].

14.4.6 Antenna Design and Technology

Antenna design for THz communication is a critical aspect, requiring novel approaches to address the high frequency and wide bandwidth of THz waves. Due to the short wavelengths, THz antennas can be compact, allowing for integration into smaller devices. However, designing antennas with high gain and efficiency that can operate effectively over the broad THz spectrum poses significant challenges. Innovations in materials science, such as the use of graphene and other nanomaterials, are being explored to develop antennas that meet these requirements. Moreover, beamforming technology, which involves directing the antenna beam towards the intended receiver, is a vital consideration for overcoming the range limitations and achieving directional data transmission in THz communication systems [53].

14.4.7 Channel Modeling

Accurate channel modeling is crucial for the design and analysis of THz communication systems. Due to the unique propagation characteristics of THz waves, existing channel models for lower frequency bands cannot be directly applied. Developing THz-specific channel models requires a comprehensive understanding of factors such as molecular absorption, scattering, multipath effects, and the influence of various environmental conditions. These models play a fundamental role in system design, allowing for the simulation and prediction of system performance under different scenarios [54].

The basics of THz communication, including the spectrum and bandwidth, how waves travel, techniques for modulation and coding, designing antennas, and modelling channels, are essential for the future development of this promising technology. It is crucial to have a thorough understanding of these principles in order to fully harness the potential of THz communication in transforming wireless systems.

Path Loss: The transmission within the THz band encounters varying molecular absorption characteristics along the channels during signal propagation. Pathloss can be formulated as follows:

$$L_{fs}(d, f) = \left(\frac{4\pi df}{c}\right)^2$$

where L_{fs} is the free-space path loss, d is the distance, f is the frequency, and c is the speed of light. The equation is derived from the Friis transmission formula, which is fundamental in electromagnetic theory and widely used in wireless communications to predict the power loss as electromagnetic waves propagate through free space. The quadratic dependence of the path loss on frequency (f) is particularly noteworthy in the THz band [2]. Due to the extremely high frequencies involved, even over relatively short distances, the path loss can be substantial compared to lower frequency bands.

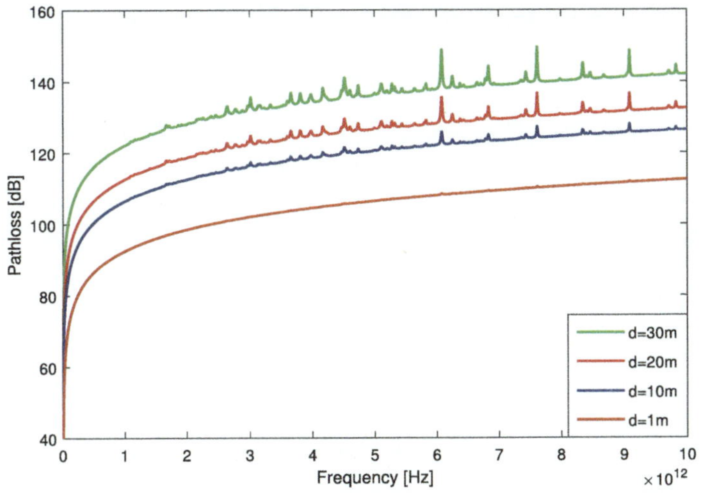

Fig. 14.6 Path loss in the THz band with different distances

This high path loss is a consequence of the spread of electromagnetic energy over a larger area as the distance from the source increases, following the inverse square law. Additionally, the shorter wavelengths associated with THz frequencies lead to increased sensitivity to obstacles and surface roughness, further exacerbating path loss [55]. The impact of pathloss for different distances is shown in Fig. 14.6.

Molecular Absorption Loss: THz waves are absorbed by atmospheric molecules, especially water vapour, oxygen, and carbon dioxide leading to molecular absorption loss [55]. This can be expressed as:

$$L_{ma}(f, d) = e^{k(f)d}, \tag{14.3}$$

where L_{ma} is the molecular absorption loss, $k(f)$ is the absorption coefficient at frequency f, and d is the distance between transmitter and receiver. The key component in this equation is $k(f)$, the absorption coefficient, which is a function of frequency and varies significantly across the THz band due to molecules present in the environment. The absorption coefficient $k(f)$ is influenced by the interaction of THz waves with atmospheric gases, primarily water vapour, oxygen, carbon dioxide. It can be further detailed as:

$$k(f) = \sum_i k_i(f) \tag{14.4}$$

where $k_i(f)$ represents the absorption coefficient for the ith atmospheric gas component. The summation accounts for the contributions of various gases present in the atmosphere, with water vapour often being the dominant factor in the THz range. The impact of water molecules in the THz frequency range is shown in Fig. 14.7.

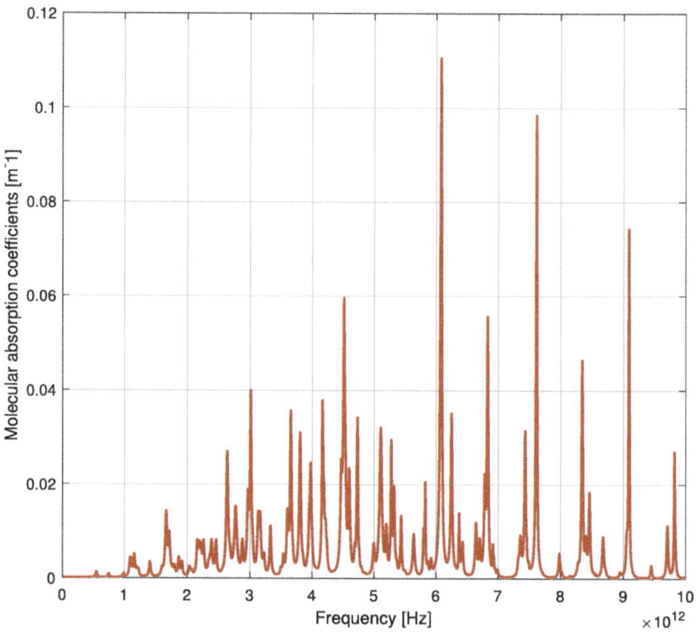

Fig. 14.7 Water molecular absorption in the THz band

The HITRAN (High-Resolution Transmission Molecular Absorption) database is a comprehensive resource that provides detailed spectral line information for a wide range of gases. This database is instrumental in calculating the absorption coefficients for various gases at THz frequencies. By offering high-resolution data, the Hitran database enables precise modeling of molecular absorption phenomena in the THz band, which is essential for accurate prediction of wave propagation behavior in different atmospheric conditions.

The molecular absorption loss, as quantified by the above equations, has profound implications for the design and operation of THz communication systems. It necessitates careful consideration of frequency selection, as certain frequencies may experience significantly higher absorption, leading to reduced range and efficiency. This characteristic also influences the choice of transmission power, antenna design, and overall system architecture to compensate for the high attenuation rates. Moreover, understanding molecular absorption is crucial for developing adaptive communication strategies that can dynamically respond to varying atmospheric conditions, such as changes in humidity and temperature, which directly affect the absorption characteristics of the THz band.

14.5 Technological Challenges and Innovations in THz Communication

The advancement of THz communication is contingent upon overcoming a series of technological challenges. Each challenge demands innovative solutions that push the boundaries of current wireless communication technologies. This section articulates these challenges and the corresponding innovations that are shaping the future of THz communication.

14.5.1 Transmitter and Receiver Technology

Challenge: The development of efficient and compact THz transmitters and receivers is a fundamental challenge. Existing technologies for generating and detecting THz waves are often bulky, energy-intensive, and lack the efficiency required for practical communication systems.

Innovations: Research is focused on developing solid-state devices, such as silicon-germanium (SiGe) heterojunction bipolar transistors (HBTs) and complementary metal-oxide-semiconductor (CMOS) technologies, for THz applications. These technologies offer the potential for integration into compact, low-power, and cost-effective systems. Additionally, quantum cascade lasers (QCLs) and photonic-based methods are being explored for their high-power and frequency agility in THz wave generation and detection.

14.5.2 Propagation Medium and Range

Challenge: THz waves face significant propagation losses, particularly due to atmospheric absorption and scattering, which limit their range and reliability. This issue is compounded in urban environments with numerous obstructions and dynamic conditions.

Innovations: To combat these challenges, research is underway to develop adaptive communication techniques, such as intelligent reflective surfaces (IRS) that can dynamically alter the propagation environment to enhance signal strength and reduce attenuation. Additionally, the development of advanced channel models that accurately reflect the THz propagation characteristics under various environmental conditions is critical for system design. The path gain of the THz channel impulse response can be expressed as

$$h_l(d,t) = \mathcal{F}^{-1}\left[H_{\text{spr}}(d,f)H_{\text{abs}}(d,f)e^{-j2\pi f\frac{d}{c}}\right], \tag{14.5}$$

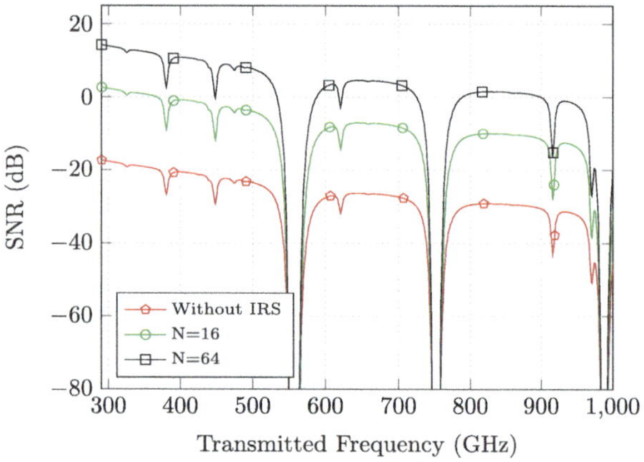

Fig. 14.8 Average received SNR versus THz frequencies plot [56]

where $\mathcal{F}^{-1}$ is the inverse Fourier transform. The $H_{spr}(d, f)$ denotes the free-space spreading loss and it is denoted as $H_{spr}(d, f) = \frac{c}{4\pi f d}$. The $H_{abs}(d, f)$ is the molecular absorption loss. Parameters f, d, c, and t are the transmitted signal frequency, signal transmission distance, the speed of light, and time, respectively. Molecular absorption loss according to Beer-Lambert Law is given as [55]

$$H_{abs}(d, f) = e^{-\frac{1}{2}k(f)d} \tag{14.6}$$

where k is the molecular absorption coefficient, which is highly depend on the transmitted signal frequency. The value of k with respect to different frequencies depends on various environmental parameters such as temperature, pressure, and the component of molecules along the propagation path.

Figure 14.8 depicts the average received signal-to-noise ratio (SNR) corresponding to a transmit power of one Watt. Considering a noise power spectral density (PSD) of −173 dBm per Hertz and a system bandwidth of 20 GHz, it's evident that the received SNR of the THz system assisted by IRS surpasses that of the conventional THz system lacking IRS, as indicated in Fig. 14.8. Moreover, Fig. 14.8 showcases that an increased number of reflective elements on the IRS panel leads to a more notable enhancement in the SNR. Additionally, higher transmitted signal frequencies result in increased losses attributed to higher spreading loss.

14.5.3 Beamforming and MIMO Technology

Challenge: The high path loss at THz frequencies necessitates highly directional beamforming and the use of MIMO systems. However, implementing these technologies at THz frequencies is complex due to the short wavelengths and the need for precise alignment between transmitters and receivers.

Innovations: Advances in antenna technology, including the development of electronically steerable phased arrays and metamaterial-based antennas, are paving the way for effective beamforming at THz frequencies. These technologies allow for rapid and precise beam steering, essential for maintaining alignment in mobile environments. Furthermore, research in THz MIMO systems is exploring the use of spatial multiplexing to significantly increase the channel capacity.

14.5.4 Near-Field Communication

Near-field communication (NFC) in the THz band faces several challenges that are distinct from those encountered in far-field THz applications. These challenges are primarily due to the short-range nature of NFC and the unique propagation properties of THz waves [57]. One of the foremost challenges is the coupling efficiency between THz devices [58]. Due to the short wavelengths involved, small misalignments can lead to significant losses in signal strength, which requires precise positioning and alignment mechanisms for optimal performance [59]. Additionally, the THz band's high-frequency signals are more susceptible to absorption and reflection by materials commonly found in the operating environment, which can lead to signal attenuation and phase distortions [60].

Another challenge is the design of antennas and transceivers for THz NFC that are not only highly directive but also capable of maintaining efficient communication over the short distances typical of NFC interactions. The integration of such high-frequency components into compact form factors suitable for consumer devices presents additional hurdles in terms of manufacturing complexity and cost. Moreover, THz NFC must contend with the issue of power consumption. The generation and detection of THz waves typically require significant energy, which can be prohibitive for battery-powered devices such as smartphones. Thus, power-efficient design is crucial for the practical adoption of THz NFC [61]. Expected solutions to these challenges are multifaceted. For instance, advancements in metamaterials and plasmonic technologies offer promising avenues for improving coupling efficiency and developing compact antenna systems. These materials can be engineered to have unique electromagnetic properties that could mitigate the effects of misalignments and environmental interference. Research into new modulation and error-correction schemes tailored to the THz band's specific propagation characteristics could enhance data integrity and communication robustness in NFC applications. Additionally, leveraging the development of semiconductor devices capable of operating at THz frequencies could address the power consumption issue. Silicon-based technologies, for instance, are being refined to create more power-efficient transceivers suitable for portable electronics [62]. As THz technology matures, the symbiosis of computational modeling, innovative materials, and advanced manufacturing techniques is expected to overcome the current limitations of THz NFC, unlocking its potential for a myriad of applications in secure data transfer, wireless power transmission, and beyond.

14.5.5 Power Consumption and Device Integration

Challenge: High power consumption is a significant barrier for THz devices, particularly for portable and IoT applications. Additionally, integrating THz components into existing communication infrastructures and devices poses substantial technical challenges.

Innovations: Researchers are working on low-power THz transceiver designs, employing novel materials and circuit topologies to reduce power consumption. Integration efforts are focused on developing THz systems compatible with existing communication standards and infrastructures, such as 5G networks, ensuring seamless coexistence and interoperability [63].

14.5.6 Standardization and Regulation

Challenge: The lack of standardized protocols and regulations for THz communication is a hurdle for its widespread adoption and commercialization. This includes standardization in spectrum allocation, transmission protocols, and safety standards.

Innovations: Efforts are underway to establish international standards and regulatory frameworks for THz communication. Organizations such as the International Telecommunication Union (ITU) and the IEEE are actively working to define standards that address the operational parameters, interoperability, and security aspects of THz communication systems.

14.5.7 Security and Privacy

Challenge: The high bandwidth and unique propagation characteristics of THz waves introduce novel security and privacy challenges. The susceptibility to eavesdropping and interception in highly directional THz links necessitates the development of robust security protocols.

Innovations: Advanced encryption methods and secure key distribution mechanisms specific to THz communication are being developed. Furthermore, leveraging the inherent physical layer security features of THz communication, such as beamforming, provides opportunities for enhancing security through spatial and temporal confinement of the signal.

The technological challenges in THz communication are significant yet surmountable with ongoing research and innovation. Each challenge presents an opportunity to rethink traditional approaches and develop breakthrough technologies that not only address the immediate issues but also lay the groundwork for future advancements in wireless communication. As these challenges are progressively addressed, THz communication is poised to become a key enabler of ultra-high-speed, high-capacity wireless networks in the coming years.

14.5.8 Empowering THz Protocol Testing

In pursuit of augmenting wireless data transmission rates, communication within the THz band is anticipated to emerge as a pivotal wireless technology in the ensuing decade. Concurrently with the advancement of THz transceivers and antennas, there is a pressing need for simulation tools to accelerate the development of communication and networking protocols bespoke to this nascent networking paradigm, and to achieve this at a markedly reduced cost.

In alignment with this objective, the UN Lab [64] has meticulously developed and is honored to present to the research community, TeraSim: an open-source network simulation platform tailored for THz communication networks. TeraSim is architected as an extension to ns-3, which stands as one of the most prevalent network simulation software utilized in educational and academic spheres. This simulator has been crafted with a focus on two primary application scenarios-nanoscale communication networks (with an average transmission range typically under one meter) and macroscale communication networks (encompassing distances exceeding one meter). TeraSim encapsulates a common channel module, distinct physical and link layers designated for each scenario, supplemented by two auxiliary modules, namely, the THz antenna module and the energy harvesting module, each originally conceived for the macroscale and nanoscale scenarios, respectively. Rigorous testing has been undertaken to authenticate the functionalities embodied in the implemented modules [65, 66].

Through disseminating TeraSim is to empower the networking community to evaluate THz networking protocols, obviating the necessity to delve into the intricacies of the channel and physical layers. TeraSim is officially accessible in the ns-3 App Store. Additionally, the ongoing development and experimental features of TeraSim can be tracked on the GitHub TeraSim project website. This initiative aims to cultivate an environment conducive to the rigorous exploration and advancement of THz communication networks, thereby significantly contributing to the broader narrative of wireless communication technology evolution [65, 66].

14.6 Applications of THz Communication

The advent of THz communication technology heralds a new era in wireless communication, with its potential to unlock unprecedented data rates and bandwidth capabilities. This section delves into the diverse and transformative applications that

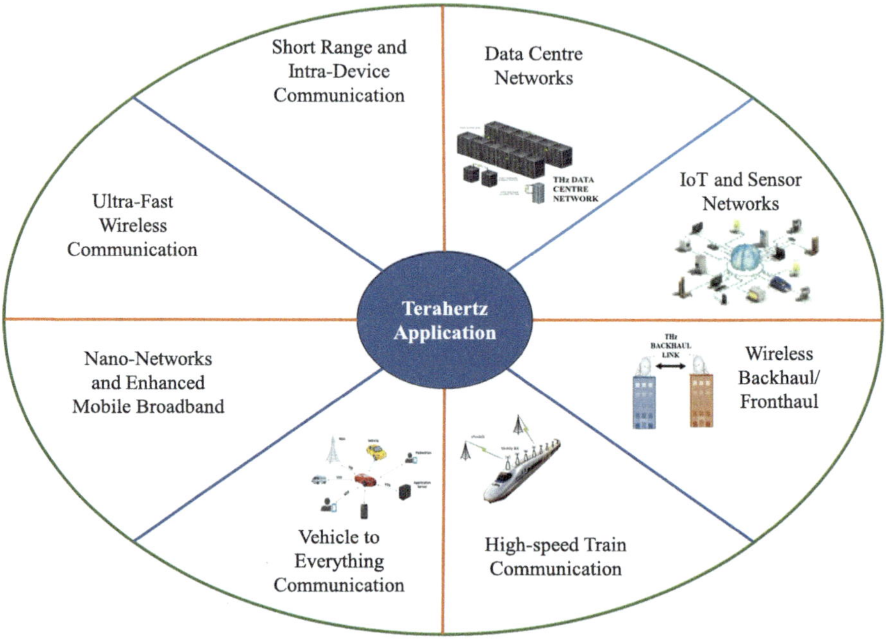

Fig. 14.9 THz Applications

THz communication is poised to revolutionize, spanning from high-speed wireless networks to innovative industrial and medical uses (Fig. 14.9).

14.6.1 Short-Range Communications

Short-range communications involve scenarios such as file transfers between user devices and kiosks located in public spaces like airports and train stations, as well as connections between routers and laptops in a stationary setting. These communications typically occur within a direct LoS over a short distance, not exceeding 1 m. This method is poised to deliver extremely high data rates, potentially reaching up to 100 Gbps, particularly at frequencies of 300 GHz or higher. This makes it an excellent candidate for creating "information showers," a term denoting rapid data transmission [67]. However, one notable challenge in this context is the precise alignment of user equipment antennas. For optimal performance, it is essential for the receiving device to be accurately oriented towards the transmission source.

14.6.2 Data Center Networks

Centralized data centers house server racks that communicate via wired links [68]. The use of THz waves at frequencies ≥300 GHz in these settings is promising due to the high bandwidth they offer, alongside the benefits of hardware flexibility in configuration, cost-effectiveness, and enhanced environmental control through cooling technologies. Data center racks are typically arranged symmetrically, allowing for both line-of-sight (LoS) and directed-non-line-of-sight (D-NLoS) propagation, covering short ranges under 20 m.

14.6.3 Intra-Device Communications

The increasing complexity of modern devices, characterized by intricate layouts and high energy consumption, necessitates efficient data transfer between components. THz wireless communication links within devices, operating at high data rates, present a promising solution. The propagation environment for such communications is confined to a few centimeters due to the device's casing. Antenna design for these applications can be less directive, accommodating both LoS and significant NLoS multipath scattering within the device's enclosure. This approach is particularly suited for higher frequency bands in the THz spectrum [69].

14.6.4 Wireless Backhaul/Fronthaul

THz communications are a potential solution for the limitations of current wireless backhaul links, which connect base stations to core networks over kilometers. These links, aiming for data rates up to 100 Gbps at 300 GHz, could serve as wireless extensions to optical fibers in both rural and urban areas, enhancing high-speed connectivity in a cost-effective manner. THz fronthaul links, in contrast, focus on providing broadband access to end-users and coexist with lower frequency bands to maximize coverage, typically involving shorter distances than backhaul links.

14.6.5 Wireless Local Area Network (WLAN)

In indoor settings like offices and conference rooms, THz wireless connectivity could replace fiber-optic cables for interconnecting access points and providing high data rates. These environments necessitate consideration of LoS, link blockage, and scattering as primary propagation mechanisms, with the communication range limited to a few meters within these enclosed spaces.

14.6.6 High-Speed Train (HST) Communications

For high-speed broadband access on trains, THz communications are poised to play a key role. The train-to-infrastructure (T2I) channel is especially promising for delivering data rates of 100 Gbps. In-train communication, like links between passengers and access points, could also benefit from the large bandwidth at THz frequencies, accommodating a higher user density in conjunction with millimeter-wave bands. The required propagation distances and antenna designs will vary based on the specific application scenario.

14.6.7 Vehicle-to-Everything (V2X) Communications

Envisioned as a highly intelligent network for 6G V2X communications, THz band communications can meet the demands for high data rates and ultra-low latency in scenarios like cooperative driving among semi-autonomous vehicles and data relaying in congested networks. These communications are particularly relevant in short-range scenarios, such as vehicle-to-vehicle exchanges at slow speeds or while stationary, involving LoS and NLoS propagation influenced by the surrounding environment.

14.6.8 Nano-Networks

In the medical field, THz communications are being explored for nano-machine enabled communications, particularly for in-vivo applications within sub-millimeter regions of the human body. Channel modeling for these applications must consider human tissue properties, including path loss calculations based on molecular characteristics, tissue scattering, the distance between nano-machines, molecular noise, and operating frequency. Nano-network communications are suitable across the THz band due to the short communication distances and the wavelength's compatibility with molecular structures in human tissue, enabling unique spectral identification not possible in lower frequency bands.

14.6.9 Ultra-Fast Wireless Communication

Wireless Fiber Optic Networks: THz communication is envisioned as a key technology in achieving wireless networks with data transmission speeds comparable to fiber optics. In urban and densely populated areas where laying physical fiber cables is challenging or cost-prohibitive, THz wireless links can provide a viable alternative.

This application is particularly crucial for the last-mile connectivity in broadband services, offering Gbps to Tbps speeds to end-users.

Data Centers and High-Performance Computing: The high data rate capability of THz communication can revolutionize data center operations. THz links can facilitate ultra-fast intra- and inter-data center communications, enabling quicker data transfer, reduced latency, and enhanced connectivity between servers. This can significantly boost the efficiency and performance of cloud services and high-performance computing applications.

14.6.10 Enhanced Mobile Broadband (eMBB) and 6G Networks

THz communication is a frontrunner technology for future 6G mobile networks. It can provide exceptionally high data rates and low-latency communication, essential for eMBB applications. This will enable seamless streaming of UHD content, augmented reality (AR), virtual reality (VR), and other bandwidth-intensive services on mobile devices [70].

14.6.11 IoT and Sensor Networks

The IoT ecosystem, comprising billions of interconnected devices, requires a robust and high-capacity communication framework. THz technology, with its high bandwidth, can support the massive data exchange requirements of IoT applications. It is particularly well-suited for high-data-rate IoT environments like smart cities, industrial IoT (IIoT), and intelligent transportation systems, enabling real-time data collection, processing, and transmission. The integration of the IoT and sensor networks with THz communication is a burgeoning area of research, poised to revolutionize how data is collected and transmitted in these networks. Operating in the THz frequency range, THz communication offers exceptionally high data rates, potentially in the terabits per second, which is a game-changer for IoT and sensor networks that are increasingly becoming data-intensive [71]. One of the most compelling advantages of using THz communication in IoT and sensor networks is the ability to transmit large volumes of data with minimal latency. This capability is particularly crucial in applications requiring real-time data processing and rapid decision-making, such as autonomous vehicles, smart cities, and industrial automation. The high bandwidth of THz bands allows for the simultaneous connection of multiple devices, supporting the dense deployment of sensors and IoT devices without the risk of network congestion. Moreover, the unique propagation characteristics of THz waves, such as their short range and susceptibility to blockage, make them well-suited for secure and private communication. This is a significant advantage for IoT applications in sensitive environments, where data security and privacy are paramount. The directional nature of THz waves also enables more precise localization and tracking capabilities, enhancing the functionality of sensor networks in applications like indoor navigation and asset tracking. However, integrating THz communication into IoT and sensor networks also presents challenges. The high path loss and sensitivity to atmospheric conditions in the THz band necessitate the development of advanced beamforming and error correction techniques to ensure reliable communication. Additionally, the design of energy-efficient THz transceivers is crucial, given the typically limited power resources of IoT and sensor devices. The potential of THz communication in transforming IoT and sensor networks is immense, offering a leap forward in data transmission capabilities. As research progresses in overcoming the technical challenges, THz-enabled IoT and sensor networks are expected to play a pivotal role in driving the next wave of innovations in smart technologies and connected devices.

14.6.12 Healthcare and Medical Imaging

THz waves hold significant promise in the field of medical imaging and diagnostics. Due to their non-ionizing nature and high resolution, THz waves can be used for safe and detailed imaging of soft tissues in the human body. This can be instrumental in applications such as skin cancer detection, dental imaging, and burn wound assessment [46]. Furthermore, the ability of THz waves to penetrate various materials can be utilized in non-invasive scanning techniques for pharmaceutical quality control and biological sample analysis. The application of THz communication in healthcare and medical imaging represents a significant leap forward in medical technology, offering new and improved methods for diagnosis and treatment. One of the key advantages of THz waves in medical imaging is their ability to penetrate a variety of materials, including clothing and biological tissues, without the harmful ionizing radiation associated with X-rays. This makes THz imaging a safer alternative for patient diagnostics. THz waves are particularly effective in differentiating between different types of tissues and cells, due to their sensitivity to water content. This property can be exploited in applications such as skin cancer detection, where THz imaging can help in identifying malignant cells, which typically have different water content compared to normal cells. Furthermore, THz technology holds promise in the field of dental imaging, offering a potential alternative to traditional X-ray imaging with the advantage of being non-ionizing. In pharmaceuticals, THz spectroscopy can be used for precise analysis of pharmaceutical compounds, aiding in quality control and drug development. Another exciting application is in the field of endoscopy. Miniaturized THz sensors could be developed for endoscopic procedures, providing high-resolution imaging of internal tissues, which can greatly aid in the diagnosis of gastrointestinal and other internal diseases. However, the practical implementation of THz technology in healthcare faces challenges, primarily related to the development of compact, portable, and affordable THz imaging systems. Current THz imaging equipment tends to be bulky and expensive, limiting its widespread use in clinical settings. Additionally, there is a need for further research to understand the interaction of THz waves with different biological tissues and to establish clear safety guidelines for their use in medical applications.

14.6.13 Security and Surveillance

The unique penetrating ability of THz radiation can be harnessed for security and surveillance applications. THz imaging systems can detect concealed weapons or contraband in security checks without the health risks associated with X-rays. Additionally, the high-resolution imaging capability of THz waves can be employed in biometric systems and for identifying chemical substances, making them useful in anti-terrorism and forensic applications.

14.6.14 Space and Atmospheric Research

Beyondground potential exists in the domains of atmospheric science and space exploration for the utilisation of THz communication and spectroscopy. The transmission in THz frequency band offers a considerably wider bandwidth in comparison to the conventional microwave frequencies that are employed in space communication. In deep-space missions, where the high data rate capacity of THz frequencies can expedite the return of massive amounts of scientific data to Earth, this characteristic is especially advantageous. Such efficiency is critical for the timely analysis of data from deep space missions and spacecraft, including high-resolution images and complex scientific measurements [72].

THz spectroscopy is further recognised as an indispensable instrument when it comes to investigating planetary atmospheres and interstellar space. A considerable number of significant molecules, particularly those that are critical for comprehending planetary atmospheres and the interstellar medium, demonstrate spectral lines in the THz range. Scientists can deduce vital information regarding the chemical composition, temperature, and other physical properties of these environments through the analysis of these lines. A comprehensive understanding of planetary formation, climate, and the possibility of life on other celestial bodies requires this knowledge [73]. In addition, there is potential for THz technology to significantly improve satellite communication, specifically for Earth observation satellites. The capacity of THz communication to process high data rates confers a benefit when it comes to relaying significant volumes of information, including high-definition videos and images, from Earth observation satellites. The implementation of this capability has the potential to greatly enhance weather monitoring, environmental surveillance, and disaster management applications.

Nevertheless, the integration of THz technology into the study of the atmosphere and space presents certain obstacles. The THz frequency band is subject to greater atmospheric attenuation, primarily attributable to water vapour, which in certain circumstances may reduce its efficacy. Furthermore, the development of miniaturised and durable THz components that are appropriate for space missions poses a significant engineering obstacle. Ongoing developments in THz sources, detectors, and components are progressively surmounting these challenges [74]. In brief, spectroscopy and THz communication are positioned to have a revolutionary impact on space and atmospheric research. In order to advance space exploration and comprehension of the Earth's atmosphere, their capacities for rapid data transmission and comprehensive spectral analysis are indispensable. It is anticipated that as technology advances, THz systems will play an ever-increasing role in the investigation of the cosmos and the expansion of our understanding of the Earth's environmental dynamics.

14.6.15 Industrial Applications

In industrial settings, THz technology can be applied for non-destructive testing and quality control. It can be used to detect defects in materials, monitor production processes, and ensure the integrity of products, ranging from automotive parts to semiconductor wafers.

The applications of THz communication are vast and diverse, offering transformative potential across multiple domains. From revolutionizing high-speed wireless networks to enabling new capabilities in healthcare and industrial processes, THz technology stands at the cusp of a new communication paradigm. As research and development in this field continue to advance, the integration of THz communication into these applications will significantly impact society, driving technological innovation and opening new frontiers in wireless communication.

14.7 Emerging Research Areas in THz Technology: A Detailed Overview

The exploration of THz technology is unlocking a myriad of possibilities across diverse fields. Here is a detailed look into the key emerging research areas:

14.7.1 THz Communication Systems

Ultra-High-Speed Wireless Networks: Research is intensifying to harness the THz band for future wireless communication systems, aiming to support data-intensive applications like 3D holographic communication, ultra-high-definition video streaming, and cloud-based services. The focus is on overcoming challenges like high path loss and molecular absorption to achieve stable and efficient terabit-per-second transmissions.

Device and Material Development: Innovations are underway in creating new semiconductor materials and nano-scale devices. Graphene, for instance, shows promise due to its high electron mobility and broadband response. Research also extends to novel antenna designs, such as phased arrays and metamaterial-based antennas, to enhance signal directionality and range.

14.7.2 THz Imaging and Sensing

THz imaging and sensing represent a significant and rapidly evolving area within the field of applied physics and engineering. THz imaging is a technique that allows for the capture of images based on the THz radiation emitted or reflected by objects. Unlike visible light, THz radiation can penetrate a variety of non-conducting materials such as clothing, paper, wood, and plastic, making it particularly useful for non-destructive testing and security screening. In medical diagnostics, THz imaging is being explored for its potential in detecting skin or breast cancer, as THz waves can differentiate between healthy and diseased tissue without the harmful effects associated with ionizing radiation like X-rays [75]. In addition to imaging, THz sensing technology is gaining prominence due to its unique capabilities. THz sensors are capable of detecting and identifying chemicals and materials based on their spectral fingerprints in the THz range. This is particularly useful in applications such as hazardous material detection, pharmaceutical quality control, and environmental monitoring. The ability of THz waves to detect both chemical composition and physical structures makes these sensors highly versatile [6].

Moreover, THz sensing is being explored in the field of wireless communication for high-speed data transmission, as THz waves can carry more data than microwaves or radio waves currently used in wireless communication. The development of THz imaging and sensing technologies is, however, not without challenges. These include the creation of efficient and compact THz sources and detectors, as well as the development of materials and technologies for manipulating THz waves. Despite these challenges, the potential applications of THz imaging and sensing are vast and varied, promising significant advancements in a range of fields from security and medicine to communication and environmental monitoring [6].

14.7.3 Quantum Computing and Information Processing

Advancements in the field of quantum computing and information processing are increasingly evident, particularly in the development of THz quantum cascade lasers. These lasers have undergone significant improvements in terms of power output and efficiency. As a result, they are becoming increasingly suitable for a diverse array of applications. One notable area of application is spectroscopy, where their enhanced capabilities allow for more precise and detailed analysis. Additionally, these lasers are proving to be highly effective in free-space communication, a domain where the need for high-speed and reliable data transmission is paramount [76].

Furthermore, THz technology is at the forefront of pioneering ultra-secure quantum communication networks. This exploration is primarily focused on leveraging the unique properties of quantum entanglement for data transmission and encryption. Quantum entanglement, a fundamental principle of quantum mechanics, involves a pair or group of particles interacting in such a way that the quantum state of each particle cannot be described independently of the state of the others, even when the particles are separated by a large distance. This property is being harnessed to create communication networks that are not only faster but also inherently more secure than current technologies. The use of quantum entanglement in THz communication networks promises a revolutionary leap in the way secure communication is handled, potentially leading to a paradigm shift in data encryption and transmission technologies [77].

14.7.4 Integrated THz Systems and Networks

The integration of THz components into compact, energy-efficient systems represents a critical area of research within the domain of Integrated THz Systems and Networks. This endeavor primarily focuses on the development of specialized integrated circuits (ICs) and system-on-chip (SoC) solutions, tailored to operate efficiently at THz frequencies. These advancements are essential for miniaturizing THz systems, thereby enhancing their practicality and accessibility for a wide range of applications. Concurrently, there is a significant emphasis on developing network architectures that can effectively incorporate THz frequencies. This involves the creation of novel communication protocols and algorithms specifically designed for the THz band, ensuring these new systems are compatible with existing wireless networks. Additionally, it necessitates addressing complex challenges such as handover and interference management, which are critical for maintaining continuous and reliable communication. The integration of THz technology into wireless networks is a multifaceted task, requiring innovations in electronics design, network architecture, and communication protocols, all of which are pivotal for the advancement and application of THz technology in wireless communication [78, 79]. A summary of research directions is summarized in Fig. 14.10.

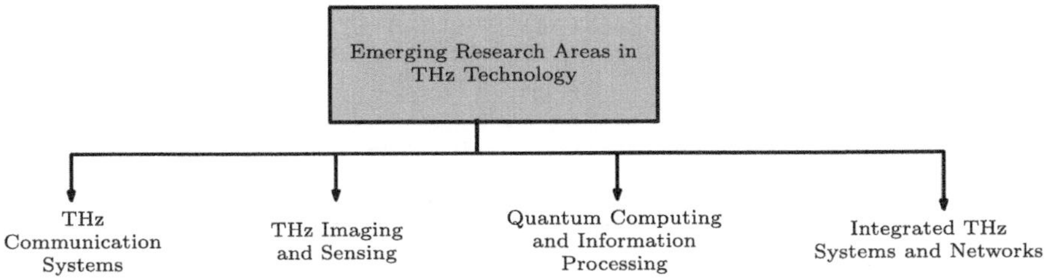

Fig. 14.10 Possible research direction in THz communication

14.8 Conclusion

This chapter has explored THz communication, covering its fundamental concepts, challenges, and extensive applications. Positioned at the forefront of wireless innovation, THz communication offers promise for elevated data transfer rates and enhanced connectivity in next-generation infrastructures. It presents untapped potential for revolutionizing high-speed communications, catering to data-intensive applications like virtual reality, ultra-high-definition content, and scientific data transfers. Delving into the complexities of harnessing THz spectrum, the chapter highlights both potential and hurdles, from hardware development to communication protocols. Despite challenges, it underscores groundbreaking applications, projecting THz's pivotal role in future wireless ecosystems. As we approach 6G, the future of wireless communication is intertwined with THz progress, fueled by ongoing research and demands for improved data rates and connectivity. This chapter affirms the relentless pursuit of innovation in wireless communication, emphasizing THz's capabilities and complexities. It enriches discourse on next-gen systems, emphasizing the concerted effort needed to realize THz communication's vision. THz technology emerges not only as a bandwidth solution but as a strategic enabler of a deeply interconnected future, shaping the wireless world through research, standardization, and technical refinement.

References

1. Tarboush S, Ali A, Al-Naffouri TY (2023) Compressive estimation of near field channels for ultra massive-MIMO wideband THz systems. In: Proceeding of the IEEE international conference on acoustics, speech, and signal processing. IEEE, pp 1–5
2. Akyildiz IF, Jornet JM, Han C (2014) Terahertz band: next frontier for wireless communications. Phys Commun 12:16–32
3. Guo F, Yu FR, Zhang H, Li X, Ji H, Leung VC (2021) Enabling massive IoT toward 6G: a comprehensive survey, pp 11 891–11 915
4. Faisal A, Sarieddeen H, Dahrouj H, Al-Naffouri TY, Alouini M-S (2019) Ultra-massive MIMO systems at terahertz bands: prospects and challenges. IEEE Veh Technol Mag

5. Vaezi M, Azari A, Khosravirad SR, Shirvanimoghaddam M, Azari MM, Chasaki D, Popovski P (2022) Cellular, wide-area, and non-terrestrial IoT: a survey on 5G advances and the road toward 6G. IEEE Commun Surv Tutor 24(2):1117–1174
6. Chaccour C, Soorki MN, Saad W, Bennis M, Popovski P, Debbah M (2022) Seven defining features of terahertz (THz) wireless systems: a fellowship of communication and Sensing. IEEE Commun Surv Tutor 24(2):967–993
7. Elbir AM, Mishra KV, Chatzinotas S (2021) Terahertz-band joint ultra-massive MIMO radar-communications: Model-based and model-free hybrid beamforming. IEEE J Sel Top Signal Process 15(6):1468–1483
8. Wang Z, Mu X, Liu Y (2023) Terahertz near-field communications and sensing. arXiv:2306.09723
9. Sharma V, Keshari P, Sharma S, Deka K, Jain S, Bhatia V (2024) Deep learning-based mitigation of nonlinear hardware impairments for THz communication. In: IEEE 35th international symposium on personal, indoor and mobile radio communications (PIMRC). IEEE, pp 1–7
10. Siegel PH (2002) Terahertz technology. IEEE Trans Microw Theory Tech 50(3):910–928
11. Fitch MJ, Osiander R (2004) Terahertz waves for communications and sensing. Johns Hopkins APL Techn Dig 25(4):348–355
12. Piesiewicz R et al (2007) Short-range ultra-broadband terahertz communications: concepts and perspectives. IEEE Antennas Propagat Mag 49(6):24–39
13. Hosako I et al (2007) At the dawn of a new era in terahertz technology. Proc IEEE 95(8):1611–1623
14. Tonouchi M (2007) Cutting-edge terahertz technology. Nat Photon 1(2):97–105
15. Federici J, Moeller L (2010) Review of terahertz and subterahertz wireless communications. J Appl Phys 107(11):6
16. Huang KC, Wang Z (2011) Terahertz terabit wireless communication. IEEE Microw Mag 12(4):108–116
17. Kurner T (2012) Towards future terahertz communications systems. Terahertz Sci Technol 5:11–17
18. Akyildiz IF, Jornet JM, Han C (2014) Teranets: ultra-broadband communication networks in the terahertz band. IEEE Wireless Commun 21(4):130–135
19. Hirata A, Yaita M (2015) Ultrafast terahertz wireless communications technologies. IEEE Trans THz Sci Technol 5(6):1128–1132
20. Lin C, Li GY (2016) Terahertz communications: an array-of-subarrays solution. IEEE Commun Mag 54(12):124–131
21. Hasan M et al (2016) Graphene terahertz devices for communications applications. Nano Commun Netw 10:68–78
22. Federici JF, Ma J, Moeller L (2016) Review of weather impact on outdoor terahertz wireless communication links. Nano Commun Netw 10:13–26
23. Mumtaz S et al (2017) Terahertz communication for vehicular networks. IEEE Trans Veh Technol 66(7):5617–5625
24. Petrov V et al (2018) Last meter indoor terahertz wireless access: performance insights and implementation roadmap. IEEE Commun Mag 56(6):158–165
25. Boulogeorgos AAA et al (2018) Terahertz technologies to deliver optical network quality of experience in wireless systems beyond 5G. IEEE Commun Mag 56(6):144–151
26. Jamshed MA et al (2020) Antenna selection and designing for thz applications: suitability and performance evaluation: a survey. IEEE Access 8:113 246–113 261
27. Wang CX et al (2020) 6g wireless channel measurements and models: trends and challenges. IEEE Veh Technol Mag 15(4):22–32
28. Sarieddeen MSAH, Al-Naffouri TY (2019) Terahertz-band ultramassive spatial modulation MIMO. IEEE J Sel Areas Commun 37(9):2040–2052
29. Faisal A et al (2020) Ultramassive mimo systems at terahertz bands: prospects and challenges. IEEE Veh Technol Mag 15(4):33–42
30. Akyildiz IF, Kak A, Nie S (2020) 6G and beyond: the future of wireless communications systems. IEEE Access 8:133 995–134 030

31. Chowdhury MZ, Shahjalal M, Ahmed S, Jang YM (2020) 6G wireless communication systems: applications, requirements, technologies, challenges, and research directions. IEEE Open J Commun Soc 1:957–975
32. Ning B, Chen Z, Chen W, Du Y, Fang J (2021) Terahertz multi-user massive MIMO with intelligent reflecting surface: beam training and hybrid beamforming. IEEE Trans Veh Technol 70(2):1376–1393
33. Candes EJ, Wakin MB (2008) An introduction to compressive sampling: a sensing/sampling paradigm that goes against the common knowledge in data acquisition. IEEE Signal Process Mag 25(2):21–30
34. Sarieddeen H, Alouini M-S, Al-Naffouri TY (2020) An overview of signal processing techniques for terahertz communications. arXiv:2005.13176
35. Yang S, Zhang Z, Hu Z, Song N, Liu H (2021) Efficient multi-beam training for terahertz wireless communications. arXiv:2112.15346
36. Do H et al (2021) Terahertz line-of-sight MIMO communication: theory and practical challenges. IEEE Commun Mag 59(3):104–109
37. Yan L, Han C, Yuan J (2022) Energy-efficient dynamic-subarray with fixed true-time-delay design for terahertz wideband hybrid beamforming. IEEE J Sel Areas Commun 40(10):2840–2854
38. Han C, Yan L, Yuan J (2021) Hybrid beamforming for terahertz wireless communications: challenges, architectures, and open problems. IEEE Wireless Commun 28(4):198–204
39. Han LYC, Yuan J (2021) Hybrid beamforming for terahertz wireless communications: challenges, architectures, and open problems. IEEE Wireless Commun 28(4):198–204
40. Badarneh OS (2022) Performance analysis of terahertz communications in random fog conditions with misalignment. IEEE Wireless Commun Lett 11(5):962–966
41. Papasotiriou EN, Boulogeorgos AAA, Alexiou A (2020) Performance analysis of THz wireless systems in the presence of antenna misalignment and phase noise. IEEE Commun Lett 24(6):1211–1215
42. Sarieddeen MSAH, Al-Naffouri TY (2021) An overview of signal processing techniques for terahertz communications. Proc IEEE 109(10):1628–1665
43. Qiao J, Alouini MS (2020) Secure transmission for intelligent reflecting surface-assisted mmWave and terahertz systems. IEEE Wireless Commun Lett 9(10):1743–1747
44. Chen Y, Li R, Han C, Sun S, Tao M (2023) Hybrid spherical- and planar-wave channel modeling and estimation for terahertz integrated UM-MIMO and IRS systems. IEEE Trans Wireless Commun
45. Wu Q, Zhang S, Zheng B, You C, Zhang R (2021) Intelligent reflecting surface-aided wireless communications: a tutorial. IEEE Trans Commun 69(5):3313–3351
46. Banerjee A, Chakraborty C, Rathi Sr M (2020) Medical imaging, artificial intelligence, internet of things, wearable devices in terahertz healthcare technologies. In: Terahertz biomedical and healthcare technologies. Elsevier, pp 145–165
47. Sarieddeen H et al (2020) Next generation terahertz communications: a rendezvous of sensing, imaging, and localization. IEEE Commun Mag 58(5):69–75
48. Hoseini SA, Ding M, Hassan M (2017) Massive MIMO performance comparison of beamforming and multiplexing in the terahertz band. IN: IEEE Globecom Workshops (GC Wkshps), pp 1–6
49. Wu Y, Song G, Liu H, Xiao L, Jiang T (2023) 3-D hybrid beamforming for terahertz broadband communication system with beam squint. IEEE Trans Broadcast 69(1):264–275
50. Gao F, Wang B, Xing C, An J, Li GY (2021) Wideband beamforming for hybrid massive MIMO terahertz communications. IEEE J Sel Areas Commun 39(6):1725–1740
51. Yuan H, Yang N, Yang K, Han C, An J (2020) Hybrid beamforming for terahertz multi-carrier systems over frequency selective fading. IEEE Trans Commun 68(10):6186–6199
52. Ning B, Tian Z, Mei W, Chen Z, Han C, Li S, Yuan J, Zhang R (2023) Beamforming technologies for ultra-massive MIMO in terahertz communications. IEEE Open J Commun Soc 4:614–658
53. Zhang H, Shlezinger N, Guidi F, Dardari D, Imani MF, Eldar YC (2021) Beam focusing for multi-user mimo communications with dynamic metasurface antennas. In: ICASSP, IEEE

international conference on acoustics, speech and signal processing - proceedings, vol 2021. Institute of Electrical and Electronics Engineers Inc., pp 4780–4784
54. Kim S, Zajic A (2016) Statistical modeling and simulation of short-range device-to-device communication channels at Sub-THz frequencies. IEEE Trans Wireless Commun 15(9):6423–6433
55. Jornet JM, Akyildiz IF (2011) Channel modeling and capacity analysis for electromagnetic wireless nanonetworks in the terahertz band. IEEE Trans Wireless Commun 10(10):3211–3221
56. Sharma S, Deka K, Mishra A (2024) Reconfigurable intelligent surfaces aided terahertz communication for 6G: challenges and potential solutions. Wireless Personal Commun 1–15
57. Jiang H, Xiong B, Zhang H, Basar E (2023) Hybrid far- and near-field modeling for reconfigurable intelligent surface assisted V2V channels: a sub-array partition based approach. IEEE Trans Wireless Commun
58. Cui M, Wu Z, Lu Y, Wei X, Dai L (2023) Near-field MIMO communications for 6G: fundamentals, challenges, potentials, and future directions. IEEE Commun Mag 61(1):40–46
59. Chen BJ, Chan CH et al (2016) High-selectivity bandpass frequency-selective surface in terahertz band. IEEE Trans Terahertz Sci Technol 6(2):284–291
60. Jansen C, Piesiewicz R, Mittleman D, Kurner T, Koch M (2008) The impact of reflections from stratified building materials on the wave propagation in future indoor terahertz communication systems. IEEE Trans Antennas Propag 56(5):1413–1419
61. Karim R, Iftikhar A, Ijaz B, Mabrouk IB (2019) The potentials, challenges, and future directions of on-chip-antennas for emerging wireless applications-a comprehensive survey. IEEE Access 7:173 897–173 934
62. Liu W, Pan C, Ren H, Shu F, Jin S, Wang J (2023) Low-overhead beam training scheme for extremely large-scale RIS in near field. IEEE Trans Commun
63. Chen Z, Han C, Wu Y, Li L, Huang C, Zhang Z, Wang G, Tong W (2021) Terahertz wireless communications for 2030 and beyond: a cutting-edge frontier. IEEE Commun Mag 59(11):66–72
64. https://unlab.tech/nano_downloads/teranova/
65. Sen P, Ariyarathna V, Madanayake A, Jornet JM (2021) A versatile experimental testbed for ultrabroadband communication networks above 100 GHz. Comput Netw 193:1–11
66. Sen P, Pados DA, Batalama SN, Einarsson E, Bird JP, Jornet JM (2020) The TeraNova platform: an integrated testbed for ultra-broadband wireless communications at true Terahertz frequencies. Comput Netw 179:1–11
67. Petrov V, Moltchanov D, Koucheryavy Y (2016) Applicability assessment of terahertz information showers for next-generation wireless networks. IEEE Intern Confer Commun (ICC) 1–7
68. Hamza AS, Deogun JS, Alexander DR (2016) Wireless communication in data centers: a survey. IEEE Commun Surv Tutor 18(3):1572–1595
69. Boulogeorgos A-AA, Alexiou A, Kritharidis D, Katsiotis A, Ntouni G, Kokkoniemi J, Lethtomaki J, Juntti M, Yankova D, Mokhtar A et al (2018) Wireless terahertz system architectures for networks beyond 5G. arXiv:1810.12260
70. Serghiou D, Khalily M, Brown TW, Tafazolli R (2022) Terahertz channel propagation phenomena, measurement techniques and modeling for 6G wireless communication applications: a survey, open challenges and future research directions. IEEE Commun Surveys Tuts
71. Nguyen DC, Ding M, Pathirana PN, Seneviratne A, Li J, Niyato D, Dobre O, Poor HV (2021) 6G Internet of Things: a comprehensive survey. IEEE Internet Things J 9(1):359–383
72. Siegel PH (2007) THz instruments for space. IEEE Trans Antennas Propag 55(11):2957–2965
73. Saeed A, Gurbuz O, Akkas MA (2020) Terahertz communications at various atmospheric altitudes. Phys Commun 41
74. Solyman A, Elhaty I (2021) Potential key challenges for terahertz communication systems. Int J Electr Comput Eng 11(4)
75. Hillger P, Grzyb J, Jain R, Pfeiffer UR (2018) Terahertz imaging and sensing applications with silicon-based technologies. IEEE Trans Terahertz Sci Technol 9(1):1–19

76. Latva-aho M, Leppänen K et al. (2019) Key drivers and research challenges for 6G ubiquitous wireless intelligence. University of Oulu
77. Vitiello MS, De Natale P (2022) Terahertz quantum cascade lasers as enabling quantum technology. Adv Quan Technol 5(1)
78. Lu X, Venkatesh S, Saeidi H (2021) A review on applications of integrated terahertz systems. China Commun 18(5):175–201
79. Huq KMS, Busari SA, Rodriguez J, Frascolla V, Bazzi W, Sicker DC (2019) Terahertz-enabled wireless system for beyond-5G ultra-fast networks: a brief survey. IEEE Network 33(4):89–95

Chapter 15
Ambulatory ECG Measurement Denoising and Implementation for Smart Healthcare Applications

Vinit Kumar, Jayesh Jayarajan, Maulesh N. Gadani, Payal Sharma, and Priya Ranjan Muduli

Abstract Wireless health monitoring devices are proliferating in the current era. Ambulatory electrocardiogram (ECG) devices are one of the tools that help in long-term ECG measurement. Ambulatory electrocardiogram is a recent technology for analyzing heart activity from extensive ECG measurements. Generally, the ECG signals get tempered through different noises during measurement. Among the noises, motion artifacts and baseline wander are the predominant noises present in ambulatory ECG signals. It is a challenging task to remove these noises from ambulatory ECG recordings. This chapter presents a robust approach for motion artifact and baseline wander suppression using empirical wavelet transform (EWT) and non-local means (NLM) estimation. This denoising technique is tested using the Xilinx PYNQ-Z2 board. The efficacy is assessed using different parameters, i.e., increment in signal-to-noise ratio, correlation coefficients, and root mean square error. The efficacy of the proposed method is validated using MIT-BIH arrhythmia and MIT-BIH NSTDB database. The denoising technique discussed in this chapter performs better in suppressing motion artifacts and baseline wander from the noisy ambulatory

V. Kumar (✉) · P. R. Muduli
Department of Electronics Engineering, Indian Institute of Technology (BHU), Varanasi, India
e-mail: vinitkumar.rs.ece21@iitbhu.ac.in

P. R. Muduli
e-mail: prmuduli.ece@iitbhu.ac.in

J. Jayarajan · M. N. Gadani · P. Sharma
Indian Space Research Organisation, Bengaluru, India
e-mail: jayeshj@sac.isro.gov.in

M. N. Gadani
e-mail: drmaulesh@sac.isro.gov.in

P. Sharma
e-mail: payal@sac.isro.gov.in

S. Sharma et al. (eds.), *Next-Generation Wireless Systems*,
Transactions on Computer Systems and Networks,
https://doi.org/10.1007/978-981-96-3185-8_15

ECG measurements. The verified hardware compatibility for the proposed algorithm with the PYNQ-Z2 board substantiates that the proposed technique can be implemented over an ambulatory ECG recording device for wireless ECG monitoring.

Keywords Ambulatory ECG · Denoising · Motion artifact · Baseline wander

15.1 Introduction

The Electrocardiogram (ECG) signal analysis is an extensively employed method for diagnosing various cardiac issues [1]. Wireless ECG systems are utilized in numerous application domains, including telemonitoring, sports applications, home health care for elderly individuals, fetal ECG, wearable technology, and ambulatory monitoring. Cables frequently impede users' freedom of movement in long-term ambulatory ECG monitoring [2]. For continuous long-time monitoring of cardiac activities, doctors require long-duration ECG recordings. Keeping the patient still during the continuous recording is an exceedingly difficult task. Hence, ambulatory ECG devices are used to acquire the ECG signal [3]. Over the last few decades, the evolution of ambulatory ECG monitoring systems has taken place. Various ambulatory ECG acquisition devices are manufactured to allow us to record and transmit ECG measurements wirelessly. The major advantage of ambulatory ECG monitoring is its ability to monitor the signal continuously with the help of a small recorder, which can be carried anywhere along with the patient, and many ambulatory ECG devices allow us to share the recorded data using wireless channel [4]. The evolution in ECG monitoring from conventional method to ambulatory ECG devices is shown in Fig. 15.1. The presence of artifacts in ECG recordings is one of the primary concerns. Some prominent artifacts originated during the ECG acquisition process are as follows [5, 6].

- Motion artifact (originated by the movement of patient or electrode)
- Baseline wander (originated by the patient's movement or respiration)

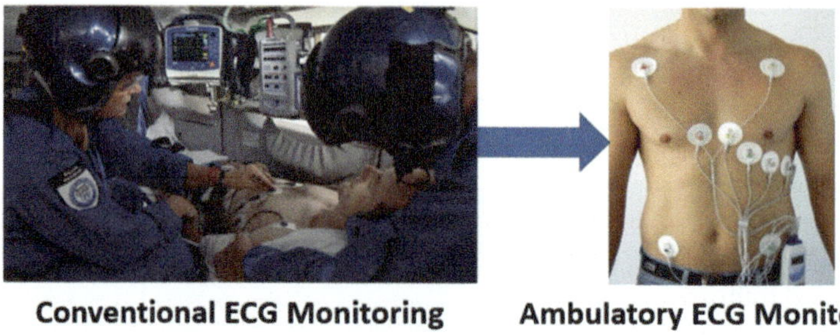

Fig. 15.1 Evolution in ECG monitoring [8, 9]

- Electromyogram noise (originated by the muscle contraction and relaxation)
- Powerline interference (originated by the power supply)

Out of these artifacts, motion artifacts and baseline wander are low-frequency noises, whereas powerline interference and electromyogram are high-frequency [5]. The two highly noticeable noises during signal acquisition through ambulatory ECG devices are motion artifacts and baseline wander. These artifacts affect the ECG waveform's morphology, making the signal analysis challenging for a clinician [7]. The suppression of this noise in ambulatory ECG recordings is an inevitable task. In the literature, several approaches are proposed for denoising ECG signals. In [10], the authors have proposed the total variation-based denoising technique. In [11], an NLM-based estimation technique is proposed for ECG signal denoising. In [12], authors combined discrete wavelet transform (DWT) [13] with NLM for effectively suppressing the noise from ECG signals. In [14], a weighted empirical mode decomposition (EMD) based technique is presented for denoising the ECG signal. Similarly, compressed sensing-based methods play a vital role in restoring the source signals from noisy measurements [15, 16]. In [17], a Moreau envelop-based nonlinear filtering approach is presented for biomedical denoising. In [18], a hybrid adaptive filtering technique based on orthogonal wavelet is used to suppress motion artifacts and baseline wander from ECG signals. In [19], authors propose a compressive sampling Taylor-Fourier multifrequency-based technique for ECG signal denoising. A proximity operator-based approach is presented for biomedical signals denoising, including ECG in [20]. In [21], a hybrid approach based on infimal convolution and AM-GM-based majorization technique is presented for denoising the ECG signals. A non-linear filtering technique is proposed in [22] for denoising of biosignals. In [23], a stationary wavelet transform (SWT) [24] with a level thresholding-based approach is proposed. In [5], the authors have presented a SWT [24] based approach for artifact suppression from ECG signals. In [25], the authors have proposed a hybrid approach using SWT and weighted total variation (WTV) [26]. In [27], a hybrid wavelet transform-based technique is presented for motion artifact suppression from ECG recordings. In [28], authors use a combination of ensemble empirical mode decomposition and SWT for denoising ECG signals, whereas in [29], an eigenvalue decomposition-based approach is used for baseline wander suppression. In [30], a disentangled autoencoder is proposed for ECG denoising. In [31], a DWT decomposition-based convolutional neural network is presented for baseline wander and motion artifact suppression. In most approaches mentioned above, DWT or SWT is used for multi-resolution-based signal decomposition. Both of these techniques require the proper selection of wavelets, which may vary in the case of different applications or even different databases and have higher computational complexity. The empirical wavelet transform (EWT) has many advantages, such as low computational time, data-adaptive, and suitability for closely spaced low and high-frequency data [32, 33]. Several authors have used the DWT, EMD [34], and SWT-based multi-resolution techniques in ECG signal denoising. However, the EWT has still not been explored well in the case of ECG applications, especially for removing motion artifacts and baseline wander. In this chapter, we incorporate the EWT-based multi-resolution technique for extracting the multi-resolution (MR) components of different frequency bands [35]. The wavelet domain non-local means (NLM) [11] estimation effectively suppresses motion artifacts and baseline wander from ECG recordings. The main takeaways from this chapter are as follows.

- In this chapter, a comprehensive discussion regarding ambulatory ECG monitoring and denoising is presented.
- Further, a new wavelet-domain non-local means estimation-based denoising technique is proposed for motion artifacts and baseline wander suppression from ECG recordings.
- To verify the hardware compatibility, the proposed algorithm is implemented over the Xilinx PYNQ-Z2 board.
- The effectiveness of the proposed technique is demonstrated using both motion artifact and baseline wander contaminated ECG database with a comparison with recent techniques using various distortion-measuring parameters.

The remainder of the chapter is structured as follows. Section 15.2 contains the proposed approach. The experimental evaluations are presented in Sect. 15.3. Finally, Sect. 15.4 contains the conclusion of this chapter.

15.2 Methodology

The proposed baseline wander and motion artifact suppression technique is presented in this section. The noisy (motion artifact or baseline wander contaminated) signal, Y=$[Y(1), Y(2), \cdots, Y(N)]^T \in \mathbb{R}^N$, which is the combination of ECG signal and noise is represented as follows.

$$Y = S + W \tag{15.1}$$

Here, S=$[S(1), S(2), \cdots, S(N)]^T \in \mathbb{R}^N$ is the clean ECG signal that is corrupted by motion artifact or baseline wander noise, $W \in \mathbb{R}^N$, of additive in nature.

This work deals with the motion artifact and baseline wander noises present in ECG measurements. The welch power spectral estimate [36] for motion artifact and baseline wander data present in MIT-BIH NSTDB database [37] are shown in Figs. 15.2 and 15.3, respectively. The signal power as a function of frequency is described by the power spectral density (PSD). Figures 15.2 and 15.3 show that most of the power of both of these artifacts is present in the lower-frequency bands. Section 15.2.1 presents the proposed method for eliminating baseline wander and motion artifacts.

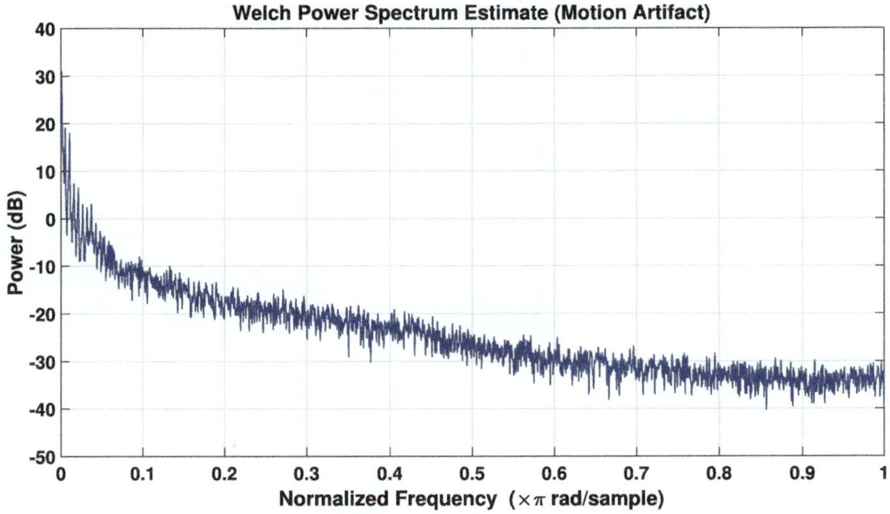

Fig. 15.2 Power spectral estimate of motion artifact

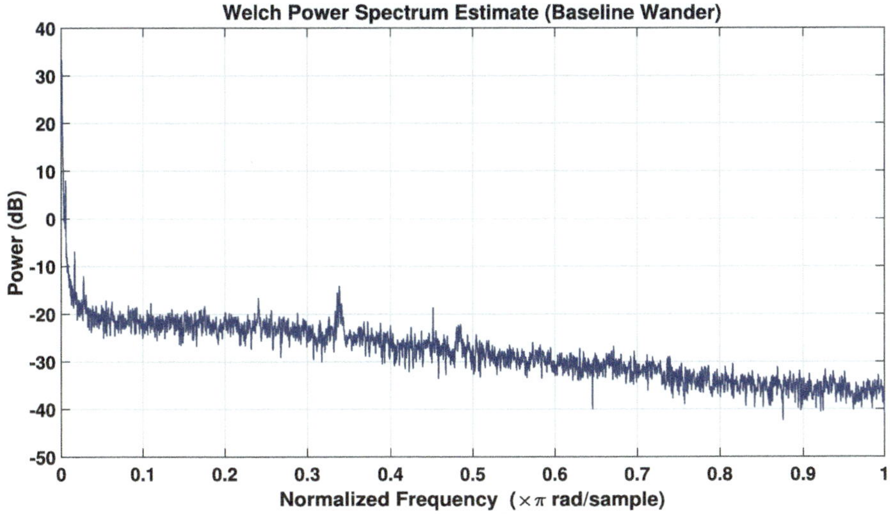

Fig. 15.3 Power spectral estimate of baseline wander

15.2.1 Denoising Technique

The proposed approach for removing motion artifacts and baseline wander is depicted in Fig. 15.4. In this approach, initially, noisy ECG signals $Y(n)$ are decomposed into their multi-resolution (MR) components using EWT technique [35]. The EWT signal decomposition technique works on the principle of adaptive wavelet utilizing the

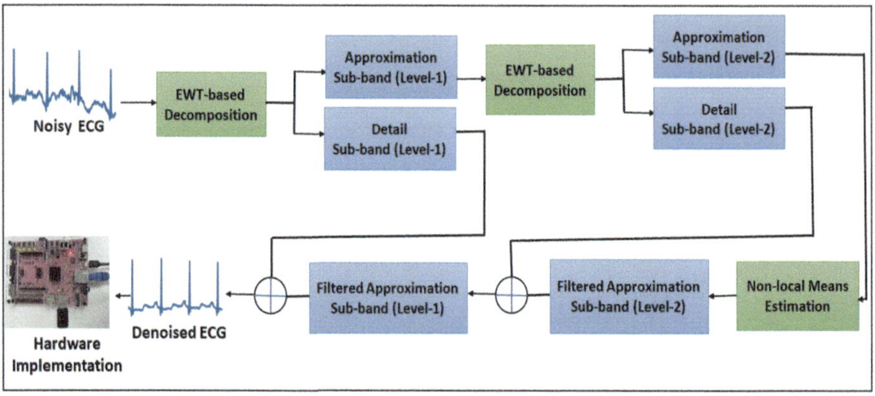

Fig. 15.4 Block diagram of the proposed noise suppression technique

information content of the signal. In EWT, the signal's local maxima are detected. Based on the detected maxima, the spectrum is sectioned, and finally, the wavelet filter bank is generated. The empirical wavelets are implemented as a band-pass filter over each segment. A normalized Fourier axis with a periodicity of 2π is considered. However, the axis is constrained to $\omega \in [0, \pi]$ to meet Shannon's requirements. The empirical scaling function $\phi(\omega)$ and empirical wavelets $\psi(\omega)$ are as follows.

$$\phi_n(\omega) = \begin{cases} 1, & \text{if } |\omega| \le \omega_n - \tau_n \\ \cos[\frac{\pi}{2}\Gamma(\frac{1}{2\tau_n}(|\omega| - \omega_n + \tau_n))], \\ \text{if } \omega_n - \tau_n \le |\omega| \le \omega_n + \tau_n \\ 0, & \text{otherwise} \end{cases} \tag{15.2}$$

$$\psi_n(\omega) = \begin{cases} 1, & \text{if } \omega_n + \tau_n \le |\omega| \le \omega_{n+1} - \tau_{n+1} \\ \cos[\frac{\pi}{2}\Gamma(\frac{1}{2\tau_{n+1}}(|\omega| - \omega_{n+1} + \tau_{n+1}))], \\ \text{if } \omega_{n+1} - \tau_{n+1} \le |\omega| \le \omega_{n+1} + \tau_{n+1} \\ \sin[\frac{\pi}{2}\Gamma(\frac{1}{2\tau_n}(|\omega| - \omega_n + \tau_n))], \\ \text{if } \omega_n - \tau_n \le |\omega| \le \omega_n + \tau_n \\ 0, & \text{otherwise} \end{cases} \tag{15.3}$$

Here, the function $\Gamma(t)$ is an arbitrary $C^k([0, 1])$ function such that

$$\Gamma(t) = \begin{cases} 0, & \text{if } t \le 0 \\ \text{and } \Gamma(t) + \Gamma(1-t) = 1 \ \forall t \in [0, 1] \\ 1, & \text{if } t \ge 1 \end{cases} \tag{15.4}$$

various functions may satisfy the properties mentioned earlier. Still, the most used function in the literature [35] is given below.

$$\Gamma(t) = t^4(35 - 84t + 70t^2 - 20t^3) \tag{15.5}$$

Here, $\tau_n = \beta\omega_n$, and $0 \leq \beta \leq 1$. Further, the detail coefficients $D_c(n, k)$ and approximation coefficients $A_c(n, k)$ are computed using the inner products of the signal with empirical wavelets $\psi_n(\omega)$ and scaling functions $\phi_n(\omega)$ respectively. For our given noisy ECG signal $Y(n)$, the first level detail and approximation coefficients $D_{c1}(n, k)$ and $A_{c1}(n, k)$ are given as follows.

$$D_{c1}(n, k) = \sum_{k=-\infty}^{+\infty} Y(k)\psi_n(n - k) = Y(n) \star \psi_n(n)$$

$$A_{c1}(n, k) = \sum_{k=-\infty}^{+\infty} Y(k)\phi_n(n - k) = Y(n) \star \phi_n(n) \tag{15.6}$$

The first level decomposed detail and approximation empirical mode functions (referred to as D_1 and A_1 in this chapter) are defined as given below, where D_1 denotes the higher frequency band and A_1 denotes the lower frequency band.

$$D_1(n) = D_{c1}(n, k) \star \psi_n(n)$$

$$A_1(n) = A_{c1}(n, k) \star \phi_n(n) \tag{15.7}$$

To get the second level decomposed coefficients of the noisy ECG, the first level decomposed approximation sub-band signal A_1 is subsequently decomposed using EWT. The second level detail and approximation $D_{c2}(n, k)$ and $A_{c2}(n, k)$ are given as follows.

$$D_{c2}(n, k) = \sum_{k=-\infty}^{+\infty} A_1(k)\psi_n(n - k) = A_1(n) \star \psi_n(n)$$

$$A_{c2}(n, k) = \sum_{k=-\infty}^{+\infty} A_1(k)\phi_n(n - k) = A1(n) \star \phi_n(n) \tag{15.8}$$

Further, the second level decomposed detail and approximation empirical mode functions D_2 and A_2 are defined below.

$$D_2(n) = D_{c2}(n, k) \star \psi_n(n)$$

$$A_2(n) = A_{c2}(n, k) \star \phi_n(n) \tag{15.9}$$

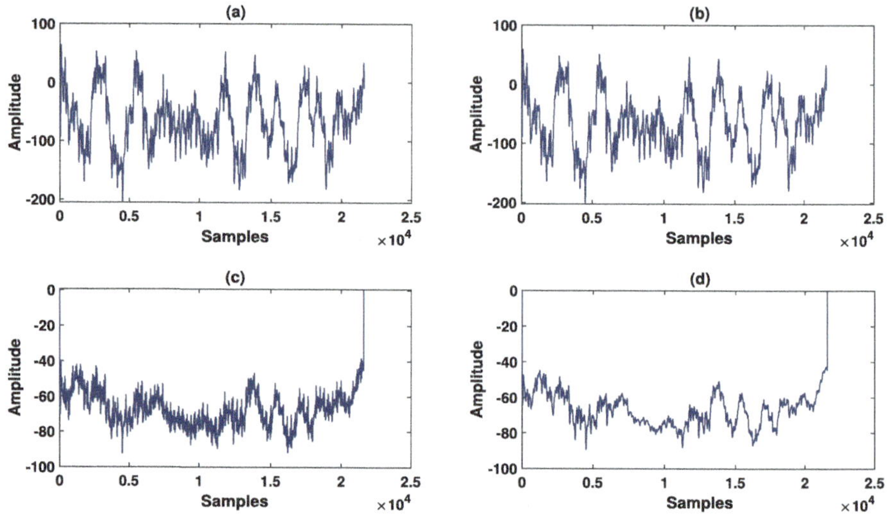

Fig. 15.5 **a** Level 1 detail coefficient, **b** Level 1 approximation coefficients, **c** Level 2 detail coefficient, **d** Level 2 approximation coefficients of EWT-based decomposed motion artifact contaminated the ECG signal

Both first-level and second-level decomposed detail and approximation sub-band signals for motion artifact and baseline wander contaminated signals are depicted in Figs. 15.5 and 15.6, respectively. As shown in Fig. 15.4, after the second level of EWT-based signal decomposition, the non-local means estimation [11] is applied over the second level approximation sub-band, i.e., A_2 as given below. For any sample, n of the second-level empirical approximation mode function A_2, its estimate after non-local means estimation is $\hat{A}_2(n)$ which is given as the weighted sum of values at other points t which are located inside a search neighborhood $N(n)$ containing L_Δ samples. It is mathematically given as follows [11].

$$\hat{A}_2(n) = \frac{1}{P(n)} \sum_{t \in N(n)} w(n,t) A_2(t) \tag{15.10}$$

where $P(n) = \sum_t w(n,t)$ and the weights $w(n,t)$ are given as below [38].

$$\begin{aligned} w(n,t) &= exp\Big(-\frac{\sum_{\delta \in \Delta}(A_2(n+\delta) - A_2(t+\delta))^2}{2L_\Delta \lambda^2}\Big) \\ &= exp\Big(-\frac{d^2(n,t)}{2L_\Delta \lambda^2}\Big) \end{aligned} \tag{15.11}$$

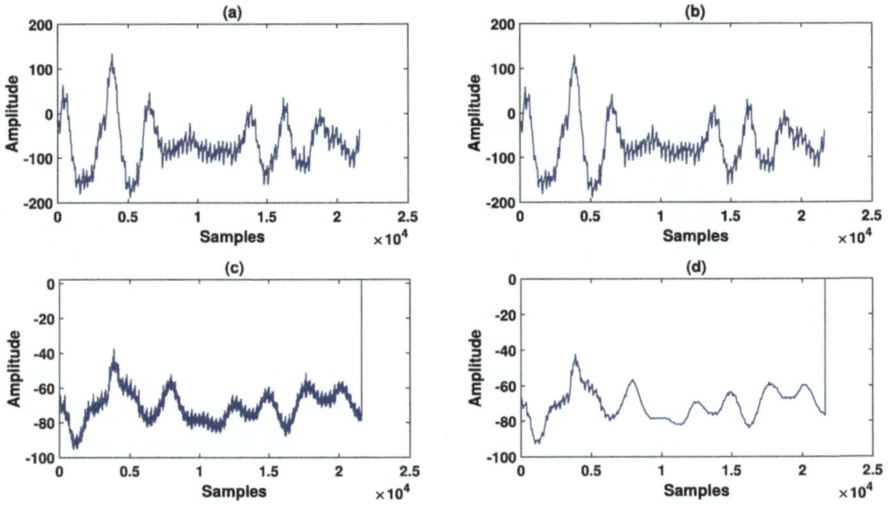

Fig. 15.6 **a** Level 1 detail coefficient, **b** Level 1 approximation coefficients, **c** Level 2 detail coefficient, **d** Level 2 approximation coefficients of EWT-based decomposed baseline wander contaminated ECG signal

Here, λ is the bandwidth, and other key parameters are neighborhood half-width M and patch half-width K, i.e., given as follows.

$$K = \frac{L_{\Delta} - 1}{2} \tag{15.12}$$

Further, the first level reconstructed approximation sub-band signal $A_1\hat{(n)}$ is given as follows.

$$\hat{A}_1(n) = D_2(n) + \hat{A}_2(n) \tag{15.13}$$

Finally, the denoised ECG signal $Z(n)$ is obtained using the following equation.

$$Z(n) = D_1(n) + \hat{A}_1(n) \tag{15.14}$$

15.2.2 Optimal Parameter Selection

In non-local means estimation, the selection of optimal values for the parameters, patch half-width (K), search neighborhood half-width (M), and bandwidth (λ) is essential. The bandwidth (λ) is generally given as the product of the smoothness parameter, α, and the standard deviation of noise σ_n, i.e.,

$$\lambda = \alpha.\sigma_n \tag{15.15}$$

Here, α determines the amount of smoothness, and σ_n represents the standard deviation of a noisy ECG signal. A very high value of α leads to an incorrect patch. Similarly, a small value of α leads to an over-smoothing problem [12]. Hence, a correct selection of α is required depending upon the application. The patch half-width K determines the size of the smallest features in samples. It is experimentally observed that the small sample size is better than the smallest feature size. The wider the neighborhood search width M, the more chance there is to find similar patches, but at the same time, computational complexity is higher. Herein, we deal with two varieties of noises, i.e., baseline wander and motion artifact. The exhaustive experiments using these noises found that the better choice of K and M are $K = 2$ and $M \geq 1900$. However, we set $M = 1900$ for all the numerical experiments to reduce the computational complexity. The proper bandwidth selection λ depends upon the correct choice of α. It is experimentally observed that for motion artifact suppression and $\alpha = 1.7$ for baseline wander suppression, $\alpha = 2.5$ are the best choices.

15.2.3 Hardware Implementation

For the real-time application of any algorithm, hardware compatibility is crucial [39]. Hence, to check the hardware compatibility of the proposed algorithm, the PYNQ-Z2 board developed by Xilinx is used. Figures 15.7 and 15.8 present the PYNQ board and its connection with the computer interface. In Fig. 15.8, an ethernet cable connects the computer with the PYNQ-Z2 board. A micro-USB cable that also serves as a UART connection via the board's PROG-UART micro-USB connector is used to provide the power supply to the board. An SD card stores the data in the PYNQ-Z2

Fig. 15.7 PYNQ-Z2 board

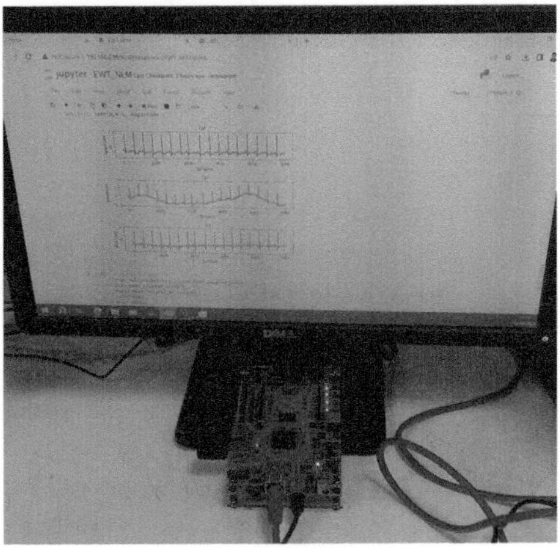

Fig. 15.8 Computer interface and hardware configuration

board, whereas a Wireless USB Adapter provides internet access. On the computer side, a Jupyter notebook and serial terminal are provided for interaction with the PYNQ-Z2 board. The major steps involved in the hardware implementation are as follows.

- Firstly, establish the connection between the PYNQ board and the computer.
- Login to the PYNQ's Jupyter Notebook interface.
- Implement the algorithm's code on the PYNQ board.

The detailed information about PYNQ-Z2 can be accessed on [40]. The proposed algorithm is implemented successfully over the PYNQ-Z2 board, which supports the fact that our algorithm is also hardware-compatible. With some necessary adjustments, it can be used as the post-processing unit in any ambulatory ECG acquisition device, system on chip (SoC), or Internet of Things (IoT) device to suppress motion artifacts and baseline wander from the device's collected ECG signal.

15.3 Experimental Evaluations

In this section, the experimental details, along with the details of the dataset used for experimentation, are discussed in detail. Further, the quantitative and qualitative experimental results are presented, showing the proposed technique's efficacy.

15.3.1 Data-Set Details

The MIT-BIH arrhythmia [41], and MIT-BIH NSTDB [37] databases available online on physionet ATM [42] are used in the experimentation process. The MIT-BIH arrhythmia database consists of 48 two-channel ambulatory ECG recordings, and the MIT-BIH NSTDB database contains motion artifacts and baseline wander noise data. The sampling rate for both datasets is 360 Hz with 11-bit resolution over a 10 mV range. We considered one-minute epochs for our experiment's ECG recordings and noises. This work considers two types of noises, i.e., motion artifact and baseline wander. The noisy ECG recording is obtained by adding the noise templates obtained from the MIT-BIH NSTDB database with the clean ambulatory ECG recordings taken from the MIT-BIH arrhythmia database. Due to the availability of separate reference signal and noise components, the proposed approach is evaluated with reference signals using various evaluation parameters.

15.3.2 Evaluation Parameters

Here, the effectuality of the proposed technique is validated using the following three different parameters.

- Increment in signal-to-noise ratio (ΔSNR)
- Percentage increment in correlation coefficient with the ground truth signal, which is (η)
- Root mean square error (RMSE)

Increment in Signal-to-Noise Ratio (ΔSNR): The ΔSNR is given by [7].

$$\Delta SNR = SNR_A - SNR_B \tag{15.16}$$

where SNR_A and SNR_B denote the signal-to-noise ratio post and before the denoising operation. SNR_A and SNR_B are mathematically given as follows.

$$SNR_B = 10 \log \frac{|||S||_2^2}{||S - Y||_2^2}$$

$$SNR_A = 10 \log \frac{|||S||_2^2}{||S - Z||_2^2} \tag{15.17}$$

Here, $S \in \mathbb{R}^N$ represents the clean signal, $Y \in \mathbb{R}^N$ represents the noisy ECG signal, whereas $Z \in \mathbb{R}^N$ represents the denoised signal.

Percentage Increment in Correlation Coefficient with the Ground Truth Signal which is (η) The mathematical formulation of the increment in correlation coefficient (η) (in percent) is as follows [7].

$$\eta(\%) = \frac{CC_A - CC_B}{CC_B} \times 100 \tag{15.18}$$

$CC_A = CC(S, Z)$ and $CC_B = CC(S, Y)$ represent the correlation coefficients of denoised ECG and noisy ECG, respectively, with the ground truth ECG. Further, the correlation coefficient of the denoised signal with the ground truth signal ($CC_A = CC(S, Z)$) is mathematically formulated as

$$CC_A = \frac{1}{K} \sum_{i=1}^{K} \frac{S(i) - \mu_{S(i)}}{\sigma_{S(i)}} \frac{Z(i) - \mu_{Z(i)}}{\sigma_{Z(i)}} \tag{15.19}$$

where μ and σ denote the mean and standard deviation, respectively. The correlation coefficient of the noisy signal with the ground truth signal ($CC_B = CC(S, Y)$) is computed using a similar way as in Eq. 15.19.

Root Mean Square Error (RMSE): The mathematical formulation for root mean square error (RMSE) is given as follows [21].

$$RMSE = \sqrt{\frac{1}{K} \sum_{n=1}^{K} (Z(n) - S(n))^2} \tag{15.20}$$

Here, K represents the number of samples in an ECG recording.

Higher values of ΔSNR and η represent the better denoising performance, and the lower value of $RMSE$ depicts the superior denoising performance of the proposed approach.

15.3.3 Results

This subsection presents the numerical and visual analyses of proposed motion artifacts and baseline wander noise removal techniques. Figure 15.9 presents the visual analysis of motion artifact suppression using the proposed approach, whereas, in Fig. 15.10, the visual analysis of baseline wander suppression using the proposed approach is shown. It can be observed that the proposed technique effectively suppresses motion artifacts and baseline wander from ECG measurements. In addition to the visual analysis, a comparison analysis is provided for both types of noises using the metrics ΔSNR, η, and RMSE. The aforementioned parameters' values are presented in Tables 15.1 and 15.2 for motion artifact removal and baseline wander suppression, respectively. Both the tables contain the values in the format of

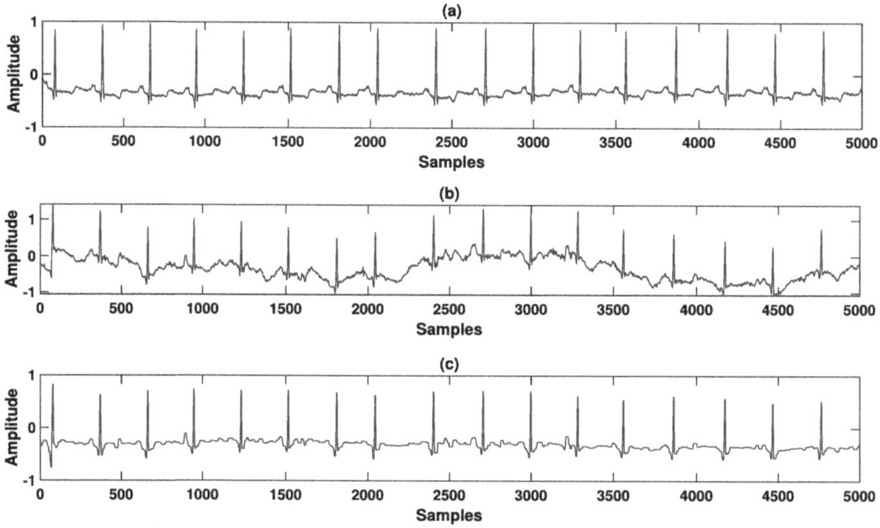

Fig. 15.9 **a** Clean ECG measurement, **b** ECG measurement corrupted by motion artifact (SNR = 4.19), **c** ECG measurement denoised using proposed approach (SNR = 13.33)

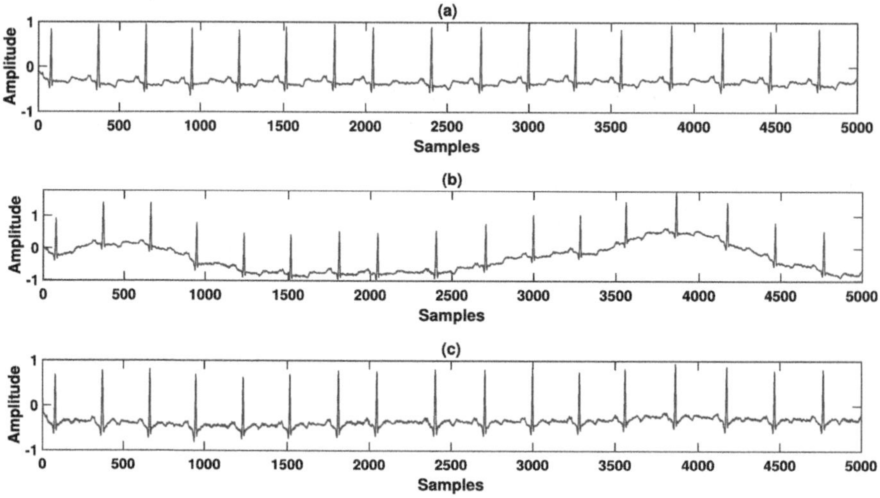

Fig. 15.10 **a** Clean ECG measurement, **b** ECG measurement corrupted by baseline wander (SNR = 3.05), **c** ECG measurement denoised using proposed approach (SNR = 13.76)

Table 15.1 Comparative analysis of the proposed technique with the existing methods for motion artifact suppression. Here μ denotes mean value whereas σ denotes standard deviation

Methods	$\Delta SNR(dB)(\mu \pm \sigma)$	$\eta(\%)(\mu \pm \sigma)$	$RMSE(\mu \pm \sigma)$
TVD [10]	1.23 ± 0.27	2.79 ± 1.21	0.195 ± 0.027
MMNF-Atan [17]	1.63 ± 0.21	3.02 ± 1.25	0.180 ± 0.026
MMED [20]	2.71 ± 1.02	5.94 ± 1.10	0.177 ± 0.024
NLM [11]	2.72 ± 0.35	6.23 ± 2.62	0.177 ± 0.025
DWT-NLM [12]	3.81 ± 0.41	9.84 ± 3.15	0.165 ± 0.029
EWT-WT [27]	4.28 ± 1.15	10.75 ± 4.82	0.148 ± 0.021
SWT-Thresholding [5]	3.90 ± 0.45	9.97 ± 3.97	0.151 ± 0.019
Proposed technique	$\mathbf{7.08 \pm 1.48}$	$\mathbf{25.67 \pm 10.09}$	$\mathbf{0.107 \pm 0.017}$

Table 15.2 Comparative analysis of the proposed technique with the existing methods for baseline wander suppression. Here, μ denotes mean value whereas σ denotes standard deviation

Methods	$\Delta SNR(dB)(\mu \pm \sigma)$	$\eta(\%)(\mu \pm \sigma)$	$RMSE(\mu \pm \sigma)$
TVD [10]	1.04 ± 0.24	1.99 ± 1.03	0.197 ± 0.024
MMNF-Atan [17]	2.08 ± 0.27	3.74 ± 1.69	0.194 ± 0.025
MMED [20]	2.37 ± 1.33	5.94 ± 1.10	0.185 ± 0.029
NLM [11]	2.39 ± 0.31	4.92 ± 3.01	0.185 ± 0.034
DWT-NLM [12]	3.13 ± 0.34	7.02 ± 3.85	0.172 ± 0.035
EWT-WT [27]	4.56 ± 1.02	9.86 ± 4.13	0.143 ± 0.022
SWT-Thresholding [5]	3.62 ± 0.49	8.31 ± 3.05	0.152 ± 0.021
Proposed technique	$\mathbf{8.10 \pm 1.27}$	$\mathbf{35.91 \pm 13.39}$	$\mathbf{0.106 \pm 0.014}$

$(\mu \pm \sigma)$, where μ and σ depicts the mean values for all the ECG recordings used in the experimental process and the standard deviation, respectively. Here η is expressed in percentage, whereas ΔSNR is expressed in decibels (dB). From the observation of Tables 15.1 and 15.2 it is evident that the proposed technique outperforms several state-of-the-art techniques.

15.4 Conclusion

For continuous monitoring of cardiac activities, doctors require long-duration ECG recording. For long-duration ECG recording, ambulatory ECG monitoring devices are extensively used due to their ability to continuously monitor signals with the help of a small recorder and wireless transmission of the recorded data. The major problem with ambulatory ECG monitoring is the addition of motion artifacts and baseline wander. This chapter presents an efficient approach for removing motion artifacts and baseline wander from ECG measurements through an EWT decomposition-based non-local means estimation filtering technique. As per the results, this technique effectively suppresses both types of noise. A comparative study uses recent state-of-the-art techniques regarding all three distortion measuring parameters: ΔSNR, η, and RMSE. The proposed technique outperforms various state-of-the-art techniques and effectively suppresses the motion artifacts and baseline wander from ambulatory ECG signals. The proposed method is also successfully implemented over the PYNQ-Z2 board to substantiate the hardware's suitability and relevancy in practical scenarios. The denoising technique presented in this chapter can be implemented as a post-processing unit in any ambulatory ECG acquisition device to suppress the ill effects of both types of noises considered in this work in wireless ECG monitoring.

Acknowledgements We thank SAC, the Indian Space Research Organization (ISRO), Department of Space, Govt. of India, for supporting the research work presented in this chapter.

References

1. An X, Stylios GK (2020) Comparison of motion artefact reduction methods and the implementation of adaptive motion artefact reduction in wearable electrocardiogram monitoring. Sensors 20(5):1468
2. Cosoli G, Spinsante S, Scardulla F, D'Acquisto L, Scalise L (2021) Wireless ECG and cardiac monitoring systems: State of the art, available commercial devices and useful electronic components. Measurement 177:109243
3. Kennedy HL (2013) The evolution of ambulatory ECG monitoring. Prog Cardiovasc Dis 56(2):127–132
4. Simunic D, Tomac S, Vrdoljak I (2009) Wireless ECG monitoring system. In: 2009 1st international conference on wireless communication, vehicular technology, information theory and aerospace & electronic systems technology. IEEE, pp 73–76
5. Kumar A, Tomar H, Mehla VK, Komaragiri R, Kumar M (2021) Stationary wavelet transform based ECG signal denoising method. ISA Trans 114:251–262
6. Sörnmo L, Laguna P (2005) Bioelectrical signal processing in cardiac and neurological applications. Academic, Chap. ECG Signal Processing, pp 453–566
7. Kumar V, Muduli PR (2023) Attentive Bi-LSTM based method for noise suppression in ambulatory ECG measurements. IEEE Trans Instrum Meas 72:1–9
8. Clinic M, Electrocardiogram (ECG or EKG). https://www.mayoclinic.org/tests-procedures/ekg/about/pac-20384983
9. Vitalograph, Vitalograph (COMPACT): sophisticated, integrated cardio-respiratory diagnostic workstation. https://edgemedsol.com/site/wp-content/uploads/compact-leaflet.pdf
10. Condat L (2013) A direct algorithm for 1-d total variation denoising. IEEE Signal Process Lett 20(11):1054–1057
11. Tracey BH, Miller EL (2012) Nonlocal means denoising of ECG signals. IEEE Trans Biomed Eng 59(9):2383–2386
12. Singh P, Pradhan G, Shahnawazuddin S (2017) Denoising of ECG signal by non-local estimation of approximation coefficients in DWT. Biocybern Biomed Eng 37(3):599–610
13. Mallat SG (1989) A theory for multiresolution signal decomposition: the wavelet representation. IEEE Trans Pattern Anal Mach Intell 11(7):674–693
14. Lahmiri S, Boukadoum M (2015) A weighted bio-signal denoising approach using empirical mode decomposition. Biomed Eng Lett 5:131–139

15. Muduli PR, Mukherjee A (2016) A subspace projection-based joint sparse recovery method for structured biomedical signals. IEEE Trans Instrum Meas 66(2):234–242
16. Muduli PR, Mandal AK, Mukherjee A (2017) An antinoise-folding algorithm for the recovery of biomedical signals from noisy measurements. IEEE Trans Instrum Meas 66(11):2909–2916
17. Muduli PR, Mukherjee A (2019) A moreau envelope-based nonlinear filtering approach to denoising physiological signals. IEEE Trans Instrum Meas 69(4):1041–1050
18. Das M, Sahana BC (2022) Optimized orthogonal wavelet-based filtering method for electrocardiogram signal denoising. J Inst Eng (India): Ser B 1–14
19. Galli A, Frigo G, Chindamo D, Depari A, Gadola M, Giorgi G (2019) Denoising ECG signal by CSTFM algorithm: monitoring during motorbike and car races. IEEE Trans Instrum Meas 68(7):2433–2441
20. Muduli PR, Kumar V (2023) A proximity operator-based method for denoising biomedical measurements. Circuits Syst Signal Proc 1–25
21. Kumar V, Muduli PR (2024) Infimal convolution and AM-GM majorized total variation-based integrated approach for biosignal denoising. SIViP 18(2):1919–1927
22. Muduli PR, Mukherjee A (2019) A robust estimator-based nonlinear filtering approach to piecewise biosignal reconstruction. IEEE Trans Instrum Meas 69(2):362–370
23. Berwal D, Vandana C, Dewan S, Jiji C, Baghini MS (2019) Motion artifact removal in ambulatory ECG signal for heart rate variability analysis. IEEE Sens J 19(24):12 432–12 442
24. Nason GP, Silverman BW (1995) The stationary wavelet transform and some statistical applications, pp 281–299
25. Madan P, Singh V, Singh DP, Diwakar M, Kishor A (2022) Denoising of ECG signals using weighted stationary wavelet total variation. Biomed Signal Process Control 73:103478
26. Sharma T, Sharma KK (2017) Qrs complex detection in ECG signals using locally adaptive weighted total variation denoising. Comput Biol Med 87:187–199
27. Xu X, Liang Y, He P, Yang J (2019) Adaptive motion artifact reduction based on empirical wavelet transform and wavelet thresholding for the non-contact ECG monitoring systems. Sensors 19(13):2916
28. Dwivedi AK, Ranjan H, Menon A, Periasamy P (2021) Noise reduction in ECG signal using combined ensemble empirical mode decomposition method with stationary wavelet transform. Circuits Syst Signal Process 40:827–844
29. Sharma RR, Pachori RB (2018) Baseline wander and power line interference removal from ecg signals using eigenvalue decomposition. Biomed Signal Process Control 45:33–49
30. Lin H, Liu R, Liu Z (2023) ECG signal denoising method based on disentangled autoencoder. Electronics 12(7):1606
31. Jin Y, Qin C, Liu J, Liu Y, Li Z, Liu C (2024) A novel deep wavelet convolutional neural network for actual ECG signal denoising. Biomed Signal Process Control 87:105480
32. Ranjan R, Sahana BC, Bhandari AK (2022) Cardiac artifact noise removal from sleep EEG signals using hybrid denoising model. IEEE Trans Instrum Meas 71:1–10
33. Kumar V, Muduli PR (2023) EWT-Domain optimized modified log filtering for motion artifact suppression from EEG signals. In: 2023 international conference on microwave, optical, and communication engineering (ICMOCE). IEEE, pp 1–4
34. Huang NE, Shen Z, Long SR, Wu MC, Shih HH, Zheng Q, Yen N-C, Tung CC, Liu HH (1998) The empirical mode decomposition and the hilbert spectrum for nonlinear and non-stationary time series analysis. Proc R Soc Lond. Ser A: Math Phys Eng Sci 454(1971):903–995
35. Gilles J (2013) Empirical wavelet transform. IEEE Trans Signal Process 61(16):3999–4010
36. Stoica P, Moses RL et al (2005) Spectral analysis of signals. Pearson Prentice Hall Upper Saddle River, NJ, vol 452
37. Moody GB, Muldrow WK, Mark RG (1984) A noise stress test for arrythmia detectors. Comput Cardiol 11:381–384
38. Van De Ville D, Kocher M (2009) SURE-based non-local means. IEEE Signal Process Lett 16(11):973–976
39. Meena YK, Kumar V, Muduli PR (2023) Schizophrenia detection using extreme gradient boosting with enhanced features on PYNQ-Z2. In: 2023 IEEE 9th international conference on smart instrumentation, measurement and applications (ICSIMA). IEEE, pp 1–5

40. Xilinx: PYNQ: Python productivity. http://www.pynq.io/
41. Moody GB, Mark RG (2001) The impact of the mit-bih arrhythmia database. IEEE Eng Med Biol Mag 20(3):45–50
42. Goldberger AL, Amaral LA, Glass L, Hausdorff JM, Ivanov PC, Mark RG, Mietus JE, Moody GB, Peng C-K, Stanley HE (2000) Physiobank, physiotoolkit, and physionet: components of a new research resource for complex physiologic signals. Circulation 101(23):e215–e220

GPSR Compliance

The European Union's (EU) General Product Safety Regulation (GPSR) is a set of rules that requires consumer products to be safe and our obligations to ensure this.

If you have any concerns about our products, you can contact us on ProductSafety@springernature.com

In case Publisher is established outside the EU, the EU authorized representative is:

Springer Nature Customer Service Center GmbH
Europaplatz 3
69115 Heidelberg, Germany

Batch number: 08551983

Printed by Printforce, the Netherlands